FEMINIST THEORY AND THE BODY

A READER

Edited by
Janet Price and Margrit Shildrick

EDINBURGH UNIVERSITY PRESS

Edinburgh University Press
22 George Square, Edinburgh

Typeset in Sabon and Gill Sans
by Bibliocraft Ltd, Dundee, and
printed and bound in Great Britain
at The University Press, Cambridge

A CIP record for this book is available from the British Library

ISBN 0 7486 1090 1 (hardback)
ISBN 0 7486 1089 8 (paperback)

Feminist Theory and the Body

CONTENTS

ACKNOWLEDGEMENTS

Our thanks are due to all those who helped with the production of this Reader. Lynda Birke, Rosi Braidotti, Ailbhe Smyth, and Jocelyn Wogan-Browne all took time to make suggestions for pieces to include; Jackie Jones at Edinburgh University Press was the instigator of the initial idea, and remained involved throughout; and Nicci Jones helped with photostatting. Thanks too to Grindl Dockery and Lis Davidson for their continuing support.

OPENINGS ON THE BODY: A CRITICAL INTRODUCTION

Margrit Shildrick with Janet Price

The status of the body within the dominant Western intellectual tradition has largely been one of absence or dismissal. Despite the necessary ubiquity of the body, and its apparent position as the secure grounding of all thought, the processes of theorising and theory itself have proceeded as though the body itself is of no account, and that the thinking subject is in effect disembodied, able to operate in terms of pure mind alone. At the end of the twentieth century, however, that familiar form of incorporeal abstraction is a site of serious contestation, emanating not least from the advent of a substantial corpus of feminist theory. As we shall trace briefly in this essay, and in the reader at large, feminism has long seen its own project as intimately connected to the body, and has responded to the masculinist convention by producing a variety of oftimes incompatible theories which attempt to take the body into account. For some, the materiality of the body, and particularly in its female form, comes to the fore only to be once more bracketed out of consideration, but what is different between this and dominant malestream approaches is that rather than a thoroughgoing disregard for things corporeal, feminism starts at least from a position of acknowledgement. More positively, other feminist writers have developed theory that is explicitly embodied and insistent on the centrality of the material body; while yet others, influenced by poststructuralism and postmodernism in particular, have put into question the giveness and security of the so-called natural body, positing instead a textual corporeality that is fluid in its investments and meanings. The purpose of the Reader, then, is not to present a single unified theory of the body, nor yet to reconcile diverse positions. Rather it is to uncover the very richness of feminist responses to the body during what is now known as the second wave. If there is a unifying theme, it is simply that the body matters – and not just to women, though gender is a persistent theme, but to all forms of theory.

How then did the body become the unspoken of Western abstract theory? In terms of intellectual activities, the body seems to have been regarded always with suspicion as the site of unruly passions and appetites that might disrupt the pursuit of truth and knowledge. At the risk of misleading simplification, it can be argued that the denial of corporeality and the corresponding elevation of mind or spirit marks a transhistorical desire to access the pure Intelligible as the highest form of Being. What makes the mind/body binary of the Enlightenment particularly characteristic is not the split itself, but a change of emphasis. Where the intellectual traditions of Judeo-Christianity for example saw the body as the mundane path to a higher, valorised, spirituality, the post-Cartesian modernist period is marked by a rejection of the body as an obstacle to pure rational thought. As such, the body occupies the place of the excluded other, and can be dismissed from consideration altogether. The distinctions made by Descartes between the *res cogitans* (the powers of intelligence and site of selfhood) and the *res extensa* (the machine-like corporeal substance) are far more complex than is generally recognised, but what has mattered to feminism is the cultural take-up of the mind/body split, and the enduring association of the devalued term with the feminine. In the post-Cartesian schema, the body has been seen simply as a material and unchanging given, a fixed biological entity obeying mathematical–causal laws, that must be transcended in order to free the mind for the intellectual pursuits of fully rational subjectivity. But if the well-ordered, self-contained and functional body may be safely forgotten in the face of higher concerns, there are nonetheless other bodies whose disorder frustrates the project of transcendence.

In its early analyses, the focus of much feminist theory was on the characterisation of the feminine as less than fully human. There has been, however, growing recognition of the need to address the ways in which the mind/body split is not only gendered in its resonances. The association of the body with gross, unthinking physicality marks a further set of linkages – to black people, to working class people, to animals, and to slaves. Feminist theory has faced the challenge of addressing the ways in which these dimensions – of race and of class, for example – intersect to constitute particular ways of seeing, and of devaluing, bodies. Whilst all such marginalised bodies are potentially unsettling, what is at issue for women specifically is that, supposedly, the female body is intrinsically unpredictable, leaky and disruptive. Not surprisingly, then, the ability to effect transcendence and exercise rationality has been gender marked as an attribute of men alone – and further of only some men, i.e. those who are white, middle/upper class, healthy and heterosexual – such that women remain rooted within their bodies, held back by their supposedly natural biological processes.[1] In consequence, feminism has from the start been deeply concerned with the body – either as something to be rejected in the pursuit of intellectual equality according to a masculinist standard, or as something to be reclaimed as the very essence of the female. A third, more recent alternative, largely associated with feminist postmodernism, seeks to emphasise the importance and inescapability of embodiment as a differential and fluid construct, the site of potential, rather than as a fixed given.

Where the body is viewed through conventional biological and racial taxonomies that make appeal to a given nature, it is taken for granted that sexual and racial difference are inherent qualities of the corporeal, and, moreover, that male and female bodies, black and white bodies, may each respectively fit a universal category. In terms of sex, the actual occurrence of bodily forms that are not self-evidently of either sex is conveniently overlooked in the interests of establishing a set of powerful gendered norms to which all bodies are supposed to approximate without substantial variation. The effect for women has been highly deleterious, for in terms of our historical oppression and disempowerment, a series of justificatory strategies are founded in the linking of the feminine to a body that is curiously and uniquely unreliable, most evidently in the female reproductive processes. The very fact that women are able in general to menstruate, to develop another body unseen within their own, to give birth, and to lactate is enough to suggest a potentially dangerous volatility that marks the female body as out of control, beyond, and set against, the force of reason. In contrast to the apparent ordered self-containment of the male body, which may then be safely taken for granted and put out of mind, the female body demands attention and invites regulation. The age-old relation between hysteria and the womb (called *hystera* in Greek) is just one example of how femininity itself becomes marked by the notion of an inevitable irrationality. In short, women just are their bodies in a way that men are not, biologically destined to inferior status in all spheres that privilege rationality. At the same time, however, that women are seen as more wholly embodied, and hence naturally disqualified from equality with the male, the boundaries of that embodiment are never fixed or secure. As the devalued processes of reproduction make clear, the body has a propensity to leak, to overflow the proper distinctions between self and other, to contaminate and engulf. Thus women themselves are, in the conventional masculinist imagination, not simply inferior beings whose civil and social subordination is both inevitable and justified, but objects of fear and repulsion. Coincident with its marginalisation, the devalued body is capable of generating deep ontological anxiety.

So powerful are such ideas, that many feminists themselves have been reluctant to engage with the female body, or have found it difficult to provide a positive theorisation of it. Given that the putative links between a natural femininity and incapacity have worked so efficiently in the interests of patriarchal oppression, many first wave writers of the eighteenth and nineteenth centuries, and those of the early years of contemporary feminism, saw equality as predicated on the need to go beyond the influence of biology, to stress instead the potential of women for intellectual achievement regardless of their troublesome bodies.[2] And even those who more closely contested the determinism of biology, saw the corporeal in a decidedly negative light. At the beginning of the second wave of feminism, Simone de Beauvoir (1953) famously likened female genital sexuality to 'a carnivorous swamp', while Shulamith Firestone (1979) looked forward with optimism to a time when the then incipient advanced reproductive technologies might free woman from the 'oppressive "natural" conditions' of procreation. The way forward was not to reclaim and revalorise the body, but to argue that the ideal standard of disembodied subjecthood was as appropriate to, and attainable by, women as it was to men. Such somataphobia, which to an extent mimics the masculinist fear and rejection of the body, is widely evident in emergent feminist theory, but at the same time the body became a central focus of more practical concerns, which in turn led to a more positive theorisation. This is especially evident in writing around the Women's Health Movement which was as engaged with the healthy as with the sick body, and was concerned to urge women to take control of their own bodies in the face of a patriarchal medical establishment. The handbook *Our Bodies, Ourselves*, first produced in 1976 (Boston Women's Health Book Collective 1976) is an early example of women's celebration of their bodyliness, particularly with regard to reproduction.

In contradistinction, then, to the somataphobia which was a characteristic feature – though not exclusively – of liberal feminist perspectives that see equality as the goal for women, other accounts began to focus on the body in a move to reclaim and give positive meaning to the feminine as a theoretical concept. From its inception, radical feminism in particular emphasised the importance of sexuality, not simply as the ground of women's oppression, but equally as the take-off point for an account that valorised bodily difference. The huge importance given then and now to the distinctive forms of female sexuality was paralleled by a perhaps more surprising turn, at least for radical feminists. Despite an initially widespread emphasis on the need for women to escape the relations of reproduction – Shulamith Firestone was after all read as a proto-radical feminist – the reproductive body too seemed to offer a site for the reconceptualisation of the feminine. The uniquely female capacity to give birth 'naturally' has been taken up as the centre of women's power, simultaneously to be jealously guarded against the incursions of biotechnology, *and* celebrated in its own right. In the cases of both sexuality and reproduction, the body retains something of its uncomfortable status as a place of ambush, of its vulnerability to male power, and yet it grounds an affirmation of the feminine that goes beyond the negative connotations given that term in masculinist discourse. Adrienne Rich's book, *Of Woman Born* (1979), is a characteristic example. In addition, for many feminists, the maternal body has come to figure the claim that women have a unique ethical sense that lays stress on caring, relationality and responsibility – an ethical sense that is more adequate not simply to women themselves but to all humanity. In contrast, the masculinist goods of autonomous rights and duties – the privileged terms of postEnlightenment morality –

are founded in the (impossible) separation of body and mind, and of body and body that underpins a series of hierarchical oppressions. The stress given to the embodied nature of sexual difference has been, then, a powerful advance for feminism, but nonetheless in its unproblematised form it runs two related risks: on the one hand it may uncritically universalise the male and female body, while on the other it appears to reiterate the biological essentialism that historically has grounded women's subordination.

As we noted before, moreover, the focus of feminism on sexual difference as the primary and often only distinguishing category of the body erases other equally important markers. Compare bell hooks, writing in the early 1980s, who touches on the question of the embodiment of black women, working as slaves. She argues that they were regarded simultaneously as 'masculinized sub-human creatures' for their ability to perform tasks that were perceived as male, and also as sexual temptresses, 'the embodiment of female evil and sexual lust' (hooks 1981: 33) before whom no white man was safe from moral corruption. Whilst historically black women's bodies were essential property to slave owners, functioning as both labour and capital, involved within the spheres of production and reproduction, writers such as Angela Davis (1982) and Patricia Hill Collins (1991) have taken up similar themes with respect to the position of many black women in society today. Feminist theorists such as hooks (1991), Patricia Hill Collins, and Audre Lorde (1984, 1988) have addressed the body, both directly and indirectly, in ways that emphasise it can no longer function in its unmarked colourlessness. Clearly, a strategy of transcendence has little relevance in such a context for rejection of the corporeal speaks to a clear disregard for women whose lives and very survival are intimately bound up with the physical potential of their bodies. What is required, and what has emerged over the subsequent years, is a theory of embodiment that could take account not simply of sexual difference but of racial difference, class difference, and differences due to disability; in short the specific contextual materiality of the body.

For all these reasons, the objective of taking up the issue of female embodiment as a constructive force has, in consequence, been highly controversial, and nowhere more so than in the response to the work of Luce Irigaray. As more representative of a poststructuralist feminism with its roots in recent Continental philosophy, than of the predominantly materialist analyses associated with the Anglo-American tradition, her approach is concerned primarily with discourse. Nonetheless, Irigaray's project of rewriting sexual difference beyond the binary of masculine:feminine – a binary which positions the sexed body, male or female, as static, ahistorical and determinate – makes constant reference to the anatomical differences between the sexes. Her concern is to revalue the way in which femininity is inscribed on to the female form in a culture in which masculinity is in retreat from the body and where disembodiment is privileged. In her 'labial politics', she places great emphasis on the multiple forms of female embodiment, the self-touching 'two lips' that characterise female morphology (1985), and on the fluidity that marks the inherent excess of the feminine that is uncontainable within binary sexual difference:

> Between our lips, yours and mine, several voices, several ways of speaking resound endlessly, back and forth. One is never separable from the other. You/I: we are always several at once. (Irigaray 1985: 209)

For many critics, such as Moi (1985) and Weedon (1987), Irigaray's own method is uncomfortably essentialist and ahistorical, appearing to play into the hands of those for whom existing social relations are determined by a fixed and differential biology. But hers is not a 'real' biology, so much as the discursive reconfiguring of a contested terrain that takes on board the force of psychic investments, notably that of desire. Irigaray's insistence on the sexed specificity of corporeality speaks not to the female body as such, but to a feminine morphological imaginary, the body that is never one. In being strongly influenced by psychoanalysis, Irigaray engages with the supposedly biological body in a highly complex way that seeks to disrupt the fixed parameters of sex and gender. The material forms of which she speaks are never given, but are filtered through and constructed by a set of discursive strategies. Although the characteristic poetic style, in which she and Hélène Cixous in particular seek to write the body, is rarely imitated, Irigaray's work has been highly influential on contemporary feminist theory.

The concern with the irreducible interplay of text and physicality which posits a body in process, never fixed or solid, but always multiple and fluid is one that resonates, implicitly and explicitly, in the work of theorists such as Elizabeth Grosz and Judith Butler. A further related element that links together many contemporary feminist theories of the body, is an interest in and use of psychoanalysis. Although Freud is often characterised as a proponent of biological determinism, which implies at least a given corpus, both he and much more clearly Lacan are, like Irigaray, concerned with a morphological imaginary, a psychic body map which may or may not correspond with the flesh and blood body. The feminist take-up of psychoanalysis is seen at its strongest in the understanding of how one becomes a sexed/gendered subject in the Symbolic, but it is in corporeality itself that the roots of explanation lie. Given that the early infant is unable to distinguish between its own body and that of the mother, a whole host of implications follow if the infant is to successfully negotiate the acquisition of a separate sense of self. Although this is not the place to discuss such psychic/psychological processes as the Freudian oedipal crisis, the Lacanian mirror stage, or feminist object relations theory, what is clear is that all of them involve a degree of alienation from the maternal body. The masculinist fear and repulsion, mentioned above, which is a transhistorical response to the female body and most particularly to the maternal body, is then consistent with the male child's investment in establishing psychosomatic difference.

For the feminist psychoanalyst, Julia Kristeva, the turn against the female body constitutes a process of psychic violence which she calls abjection:

> The abject confronts us ... within our personal archeology, with our earliest attempts to release the hold of *maternal* entity even before existing outside of her ... It is a violent, clumsy breaking away with the constant risk of falling back under the sway of a power as securing as it is stifling. (Kristeva 1982: 13)

In Kristeva's schema, the abject is always ambiguous: desirable and terrifying, nourishing and murderous; and, moreover, the process is never simply one of repudiaton: 'It is something rejected from which one does not part' (1992: 4). That sense of the abject as both the alien other who threatens the corporeal and psychic boundaries of the embodied self, and as an intrinsic, but unstable, part of the self resonates with the widespread cultural unease with bodily, and especially female bodily, fluids. In the effort to secure the 'clean and proper' male body, the body that is sealed and self-sufficient, it is women who are marked by the capacity of that which leaks from the body – menstrual blood is the best exemplar – to defile and contaminate. In short, women, are both dangerous and excluded others, but also, as mothers, an originary presence. Where Kristeva explicitly looks back to the anthropological work of Mary Douglas (1969), her own concept of the abject is carried forward implicitly in Judith Butler's more abstract theorisation of the 'constitutive outside' that marks the self (Butler 1993).

It is somewhat ironic that in large part the enormous proliferation of feminist theorisations of the body has been mobilised by the response to the insights of poststructuralism and postmodernism, which in their masculinist forms have been often accused of an indifference to materiality. There is no doubt that in the hands of some practitioners, the potentially endless textual play of many postconventional strategies has seemed to preclude any engagement with the day-to-day lived body, or has at best emptied it out (the body-without-organs) leaving only an undifferentiated surface of inscription. But such an impasse – at least in terms of the feminist project – is not inevitable, and many have taken up the same theoretical challenge to the belief in a given reality to pursue quite disparate outcomes that engage fully with the concept of sexual difference, and deliver a politics and ethics appropriate to the feminist agenda. To say that the body is a discursive construction is not to deny a substantial corpus, but to insist that our apprehension of it, our understanding of it, is necessarily mediated by the *con*texts in which we speak. As Judith Butler succinctly puts it: 'there is no reference to a pure body which is not at the same time a further formation of that body' (1993: 10). It is then the forms of materialisation of the body, rather than the material itself, which is the concern of a feminism that must ask always what purpose and whose interests do particular constructions serve. And what that question entails is the recognition that if the body itself is not a determinate given, then the political and social structures that take it as such are equally open to transformation. Moreover, it is not simply that we can vary the meaning and significance of the body, but that the very notion of 'the' body is untenable. As Gayatri Spivak asserts: 'There are thinkings of the systematicity of the body, there are value codings of the body. The body as such cannot be thought' (1989: 149). What this means is that there are only multiple bodies, marked not simply by sex, but by an infinite array of differences – race, class, sexuality, age, mobility status are those commonly invoked – none of which is solely determinate. In such a model the universal category of the body disappears, not as the result of the disembodiment characteristic of masculinist discourse, but in favour of a fluid and open embodiment. At any given moment we are always marked corporeally in specific ways, but not as an unchanging or unchangeable fixture.

Perhaps the area of greatest take-up of postconventional modes of analysis is evident in the work of those feminists who have been inspired, directly or indirectly, by Foucauldian theory. Of all the Continental theorists emerging during the last few decades who break radically with the philosophical tradition that has dominated epistemology and ontology – names like Lacan, Derrida, Deleuze, Kristeva spring to mind – it is Foucault, particularly in his focus on the discursive construction of the body, who has been most accessible, and most easily adapted to the feminist agenda. Where Foucault famously sees his task as one 'to expose a body totally imprinted by history and the process of history's destruction of the body' (1977b: 148) – which he indicates has operated most recently in the interests of capitalism – feminists have undertaken to extend that explication to take account of patriarchy. Foucault's sustained analysis, in works such as *Discipline and Punish* (1977a) and *The History of Sexuality, Vol. 1* (1979), of the discursive operations that construct the useful, manipulable body – what he calls the 'docile body' – have been a fertile ground for feminist understandings that make clear the links between the everyday body as it is lived, and the regime of disciplinary and regulatory practices that shape its form and behaviour. Although gender is of subsidary interest to Foucault, it is nonetheless clear even in his own work that it is the female body above all that is constructed and marked in such a way, a point that feminists have been quick to develop. Theorists such as Susan Bordo (1993) and Sandra Bartky (1988) have been in the forefront in analysing how the processes of surveillance and self-surveillance are deeply implicated in constituting a set of normativities towards which bodies intend. The practices of diet, keep-fit, fertility control, fashion, health care procedures and so on are all examples of disciplinary controls which literally produce the bodies that are their concern. Given the pre-eminent position of the discourse of biomedicine in such a schema, it is incumbent on feminists, and particularly those associated with the longstanding women's health movement, to rethink the traditional claims of medical practice to cure and care. But the point is not only to make clear that notion of control, but to uncover its part in the constitution of the body: 'the discursive power of biomedicine does not simply direct choice among alternative models of the body ... it actively and continuously constructs the body' (Shildrick and Price 1998).

It is not our purpose to suggest that feminist theory is merely derivative of pre-existing masculinist analyses, simply adding in a sensitivity to sexual difference or some such where that is missing. In any case, to forefront the notion of the body as sexed is not just another consideration to be taken into account, but a factor that changes everything. Moreover, many writers, such as Irigaray who deconstructs the deconstruction of a Derrida or Lacan, are impressive theorists in their own right, moving the argument into highly original forms. In a similar way, Judith Butler's initial take-up of the Foucauldian concept of the disciplinary body has produced the notion of performativity – as distinct from the less nuanced term performance – to explain how the deployments of the body through acts and gestures, especially in terms of gendered sexuality, are, through a process of reiteration, productive of a discursive identity that is both open and constrained. And although in *Gender Trouble* (1990) her emphasis is relatively limited to sexual difference, she goes on to develop the idea in a wider field. As Butler sees it, a whole range of normative binaries that are used to characterise embodiment – male/female, health/ill-health, heterosexual/homosexual, black/white and so on – may be exposed in their instability – but also paradoxically confirmed – by the performativity of abject bodies. The precariousness of all bodies is exposed by paying attention to their 'constitutive outside', from which they are alienated and yet on which they are dependent. What seems important is that although the female body may provide the primary point of interest, the analyses offered here and elsewhere tell us something new about all bodies, that throws light not only on those that are marginalised by sex, sexuality, skin colour, disability, age, and all the other conditions of difference, but on the fully normative body itself. As we have discussed briefly, Irigaray herself is somewhat open to the charge of an ahistorical essentialism in her concentration on the female, and particularly maternal, body, but for other theorists such as Butler or Liz Grosz the commitment to feminism is fully consonant with an analysis that lends itself to a fuller understanding of corporeality in whatever form it might take.

The arguments traced out by postmodernist feminists are at times highly abstract, but as the work has moved further away from its often matter-indifferent theoretical sources it is less likely to lose sight of the body as a lived entity, or of the notion that 'language and materiality are fully embedded in each other' (Butler 1993: 69), as Butler herself puts it. That move away from a purely textual analysis is even more evident in the feminist take-up of the phenomenology of Merleau-Ponty (1962), in which the being-in-the-world of the subject is intricately and irreducibly bound up with the constitution and extension of the body. It is a model in which the Cartesian mind/body split has no place. In the phenomenological tradition, the structure of the self is indivisible from its corporeal capacities, but what feminist theorists importantly add is an emphasis on the differential forms of embodiment that confound normative boundaries. Where Merleau-Ponty seems to have in mind a universalised gender-neutral, and thus effectively male body, in which the self is disrupted by radical changes to bodily potential and comportment – what would broadly be characterised negatively as a loss of health – feminists have been quick to point out that the female body for one is subject to a life-long series of changes in form – menstruation, pregancy, lactation, menopause – that are naturally occuring life-events rather than exceptions. The work of Iris Marion Young (1990), Ros Diprose (1994), and increasingly Liz Grosz (1994) is deeply concerned with those processes of embodied subjectivity as they evolve within temporal and spatial parameters. That current feminist enthusiasm for phenomenology is not, however, entirely new in that Irigaray's emphasis on touch as the mediating force between embodied selves speaks to similar parameters. And like more recent theorists, Irigaray (1993) too challenges Merleau-Ponty, specifically with respect to his analysis of seeing/being seen (*The Visible and the Invisible* 1968) as offering a model for the relationship between self and other. For her, the privileging of sight is merely a continuation of a long-standing masculinist denial of maternal origins, and of the tactile as the substratum of all the senses.

Given the immense significance of the maternal body in both material and theoretical configurations for writers as diverse as Irigaray, Rich, Firestone, Kristeva and many others, it is not surprising that contemporary and projected technologies that impact on reproduction have stimulated new considerations for feminism. The actual and potential variation of the relation between mother and child alone has necessitated a rethinking of previously familiar terrain, but the challenge posed by the advanced reproductive technologies and genetic engineering goes much wider. What is at stake is no longer simply the maternal body as such, but questions of what constitutes human being, human individuality, human corporeality in general. Clearly the concerns thrown up by processes such as germ-line therapy, *ex-utero* gestation, cloning, trans-species generation and the like are not peculiar to feminism, and yet the question of sexual difference is one that must always be posed. As the certainties of biology become increasingly contested, mirroring in substance the deconstruction of postEnlightenment epistemology and ontology, feminist theory has pulled together the interconnected leakiness of bodies of matter and bodies of knowledge (Shildrick 1997). Just as postmodernist theory destabilises the foundational claims and grand narratives of western discourse, so too postmodern empiricism throws into doubt the substantive nature of flesh and blood itself. But it is not only in the field of reproduction that feminists have responded to the challenge. Theorists such as Donna Haraway have long argued that the body of the late twentieth century and beyond is determined neither by biological givens and boundaries, nor by discursive regimes of power on a Foucauldian model, but constitutes a field of conflicting and unstable flows that partake of the revolution in informatics. As she puts it:

'sociobiological stories depend on a high-tech view of the body as a biotic component or cybernetic communications system' (Haraway 1991: 169). Haraway's now famous 'Cyborg Manifesto', first published in the 1980s, heralded a feminist engagement with the body where difference could no longer be adequately construed according to the conventional binaries of gender, sexuality, race, class and similarly 'humanist' paradigms.

Haraway's speculative and ironic figuration of the half-organic, half-machine cyborg has been taken up and endlessly reworked in subsequent theory, much of which has been named as cyberfeminism. In that field the dispersal of the normative body is taken for granted, and the distinctions between human and machine, between male and female, between actual and virtual, lose currency. In cyberspace bodies are either of no consequence or may endlessly morph into new and uncategorisable forms that frustrate the modernist desire for hierarchy and order. Many feminists have seen such adventures as offering new freedoms to women to escape existing biological constraints, but it is not clear to what extent the move replays somataphobia and the ideals of transcendence rather than radically varies forms of embodiment. Women have good reason to be suspicious both of any reiteration of the mind/body split and of the supposed 'neutrality' of cyberspace, and as a site of feminist theory it remains controversial. In any case the understandable desire for corporeal transgression may be more adequately addressed through a set of tangible, albeit fluid, materialities. A large part of Haraway's own project has been not simply to deconstruct the boundaries between the organic and the inorganic, but to problematise the category not just of the female but of the human itself. For her the interconnectedness of human and animal bodies, the sense of nature as agentic rather than as a passive locus of action, the 'promises of monsters' (Haraway 1992), all speak to forms of posthuman embodiment. At the same time, technological advances such as xenotransplantion, genetic hybridisation, transsexual surgery and even cloning blur in practice the boundaries of the closed and proper human body. And yet there is nothing new about transgressive morphology: whether we see normative standards as natural or discursively constructed, there have always been those bodies which have failed to conform. All those who are monstrous by reason classically of excess (concorporate twins, superabundant hair), displacement (reversed limbs, intersexed bodies), or lack (too few limbs, unusually small stature) have been pushed to the margins and effectively othered. Although Haraway herself has a specific project of 'queering what counts as nature' (1992: 300), and more recent Queer Theory is deliberately transgressive, the inappropriate/d body has always cast doubt on the stability of the norm. The postmodern as a set of unstable conditions and practices is embedded already within the structures of modernity.

These and many other issues with regard to the body are taken up in the course of the Reader. Although the material is organised into seven discrete sections, none should be read as closed or complete. Feminist theories of the body are evolving very rapidly, and it is highly unlikely that we could ever reach a point of having exhaustively covered the ground. Nonetheless, the sections are intended to give an indication of the more important areas of debate, together with some sense of how those areas have developed during the second wave of feminism. Inevitably the choice of material will be contested by some, but we hope that the omissions as much as the inclusions will stimulate further discussion, and towards that end each section is followed by suggestions – again just a selection – for additional reading. The process of refining a book length selection to represent such a rich and diverse field is inevitably both a delight and a frustration which has been with us through our endless discussions and list-making over a period of several months. To a certain extent, the choices we have made reflect our own interests, but we have also tried to avoid too clear an overlap with other collections. Rather surprisingly, there have been in fact very few previous publications that bring together existing essays that focus on both the body and feminist theory. The US publication of *Writing on the Body* (Conboy et al. 1996) has similar aims to our own, and in view of its own bias towards literature and the arts, we have given less emphasis to those fields in our own work. Similarly, we have included very little that directly addresses questions of sexuality – unless the issue of the body is to the forefront – and would direct readers to a companion volume, *Feminism and Sexuality – a Reader* (Jackson and Scott 1996). Some of the material used is already extremely well-known, even seminal in its influence on subsequent theory – which is not to say, of course, that all of us will have already read it first hand; other pieces are relatively unknown, or inaccessible, but offer positive contributions, we think, to the debate in hand. In the majority of cases, the pieces reprinted here are either given in total or in sufficient length to stand on their own without need for immediate recourse to the original, though we urge readers to both pick up those originals and trace the connections with other pieces when time allows.

For all such practical considerations and organisation, however, it remains to stress that the body does not divide up neatly into sections. The whole force of modernist tabulations and taxonomies has been directed towards achieving mastery: completing knowledge and closing down on the unruly possibilities of the corporeal. For us, the central point of feminist theories of the body is that they reject that easy categorisation, or any striving for a false unity that belies the sometimes confusing but always productive tensions of disparate starting points, perspectives, and aims. Even with our own schema, the leaks and flows between the diverse figurations of the body should be apparent, and in many, if not all cases, a particular article might just as easily have appeared in several other of the sections. The body, then, has become the site of intense inquiry, not in the hope of recovering an authentic female body unburdened of patriarchal assumptions, but in the full acknowledgment of the multiple and fluid possibilities of differential embodiment. As Grosz puts it: 'the stability of the unified body image, even in the so-called normal subject, is always precarious. It cannot be simply taken for granted as an accomplished fact, for it must be continually renewed' (Grosz 1994: 43–4).

NOTES

1. A similar move with regard to the black body is used to justify claims of racial inferiority.
2. This was a move that spoke primarily to the needs of white middle class and upper class women. It failed to recognise the very differing constructions, and lived experiences, of black women and working class women, for example, whose lives were marked not by the fragility of their bodies but by their labour.

REFERENCES AND FURTHER READING

Bartky, Sandra Lee (1988) 'Foucault, Femininity and Patriarchal Power' in Irene Diamond and Lee Quinby (eds), *Feminism and Foucault*, Boston: Northeastern University Press.

Beauvoir, Simone de (1953) *The Second Sex*, London: Jonathan Cape.

Bordo, Susan (1993) *Unbearable Weight. Feminism, Western Culture and the Body*, Berkeley: University of California Press.

Boston Women's Health Book Collective (1976) *Our Bodies, Ourselves*, New York: Simon & Shuster.

Butler, Judith (1990) *Gender Trouble: Feminism and the Subversion of Identity*, London: Routledge.

Butler, Judith (1993) *Bodies that Matter. On the Discursive Limits of 'Sex'*, London: Routledge.

Collins, Patricia Hill (1991) *Black Feminist Thought. Knowledge, Consciousness and the Politics of Empowerment*, New York: Routledge.

Conboy, Katie, Medina, Nadia and Stanbury, Sarah (1996) *Writing on the Body: Female Embodiment and Feminist Theory*, New York: Columbia University Press.

Davis, Angela (1982) *Women, Race and Class*, London: The Women's Press.

Diprose, Ros (1994) *The Bodies of Women. Ethics, Embodiment and Sexual Difference*, London: Routledge.

Douglas, Mary (1969) *Purity and Danger*, London: Routledge and Kegan Paul.

Eisenstein, Zillah (1989) *The Female Body and the Law*, Berkeley: University of California Press.

Firestone, Shulamith (1979) *The Dialectic of Sex*, London: The Women's Press.

Foucault, Michel (1977a) *Discipline and Punish: the Birth of the Prison*, trans. Alan Sheridan, London: Allen Lane.

Foucault, Michel (1977b) 'Nietzsche, genealogy, history' in Donald Bouchard (ed.), *Language, Counter-Memory, Practice: Selected Essays and Interviews*, Ithaca, NY: Cornell University Press.

Foucault, Michel (1979) *History of Sexuality, Vol. 1*, trans. R. Hurley, London: Allen Lane.

Grosz, Elizabeth (1994) *Volatile Bodies. Towards a Corporeal Feminism*, London: Routledge.

Haraway, Donna (1991) 'A Cyborg Manifesto: science, technology and socialist feminism in the late twentieth century' in *Simians, Cyborgs and Women: The Reinvention of Nature*, London: Free Association Books.

Haraway, Donna (1992) 'The promises of monsters: a regenerative politics for inappropriate/d others' in L. Grossberg, C. Nelson and P. Treichler (eds), *Cultural Studies*, London: Routledge.

hooks, bell (1981) *Ain't I a Woman. Black Women and Feminism*, Boston: Southend Press.

hooks, bell (1991) 'Postmodern Blackness' in *Yearning. Race, Gender and Cultural Politics*, Boston: Southend Press.
Irigaray, Luce (1985) *This Sex Which Is Not One*, Ithaca: Cornell University Press.
Irigaray, Luce (1993) *An Ethics of Sexual Difference*, Ithaca: Cornell University Press.
Jackson, Stevi and Scott, Sue (1996) *Feminism and Sexuality – a Reader*, Edinburgh: Edinburgh University Press.
Kristeva, Julia (1982) *Powers of Horror: An Essay on Abjection*, New York: Columbia University Press.
Lorde, Audre (1984) *Sister Outsider. Essays and Speeches*, Freedom, CA: The Crossing Press.
Lorde, Audre (1988) *A Burst of Light*, New York: Firebrand Books.
Merleau-Ponty, Maurice (1962) *Phenomenology of Perception*, trans. C. Smith, London: Routledge and Kegan Paul.
Merleau-Ponty, Maurice (1968) *The Visible and the Invisible*, trans. A. Lingis, Evanston: Northwestern University Press.
Moi, Toril (1985) *Sexual/Texual Politics*, London: Methuen.
Rich, Adrienne (1979) *Of Woman Born: Motherhood as Experience and Institution*, London: Virago.
Shildrick, Margrit (1997) *Leaky Bodies and Boundaries: Feminism, Postmodernism, and (Bio)ethics*, London: Routledge.
Shildrick, Margrit and Price, Janet (1998) *Vital Signs: Feminist Reconfigurations of the Bio/logical Body*, Edinburgh: Edinburgh University Press.
Spivak, Gayatri Chakravorty (1989) 'In a Word', interview with Ellen Rooney, *differences* 1 (2): 124–56.
Suleiman, Susan Rubin (ed.) (1986) *The Female Body in Western Culture*, Cambridge, MA: Harvard University Press.
Terry, Jennifer and Urla, Jacqueline (eds) (1995) *Deviant Bodies*, Bloomington: Indiana University Press.
Weedon, Chris (1987) *Feminist Practice and Poststructuralist Theory*, Oxford: Blackwell.
Wendell, Susan (1996) *The Rejected Body. Feminist Philosophical Reflections on the Disabled Body*, London: Routledge.
Young, Iris Marion (1990) *Throwing Like a Girl*, Bloomington: Indiana University Press.

SECTION 1
WOMAN AS BODY?

INTRODUCTION

This section serves to introduce questions of how we analyse, understand and theorise our bodies as women, particularly within the values of modernism which grounded early second wave feminism. Broadly speaking the move has been from an initial disregard for body matters – except in the context of sexuality and reproduction – through a recognition of the force of body politics, to a contemporary focus on the inescapable relationship between embodiment, power and knowledge.

The identification of women with the body is a familiar idea in the Western tradition from Aristotle to postCartesian modernism, although not necessarily one that traces an unbroken continuum. It is not the purpose of this section to present the link as an ahistorical truth, but rather to uncover the complex components and implications that are inherent in the apparent straightforwardness of the woman/body/nature association. Although somatophobia, especially in association with the feminine, is a recurring feature, it must be contextualised in terms of very different paradigms of the body over time. The moves, for example, from a Platonic concern with the state of the embodied soul, to the medieval concentration on pain, death and decay, through Enlightenment concerns with the transcendence of the mind over the body, to our current focus on pleasure, sex and life do little to suggest continuity. Even in the light of such broad and relatively crude summations, it is clear that the meaning of the body changes dramatically over time. This in turn suggests that the questions feminists ask about the body should be ones set within the context and concerns of particular historical moments. Nonetheless, although the relationships between such studies and the questions pertinent to the present are not direct ones, they are ones, as Bynum (1995) suggests, of analogy and proportion. The relevance for today is that the past has shaped our own context, and allows us to draw broader lessons from the specificity of late twentieth-century studies.

With all that in mind, the extracts here come from a range of disciplinary areas – history, biology, philosophy, sociology – and examine the body in a variety of ways – as pure physicality, as the carrier of self/identity, as an indicator of position in hierarchies of power. The questions of sexual difference that are so often the focus of feminist theory are complicated by the problems of applying current categories to the past. For one thing, the specificity of sexual difference cannot be taken as a singular constant, but is often intimately tied up with other operative differences such as those of race. Londa Schiebinger, for example, traces the complexity of how histories of the measurement and documentation of bodies are bound up with colonialism, with scientific sexism and racism, with a need to prove the irrationality of women, and with the desire to exclude them from political life. As she puts it, '(w)oman, considered a monstrous error of nature, was studied for her deviation from (the) male norm', but, nonetheless, not all similarly sexed bodies had the same value. Schiebinger clearly shows how studies of sex and race in the eighteenth century were linked to explicit political questions of rights and equality, and she points out that gender is not, in all situations, the primary distinguishing characteristic between humans. The corporeal mapping of race and sex is never simple, and feminists need to ask more complex questions about the interrelationship of those categories and the body.

It is no surprise, then, that the devaluation of women through their putative identification with bodily matters is mirrored by a similarly grounded devaluation of slaves, of children, of animals, and of black people. Work in both philosophy and science studies, as the pieces by Elizabeth Spelman and Lynda Birke show, outlines a move to reduce marginalised others to their bodies. The process of categorisation that works to set in place boundaries between the normal and the abnormal, between the fully human and the sub-human, is as pertinent to the classical institution of slavery as it is to more recent biological reductionism that ties women to their reproductive functions. As Spelman eloquently points out, one heritage of the Platonic distinction between bodies and souls, which ranged women, slaves and animals on the side of the former, is a somatophobia both in the philosophies that feminists critique, and in work of feminists themselves who wish to deny the 'taint' of immanence. Rather than become mired in what is seen as the gross giveness of physicality in the feminine – especially with respect to pregnancy and childbirth – it seemed more liberating to claim that women too could transcend their bodies. Even Simone de Beauvoir who took women's bodies seriously enough to devote a major work, *The Second Sex*, to them was unable or unwilling to reconsider the value split between transcendence and corporeality. Although she did not believe that biological facts establish a fixed or inevitable destiny – as she says, 'the facts of biology take on the values that the existent bestows on them' (de Beauvoir 1972: 69) – she displays nonetheless an apparently deep aversion to some everyday experiences of the female body. On reproduction she remarks scathingly: 'giving birth and suckling are not *activities*, they are natural functions; no project is involved; and that is why woman found in them no reason for a lofty affirmation of her existence' (de Beauvoir 1972: 94). The answer, then, is clear: women too must leave behind the materiality of their bodies.

In her emphasis on the possibilities of women's becoming, Simone de Beauvoir holds out hope to women, but it is a different move to 'the body becoming' by means of which Birke signals the possibility of transformation within the biological itself. For all her ambivalence, de Beauvoir is to be associated with the phenomenological approach that sees the body and its social/cultural world in a mutually constitutive relationship. Instead of the body being positioned as a bar to knowledge, knowledge is produced through the body and embodied ways of being in the world. Both Felly Nkweto Simmonds and Helen Marshall utilise such an approach, which entails reflection on the habitus of the body, and on the image of, and experiential changes to, the body itself. Clearly, while being female is a major parameter of experience, it is only one of a number of possible aspects of bodily comportment. For Simmonds, her position as a Black woman is both a defining part of her social reality, a fully evident physical 'fact', but at the same time as being asked to speak on race issues, she feels that to 'talk about the body is to invite derision'. Simmonds takes the risk, however, in discussing her embodied reality not only in terms of her blackness, but also in terms of her experience of illness. In a similar way, Marshall utilises the the bodily transitions of her own pregnancy, to offer a phenomenology of the everyday body. What both extracts speak to, in a sense that is scarcely apparent in de Beauvoir's work, is a way of theorising the physicality of the body – including the body that bleeds, that sweats, that feels pain, that is marked both externally and internally – a way which moves beyond the dis-ease with flesh and blood experience that has often characterised feminist theory.

Where the selections in this section reject the crudity of the traditional identification of women with their bodies, they do not reject the body itself, but seek to find ways to reconceptualise its nature as an organism with both integrity and transformativity. For a very different set of theorisations, see Section 4, 'After the Binary' in which the body as such is a much more slippery concept, evading all attempts at categorisation.

REFERENCES AND FURTHER READING

de Beauvoir, Simone (1972) *The Second Sex*, Harmondsworth: Penguin Books.

Bynum, Caroline (1995) 'Why All the Fuss about the Body? A Medievalist's Perspective' in *Critical Inquiry* 22: 1–33.

Doy, Gen (1996) 'Out of Africa: Orientalism, "Race" and the Female Body' in *Body & Society* 2 (4): 17–44.

Gatens, Moira (1991) 'Woman as the Other' in *Feminism and Philosophy. Perspectives on Difference and Equality* Cambridge: Polity Press.
hooks, bell (1981) 'Sexism and the Black Female Slave Experience' in *Ain't I a Woman*, Boston: Southend Press.
Jordanova, Ludmilla (1989) *Sexual Visions: Images of Gender in Science and Medicine between the Eighteenth and Twentieth Centuries*, London: Harvester Wheatsheaf.
Kaplan, Gisela and Rogers, Lesley (1990) 'The Definition of Male and Female. Biological Reductionism and the Sanctions of Normality' in Sneja Gunew (ed.), *Feminist Knowledge: Critique and Construct*, New York: Routledge.
Kappeler, Susanne (1994/95) 'From Sexual Politics to Body Politics' in *Trouble and Strife* 29 (30): 73–70.
Oakley, Ann (1982) 'Genes and Gender' in *Subject Woman*, London: Fontana.
Ortner, Sherry B. (1974) 'Is Female to Male as Nature is to Culture' in M. Z. Rosaldo and L. Lamphere (eds), *Women, Culture and Society*, Stanford: Stanford University Press.
Spanier, Bonnie B. (1991) '"Lessons" from "Nature": Gender Ideology and Sexual Ambiguity in Biology' in J. Epstein and K. Straub (eds), *Body Guards. The Cultural Politics of Ambiguity*, New York: Routledge.
Scheman, Naomi (1993) 'The Body of Privilege' in *Engenderings. Constructions of Knowledge, Authority and Privilege*, New York: Routledge
Wittig, Monique (1992) 'The Category of Sex' and 'One is Not Born a Woman' in *The Straight Mind and Other Essays*, Hemel Hempstead: Harvester Wheatsheaf.

1.1

THEORIES OF GENDER AND RACE

Londa Schiebinger

> The volcanic eruption of freedom in France will soon bring on a larger explosion, one that will transform the destiny of humankind in both hemispheres. (Henri Grégoire, *De la littérature des nègres*, 1808)

The expansive mood of the Enlightenment – the feeling that all men are by nature equal – gave middle- and lower-class men, women, Jews, Africans, and West Indians living in Europe reason to believe that they, too, might begin to share the privileges heretofore reserved for élite European men. Optimism rested in part on the ambiguities inherent in the word 'man' as used in revolutionary documents of the period. The 1789 *Declaration of the Rights of Man and Citizen* said nothing about race or sex, leading many to assume that the liberties it proclaimed would hold universally. The future president of the French National Assembly, Honoré-Gabriel Riqueti, comte de Mirabeau, declared that no one could claim that 'white men are born and remain free, black men are born and remain slaves'.[1] Nor did the universal and celebrated 'man' seem to exclude women. Addressing the convention in 1793, an anonymous woman declared: 'Citizen legislators, you have given men a constitution ... as the constitution is based on the rights of man, we now demand the full exercise of these rights for ourselves'.[2]

Within this revolutionary republican framework, an appeal to natural rights could be countered only by proof of natural inequalities. The marquis de Condorcet wrote, for instance, that if women were to be excluded from the

From: L. Schiebinger, *Nature's Body: Gender in the Making of Modern Science*, Boston: Beacon Press, 1993.

polis, one must demonstrate a 'natural difference' between men and women to legitimate that exclusion.[3] In other words, if social inequalities were to be justified within the framework of Enlightenment thought, scientific evidence would have to show that human nature is not uniform, but differs according to age, race, and sex.

Scientific communities responded to this challenge with intense scrutiny of human bodies, generating countless examples of radical misreadings of the human body that scholars have described as scientific racism and scientific sexism.[4] These two movements shared many key features. Both regarded women and non-European men as deviations from the European male norm. Both deployed new methods to measure and discuss difference. Both sought natural foundations to justify social inequalities between the sexes and races. Eighteenth-century anthropologists, though, did not always perceive that what they said about sex had a bearing on race and vice versa. Leading theories underlying scientific racism (the doctrine of a great chain of being, for example) did not incorporate new views on sexual difference, while leading theories explaining sexual divergence (the doctrine of sexual complementarity being a prime example) applied only to Europeans.

Here we explore the paradoxes and incompatibilities plaguing eighteenth-century theories of sexual and racial difference. Where did naturalists place women along the great chain of being? To what extent did the theory of sexual complementarity reach beyond Europe? How, in other words, did notions of gender influence the study of race, and how did notions of European superiority influence studies of sex? The anatomy of sex and race was caught up, as we shall see, in eighteenth-century politics of participation – struggles over who should do science and who should be actively involved in affairs of the state. Eighteenth-century male anatomists in Europe were obsessed with black men (the dominant sex of an inferior race) and white women (the inferior sex of the dominant race). It was these two groups, and not primarily women of African descent, who challenged European male élites in their calls for equal rights and political participation.

WERE WOMEN ON THE CHAIN?

One of the most powerful doctrines governing theories of race in the eighteenth century was the great chain of being. This doctrine postulated that species were immutable entities arrayed along a fixed and vertical hierarchy stretching from God above down to the lowliest sentient being. The historian Winthrop Jordan has shown that the notion of a chain of being became the darling of eighteenth-century conservatives in their attempts to stem the levelling tide of democracy and abolitionism.[5] The conservative British naturalist William Smellie, for example, taught that social hierarchies issued from natural hierarchies. 'Independently of all political institutions', Smellie wrote in his 1790 *Philosophy of Natural History*, 'Nature herself has formed the human species into castes and ranks'.[6]

Europe's anatomists dissected and analyzed the skeletons of animals and humans from every corner of the world in their attempts to substantiate the notion that nature shades continuously from one form to another. Of special interest were the transitional forms bridging the gap between animals and humans. Although different animals vied for a time as the 'missing link' (elephants, for their intelligence, and parrots, for their ability to talk), by the eighteenth century naturalists had settled on the ape, and especially the orangutan (still commonly used as a generic name for both chimpanzees and orangutans), as the animal most resembling humankind. What, though, was the 'lowest' sort of human? Voyagers, coming into contact with Africans in the course of colonial expansion and the slave trade, had already suggested that the people of this continent resembled the apes who inhabited this same region.[7] (Some went so far as to suggest that the black race originated from whites copulating with apes.)[8] Within this context arose a project central to eighteenth-century anatomy: investigation into the exact relationship among apes, Africans, and Europeans.

Much has been written about the racist implications of the chain of being.[9] What has not been investigated, however, is the place of females in that hierarchy. The notion of a single chain of being stretching throughout nature (and society) created a problem of where to fit women. Scientific racism and scientific sexism both taught that proper social relations between the races and the sexes existed in nature. Many theorists failed to see, however, that their notions of racial and sexual relations rested on contradictory visions of nature. Scientific racism depended on a chain of being or hierarchy of species in nature that was inherently unilinear and absolute. Scientific sexism, by contrast, depended on radical biological divergence. The theory of sexual complementarity attempted to extract males and females from competition with or hierarchy over each other by defining them as opposites, each perfect though radically different and for that reason suited to separate social spheres.[10] Thus the notion of a single chain of being worked at odds with the revolutionary view of sexual difference which postulated a radical incommensurability between the sexes (of European descent).

Before investigating further women's place on the chain, we must turn to the glaring asymmetries in studies of race and sex in this period. Most strikingly, racial science interrogated males and male physiology, while sexual science scrutinized European subjects. As one might imagine, eighteenth-century comparative anatomists and anthropologists were overwhelmingly male. What is especially revealing, however, is that they developed their theories about race by examining male bodies. Females were studied, but only as a sexual subset of any particular race. Consider the work of the German anatomist Samuel Thomas von Soemmerring in this regard. His 1785 book on race, *Über die körperliche Verschiedenheit des Negers vom Europäer*, compared the bodies of Africans and Europeans, most of which were male.[11] His preference for male bodies was not simply an artifact of availability. He had dissected at least one female

African in Kassel, observed 'dozens' of blacks (including females) at the public baths,[12] and had at least part of a female African skeleton (probably from the dissected female) in his anthropological collection.[13] Soemmerring also knew black women personally, having arranged transportation to Amsterdam for one of 'his Mohrin' (he commonly referred to Africans as Moors).[14] It is not clear, though, what was meant by calling her 'his' Mohrin – she may have been his servant or perhaps simply one of his objects of study.

At the same time, when anatomists turned their attention to sexual difference, they tended to confine their studies to middle-class Europeans. While Soemmerring's study of race focused on males, his study of sex, epitomized in his classic illustration of the female skeleton undertaken eleven years after his anatomy of race, treated only Europeans.[15] Indeed, anatomists' portrayals of distinctively female skeletons, ushering in the eighteenth-century revolution in views of sexual difference, were all of Europeans. The canonical texts of that revolution – by Rousseau, Roussel, Ackermann, and Moreau de la Sarthe – similarly compared males and females of undifferentiated European origin.[16] Females were rarely compared across racial lines in the eighteenth century; or, if they were, it was commonly in relation to their sexual parts. Only in the nineteenth century did Virey's *De la femme*, written to complement his work on race published some twenty years earlier, explore the 'natural history of woman' with some attention to race.[17]

This assumption that the racial subject was male and that sexual differentiation was primarily about Europeans had deep roots. At least since Aristotle natural historians had given preference to the study of male bodies, or more precisely, the bodies of male citizens. Woman, considered a monstrous error of nature, was studied for her deviation from this male norm. The philosopher Elizabeth Spelman has recently shown that when Aristotle spoke about 'women', he referred, in fact, only to free women (the wives of citizens). She goes on to show that when Aristotle spoke about slaves, he assumed a male subject, even though he recognized that there were significant numbers of women among slave populations. Aristotle's political philosophy drew no distinction between male and female slaves; it was also neutral with respect to the relation between slaves and free women because both women and slaves found their place in the polis through the services they rendered to male citizens.[18]

Physical anthropologists in the eighteenth century did not consciously center their studies of race on the male body. For them, the 'European', 'African', and 'American' were assumed to be generic, sexless universal types. While men of science argued over whether a skull, arriving from overseas stripped of skin and hair, was genuinely that of an African, Carib, or Brazilian (suppliers were often unreliable concerning the provenance of their specimens), they rarely queried the sex of the specimen. Inventories of their collections of hair, skulls, skin, genitalia, and other soft tissues reveal that, where the sex of the specimen was known and recorded, it was most often male.[19] Even Buffon, who drew his information from travelogues (not dissection tables) and had quite a lot to say about women, assumed that the person described was male unless specifically labelled female.[20]

One might suppose that anatomists lavished attention on males in this period because there were simply more men among foreign populations living in Europe, and their bodies were more readily available for study. Three quarters of the 765 *gens de couleur* living in Paris between the years 1777 and 1790, for instance, were male.[21] Men (both European and non-European) travelled more freely and were active in foreign armies. But the attention given to males in this period also reveals a recalcitrant preference for them. Charles White collected data primarily from men (at a ratio of about six to one) in settings where women were equally available. When measuring the upper arms and forearms of living subjects in Manchester, he used his butler, gardener, coachman, and footman, not his maid or (female) cook. He also sought out measurements from a male apothecary at the local lying-in hospital – an environment abounding with potential female subjects whom he investigated only when studying distinctively female features, such as changes in pigmentation during pregnancy.[22]

That males were taken as universal racial subjects can best be seen in anthropological studies of skulls. In the eighteenth century, skulls were used to forge a new, racial chain of being. Interest in the faculty of reason, considered by many the generic characteristic responsible for humankind's vast superiority over all other animals, and increasingly important for citizenship in European society, brought to the fore the study of skulls as measures of intelligence. Reason had long been considered a masculine trait in Western cultures, and the skulls used in these studies were overwhelmingly male.[23]

[...]

When anthropologists did compare women across cultures, their interest centered on sexual traits – feminine beauty, redness of lips, length and style of hair, size and shape of breasts or clitorises, degree of sexual desire, fertility, and above all the size, shape, and position of the pelvis. For the anatomists among them, it was the pelvis (and its procreative virtues) that ultimately emerged as the universal measure of womanliness.[24] The female pelvis, however, never achieved the symbolic standing of the male skull. No chart of pelvises took on the mythic proportions of Camper's skulls and their successors. Skulls, and indeed male skulls, remained the central icon of racial difference until craniometry was replaced by intelligence testing in the late-nineteenth and early-twentieth centuries.[25]

Initially, anatomists steered away from the pelvis when ordering the races because it posed problems for their supposed chain of racial being. While their rankings of skulls were devised to show the proximity of apes and Negroes, pelvises threatened to tell a different story. African women's alleged extraordinary ease in parturition seemed to indicate pelvises more capacious than European women's – bigger in this case being better. Voyagers and ships' surgeons had long reported that women of African descent gave birth easily (this was also assumed to be true of apes and other quadrupeds). Buffon assured his eighteenth-century readers that African women 'bring forth their children with great ease, and require no assistance. Their labors are followed by no troublesome consequences; for their strength is fully restored by a day or, at most, two days' repose'.[26] The historian Barbara Bush has pointed out that this myth justified working slave women throughout their pregnancies and returning them to the fields shortly after parturition.[27]

Camper made sporadic attempts to measure female pelvises beginning in the 1750s. His initial results seemed to confirm the 'well-known fact' that women of warm climates gave birth easily. Devising a pelvic angle from various measurements of the pelvic aperture, he set the optimum for the well-built (European) woman at one hundred degrees. He then procured 'not without much trouble and great cost' pelvic bones from a female African, Asian, and American. In each of these subjects he discovered the pelvic opening to be 'wider and noticeably rounder' so that the child's head could easily 'shoot through'.[28] Elsewhere, however, Camper drew the conclusion that Africans had *narrower* pelvises than Europeans from his comparison of the ratio of the length to the breadth of the pelvis in seven humans (one European female and probably two others, three European males, and one male Negro) and from his examination of pelvic measurements from three statues (the Farnese Hercules, Pythian Apollo, and Venus de Medici).[29] His explanation in this text was that Africans' pelvises were narrowed during childhood by virtue of the fact that the back part of their skulls were proportionally heavier than whites'. In order to keep their balance, Africans tended to throw their heads back and jut their necks forward in such a manner that their pelvises gradually bent inward.[30] Soemmerring confirmed Camper's conclusion that Africans had narrower pelvises than Europeans after measuring four male subjects (but not the African female he had dissected).[31] Prichard and White suggested that Camper and Soemmerring's conclusions were flawed because too many of their subjects were male.[32]

While the study of skulls emerged prominently in the late eighteenth century, systematic study of the racial pelvis did not begin until the 1820s. Moritz Weber, a pioneer in this respect, proclaimed in his 1830 *Theory of Fundamental and Racial Forms of the Skull and Pelvis in Humankind*: 'That racial skulls exist is now a matter of fact; that racial pelvises also exist ... has only very recently been proven'.[33] When pelvises were studied systematically, they were ranked superior or inferior, lower or higher in accord with the principles of the supposed chain of being. Studies of the 'racial pelvis' brought the female pelvis into line with the chain. Differences in sex, which had long been discussed, were subordinated to racial types which, in turn, were arrayed along the chain. In Africans, the female pelvis, by contrast with the male, was said to be light and delicate; the male pelvis was said to be so dense that it resembled the pelvis of a wild beast. When compared to the pelvis of the European female, however, the African woman's was described as entirely destitute of the transparent delicacy characteristic of the female European. The Dutch anatomist Willem Vrolik made a point of saying that no matter how light or fine the pelvis of African females, they were of the same race as African males. Their pelvises shared with the males the *Urform* of that race: an elongated form said to recall the shape of the pelvis in apes.[34]

In the 1830s, Weber sorted pelvises into distinct racial types. The oval pelvis was found most commonly in Europeans; the round pelvis in Americans; the square in Mongolians; and the oblong (the most narrow, bestial type) in Africans. The seeming contradiction of the African pelvis had been resolved. Ease of parturition no longer suggested a larger African pelvis, but a smaller African skull (there was assumed to be a natural complementarity between head and pelvis). Effortless birthing was also assumed to result from their primitive way of life (simple diets and constant labor).[35] The notion that Africans had crudely narrow pelvises spawned the absurd notion current in the late nineteenth century that steatopygia (the enlarged buttocks characteristic of Hottentot females) was a natural adaptation mimicking the large pelvis of the 'higher races'. Their large buttocks were seen, in other words, as nature's way of compensating for racial deficiencies.[36]

This brings us back to the question: Where were women on the postulated chain of being? Recent scholars, discussing nineteenth-century racial science, have argued that scientists of that period saw sex and race as two aspects of the same problem. Women and Africans were seen as sharing similar deficiencies when measured against a constant norm – the élite European man. Women and black males had narrow, childlike skulls; both were innately impulsive, emotional, and imitative. European women shared the apelike jutting jaw of the lower races, while males of the lower races had prominent bellies similar to those of Caucasian women who had borne many children.[37]

This assessment, however, does not take into account the contrasting ways in which European naturalists described women of different ethnic and cultural backgrounds. The eighteenth-century revolution in views of sexual difference applied only to Europeans. The theory of sexual complementarity born of that revolution offered a new picture of the middle-class European female. Fragments of the old picture of woman as man *manqué* did persist: she did not measure up to the male in terms of physical and intellectual strength, for example. But her portrait was no longer entirely negative. In this era of the polarization of public and private spheres, the ideal of republican motherhood saw woman as delicate, pure, and passionless, a bastion of moral and spiritual virtue – 'the better half' of her robust, solid, and assertive companion.

The black female (and by various degrees of skin hue, other non-European women) did not fit this vision of womanhood. African women were seen as wanton perversions of sexuality, not paragons of piety and purity. They served as foils to the Victorian ideal of the passionless woman, becoming, as Sander Gilman has written, the central icon for sexuality in the nineteenth century.[38] Women of African descent were not idealized as angels of the household, but forced to labor in the fields like beasts whom they were said to resemble. Elite European naturalists who set such store by complementarity when describing their own mothers, wives, and sisters refused to include African women in their new definitions of femininity.

Blumenbach, who ridiculed attempts to identify a natural scale of being as 'metaphorical and allegorical amusements', discussed the impossibility of building a single sexual and racial hierarchy. In animals such as silkworms, he suggested, there was so great a difference in the appearance of either sex, that if you wanted to refer them to a scale of that kind, 'it would be necessary to separate the males as far as possible from their females, and to place the different sexes of the same species in the most different places'.[39] Such a separation would have been, in his view, absurd.

To the extent, then, that comparative anatomists in this period devised a scale of being, it emerged from the comparison of male virtues across races, especially the virtues of male skulls. In most instances, sexual differences were considered secondary to racial differences. As Camper wrote, 'men, women, and children bear the characteristic marks of their race from their births', for, as Virey explained a half century later, the two sexes are submitted to the same general laws of nature.[40] Europeans were not particularly interested in whether African females were physically and morally superior or inferior to African males, rather both sexes were compared to Europeans. Females in general were considered a sexual subset of their race; unique female traits only served to confirm their racial standing. In eighteenth-century Europe, the male body remained the touchstone of human anatomy.

NOTES

1. Cited in Charles Hardy, *The Negro Question in the French Revolution* (Menasha, Wisc.: George Banta Publishing Co., 1919), p. 15. Condorcet expressed similar sentiments in his 'Lettres d'un bourgeois de New Haven à un citoyen de Virginie' (1787), in *Oeuvres de Condorcet*, ed., A. Condorcet O'Connor and M. F. Arago (Paris, 1847), vol. 9, pp. 15–19.
2. Cited in Jane Abray, 'Feminism in the French Revolution', *American Historical Review* 80 (1975): 48.
3. Marie-Jean-Antoine-Nicolas de Caritat, marquis de Condorcet, 'Sur l'admission des femmes au droit de cité' (1790), in *Oeuvres* (Stuttgart: F. Frommann, 1968), vol. 10, p. 129. See Maurice Bloch and Jean Bloch, 'Women and the Dialectics of Nature in Eighteenth-Century French Thought', in *Nature, Culture and Gender*, ed. Carol P. MacCormack and Marilyn Strathern (Cambridge: Cambridge University Press, 1980), pp. 25–41; Steven Rose, Leon Kamin, and Richard Lewontin, *Not in Our Genes: Biology, Ideology and Human Nature* (New York: Pantheon, 1984), pp. 63–81; Christine Fauré, *Democracy without Women: Feminism and the Rise of Liberal Individualism in France*, trans. Claudia Gorbman and John Berks (Bloomington: Indiana University Press, 1991); and Michèle Crampe-Casnabet, 'Saisie dans les oeuvres philosophiques (XVIIIe siècle)', in *Histoire des femmes en Occident*, ed. Natalie Davis and Arlette Farge (Paris: Plon, 1991), vol. 3, pp. 327–358.
4. For studies of race see, for example, Maragaret Hodgen, *Early Anthropology in the Sixteenth and Seventeenth Centuries* (Philadelphia: University of Pennsylvania Press, 1964); Philip D. Curtin, *The Image of Africa: British Ideas and Actions, 1780–1850*

(Madison: University of Wisconsin Press, 1964); David Brion Davis, *The Problem of Slavery in Western Culture* (Oxford: Oxford University Press, 1966); George Stocking, Jr., *Race, Culture, and Evolution: Essays in the History of Anthropology* (New York: Free Press, 1968); also his *Bones, Bodies, Behavior: Essays on Biological Anthropology* (Madison: University of Wisconsin Press, 1988); Winthrop D. Jordan, *White over Black: American Attitudes toward the Negro, 1550–1812* (Chapel Hill: University of North Carolina Press, 1968); and Nancy Leys Stepan, *The Idea of Race in Science: Great Britain, 1800–1960* (Hamden, Conn.: Archon Books, 1982). For studies of sexual differences, see Elizabeth Fee, 'Nineteenth-Century Craniology: The Study of the Female Skull', *Bulletin of the History of Medicine* 53 (1979): 415–433; Esther Fischer-Homberger, *Krankheit Frau und andere Arbeiten sur Medizingeschichte der Frau* (Bern: Hans Huber Verlag, 1979); Londa Schiebinger, *The Mind Has No Sex? Women in the Origins of Modern Science* (Cambridge, Mass.: Harvard University Press, 1989); Ludmilla Jordanova, *Sexual Visions: Images of Gender in Science and Medicine between the Eighteenth and Twentieth Centuries* (Madison: University of Wisconsin Press, 1989); Thomas Laqueur, *Making Sex: Body and Gender from the Greeks to Freud* (Cambridge, Mass.: Harvard University Press, 1990). For studies of sex and race, see, for example, Nancy Leys Stepan, 'Race and Gender: The Role of Analogy in Science' *Isis* 77 (June 1986): 261–277; Elizabeth Spelman, *Inessential Woman: Problems of Exclusion in Feminist Thought* (Boston: Beacon Press, 1988); Bell Hooks, *Black Looks: Race and Representation* (Boston: South End Press, 1992); Henry Louis Gates, Jr., ed., *'Race', Writing, and Difference* (Chicago: University of Chicago Press, 1986); Kathy Peiss and Christina Simmons, eds., *Passions and Power: Sexuality in History* (Philadelphia: Temple University Press, 1989); David Goldberg, ed., *Anatomy of Racism* (Minneapolis: University of Minnesota Press, 1990); Dominick La Capra, ed., *The Bounds of Race: Perspectives on Hegemony and Resistance* (Ithaca: Cornell University Press, 1991); and Evelyn Brooks Higginbotham, 'African-American Women's History and the Metalanguage of Race', *Signs: Journal of Women in Culture and Society* 17 (1992): 251–274.

5. Arthur Lovejoy, *The Great Chain of Being: A Study of the History of an Idea* (1933; Cambridge, Mass.: Harvard University Press, 1964). Winthrop D. Jordan, *White over Black: American Attitudes toward the Negro, 1550–1812* (Chapel Hill: University of North Carolina Press, 1968), pp. 217–228.
6. William Smellie, *The Philosophy of Natural History* (Edinburgh, 1790), vol. 1, pp. 521–522. See also Charles White, *An Account of the Regular Gradation in Man and in Different Animals and Vegetables* (London, 1796).
7. See also Samuel Thomas von Soemmerring, *Über die körperliche Verschiedenheit des Negers vom Europäer* (Frankfurt, 1785), p. xiv.
8. Reported in Petrus Camper, *The Works of the Late Professor Camper on the Connexion between the Science of Anatomy and the Arts of Drawing, Painting, Statuary, etc.*, trans. T. Cogan (London, 1794), p. 32, though this was not his opinion.
9. One of the best discussions is found in Jordan, *White over Black*, pp. 215–265.
10. See Thomas Laqueur, *Making Sex: Body and Gender from the Greeks to Freud* (Cambridge, Mass.: Harvard University Press, 1990); and also Londa Schiebinger, *The Mind Has No Sex? Women in the Origins of Modern Science* (Cambridge, Mass.: Harvard University Press, 1989), chaps. 7, 8.
11. As Soemmerring wrote in his preface, he had studied several (*mehere*) males and one female. Soemmerring supplemented his own specimens with parts from other anatomists' collections, occasionally referring, for example, to a female African skull that one of his colleagues had prepared two years earlier. He had five complete African skulls (four male and one female) in his own collection (*Über die körperliche Verschiedenheit des Negers vom Europäer*, pp. xvii, 50). Two illustrations drawn for this book were also of males (these portraits were never published).

Soemmerring's papers, Senckenbergische Bibliothek, Frankfurt, Lfd. Nr. 2, fig. 3 and 4. He also had a drawing of a female African among his papers.

12. Soemmerring, *Über die körperliche Verschiedenheit des Negers vom Europäer*, p. 34. From records from the Chatoul Rechnungen held at the Staats-Archiv in Marburg, we know that women were among the Africans living at Wilhelmshöhe.
13. The catalogue of Soemmerring's collection done by Rudolph Wagner indicates that he had four male Africans and one black female reportedly from Java among his twenty-five skeletons. It does not list the skeleton from the Kassel female ('Katalog des Präparate welche sich in dem von Soemmerring angelegten anatomischen Museum befanden', in Rudolph Wagner, *Samuel Thomas von Soemmerring's Leben und verkehr mit sienen Zeitgenossen* [Leipzig, 1844], vol. 1, pp. lxxx–lxxxv).
14. Letter from Merck to Soemmerring, 13 August 1784, in ibid., p. 288.
15. Even though Soemmerring raised the issue of sexual subordination in his book on race (*Über die körperliche Verschiedenheit des Negers vom Europäer*, p. ix), he did not treat the issue of sexual differences among Africans.
16. Jean-Jacques Rousseau, *Emile, ou De l'éducation* (1762), in *Oeuvres complètes*, ed. Bernard Gagnebin and Marcel Raymond (Paris: Gallimard, 1959–1969), vol. 4; Pierre Roussel, *Système physique et moral de la femme, ou Tableau philosophique de la constitution, de l'état organique, du tempérament, des moeurs, & des fonctions propres au sexe* (Paris, 1775); Jakob Ackermann, *Über die körperliche Verschiedenheit des Mannes vom Weiber ausser Geschlechtstheilen*, trans. Joseph Wenzel (Koblenz, 1788); and Jacques Moreau de la Sarthe, *Histoire naturelle de la femme* (Paris, 1803).
17. Julien-Joseph Virey, *De la femme* (Paris, 1823). Virey's *Histoire naturelle du genre humain* (Paris, 1800) makes some reference to women.
18. Elizabeth Spelman, *Inessential Woman: Problems of Exclusion in Feminist Thought* (Boston: Beacon Press, 1988), pp. 37–56.
19. Soemmerring listed one partial skeleton and a skull of African females in his extensive private collection, consisting of 1,462 specimens preserved in alcohol, 2,439 dried preparations, and more than 200 skulls of people of various nations, remarkable or sick individuals (Ignaz Döllinger, *Gedächtnißrede auf Samuel Thomas von Soemmerring* [Munich, 1830], pp. 14–15). See also 'Präparate, welche Herr Hofrath liess', *Medicinisches Journal*, ed. Ernst Gottfried Baldinger, 4 (1787): 14–23. Eighteenth-century anatomists' collections were, indeed, extensive; a Mr Van Butchel kept his first wife embalmed at his home (Petrus Camper, 'Petri Camperi itinera in Angliam, 1748–85', *Opuscula selecta Neerlandicorum de arte medica* 15 (1939): 205. See also Johann Blumenbach's partial catalogue prefacing the third edition of his *On the Natural Varieties of Mankind*, trans. Thomas Bendyshe (1795; New York: Bergman, 1969). Seventy-one of the eighty-two skulls listed are male.
20. Georges-Louis Leclerc, comte de Buffon, *Histoire naturelle, générale et particulière* (Paris, 1749–1804), vol. 3, 'Variétés dans l'espèce humaine', and vol. 4, supplément.
21. Pierre Boulle, 'Coloured People in Paris on the Eve of the French Revolution', in *L'image de la Révolution française*, ed Michel Vovelle (New York: Pergamon Press, 1990), vol. 4, p. 2467. See also Paul Edwards and James Walvin, *Black Personalities in the Era of the Slave Trade* (Baton Rouge: Louisiana State University Press, 1983), p. 19.
22. White, *An Account of the Regular Gradation*, chart, p. 45.
23. Genevieve Lloyd, *The Man of Reason: 'Male' and 'Female' in Western Philosophy* (Minneapolis: University of Minnesota Press, 1984).
24. Schiebinger, *Mind Has No Sex?* pp. 197, 209, 212, and 214.
25. For Blumenbach, as for many of his era, the skull epitomized racial diversity, revealing 'the principle physical characters of humanity'. Blumenbach wrote proudly

of his vast skull collection: 'When viewing this collection hopefully no one will say with the Cynic Menippus – when he arrived in the underworld – that they all look alike'. Blumenbach, *Beyträge zur Naturgeschichte*, Göttingen (1806–1811), p. 66.
26. Buffon, *Histoire naturelle*, vol. 3, pp. 459–460; Edward Long, *The History of Jamaica* (London, 1774), vol. 2, p. 380; White, *An Account of the Regular Gradation*, pp. 71–72. Maria Nugent, *Lady Nugent's Journal (1801–1815)*, ed. Frank Cundall (London: Adam & Charles Black, 1907), p. 94.
27. Barbara Bush, *Slave Women in Caribbean Society, 1650–1838* (Bloomington: Indiana University Press, 1990), p. 15.
28. Petrus Camper, *Vermischte Schriften* (Lingen, 1801), pp. 342–343. In the 1780s, Camper measured the Negro pelvis at 125 degrees (these measurements were added by the editor from Camper's notebooks).
29. Camper, *Works*, pp. 61–63. The subjects measured were reported in Soemmerring, *Über die körperliche Verschiedenheit des Negers vom Europäer*, p. 35. Camper also compared pelvic proportions reported by Albrecht Dürer and in 'the Antonius', which I take to be another well-known European piece of statuary.
30. Camper, *Works*, pp. 60–61.
31. Soemmerring, *Über die körperliche Verschiedenheit des Negers vom Europäer*, pp. 34–35.
32. White, *An Account of the Regular Gradation*, p. 72; cited in James Prichard, *Researches into the Physical History of Mankind* (London, 1841), vol. 1, p. 323.
33. Moritz Weber, *Die Lehre von den Ur- und Racen-Formen der Schädel und Becken des Menschen* (Düsseldorf, 1830), p. 5.
34. Willem Vrolik, *Considérations sur la diversité des bassins de différentes races humaines* (Amsterdam, 1826), p. 11.
35. See, for example, Soemmerring, *Über die körperliche Verschiedenheit des Negers vom Europäer*, pp. 67–69; Lawrence, *Lectures*, p. 462.
36. Havelock Ellis, cited in Sander Gilman, *Sexuality: An Illustrated History* (New York: John Wiley & Sons, 1989), p. 295.
37. Schiebinger, *Mind Has No Sex?* pp. 212–213; Nancy Leys Stepan, 'Race and Gender: The Role of Analogy in Science', *Isis* 77 (June 1986): 261–277; and Cynthia Russett, *Sexual Science: The Victorian Construction of Womanhood* (Cambridge, Mass.: Harvard University Press, 1989).
38. Sander Gilman, *Difference and Pathology: Stereotypes of Sexuality, Race and Madness* (Ithaca: Cornell University Press, 1985), p. 83.
39. Blumenbach, *On the Natural Varieties of Mankind*, p. 151.
40. Camper, *Works*, p. 20; Virey, *Histoire naturelle du genre humain*, vol. 1, p. 317.

1.2

WOMAN AS BODY: ANCIENT AND CONTEMPORARY VIEWS

Elizabeth V. Spelman

> and what
> pure happiness to know
> all our high-toned questions
> breed in a lively animal.
> (Adrienne Rich, from 'Two Songs')

What philosophers have had to say about women typically has been nasty, brutish, and short. A page or two of quotations from those considered among the great philosophers (Aristotle, Hume, and Nietzsche, for example) constitutes a veritable litany of contempt. Because philosophers have not said much about women,[1] and, when they have, it has usually been in short essays or chatty addenda which have not been considered to be part of the central body of their work,[2] it is tempting to regard their expressed views about women as asystemic: their remarks on women are unofficial asides which are unrelated to the heart of their philosophical doctrines. After all, it might be thought, how could one's views about something as unimportant as women have anything to do with one's views about something as important as the nature of knowledge, truth, reality, freedom? Moreover – and this is the philosopher's move par excellence – wouldn't it be charitable to consider those opinions about women as coming merely from the *heart*, which all too easily responds to the tenor of the times, while philosophy 'proper' comes from the *mind*, which resonates not with the times but with the truth?

From: *Feminist Studies* 8 (1): 109–31, Spring 1982.

Part of the intellectual legacy from philosophy 'proper', that is, the issues that philosophers have addressed which are thought to be the serious province of philosophy, is the soul/body or mind/body distinction (differences among the various formulations are not crucial to this essay). However, this part of philosophy might have not merely accidental connections to attitudes about women. For when one recalls that the Western philosophical tradition has not been noted for its celebration of the body, and that women's nature and women's lives have long been associated with the body and bodily functions, then a question is suggested. What connection might there be between attitudes toward the body and attitudes toward women?

If one begins to reread philosophers with an eye to exploring in detail just how they made the mind/body distinction, it soon becomes apparent that in many cases the distinction reverberates throughout the philosopher's work. How a philosopher conceives of the distinction and relation between soul (or mind) and body has essential ties to how that philosopher talks about the nature of knowledge, the accessibility of reality, the possibility of freedom. This is perhaps what one would expect – systematic connections among the 'proper' philosophical issues addressed by a given philosopher. But there is also clear evidence in the philosophical texts of the relationship between the mind/body distinction, that is drawn, on the one hand, and the scattered official and unofficial utterances about the nature of women, on the other.

In this article, I shall refer to the conceptual connections between a philosopher's views about women and his expressed metaphysical, political, and ethical views. That is, I shall refer to conceptual relations internal to the texts themselves, and not to relations between the texts and their political and historical contexts. So my task is different from that of a historian of ideas or a social historian who might look at the relation between the political, economic, and cultural conditions under which a philosopher writes, on the one hand, and the focus and force of that philosopher's writings, on the other.[3]

My focus below is on the works of Plato, to discover what connections there are between his views about women and his views about the philosophical issues for which he is regarded with such respect. His descriptions of women's nature and prescriptions for women's proper societal niche recently have been under scrutiny by feminists.[4] What I hope to show is why it is important to see the connections between what Plato says about women and other aspects of his philosophical positions. Feminist theorists frequently have wanted to reject the kinds of descriptions of woman's nature found in Plato and other philosophers, and yet at the same time have in their own theorizing continued to accept uncritically other aspects of the tradition that informs those ideas about 'woman's nature'. In particular, by looking at the example of Plato, I want to suggest why it is important for feminists not only to question what these philosophers have said about women, but also what philosophers have had to say about the mind/body distinction.

PLATO'S LESSONS ABOUT THE SOUL AND THE BODY

Plato's dialogues are filled with lessons about knowledge, reality, and goodness, and most of the lessons carry with them strong praise for the soul and strong indictments against the body. According to Plato, the body, with its deceptive senses, keeps us from real knowledge; it rivets us in a world of material things which is far removed from the world of reality; and it tempts us away from the virtuous life. It is in and through the soul, if at all, that we shall have knowledge, be in touch with reality, and lead a life of virtue. Only the soul can truly know, for only the soul can ascend to the real world, the world of the Forms or Ideas. That world is the perfect model to which imperfect, particular things, we find in matter merely approximate. It is a world which, like the soul, is invisible, unchanging, not subject to decay, eternal. To be good, one's soul must know the Good, that is, the Form of Goodness, and this is impossible while one is dragged down by the demands and temptations of bodily life. Hence, bodily death is nothing to be feared: immortality of the soul not only is possible, but greatly to be desired, because when one is released from the body one finally can get down to the real business of life, for this real business of life is the business of the soul. Indeed, Socrates describes his own commitment, while still on earth, to encouraging his fellow Athenians to pay attention to the real business of life:

> [I have spent] all my time going about trying to persuade you, young and old, to make your first and chief concern not for your bodies nor for your possessions, but for the highest welfare of your souls. (*Apology* 30a–b)

Plato also tells us about the nature of beauty. Beauty has nothing essentially to do with the body or with the world of material things. *Real* beauty cannot 'take the form of a face, or of hands, or of anything that is of the flesh' (*Symposium* 221a). Yes, there are beautiful things, but they only are entitled to be described that way because they 'partake in' the form of Beauty, which itself is not found in the material world. Real beauty has characteristics which merely beautiful *things* cannot have; real beauty

> is an everlasting loveliness which neither comes nor goes, which neither flowers nor fades, for such beauty is the same on every hand, the same then as now, here as there, this way as that way, the same to every worshipper as it is to every other. (*Symposium* 221a)

Because it is only the soul that can know the Forms, those eternal and unchanging denizens of Reality, only the soul can know real Beauty; our changing, decaying bodies only can put us in touch with changing, decaying pieces of the material world.

Plato also examines love. His famous discussion of love in the *Symposium* ends up being a celebration of the soul over the body. Attraction to and appreciation for the beauty of another's body is but a vulgar fixation unless one can use such appreciation as a stepping stone to understanding Beauty itself. One can begin to learn about Beauty, while one is still embodied, when one notices that this body is beautiful, that that body is beautiful, and so on, and then one begins to realize that Beauty itself is something beyond any particular beautiful body or thing. The kind of love between people that is to be valued is not the attraction of one body for another, but the attraction of one soul for another. There is procreation of the spirit as well as of the flesh (*Symposium* 209a). All that bodies in unison can create are more bodies – the children women bear – which are mortal, subject to change and decay. But souls in unison can create 'something lovelier and less mortal than human seed', for spiritual lovers 'conceive and bear the things of the spirit', that is, 'wisdom and all her sister virtues' (*Symposium* 209c). Hence, spiritual love between men is preferable to physical love between men and women. At the same time, physical love between men is ruled out, on the grounds that 'enjoyment of flesh by flesh' is 'wanton shame', while desire of soul for soul is at the heart of a relationship that 'reverences, aye and worships, chastity and manhood, greatness and wisdom' (*Laws* 837c–d). The potential for harm in sexual relations is very great – harm not so much to one's body or physique, but to one's soul. Young men especially shouldn't get caught up with older men in affairs that threaten their 'spiritual development', for such development is 'assuredly and ever will be of supreme value in the sight of gods and men alike' (*Phaedrus* 241c).

So, then, one has no hope of understanding the nature of knowledge, reality, goodness, love, or beauty unless one recognizes the distinction between soul and body; and one has no hope of attaining any of these unless one works hard on freeing the soul from the lazy, vulgar, beguiling body. A philosopher is someone who is committed to doing just that, and that is why philosophers go willingly unto death; it is, after all, only the death of their bodies, and finally, once their souls are released from their bodies, these philosophical desiderata are within reach.

The offices and attributes of the body *vis-à-vis* the soul are on the whole interchangeable, in Plato's work, with the offices and attributes of one part of the soul *vis-à-vis* another part. The tug-of-war between soul and body has the same dynamics, and the same stakes, as the tug-of-war between 'higher' and 'lower' parts of the soul. For example, sometimes Plato speaks as if the soul should resist the desires not of the body, but of part of its very self (*Gorgias* 505b). Sometimes he describes internal conflict as the struggle between soul and body, and sometimes as the battle among the rational, the spirited, and the appetitive parts of the soul. The spirited part of the soul is supposed to help out the rational part in its constant attempt to 'preside over the appetitive part which is the mass of the soul in each of us and the most insatiate by nature of wealth'; unless it is watched, the appetitive part can get 'filled and infected with the so-called pleasures associated with the body' (*Republic* 442a–b).

The division among parts of the soul is intimately tied to one other central and famous aspect of Plato's philosophy that hasn't been mentioned so far:

Plato's political views. His discussion of the parts of the soul and their proper relation to one another is integral to his view about the best way to set up a state. The rational part of the soul ought to rule the soul and ought to be attended by the spirited part in keeping watch over the unruly appetitive part; just so, there ought to be rulers of the state (the small minority in whom reason is dominant), who, with the aid of high-spirited guardians of order, watch over the multitudes (whose appetites need to be kept under control).

What we learn from Plato, then, about knowledge, reality, goodness, beauty, love, and statehood, is phrased in terms of a distinction between soul and body, or alternatively and roughly equivalently, in terms of a distinction between the rational and irrational. And the body, or the irrational part of the soul, is seen as an enormous and annoying obstacle to the possession of these desiderata. If the body gets the upper hand (!) over the soul, or if the irrational part of the soul overpowers the rational part, one can't have knowledge, one can't see beauty, one will be far from the highest form of love, and the state will be in utter chaos. So the soul/body distinction, or the distinction between the rational and irrational parts of the soul, is a highly charged distinction. An inquiry into the distinction is no mild metaphysical musing. It is quite clear that the distinction is heavily value-laden. Even if Plato hadn't told us outright that the soul is more valuable than the body, and the rational part of the soul is more important than the irrational part, that message rings out in page after page of his dialogues. The soul/body distinction, then, is integral to the rest of Plato's views, and the higher worth of the soul is integral to that distinction.

PLATO'S VIEW OF THE SOUL AND BODY, AND HIS ATTITUDE TOWARD WOMEN

Plato, and anyone else who conceives of the soul as something unobservable, cannot of course speak as if we could point to the soul, or hold it up for direct observation. At one point, Plato says no mere mortal can really understand the nature of the soul, but one perhaps could tell what it resembles (*Phaedrus* 246a). So it is not surprising to find Plato using many metaphors and analogies to describe what the soul is *like*, in order to describe relations between the soul and the body or relations between parts of the soul. For example, thinking, a function of the soul, is described by analogy to talking (*Theaetetus* 190a; *Sophist* 263e). The parts of the soul are likened to a team of harnessed, winged horses and their charioteer (*Phaedrus* 246a). The body's relation to the soul is such that we are to think of the body *vis-à-vis* the soul as a tomb (*Gorgias* 493a), a grave or prison (*Cratylus* 400c), or as barnacles or rocks holding down the soul (*Republic* 611e–612a). Plato compares the lowest or bodylike part of the soul to a brood of beasts (*Republic* 590c).

But Plato's task is not only to tell us what the soul is like, not only to provide us with ways of getting a fix on the differences between souls and bodies, or differences between parts of the soul. As we've seen, he also wants to convince us that the soul is much more important than the body, and that it is to our peril that we let ourselves be beckoned by the rumblings of the body at the expense of harkening to the call of the soul. And he means to convince us of this by holding up for our inspection the silly and sordid lives of those who pay too much attention to their bodies and do not care enough for their souls; he wants to remind us of how unruly, how without direction, are the lives of those in whom the lower part of the soul holds sway over the higher part. Because he can't *point* to an adulterated soul, he points instead to those embodied beings whose lives are in such bad shape that we can be sure that their souls are adulterated. And whose lives exemplify the proper soul/body relationship gone haywire? The lives of women (or sometimes the lives of children, slaves, and brutes).

For example, how are we to know when the body has the upper hand over the soul, or when the lower part of the soul has managed to smother the higher part? We presumably can't see such conflict, so what do such conflicts translate into, in terms of actual human lives? Well, says Plato, look at the lives of women.[5] It is women who get hysterical at the thought of death (*Phaedo* 60a, 112d; *Apology* 35b); obviously, their emotions have overpowered their reason, and they can't control themselves. The worst possible model for young men could be 'a woman, young or old or wrangling with her husband, defying heaven, loudly boasting, fortunate in her own conceit, or involved in misfortune or possessed by grief and lamentation – still less a woman that is sick, in love, or in labor' (*Republic* 395d–e). He continues:

> When in our own lives some affliction comes to us you are aware that we plume ourselves ... on our ability to remain calm and endure, in the belief that this is the conduct of a man, and [giving in to grief] that of a woman. (*Republic* 605c–d)

To have more concern for your body than your soul is to act just like a woman; hence, the most proper penalty for a soldier who surrenders to save his body, when he should be willing to die out of the courage of his soul, is for the soldier to be turned into a woman (*Laws* 944e).[6] Plato believed that souls can go through many different embodied life-times. There will be certain indications, in one's life, of the kind of life one is leading now; and unless a man lives righteously now, he will as his next incarnation 'pass into a woman' and if he doesn't behave then, he'll become a brute! (*Timaeus* 42b–c, 76e, 91a).

Moreover, Plato on many occasions points to women to illustrate the improper way to pursue the things for which philosophers are constantly to be searching. For example, Plato wants to explain how important and also how difficult the attainment of real knowledge is. He wants us to realize that not just anyone can have knowledge, there is a vital distinction between those who really have knowledge and those who merely think they do. Think, for example, about the question of health. If we don't make a distinction between those who know what health is, and those who merely have unfounded and confused opinions about what health is, then 'in the matter of good or bad health ... any woman or child – or animal, for that matter – knows what is wholesome for it and is capable of curing itself' (*Theaeteus* 171c). The implication is clear: if any old opinion were to count as real knowledge, then we'd have to say that women, children, and maybe even animals have knowledge. But surely *they* don't have knowledge! And why not? For one thing, because they don't recognize the difference between the material, changing world of appearance, and the invisible, eternal world of Reality. In matters of beauty, for example, they are so taken by the physical aspects of things that they assume that they can see and touch what is beautiful; they don't realize that what one knows when one has knowledge of real Beauty cannot be something that is seen or touched. Plato offers us, then, as an example of the failure to distinguish between Beauty itself, on the one hand, and beautiful things, on the other, 'boys and women when they see bright-colored things' (*Republic* 557c). They don't realize that it is not through one's senses that one knows about beauty or anything else, for real beauty is eternal and invisible and unchangeable and can only be known through the soul.

So the message is that in matters of knowledge, reality, and beauty, don't follow the example of women. They are mistaken about those things. In matters of love, women's lives serve as negative examples also. Those men who are drawn by 'vulgar' love, that is, love of body for body, 'turn to women as the object of their love, and raise a family' (*Symposium* 208e); those men drawn by a more 'heavenly' kind of love, that is, love of soul for soul, turn to other men. But there are strong sanctions against physical love between men: such physical unions, especially between older and younger men, are 'unmanly'. The older man isn't strong enough to resist his lust (as in woman, the irrational part of the soul has overtaken the rational part), and the younger man, 'the impersonator of the female', is reproached for this 'likeness to the model' (*Laws* 836e). The problem with physical love between men, then, is that men are acting like women.

To summarize the argument so far: the soul/body distinction is integral to the rest of Plato's views; integral to the soul/body distinction is the higher worth and importance of the soul in comparison to the body; finally, Plato tries to persuade his readers that it is to one's peril that one does not pay proper attention to one's soul – for if one doesn't, one will end up acting and living as if one were a woman. We know, Plato says, about lives dictated by the demands and needs and inducements of the body instead of the soul. Such lives surely are not good models for those who want to understand and undertake a life devoted to the nurturance of the best part of us: our souls.

To anyone at all familiar with Plato's official and oft-reported views about women, the above recitation of misogynistic remarks may be quite surprising. Accounts of Plato's views about women usually are based on what he says in book 5 of the *Republic*. In that dialogue, Plato startled his contemporaries, when as part of his proposal for the constitution of an ideal state, he suggested that

> there is no pursuit of the administrators of a state that belongs to woman because she is a woman or to a man because he is a man. But the natural

> capacities are distributed alike among both creatures, and women naturally share in all pursuits and men in all. (*Republic* 455d–e)

The only difference between men and women, Plato says at this point, is that women have weaker bodies than men, but this is no sign that something is amiss with their souls.

Plato also says, in a dialogue called the *Meno*, that it doesn't make sense to talk about 'women's virtues' or 'men's virtues', because virtue as virtue is the same, whether it happens to appear in the life of a woman, a man, or a child. This view is part of Plato's doctrine of the Forms, referred to earlier. Virtue, like any other Form, is eternal and unchanging; so it can't be one thing here, another thing there; it is always one and the same thing. Virtue as virtue does not 'differ, in its character as virtue, whether it be in a child or an old man, a woman or a man' (*Meno* 73a).

Well now, what are we to make of this apparent double message in Plato about women? What are we to do with the fact that on the one hand, when Plato explicitly confronts the question of women's nature, in the *Republic*, he seems to affirm the equality of men and women[7]; while on the other hand, the dialogues are riddled with misogynistic remarks? I think that understanding the centrality and importance of the soul/body distinction in Plato's work helps us to understand this contradiction in his views about women. As we've seen, Plato insists, over and over again in a variety of ways, that our souls are the most important part of us. Not only is it through our souls that we shall have access to knowledge, reality, goodness, beauty; but also, in effect we *are* our souls; when our bodies die and decay, we, that is our souls, shall live on. Our bodies are not essential to our identity; in their most benign aspect, our bodies are incidental appendages; in their most malignant aspect, they are obstacles to the smooth functioning of our souls. If we *are* our souls, and our bodies are not essential to who we are, then it doesn't make any difference, ultimately, whether we have a woman's body or a man's body. When one thinks about this emphasis in Plato's thought, his views about the equality of women and men seem integral to the rest of his views. If the only difference between women and men is that they have different bodies, and if bodies are merely incidental attachments to what constitutes one's real identity, then there is no important difference between men and women.[8]

But as we have also seen, Plato seems to want to make very firm his insistence on the destructiveness of the body to the soul. In doing so, he holds up for our ridicule and scorn those lives devoted to bodily pursuits. Over and over again, women's lives are depicted as being such lives. His misogyny, then, is part of his somatophobia: the body is seen as the source of all the undesirable traits a human being could have, and women's lives are spent manifesting those traits.

So the contradictory sides of Plato's views about women are tied to the distinction he makes between soul and body and the lessons he hopes to teach his readers about their relative value. When preaching about the overwhelming importance of the soul, he can't but regard the kind of body one has as of no final significance, so there is no way for him to assess differentially the lives of women and men; but when making gloomy pronouncements about the worth of the body, he points an accusing finger at a class of people with a certain kind of body – women – because he regards them, as a class, as embodying (!) the very traits he wishes no one to have. In this way, women constitute a deviant class in Plato's philosophy, in the sense that he points to their lives as the kinds of lives that are not acceptable philosophically: they are just the kinds of lives no one, especially philosophers, ought to live. It is true that Plato chastises certain kinds of men: sophists, tyrants, and cowards, for example. But he frequently puts them in their place by comparing them to women! We've already seen some examples of that, such as male homosexuals being ridiculed for their likeness to women. Another example comes from the same dialogue in which Plato's argument about equality occurs. At the one point in the *Republic* (579c), Plato tries to convince us that tyranny does not pay, by saying that a tyrant is someone who 'must live for the most part cowering in the recesses of his house like a woman, envying among the other citizens anyone who goes abroad and sees any good thing'.

Plato had what I have described elsewhere[9] as a case of psychophilic somatophobia. As a psychophile who sometimes spoke as if the souls of women were not in any important way different from the souls of men, he had some remarkably nonsexist things to say about women. As a somatophobe who often referred to women as exemplifying states of being and forms of living most removed from the philosophical ideal, he left the dialogues awash with misogynistic remarks. Of course, one can be a dualist without being a misogynist, and one can be a misogynist without being a dualist. However, Plato was both a dualist and a misogynist, and his negative views about women were connected to his negative views about the body, insofar as he depicted women's lives as quintessentially body-directed.

In summary, Plato does not merely embrace a distinction between soul and body; for all the good and hopeful and desirable possibilities for human life (now and in an afterlife) are aligned with the soul, while the rather seedy and undesirable liabilities of human life are aligned with the body (alternatively, the alignment is with the higher or lower parts of the soul). There is a highly polished moral gloss to the soul/body distinction in Plato. One of his favorite devices for bringing this moral gloss to a high luster is holding up, for our contempt and ridicule, the lives of women. This is one of ways he tries to make clear that it makes no small difference whether you lead a soul-directed or a bodily directed life.

[...]

NOTES

I want to thank the Women's Resource Center at Smith College for inviting me to give an early version of parts of this essay in April 1979. I am grateful to the editors of *Feminist Studies* for some very helpful suggestions about an earlier draft.

1. There is no reason to think philosophers used 'man' or its equivalent in other languages generically. For example, in discussing the conditions of happiness for 'man'. Aristotle raises the question of whether a 'man's' being self-sufficient is compatible with his having a wife (*Nicomachean Ethics 1097b11*).

   All references to Plato are from *Collected Dialogues of Plato*, ed. Edith Hamilton and Huntington Cairns (New York: Pantheon, 1963), and are supplied in parentheses in the text.

2. For example, Alice Rossi noted in her introductory essay to *Essays on Sex Equality* by John Stuart Mill and Harriet Taylor Mill (Chicago: University of Chicago Press, 1970, p. 5 and especially note 3) that major collections of John Stuart Mill's works typically do not include *The Subjection of Women*.
3. Our tasks are not, of course, unconnected, but they are distinct. It is always conceivable that a philosopher's remarks not be motivated by nor have consequences for the historical and political events in which his life was framed. It is the task of historians to trace the extent to which in any given case such motivations existed and such consequences followed. My task here, however, is to look at the logical connections between parts of a philosopher's works, whatever the actual connection was between those works and the particular historical moment in which they were created. At the same time, one reason I find the philosophical exercise interesting and worthwhile is just because the mind/body distinction – whatever its particular place in the history of Western philosophy – appears to be so deeply connected with political and social institutions used to define and shape women's lives.
4. See, for example, Christine Garside Allen, 'Plato on Women', *Feminist Studies* 2, no. 2–3 (1975): 131–8; Julia Annas, 'Plato's *Republic* and Feminism', *Philosophy* 51. (1976): 307–21; Anna Dickason, 'Anatomy and Destiny: The Role of Biology in Plato's View of Women', in *Women and Philosophy*, ed. Carol C. Gould and Marx Wartofsky (New York: Putnam's 1976), pp. 45–53; Susan Moller Okin, *Women in Western Political Thought* (Princeton: Princeton University Press, 1979), pt. 1; Martha Lee Osborne, 'Plato's Unchanging View of Women: A Denial that Anatomy Spells Destiny', *Philosophical Forum*, 6, no. 2–3 (1975): 447–52; Sarah Pomeroy, 'Feminism in Book V of Plato's *Republic*'. *Apeiron* 8 (1974): 32–35; and my 'Metaphysics and Misogyny: Souls, Bodies and Women in Plato's Dialogues', unpublished manuscript.
5. Although Plato objects to certain types of men – sophists, tyrants, and so forth – his disdain for women is always expressed as disdain for women in general and not for any subgroup of women. Moreover, one of the ways he shows his disdain for certain types of men is to compare them to women . . .
6. In passages like this we see Plato assuming that a certain kind of body implies the presence of a certain kind of soul. This is at odds with his explicit view elsewhere that what is really important about someone is that the person has a soul, no matter what kind of body she has.
7. In contrast, see the articles by Julia Annas and others cited in note 4.
8. This line of thinking may remind us of some contemporary discussions of androgyny; it also has fascinating connections to the complicated phenomenon of transsexualism. In *Conundrum* (New York: Signet, 1974), Jan Morris insists that 'she' had always had a woman's soul housed in a man's body. That she could think about herself in this way suggests that she thought her bodily identity not to be indicative of what her soul was like. On the other hand, that she felt compelled to change her body suggests – among many other things – that she felt her body had to properly reveal the kind, the gender of soul she had.
9. Spelman, 'Metaphysics and Misogyny'.

1.3

BODIES AND BIOLOGY

Lynda Birke

WHAT IS THE 'BIOLOGICAL BODY'?

Our bodies are ourselves: yet we are also more than our bodies. In the early years of 'second-wave' feminism in the West, embodiment was acknowledged implicitly in the action of women's health groups, and campaigns for reproductive rights. But simultaneously, bodies failed to enter our theorizing. Central to theorizing then was a distinction between 'sex', (which anatomically distinguishes males and females) from 'gender' (the processes of becoming 'woman' or 'man'). Although recent feminist writing tends to decry that simple opposition, the ghost of biology still haunts us: biological sex, the biological body, remain problematic concepts for feminist theorizing.

But what does the term 'biology' connote? It can mean a particular discipline, part of the natural sciences. 'Biology' implies the study of living organisms and their processes. But the word can also be synonymous with those processes, as in 'human biology'. In this sense, the term 'biology' all too often invokes dualism, as it is taken to include bodily processes, and nature 'out there'. This sense of biology, and of 'biological', tends to be troublesome for feminism.

Biological arguments have all too often been made in ways that buttress gender divisions. Such biological determinism has, for example, been adduced to argue that women are, say, genetically predisposed toward nurturing behavior while men are inclined toward adventures and fights. Politically, then, feminists have tended to oppose biological determinism and to insist on

---

From: L. Birke, 'Biological Sciences' in A. Jagger and I. Young, *Companion to Feminist Philosophy*, Oxford: Blackwell, 1998.

some form of social constructionism of gender, or of other social categories (such as sexuality).

In this article, I will examine some of the ways in which we have analyzed arguments rooted in biological claims. In particular, I draw on the work of feminists challenging the philosophical and theoretical underpinnings of biological ideas. Mainstream philosophy of biology covers many areas: it shares with other studies in philosophy of science a concern with issues such as empiricism and positivism, with objectivism and realism, for example. But it also has its own concerns, with, for instance, theories of evolution or genetics. Here, what is at issue includes the nature of explanation and evidence when, say, biologists speak of phenomena such as natural selection or adaptation (e.g. Hull 1988; Sober 1993).

[...]

THE BODY IS GOOD TO THINK WITH? LIVING THE BODY IN 1980s FEMINISM

'The body' is a focus of growing intellectual interest, both within and without feminism. Some writers, such as Moira Gatens, employ earlier frameworks to transcend dualisms such as sex/gender. She suggests returning to Spinoza's metaphysics, in which the 'body is not part of passive nature ruled over by an active mind but rather the body is the ground of human action' (Gatens 1988; 68). This, she claims, would allow us to acknowledge cultural and historical specificities while moving beyond the traditional political assumption of bodies as given.

Although forms of social constructionism still prevail, some theorists attempt to transcend the mind/body dichotomy through phenomenological approaches emphasizing the *lived body* (e.g. Young 1990). This is a body that is not given but is both signifying and signified, historically contingent and social (see Grosz 1994). 'The body' in this theorizing becomes central to understanding women's experiences but is not fixed or presocial. It becomes instead, 'a body as social and discursive object, a body bound up in the order of desire, signification, and power' (Grosz 1994, 19; Butler 1993).

Insisting on the 'lived body' is important, and understanding how it is signified – and lived – is critical for feminist theorizing which no longer ignores the body. Yet gaps remain. First, it fails to pay much heed to the body's *interior* and its processes; and secondly, it does not sufficiently address bodily development. Both of these fall within the remit of 'biology': as such, it has fallen largely to feminist biologists to begin the task of retheorizing.

Feminist theorists continue to deconstruct texts and visual images; yet the abstractions called 'diagrams of the body' rarely merit much attention, except as historical artifacts. Thomas Laqueur (1990) notes how representations of the reproductive organs have changed during recent centuries. From carefully executed drawings, shaded to show fine details, anatomical illustration has moved to highly abstract, stylized images. These, which most of us would take for granted in biology textbooks, need more detailed feminist analyses.

Picturing the body's interior is, nevertheless, evident in two important feminist works, both dealing with immunology. Donna Haraway (1991a) and Emily Martin (1994) work with changing images of the immune system and how these are culturally mediated. Here, at least, the interior of the body enters the realms of cultural production and feminist theory. I want to emphasize two aspects of these texts. The first is the stress on cultural understandings, so that 'the immune system' can be understood principally (perhaps only) in terms of the language and images with which it is described. Thus, Emily Martin notes the ways in which narratives of immune bodies have changed dramatically from one of bodily defences, under siege from external pathogens, to bodies responding flexibly to external demands.

The permeability of the body's exterior is the second theme. The postmodern immune system, suggests Haraway, is part of a 'network-body'; it is 'everywhere and nowhere' all at once (1991 b: 218). The body – as bounded, as the quintessential individual – is threatened even by the discourses of science itself. Its boundaries become permeable, opening it up to networks of influence inside and out.

We can see in these accounts feminist insistence on the situatedness of scientific knowledge, and on biological complexity. Nevertheless, apart from these studies of the discourses of immunology, there is rather little consideration of bodies *as* biological. Indeed, it seems to me that much feminist thinking still harbours an underlying belief in the biological body as fixed – even when that belief is apparently denied by statements that we cannot understand our biological selves *except* through culture. Where does that leave (say) the action of nerves, the functioning of immune systems, or the development of embryos from fertilized eggs? That *level* of bodily working seems to remain forever outside culture, fixed into 'biology'. In important ways, that underlying assumption that some aspects of 'biology' are fixed becomes itself the grand narrative (albeit implicit) from which feminist and other social theorists are seeking to escape.

While the new focus on the body is welcome, it perpetuates an additive model. There is always a level of 'biology' that seems beyond the cultural analysis – usually, the body's interior. Interestingly, it is also the workings of the interior that largely escape the attentions of mainstream philosophers of biology, who tend to focus on areas such as evolution or genetics. Physiology seems doomed to narratives of mechanism and reductionism, unsullied by philosophical attention.

Most of the time, our physiology seems constant; it is a part of our bodily 'nature'. Now there is an important reason for assuming such constancy, which comes from the study of physiology itself. For those of us trained in the biological sciences, the body's functions – physiology – can be roughly categorised into systems: nervous system, endocrine system, immune system, and so on. A central principle of how these systems work is homoeostasis, the body's ability to maintain a constant state. So, for example, body temperature is normally around 37 degrees Celsius, and levels of sugar in the blood usually (except in some disease states) remain within certain limits.

Scientific language of separate systems maintaining constancy becomes part of a wider cultural language, assumed even within accounts of the 'socially constructed body'. Thus, health is a matter of maintenance, or keeping things constant, while disease represents perturbation. This way of thinking is paralleled by the abstraction and reductionism of the language of genetics as fixity, a language increasingly moving from laboratories and into the street. Biological bodies, within these narratives, become fixed by the parallel languages of genes (determining who we are) and homoeostasis (which ensures we stay that way). Yet isn't such language itself a social and cultural construction?

Bodily interiors need to emerge from the confines of physiological discourse into wider cultural criticism. We need to insist on thinking about the biological body as changing and changeable, as *transformable* (Birke 1986, 1994; Fausto-Sterling 1992; Hubbard 1990). All our cells constantly renew themselves, even bone (which is always remodelling, especially when we put loads on it in exercise). There are, nevertheless, constraints, imposed by one part of the body on another; as a result, our overall bodily appearance changes relatively little in adulthood.

Living the body means experiencing it *as* transformable, not only as cultural meanings/readings, but also within itself. Whatever physiology may say, I do not know whether we would experience the interior workings of our bodies in similar ways if the culture in which we live were to change dramatically. People with diseases, or some forms of physical disability, may well experience their bodily interiors differently from persons who are well or able-bodied: but part of that experience depends upon the cultural experience of living out medical definitions of pathological functioning. In that sense at least, culture shapes our internal experiencing.

Moreover, 'homoeostasis' can be turned around, decentering the 'constancy' theme and focusing instead on fine changes involved in keeping within gross limits. How might we understand potential changes in these, and how they are culturally contingent? Over time, too, bodies are transformable; the type of muscle fiber predominating in any one muscle mass is at least partly a product of the kind of stress put on the muscle in exercise. Judith Butler's concept of performativity (1993) is useful here: She focuses on gendered performance, analyzing in detail the cultural production of gender transgression. But might performativity (in the sense of iterated performance, whether or not to do with gender) itself also influence the 'way the body works', its interiority? We perform many roles, any or all of which could influence bodily workings.

Meanwhile, our internal organs and tissues also perform. Physiological language is deeply mechanistic – it speaks of control systems, and feedback loops serving to stabilize them. Yet implicit in these systems is *active* response to change and contingency, bodily interiors that constantly react to change inside or out, and act upon the world.

THE BODY BECOMING

Even fetuses enter culture, through the use of techniques of prenatal visualization and screening. Yet, human development – the processes of becoming human as we enter the world, or of becoming adult as we grow – seems to be missing from feminist insistence on 'lived bodies' or social constructionism.

We can emphasize human development in terms of transformability, in opposition to the fixity implied by some concepts of 'the gene'. The discourse of 'the' gene is gaining ground; Dorothy Nelkin and Susan Lindee argue that: 'The findings of scientific genetics – about human behavior, disease, personality and intelligence – have become a popular resource precisely because they conform to and complement existing beliefs about identity, family, gender and race' (1995: 197). So, while feminist and postmodern theorists increasingly question notions of identity, such ideas are reinforced in the wider culture within a (dangerous) discourse of 'genes'.

Within this discourse of identity and determinism, 'we' unfold from the genes laid down when sperm meets egg. It is a modern version of preformationism – the eighteenth-century idea that we unfold from a tinier version of ourselves housed comfortably in sperm (or egg). Genes as blueprints is a similar and persistent idea, as Susan Oyama (1985) pointed out.

Yet there are other ways of thinking about our becoming. Even within science, there are other positions, such as emphasizing active engagement of the embryo in its own development (see, for example, Fausto-Sterling 1989). The embryo actively makes over its environment, engaging with its own development; it is thus a self-organizing entity, rather than a passive victim of genetic inheritance (see Goodwin 1994). Its essence, if there is one, is not fixity, but transformability. The embryo/fetus in this story is like the physiological organism as I conceive it above; it is constantly changing and having agency in that change.

REINVENTING THE ORGANISM

In thinking about organisms or development as transformative, I recognize that the organism itself becomes a more fluid and permeable concept (with implications for selfhood and subjectivity). Elizabeth Grosz (1994) suggests an association with gender in notions of fluidity: 'women's corporeality is inscribed as a mode of seepage', she argues (1994: 203). Female bodies thus culturally echo themes of seeping liquids and formless flow, of uncontrollability.

Donna Haraway's vision of the cyborg (1991b) also implies fluidity. She speaks of 'polymorphous, information' systems, emphasizing rates of flow across boundaries rather than bodily integrity. While both Haraway's vision and Grosz's description of female fluidity are compatible with my insistence on transformation, I want also to retain some sense of organisms as entities. Haraway opposes holistic/organismic views (and related stories of development as progress), as fostering a kind of solipsism. In her utopia, organisms seem to disappear into webs of complexity, with entities dispersing into information; they become 'strategic assemblages ... ontologically contingent constructs' (1991a: 220).

But organisms are more than just strategic assemblages of cells/information: they are self-actualizing agents. Insisting on organisms as entities/agents returns them conceptually to the study of biology – from which whole organisms have almost disappeared in the world of genes as prime movers. If, as Haraway insists, it's problematic to think of organisms in terms of a path of (genetic) progress (a narrative deeply embedded in Western culture), then we should certainly use other metaphors. One approach retaining organismic entities is provided by Brian Goodwin (1994). Drawing on the sciences of chaos, he describes ways in which emergent order can arise from apparent chaos in nature. Organisms (bodies), in this view, are *self*-organizing; they *are* processes. And they have value, he suggests, as entities – a position arguing against the extreme reductionism that takes organisms apart to reconstitute them (as, for example, in genetic engineering or the use of animals as organ 'donors' for transplant surgery).

Ascribing agency and transformativity to organisms/bodies works against the social devaluation of the body and its interior that contributes to women's (and others') oppressions. Moreover, it works against simple dichotomous classifications of mind/self versus body, for both exemplify the same or overlapping agencies. Elizabeth Grosz similarly emphasizes the need for feminist philosophy to seek an 'embodied subjectivity' (1994: 22).

To see organisms/bodies as having agency and the ability to be self-organizing also implies that social constructions and experiences of gender can themselves be part of a process. 'Sex' cannot thus be prior to gender, but itself shaped by, and contingent to, gender. Put another way, processes involved in creating and continually recreating (sexed) bodies are partly material and partly social/experiential. Out of those are created the marked bodies, the bodies of difference, that feminist writers such as Grosz insist upon.

REMAINING UNCERTAINTIES

In examining feminist work on embodiment in relation to scientific explanations, I recognize two tensions. First, I draw on both postmodernist insistence on science as narratives *and* on belief in some form of realism. These are not necessarily incompatible, and tension between them seems inevitable – indeed, desirable, if we are to escape from such binaries as narrative versus realism itself.

Secondly, 'transformation' may not always serve feminist political ends. My insistence on transformability is for thinking about organisms. Genetic reductionism, however, also (somewhat paradoxically) permits discourses of transformation, by moving genes around, *within* the rhetoric of reductionism. Surgical transformation may not serve progressive interests, either; cosmetic or transsexual surgeries, for example, involve literally making the body over to achieve desired goals. But in neither is the material body thought of as having internal agency; rather, it is a fixed entity which is at odds with what is desired. (I am not saying here that thinking of agency will necessarily make the desired changes, simply that seeing bodies as reducible to interchangeable bits is part of the discourse of fixity.)

Yet the fact that bodies are alterable within reductionist logic is not itself an argument against transformation and complexity. We need urgently to find ways of thinking about bodily processes (or about 'biology' more generally) that move away from simple reductionism, and that simultaneously allow us to theorize bodies lived *in* culture.

The search for alternative models, for different stories to tell, lies at the heart of feminist theorizing about biology. Feminists insist on more complex, nuanced, ways of interpreting biological processes. Partly, we do so because even empiricism allows different ways of interpreting evidence. Complex models better describe how things work: they also provide alternative narratives, in the post-modern sense, which challenge Enlightenment concepts of one truth (see Hekman 1992).

A second, more clearly political, reason for feminist struggles to rename nature through complexity and transformation is that we can thus challenge persistent dualisms. Seeing gender opposed to the bedrock of sex is one example. Others include the dualisms of organism/environment, human/animal, bodily fixity/cultural lability, nature/nurture, and so on. As feminist critics often note, dualistic thought is deeply problematic – not least because it feeds dualisms of gender.

In opposing reductionism, and ensuing dualisms, feminists must insist on the uncertainty and indeterminacy of bodies. Yet we must also recognize that indeterminacy and transformability are not without limit. Bodies may constantly undergo interior change, but within apparent sameness. Perhaps nowhere is this more apparent than in the bodies of those with physical disabilities. Transformation may be the modus operandi of the body's interior, but it is unlikely to lead to sudden able-bodiness. And nor will thinking of bodies in terms of transformation alter the present cultural reproduction of disability.

While recent feminist work insists on cultural contingencies in describing bodies as marked, as signifiers of culture, it rarely goes beyond bodily surfaces. Culture is inscribed *on* those surfaces. In doing so, we run the risk, as Elizabeth Grosz rightly recognizes, of leaving the body's interior in the realm of biological fixity. Yet that risk rests on how we conceptualize biology itself. Bodies are good to think with only when we think of indeterminacy or transformation. 'Biology' is not always the ultimate limitation.

REFERENCES

Birke, L. (1986) *Women, Feminism and Biology: the Feminist Challenge*, New York: Methuen.

Birke, L. (1994) *Feminism, Animals and Science: the Naming of the Shrew*, Buckingham: Open University Press.
Butler, J. (1993) *Bodies that Matter: on the Discursive Limits of 'Sex'*, London: Routledge.
Fausto-Sterling, A. (1989) 'Life in the XY Corral', *Women's Studies International Forum*.
Fausto-Sterling, A. (1992) *Myths of Gender: Biological Theories about Women and Men*, New York: Basic Books.
Gatens, M. (1988) 'Towards a Feminist Philosophy of the Body' in B. Caine, E. A. Grosz and M. de Lepervanche (eds), *Crossing Boundaries: Feminisms and the Critique of Knowledges*, Australia: Allen and Unwin.
Goodwin, B. (1994) *How the Leopard Changed its Spots*, London: Weidenfeld and Nicolson.
Grosz, E. (1994) *Volatile Bodies: Toward a Corporeal Feminism*, Bloomington: Indiana University Press.
Haraway, D. (1989) *Primate Visions*, London: Routledge.
Haraway, D. (1991a) 'The Politics of Postmodern Bodies' in *Simians, Cyborgs and Women: the Reinvention of Nature*, London: Free Association Books.
Haraway, D. (1991b) 'The Cyborg Manifesto' in *Simians, Cyborgs and Women: the Reinvention of Nature*, London: Free Association Books.
Hekman, S. J. (1992) *Gender and Knowledge: Elements of a Postmodern Feminism*, Boston: Northeastern University Press.
Hubbard, R. (1990) *The Politics of Women's Biology*, New Brunswick: Rutgers University Press.
Hull, D. L. (1988) *Science as a Process: An Evolutionary Account of the Social and Conceptual Development of Science*, Chicago: Chicago University Press.
Laqueur, T. (1990) *Making Sex: Body and Gender from the Greeks to Freud*, Cambridge: Harvard University Press.
Martin, E. (1994) *Flexible Bodies*, Boston: Beacon Press.
Nelkin, D. and Lindee, S. (1995) *The DNA Mystique: the Gene as Icon*, Freeman: New York.
Oyama, S. (1985) *The Ontogeny of Information*, Cambridge: Cambridge University Press.
Sober, E. (1993) *Philosophy of Biology*, Oxford: Oxford University Press.
Young, I. M. (1990) *Throwing Like A Girl and Other Essays in Feminist Philosophy and Social Theory*, Indiana: Indiana University Press.

1.4

MY BODY, MYSELF: HOW DOES A BLACK WOMAN DO SOCIOLOGY?

Felly Nkweto Simmonds

> Social reality exists, so to speak, twice, in things and in minds, in fields and in habitus, outside and inside of agents. And when habitus encounters a social world of which it is the product, it is like a 'fish in water': it does not feel the weight of the water, and it takes the world about itself for granted. (Bourdieu and Wacquant 1992: 127)

INTRODUCTION

I have a particular relationship with the subject of sociology because of who I am. I am a Black woman and a sociologist. At conferences, for example, I am asked to speak as a Black female academic. Black academics (and students) are expected to talk about issues of 'race' as personal experiences. White academics, even when they are 'race' experts, are not expected to. It's as if 'race', as an experience, is only of concern to those who are 'racialized' by social theory itself. But when I use autobiographical examples to illustrate the relationship between my embodied experience, and my sociological practice, to an audience which is almost always white, the impact is always dramatic. Reflecting on my experience as a Black woman challenges the silence of those who are privileged by whiteness (MacIntosh quoted in Minas 1993). It forces them to ask themselves the questions they take for granted, to locate their own 'racial' experience as I have to every day.

As a Black woman, I know myself inside and outside myself. My relation to this knowledge is conditioned by the social reality of my habitus.[1] But my

---

From: H. S. Mirza (ed.), *Black British Feminism: A Reader*, London: Routledge, 1997.

socialized subjectivity is that of a Black woman and it is at odds with the social world of which I'm a product, for this social world is a white world. I cannot be, as Bourdieu suggests, a fish in water that 'does not feel the weight of the water, and takes the world about itself for granted'. The world that I inhabit as an academic, is a white world. This white world has a problematic relationship with blackness. Academic discourses of the social have constructed blackness as the inferior 'other', so that even when blackness is named, it contains a problem of relationality to whiteness. The British Sociological Association's guide to anti-racist language acknowledges 'white' and 'Black' (Caribbean/African/Other) as 'ethnic classifications', but fails to provide an actual definition of 'White' on its own. 'Black', however, has a detailed and problematized definition which begins with, 'This term is often used to refer to a variety of non-white ethnic groups'.[2]

Sociology gives me, even as a teacher of sociology, a 'non-white' existence, doomed to inhabit the margins of white theory. In this white world, the question becomes, How does a Black woman do sociology? As Fanon laments:

> The black man [sic] ... does not know at what moment his inferiority comes into being through the other. And then the occasion arose when I had to meet the white man's eyes. An unfamiliar weight burdened me. The real world challenged my claims. In the white world the man of color encounters difficulties in the development of his bodily schema. Consciousness of the body is solely a negating activity. It is a third-person consciousness. The body is surrounded by an atmosphere of certain uncertainty. ... A slow composition of my self as a body in the middle of a spatial and temporal world – such seems to be the schema. It does not impose itself on me; it is, rather, a definitive structuring of the self and the world – definitive because it creates a real dialectic between my body and the world. (Fanon 1986: 111)

In this white world the question becomes: What relationship can a Black woman establish between being a sociologist and being a person? I want to argue that an intellectual understanding of social reality is not enough, and that such an understanding has to critically examine the relationship between individual/personal and collective/social realities. In this white world I am a fresh water fish that swims in sea water. I feel the weight of the water ... on my body.

'CERTAIN PRIVATE INFORMATION'

To talk about the body is to invite derision. We cannot invite bodies, ours and those of others, into sociological discourse without being accused of essentialism or narcissism. But I want to risk talking about the body, my body as a strategy, in the way that Gayatri Spivak suggests, as 'persistent (de) constructive critique of theory' (Spivak 1993: 3). In this sense talking about the body, my body, becomes both a strategy and a technique, to deconstruct my positioning as a woman, an African woman (a 'third world' woman) and an academic in a western institution. It is neither essentialist nor narcissistic. I want to explore the relationship between my body as a social construct and my experience of it. I want to examine the relationship I have with my body and how I negotiate, daily, with 'embodied social situations' (Scott and Morgan 1993: 112).

I live in Newcastle-upon-Tyne, which has a significantly smaller Black population than other British cities. I am the only Black person in my department, and in fact one of only a handful in the whole institution. Currently I only have two Black students out of the nearly two hundred I teach across the university. I cannot ignore the fact of my blackness, even if I wanted to. Neither can my colleagues or students,[3] even if they wanted to.

This makes me vulnerable.[4] In the final analysis, I might be an academic, but what I carry is an embodied self that is at odds with expectations of who an academic is. I can be invited and /or dismissed as the token (Black, woman, 'Third World'), and can be expected or presumed to be taking one or more of these positions in how I teach/what I teach. I can be invited to give conference papers as Black, woman, African or 'Third World' (but not British, which is what my passport says!). In her essay 'Marginality in the Teaching Machine', Spivak illustrates this position:

> At the conference on Cultural Value at Birkbeck College, the University of London, on July 16, 1988, where this paper was first presented, the speaker was obliged to speak of her cultural identity. From what space was she speaking, in what space was the representative member of the audience placing her? What does the audience expect to hear today, here? ... To whom did they want to listen? (Spivak 1993: 54–5)

For some of us, it is impossible to escape the body and its constructions, even inside the 'teaching machine'. I am expected to not only carry my body, but to acknowledge it. I have a specific and clear relationship to the knowledge that I teach, through my body. The contradiction for me is that, whereas I can clearly be invited to speak about 'race' issues, it is only when I choose to speak about the experiences of the racialization of my body, that my authority to do this is questioned or dismissed as subjective and 'confessional'.[5] I'm expected to *be, but not to know about being*. This relationship between being and knowing exposes the fragility of theory's insistence that we can articulate truths only through a rational and objective epistemology of social reality. Ontological knowledge is suspect and at worst pathologized. This tradition is sanctioned, even by some whose practice is reflexive. For example, when Loic Wacquant asks Pierre Bourdieu the question, 'Can we do a Bourdieuan Sociology of Bourdieu? Can you explain yourself? If so, why this unwavering reticence to speak about the private person Pierre Bourdieu?' Bourdieu's response is defensive:

> It is true that I have a sort of professional vigilance which forbids me to adopt the kind of egomaniacal postures that are so approved of and even

> rewarded ... this reluctance to talk about myself has another reason. By revealing certain private information, by making bovaristic confessions about myself, my lifestyle, my preferences, I may give ammunition to the people who utilize against sociology the most elementary weapon there is – relativism. ... The personal questions that are put to me are often inspired by what Kant would call 'pathological motives'. (Bourdieu and Wacquant 1992: 202–3)

This, however, is a luxury. A white male academic has the privilege to opt for silence about 'private information'. As Scott and Morgan have observed, theory 'may admit the body', but demand that 'the theorist remains disembodied' (Scott and Morgan 1993: 112). Theory thus becomes only that knowledge which is created from outside ourselves, outside our bodies, out of our heads (as it were). It is as if 'facts' come out of our heads, and 'fictions' out of our bodies. As Anne Game observes:

> Sociological practice is conceived of as representation of the real, which for this discipline is conceived as the social. And there's nothing fictitious about the social and representations of it. Thus, the discipline is defined through oppositions, fact-fiction and theory-fiction. ... Social reality is taken as determinant; theory as reflection. But, this reflection is privileged as adequate correspondence to social reality as opposed to fictional reflection. (Game 1991: 3)

Although in this case Game uses the sociology of literature as the example of 'fictional reflection', the same can be said of experience (which in any case can be presented as literature), which reveals 'certain private information'. Bourdieu also acknowledges that literature can teach sociologists more about the 'truth of temporal experience' such as those found in biographical writings, and warns us that although 'there are ... significant differences between sociology and literature ... we should be careful not to turn them into irreconcilable antagonism' (Bourdieu and Wacquant 1992: 206). But he also adds, 'It goes without saying that sociologists must not and cannot claim to compete with writers on their own turf' (ibid.).

As an African woman my 'certain private information' is not only inscribed in disciplines such as anthropology, but also in colonial narratives, literatures, photographs, paintings and so on. Here the 'facts' created by social theory and the 'fictions' created by literature can be difficult to separate. At times social theory itself becomes a fiction.[6] Anne Game concludes, 'the sociological fiction is that it is not fiction. ... As an initial move in shifting the codes of sociology I will propose a reversal: that we think of sociological writing as fiction and fiction as social analysis' (Game 1991: 18).

One of the consequences, for a Black woman, of this insistence on the separation of the 'facts' of social reality, from the 'fictions' of experience and biographical knowledge is the creation of what Fanon has identified as 'the dialectic between my body and the world' (Fanon 1986: 111). The consequences are real enough. In academia, for example, I experience what Bourdieu himself has acknowledged; the 'feeling of being a stranger in the academic universe' (Bourdieu and Wacquant 1992: 208–9):

> In France, to come from a distant province, to be born south of the Loire, endows you with a number of properties that are not without parallel in the colonial situation. It gives you a sort of objective and subjective externality and puts you in a particular relation to the central institutions of French society and therefore the intellectual situation. There are subtle (and not so subtle) forms of racism that cannot but make you perceptive; being constantly reminded of your otherness stimulates a sort of permanent sociological vigilance. It helps you perceive things that others cannot see or feel. (Bourdieu and Wacquant 1992: 209)

But such an analysis can be a dangerous. It can glorify oppression in a way that can only be spoken by those who are privileged. 'Permanent sociological vigilance' is the consequence of oppression, a consequence of the subtle and not so subtle racism that permeates academic institutions in Britain. For a Black academic this is one of the burdens we carry, everyday. It is for this reason that we cannot and must not remain disembodied theorists. To put it simply, we cannot write a sociology of the Black experience without revealing certain private information.

As a woman, as a Black person, as an African, social theory has fed on my embodied experience. In anthropology, for example, one of the central tenets of 'defining the primitive' (Torgovnick 1990: 1–41) was the very basic idea that: 'primitives live life whole, without fear of the body' (Torgovnick 1990: 9). I have a body prescribed not only as primitive, but at the very 'heart of darkness'. In her re/interpretations of Conrad's *Heart of Darkness*, Marianna Torgovnick exposes the relationship Conrad, through Marlow, gives between the African woman and Africa itself.

> In my mind, I keep coming back to the African woman who stalks through the heart of darkness. ... That African woman, is, for me the crux of *Heart of Darkness* ... She is the representative 'native'. ... She is, the text insists, the symbol of Africa. ... Her death fulfils her role as emblem of the African landscape and makes ... explicit the hidden reference of 'the feminine' and the 'primitive' to death. The African landscape is death in the novella. It is 'the white man's grave' ... Europeans enter it but leave it either dead or ill or changed and marked for ever. (Torgovnick 1990: 154–5)

Both the nature and value of the 'primitive' body are prescribed. The Black body must remain 'voiceless' (Torgovnick 1990: 9). How then, are we to write a sociology of the Black experience in Britain, without taking on the body, and without revealing 'certain private information'?

FEAR AND DESIRE

When a young Black man is murdered by a group of young white men, we could write whole texts on the politics of race and racism, such as the collusion of the legal system in the killing of our sons.[7] But I fear such grand narratives ignore the very basic act of the killing of a Black body which is the final solution, the very logic of racism. History is littered with such bodies – Black bodies swinging from poplar trees in Alabama – Black bodies hanging from Mopane trees in Central Africa – Black bodies hanging from Flame trees in Kenya. Maybe, if we began by counting the bodies, we might arrive at a clearer picture of what the idea of race and racism as an ideology produces, socially and politically, and what the bodily experience can be.

In her essay, 'Myth of the Black Rapist', Angela Davis (Davis 1981) illustrates this with the example of the white institution of lynching (complemented by the rape of Black women). In the aftermath of the Civil War (and later, to a less extent), the lynching of Black men was used by white America as a valuable political weapon to guarantee the continued exploitation of Black labour 'and the political domination of Black people as a whole' (Davis 1981: 185). In this case, the history of the killing of Black bodies as a central political strategy cannot be separated from the social reality of how racism worked then, and continues to today, making a Black body always vulnerable to whiteness.

In this white world, the Black body, my body, is always on display. It has been documented by western disciplines such as anthropology. The essays in Elizabeth Edward's *Anthropology and Photography: 1860–1920*, chronicle how nineteenth-century anthropologists used the authority of photography to construct and display knowledge of the 'other'. A particular fascination with the female body was quite explicit in the search for anatomical landmarks of different 'races'. For example, even Colonial Office records used Thomas Henry Huxley's standardized photometric methods to collect information on colonial subjects. As Frank Spencer notes in his essay, 'Some Notes on the Attempt to Apply Photography to Anthropometry during the Second Half of the Nineteenth Century':

> In an effort to produce a photographic document that would permit the subsequent recovery of reliable comparative and morphometric data, Huxley recommended that all subjects be photographed naked, according to established and anthropometric poses. ... In particular Huxley noted the desirability that the arm in female subjects should be 'so disposed as not to interfere with the contour of the breast which is very characteristic in some races' (Huxley to Lord Granville, Dec. 8, 1869). (Spencer 1992: 100)

The spectacle of the colonized female subject in nineteenth-century writings and readings of difference between the races was also symbolically captured in the public displays of African women as curiosities. The African woman named Saartjie Baartman, also called Sarah Bartmann or Saat-Jee and known as the 'Hottentot Venus', was on public display in London and in Paris in 1810. After her death in Paris in 1810, her sexual parts, her genitalia and her buttocks were preserved and continue, to this day, to be displayed in the *Musée de l'homme* in Paris (Gilman 1992: 180–1).

These public displays of images of 'other' societies were common forms of entertainment in the nineteenth century. In his essay 'British Popular Anthropology: Exhibiting and Photographing the Other', Brian Street illustrates how exhibitions of other societies 'with their underlying associations of race, hierarchy and evolution, were most vividly experienced through exhibitions, photographs and postcards ... not simply as "entertainment" but as having educational value' (Street 1992: 122). In this case the 'facts' and the 'fictions' of 'others' were rendered one and the same thing.[8] As a Black woman, my body cannot escape this history.

It is particularly poignant for me that some of the photographs of the 'curiosities' documented in Street's chapter are of the Batwa, 'pygmies from the Ituri forest region of the Congo' (Street 1992: 128–9). The Batwa were some of the earliest settlers across most of Central Africa, including the islands of Lake Bangweulu in Zambia. My maternal ancestors are from those islands. Looking at the photographs of the Batwa (in which they are virtually naked, and includes one bare breasted woman) taken of them in London, by one Sir Benjamin Stone, in August 1905, I cannot help but take a second look to see if I can recognize myself.

Adorned and unadorned I cannot escape the fantasies of the western imagination. Robert Young illustrates this desire for colonized bodies as spectacle, as labour and so on, as essentially an extension of the 'desiring machine' of capital. This has particular implications for the female body, and is highlighted by anthropology's particular fascination with female bodies and with sexual lives. In this sense, sexuality becomes part of the political economy of desire, for money, for products and for those who produce. It becomes part of:

> 'the libidinal unconscious' [which] opens up possibilities for the analysis of the dynamics of desire in the social field. Racism is perhaps the best example through which we can immediately grasp the form of desire and its antithesis, repulsion, as a social production: thus 'fantasy is never individual: it is a group fantasy'. (Young 1995: 168–9)

It is this politics of sex, race and desire which still affects 'racial' encounters in everyday life.[9]

As Brian Street concludes:

> as in written representations of non-European peoples, nineteenth-century European discourses on race and evolution continued to frame visual

> portrayals, even at a time when anthropologists themselves were beginning to move, via field work methods, towards a more characteristically twentieth-century interest in how people might see themselves and towards a more relativist less physically based view of cultural difference. The interest in legends of little people, in little bodies as signs of little minds, in 'savage' customs as a justification and rationale for 'scientific' and business 'progress' . . . was firmly rooted in a common framework of race, evolution and hierarchy [and] served to construct and perpetuate this conceptual framework, beyond its academic life, for larger proportions of the public than could be influenced solely by the books and literature available on the subject at the time. (Street 1992: 130)

MY BODY, MYSELF

'Racial' knowledge constructed about 'the other' is what provides the contradictory experience of 'race' as an everyday reality even at the end of the twentieth century. Here I want to unearth some of these bodily and embodied experiences of My Body, My Self and of how others see me and how I experience being a 'curiosity'.

First Sketch

> One day, I walked from the bus with an Italian waiter I knew a little, I'd been in the restaurant in which he works several times. Suddenly he said, 'I bet you have a beautiful body'.
>
> In his imagination, in his fantasy, of course I have. He has seen Black female bodies. The Black female body is etched on his sexual unconscious. As a white man he also has the weight of history behind him, which tells him Black women are available to him.
>
> Suddenly conscious of my missing breast I say, 'No actually, no. I had a bad illness two years ago'. Illness? He vanished, didn't even wait for the end of the sentence, having conjured up in his imagination all the awful things I could be carrying. Blackness, dirt, disease . . . HIV AIDS, Ebola fever? He vanished.

I carry a contradictory body, so exotic and desirable, so threatening and deadly. Actually now, in my middle age, now that I've had to get in touch with my body, it feels OK. I think of my body quite often . . . pamper it with bath oils, take it to the gym twice or three times a week . . . but still too much whiskey (Irish). My body never has hangovers from whiskey.

Second Sketch

> In Dublin a rather beautiful Irish man leans across the table, takes my hand and rubs the back of my hand: 'Do you know how exotic you are? Such beautiful ebony skin, so soft, so beautiful'.
>
> Because he is so beautiful, and I suspect in love with me for the duration

> of the meal, I'm kind to him and remind him that I can only be exotic in Dublin, in my own space there are millions like me and we don't go round touching each other and telling ourselves how exotic our ebony skins are ...
> And of course what he doesn't know, cannot know is that my skin is quite dry and I have to oil it everyday.

My skin, so soft, so black, so dry, the colour of my skin so exotic to a beautiful Irish man, such a deadly cloak to wear on a dark night on the streets of London, Liverpool, Birmingham, Leeds ... Dublin.

Third Sketch

> Alice leans over and whispers; 'Barbara wants to touch your hair'. I take Barbara's hand and put it on my hair, on my dreadlocks, 'Ooh ... Just like wool. I've always wondered what it feels like. Ooh'. She coos.

I try to remember Bourdieu's phrase 'permanent sociological vigilance'. Any Black person can tell you, hair is our special thing. It is as tanning is to white people, I suspect. We have special ways of torturing our hair, twisting, braiding, straightening, curling, colouring, extending ... the perfect disguise. As this quote from the leading Black newspaper, the *Weekly Journal*, illustrates.

> One ... frivolous girl preferred white men for no other reason than that none of her white (and perhaps short sighted) lovers were ever sharp enough to work out that the cascading brown hair with blonde highlights reaching half way down her back was fake. (*Weekly Journal*, 18 May 1995)

Fourth Sketch

> Joe has his hands on my naked butt ... kneading, 'I love this, I love this'. With a Black man, I even like the idea that I have a big bum! But when I try to squeeze myself into a skirt from Warehouse, I realize that there's no life after size 10.

But I also think of Sarah Bartmann, The Hottentot Venus, whose image not only formed the 'central image of the black female throughout the nineteenth century' (Gilman 1992: 180), but has also fed countless white fantasies about the Black female form.

Fifth Sketch

> Steven says, 'You have the most beautiful eyes'. I worry. On my right eye I have a small growth which is slowly growing over my cornea. It's quite common in those who have lived in tropical climates. Ultra-violet rays damage our eyes. I have to keep watching it, if it suddenly starts to grow, or begins to affect my vision, I have to have it removed. The only problem

> is, once it's disturbed, as it were, it's likely to grow back faster than it's growing now.
> What did he say? Beautiful eyes. I think they say beauty is in the eye of the beholder.

Or is it? My experience of my body, inside and outside of myself, leaves me with more questions than answers. But I need to unearth this bodily experience for myself as an act of sorting through the fictions of theory, of realizing that for me there is a very fine line between the 'facts' and 'fictions' of my body as I experience it in the here and now, and the history of that body.

I have come to this realization through my experience of breast cancer and the 'fact' and 'fiction' of the body I live with everyday.

Sixth Sketch

From my Cancer Diary (Moss, Tuesday 24.3.92)

> I'm trying to remember my body. With two breasts. With no pain. What did it feel like to have two breasts? To touch them ... together or one at a time. To cradle a man's head between my breasts. ... It all feels so impossible now. Will I ever let a man see my lone (lonely) breast? I don't know if I can relate to a whole body again. ... This is how the surgeon broke the news to me:
> 'We have found a cancer in your right breast ... in its early stage ... a ductile carcinoma in situ ... still contained in the ducts ... has not invaded the breast tissue or the lymph nodes ... but the whole breast tissue is unstable. ... My recommendation is that we remove the breast, thus ensuring that the whole cancer is removed'.
> I think he was talking about my breast. I felt it then, and knew I hated it, wanted it off, there and then. Little did I know how much I'd miss it at first, and how much I'd forget it, in time.

It is the loss of my right breast that has made me take account of the embodied experience in the making of social reality. It's not very often we get the chance of a new body. I was 42 when I lost my breast. On the outside I carry the same body; a fact and a fiction. But I'm different, not just because the shape of my body is different, but because I have to relate to that different body. I am transformed, and the world around me is transformed also.[10] It is this new relationship with my body that has allowed me to re/think myself and my place in the social.

CONCLUSION

Being conscious of myself as a person, an embodied self, is what helps me perceive things that 'others cannot see or feel' as sociologists. This is what gives me a particular relationship with the subject of sociology. The relationship between my embodied reality and my sociological practice is at the very core of how I do sociology. I have to be equally as aware of the reality that my body imposes on my practice and of the reality that social theory imposes on that body. I cannot be silent about it. As Paulo Freire suggests:

> men [sic] are aware of their activity and the world in which they are situated. They act in function of the objectives which they propose, have the seat of their decisions located in themselves and in their relation with the world and with others, and infuse the world with their creative presence by means of the transformation they effect upon it. Unlike animals, they do not only live but exist; and their existence is historical. ... For animals, 'here' is only a habitat with which they enter into contact; for men, 'here' signifies not merely a physical space, but an historical space. (Freire 1972: 71)

The 'here' of academia is also 'an historical space'. When I teach sociology, as a Black woman in an almost all-white institution, the social reality of academia and of academic discourse is transformed. My practice is reflexive in the way that Alvin Gouldner has argued, and that is:

> A Reflexive Sociology ... is characterised not by what it studies. It is distinguished neither by the persons and the problems studied nor even by the techniques and instruments used in studying them. It is characterised, rather, by the relationship it establishes between being a sociologist and being a person, between the role and the man [sic] performing it. (Gouldner 1993: 470–1)

I have chosen to acknowledge this relationship between being a sociologist and being a person, openly, and to acknowledge the impact this has on my practice. A reflexive sociology allows me, as a person, to use embodied social realities, to do sociology and to inform theory. It is a process of uncovering embodied social reality through the practice of sociology. In this process *sociological theory has to admit the body*. The body cannot remain 'voiceless'. This is how I teach sociology.

Recently a young Black man doing a PhD on sport and identity, wrote to me after a conference:

> you suggested that I keep a personal diary of my feelings about the research alongside the more 'serious' research notes. Anyway you'll be pleased to know that I did, grudgingly, start to do this and over time (and you'll no doubt be aware of this) I found it increasingly difficult to separate the two types of notes until my field notes became increasingly reflexive and 'personalized'. I am currently at the stage of writing up my notes and attempting to theorize them.[11]

I'm aware that he has embarked on a difficult journey – toward the discovery of (an embodied) self through the practice of sociology. It is an act of transforming theory, an act of admitting the body and embodied social experiences into theory.

As Black academics (and students), one of our tasks has to be to transform theory itself, if we are not to remain permanent 'curiosities' in academia. For us, the habitus of academia is as dangerous as society at large, because we are not 'fish in water' (Bourdieu and Wacquant 1992). Our work is often marginalized and dismissed as 'not theory', because we challenge the limits of theories that will not admit our embodied realities. To have our bodies, ourselves, admitted on our own terms, will be an act of naming ourselves on this journey through the 'heart of whiteness' (Gates, quoted in Mirza 1996).

NOTES

1. As Diane Reay (1995) explains: 'Bourdieu has developed the concept of habitus to demonstrate not only the ways in which the body is in the social world but also the ways in which the social world is in the body (Bourdieu 1981):

   > The habitus as the feel for the game is the social game embodied and turned into second nature. (Bourdieu, 1990b: 63)

   Thus, one of the crucial features of habitus is that it is embodied; it is not composed solely of mental attitudes and perceptions. Bourdieu writes that it is expressed through durable ways 'of standing, speaking, walking, and thereby feeling and thinking' (Bourdieu, 1990a: 70).
2. British Sociological Association, *Anti-Racist Language: Guidance For Good Practise* (undated, unacknowledged authorship), gives this definition:

   > **Black** – This term is often used to refer to a variety of non-white ethnic groups. This term has taken on more political connotations with the rise of black activism in the USA since the 1960s and now its usage implies solidarity against racism. The idea of 'black' has thus been reclaimed as a source of pride and identity. To accept this means that we should be sensitive to the many negative connotations relating to the word 'black' in the English language (black leg, black list, etc.). However, some Asians in Britain object to the use of the word 'black' being applied to them and some argue that it also confuses a number of ethnic groups which should be treated separately – Pakistanis, Bangladeshis, Indians and so on. One solution is to refer to 'black peoples' 'black communities', etc. in the plural to imply that there are a variety of such groups. It is also important to be aware of the fact that in some contexts – such as South Africa – 'black' can also be used in a racist sense.

3. In a seminar, I was asked by one of my white students if I had come to Britain 'to better myself'. I'm quite sure she wouldn't ask a white teacher on a lucrative 'Aid' contract in Africa the same question!
4. When, for example, I pointed out the racist nature of a policy document, the response of those in authority was to call into question the validity of my assertion.
5. I first aired some of the ideas in this chapter at a meeting of the Feminist Research Group at the University of Northumbria. A man (uninvited to the meeting!) asked if 'we all had to become confessional'. My reply was that he'd used the word 'confession' not me! I refused to be drawn into having to justify (to the white man) what I was saying and how I chose to say it.
6. See for example Adam Kuper (1988) and Mary Midgley (1985).
7. Since 1969 more than one hundred Black people have died in custody in Britain, (police, psychiatric and prison custody), nearly half of them in police custody (from a special report in the *Voice* 30 January 1996).
8. This practice continues today in television documentaries. For example, 'Watching

Brief' (*Guardian* 7 January 1996) introduces the programme *Under The Sun: A Caterpillar Moon* (BBC2) thus: 'After the honey season, the caterpillar season is the favourite time of the year for the Aka pygmies of the central African rainforest. They get to gorge themselves on the juicy titbits which rain down from the tree canopy. ... Julia Simmons' fascinating film focuses on the family of Bosseke, a warm and friendly Aka who tells his son to give the first caterpillar pickings to the film crew. At first you are just relieved for the crew that the hairy, squirmy grubs are in strangely short supply.' For myself, these 'hairy, squirmy grubs' are a delicacy.

9. See, for example, Kathryn Perry, 'The Heart of Whiteness: White Subjectivity and Interracial Relationships'; Inge Blackman, 'White Girls Are Easy, Black Girls Are Studs'; Helen (charles), '(Not) Compromising: Inter-skin Colour Relations'; Felly Nkweto Simmonds, 'Love in Black and White' all in Lynne Pearce and Jackie Stacey (1995) (eds) *Romance Revisited*, London: Lawrence Wishart 1995.
10. One way I observe this transformation is when I first tell someone that I have one breast. I can literally see the bodily reaction to it ... sometimes of surprise, confusion, pity, and even fear. I've also noticed that they also 'hear' or 'read' me differently, whatever I'm talking about. In these instances I catch a glimpse of the reality of those whose embodied realities, such as disabled people, cannot be ignored.
11. Personal communication 26.10.95.

REFERENCES

Bourdieu, Pierre (1981) 'Men and Machines' in K. Knorr-Cetina and V. Cicourel (eds), *Advances in Social Theory and Methodology: Towards an Integration of Micro and Macro Sociologics*, London: Routledge & Kegan Paul.

Bourdieu, Pierre (1990a) *The Logic of Practice*, Cambridge: Polity Press.

Bourdieu, Pierre (1990b) *In Other Words: Essays Towards a Reflexive Sociology*, Cambridge: Polity Press.

Bourdieu, Pierre and Wacquant, Loic J. D. (1992) *An Invitation to Reflexive Sociology*, Cambridge and Oxford: Polity Press/Blackwell Publishers.

Davis, Angela (1981) *Women, Race and Class*, London: Women's Press.

Donald, James and Rattansi, Ali (eds) (1992) *'Race', Culture and Difference*, London: Sage Publications in association with the Open University.

Edwards, Elizabeth (ed.) (1992) *Anthropology and Photography 1860–1920*, New Haven and London: Yale University Press in association with The Royal Anthropological Institute, London.

Fanon, Frantz (1986) *Black Skin, White Masks*, London: Pluto Press.

Freire, Paulo (1972) *Pedagogy of The Oppressed*, London: Penguin.

Game, Anne (1991) *Undoing The Social*, Buckingham: Open University Press.

Gilman, Sander (1992) 'Black Bodies, White Bodies: Towards an Iconography of Female Sexuality in Late Nineteenth-Century Art, Medicine and Literature' in J. Donald and A. Rattansi (eds), *'Race', Culture and Difference*, London: Sage Publications in association with the Open University.

Gouldner, Alvin W. (1993) 'Towards a Reflexive Sociology' in C. Lemert (ed.), *Social Theory: The Multicultural and Classic Readings*, Boulder, CO: Westview Press.

Griffiths, Morwenna and Troyna, Barry (eds) (1995) *Antiracism, Culture and Social Justice in Education*, Stoke-on-Trent. Trentham Books Limited.

Kuper, Adam (1988) *The Invention of Primitive Society: Transformations of an Illusion*, London: Routledge.

Midgley, Mary (1985) *Evolution as Religion*, London: Methuen.

Minas, Anne (1993) *Gender Basics: Feminist Perspectives on Women and Men*, Belmont, CA: Wadsworth Publishing Company.

Mirza, Heidi Safia (1996) 'Black Educators: Transformative Agents for Social Change', *Adults Learning*, NIACE, vol. 7, no. 6, February.

Pearce, Lynne and Stacey, Jackie (eds) (1995) *Romance Revisited*, London: Lawrence and Wishart.

Reay, Diane (1995) 'Using "Habitus" to Look at "Race" and Class in Primary School Classrooms' in M. Griffiths and B. Troyna (eds), *Antiracism, Culture and Social Justice in Education*, Stoke-on-Trent: Trentham Books Limited.

Scott, Sue and Morgan, David (eds) (1993) *Body Matters: Essays on the Sociology of The Body*, London: The Falmer Press.

Spencer, Frank (1992) 'Some Notes on the Attempts to Apply Photography to Anthropology During the Second Half of the Nineteenth Century' in E. Edwards (ed.) *Anthropology and Photography 1860–1920*, New Haven and London: Yale University Press, in association with the Royal Anthropological Institute, London.

Spivak, Gayatri Chakravorty (1993) *Outside in The Teaching Machine*, New York and London: Routledge.

Street, Brian (1992) 'British Popular Anthropology: Exhibiting and Photographing the Other' in E. Edwards (ed.), *Anthropology and Photography 1860–1920*, New Haven and London: Yale University Press, in association with the Royal Anthropological Institute, London.

Torgovnick, Marianna (1990) *Gone Primitive: Savage Intellects, Modern Lives*, Chicago and London: University of Chicago Press.

Young, Robert J. C. (1995) *Colonial Desire: Hybridity in Theory, Culture and Race*, London: Routledge.

1.5

OUR BODIES, OURSELVES: WHY WE SHOULD ADD OLD FASHIONED EMPIRICAL PHENOMENOLOGY TO THE NEW THEORIES OF THE BODY

Helen Marshall

[...]

GETTING LIVED BODIES BACK INTO THEORY

Although I am basically sympathetic to the thrust of argument that we must reconceptualise the body, and find much of the more recent work on how to do so very exciting, I am concerned that so much attention is being paid to nomenclature and theory, and so little to the lived experiences and data. We talk endlessly about how to theorise the body. When we do research about bodies in the context of how to theorise the body, it tends to be either from a psychoanalytic perspective, or from the external approach that takes bodies as texts. We study how the body is psychically created or functions as a creator, or we study how it has been presented, colonised, dressed, and inscribed. We do not, on the whole, talk about how the body is experienced as a way of getting a better theoretical hold on the concept. In other words, the enterprise of theorising the body tends to rely on the research 'external' approach or on speculative writing from the psychoanalytic version of the 'internal'. It makes disturbingly little reference to empirical work that comes from the tradition of phenomenology.

Phenomenological approaches share a concern with collecting and, where possible, collating the understandings of experiences found amongst various populations.[1] The resulting 'second order' constructs (Schutz, 1962) can be used in a variety of ways. In the conventional social sciences, they often have no other purpose than to produce a typology and a book or paper about its

From: *Women's Studies International Forum* 19 (3): 253–65, 1996.

relationship to some theoretical paradigm. When feminists engage in the exploration of their own or other women's experiences, either in the form of consciousness raising or as formal research, it is with the explicit aim of understanding better how and why women are oppressed[2] (Stanley and Wise 1983).

There are many examples of feminist phenomenological research on how women in varied situations have experienced their different bodies' transitions. The book from which I drew the title of this article is an early example of work closer to consciousness raising than academic research (Boston Women's Health Collective, 1976/1984). There are anthropological accounts such as Martin's (1987) work on the body, and sociological work on menstruation (Laws 1990), breasts (Meckelburg 1993), and childbirth (Rothman 1982).[3]

I do not mean that philosophers should set out to do empirical studies of the body when they aim to change our thinking about it. I do, however, think that it would be useful to look at existing empirical work to test, refine, and add insights to the theoretical.

In some ways, phenomenological research into the varied meanings of bodies might even be more useful than studies of how the body is inscribed. It can keep us focused on the problem of dualism in ways that help avoid the culturalist traps of the disappearing body, which Caddick (1986, 1992) outlined. I hypothesise that experiential accounts of bodies may reveal both unities and splits – the account later in this paper certainly does so. Studies of how regimes of power inscribe the body (Grosz's 'external' approach) may do this. They may also convey an impression of simple minds regulated by other minds or by the workings of an ethereal unlocated power. The tension between messy accounts (where a narrator is a sometimes a body and sometimes not a body) and a monist theory is useful. Terry Eagleton (1993) suggests that:

> One can see the point ... of dropping talk of having a body and substituting talk of being one. If my body is something I use or possess, then it might be thought that I would need another body inside this one to do the possessing and so *ad infinitum*. But this resolute anti-dualism, though salutary enough in its way, is untrue to a lot of our intuitions about the lump of flesh we lug around ... the fact remains that the human body is indeed a material object, and this is an essential component of anything more creative we get up to. ... It is not quite true that I have a body, and not quite true that I am one either. (Eagleton 1993: 7)

Further, I would argue, that a phenomenology of the ordinary experiences of the body may in some ways be of more use to us than looking at the numerically unusual. In this, I differ somewhat from writers on the body like Turner (1984) and Grosz (1987, 1994), who tend to focus on categories of disease such as anorexia (Turner 1990), or the 'phantom limb' phenomenon (Grosz). While the study of the unusual gives insight by allowing free play to imagination, the study of bodies in transition between socially recognised and 'ordinary'[4] states offers possibilities of discovering the varieties of corporality that we may miss when we look to the extreme. Moreover, the study of the ordinary body offers glimpses of power relations with a moral immediacy sometimes lacking in work where, as Eagleton observes 'there are mutilated bodies galore, but few malnourished ones, belonging as they do to bits of the globe beyond the purview of Yale' (Eagleton 1993: 7).

I have argued that it would be helpful for theorists of the body to look at work that attempts to discern what ordinary states of embodiment mean to the embodied subject. To illustrate what I have in mind, I now turn to my own private 'phenomenology of pregnancy'. The next section of this article contains two kinds of data. First, there is a formalised second-order construction of the experience of pregnancy, written a month before the birth.[5] Next, there is an edited set of informal notes taken two hours after the event on how I experienced the birth of my daughter [*not included*].

The context of this account is that I became pregnant intentionally and easily, had a medically uneventful pregnancy with minimal professional supervision, initially from a gynaecologist and then from birth centre staff at a large teaching hospital. The brief history of my birthing experience is that I spent an afternoon 'feeling funny', eventually decided that I had had some mild contractions, probably Braxton-Hicks ('practice contractions'), woke early next morning with strong regular and frequent contractions, arrived at the birth centre at 8 a.m. and gave birth three hours later after a medically uneventful labour.

Notes at Eight Months

Getting into a state

> We might question where the state of pregnancy begins. There is a relatively clear set of biological starting points (implantation, fusion) from which a choice could be made and a time set, but that is not the same as a starting point felt, or apprehended by the female body in question. Neither is it the same as the starting point apprehended by those around this person-on-the-way-to-motherhood.
>
> My own pregnancy began biologically at some time ... but began for me at a different time. I had been trying to conceive, I had early symptoms matching the ones described in books on pregnancy, I had a positive test on a DIY kit, and I said doubtfully 'I think I'm pregnant?'
>
> My partner had no trouble with this statement – as far as he was concerned it was instantly a fact. ... I (in relation to/opposition to my body) knew I was pregnant when I wrote that fact to my closest female friend.
>
> As far as other members of my network were concerned, my pregnancy seems to have started at the time they first heard about it, so that I have been gestating at the time of writing for anything between 4 and 7 months!

A constant state?

My later experiences have raised other questions beside the one about where pregnancy begins. For example, there is the question of whether the pregnant person (as distinct for the moment from the pregnant body) remains in that state constantly or can so to speak, enter, and leave it. Since I had put a lot of effort and planning into getting pregnant, and since all the literature on mothers to be, whatever perspective is adopted, focuses on the business of being pregnant, I was more than mildly surprised to find that the experience is not of on-going unified pregnancy, but of a self living in the usual fragmented way, with pregnancy as one of the fragments. Classes, conferences, college amalgamations, the amalgamation of two households and three cats – all call forth elements of identity in which a pregnant me is involved very little or not at all. I've now been in the state long enough and often enough to avoid social blunders like asking 'what for' if I'm congratulated by someone who has just received the news, but that does not mean that I spend all my time being pregnant!

A phenomenology of pregnancy focusing on the ways in which the body is experienced could also profitably look at varying perceptions of bodily size and shape, especially as the control of the female body via fashion in clothing and via food is a theme which has already been explored by some (see Turner 1984). Perceptions of the size and shape of a pregnant body seem to me to be as variable as perceptions of the starting point of a pregnancy. At 32 weeks, the midwife's tape-measure declared my belly to be growing at an exactly standard rate and to have reached an average size. A few days later I met separately, within 10 minutes, two acquaintances who had not heard my news. Both expressed pleasure that I was pregnant and asked politely about due dates. Told 'end of January' the first, (mother of one and 11 weeks pregnant with a second) said 'Ooh you don't look it; you're so small'. The second (no children and as far as I know never pregnant) gasped in horror and said 'That long to go. You'll be like an elephant!'

That one's body should appear different to different people is a commonplace observation, and I suppose that it is also commonplace that one may experience body size differently in different situations. But being pregnant seems to me to have brought about some interesting variations on my normal range of perceptions. For one thing, it has reversed my normal experiences of size in and out of clothes.[6] The general observation remains constant, but impressions of my own body reflected in the mirror have now changed. Clothed, I appear to myself to be larger and somehow more pregnant than I do naked. And within this general perception too there are variations, depending on what I am wearing, how fit I am feeling, and so on. Helen's version of how she is growing and what size she has reached varies from a feeling that she is that logical impossibility, only a little bit pregnant, to one that has her reaching elephantine proportions and in need of a crane to get up the next flight of stairs – and this range must be set in the context on which I remarked earlier, of being pregnant intermittently so that sometimes she has no particular perception of body size at all.

I am sure that there is a lot to be learned about the social construction of the feminine body from a study of the conventions of clothing it, and that this would supplement material on experiences of body size and shape. As far as I can tell (on the basis of a rudimentary and eventually despairing search of clothing shops and pattern books) there are four varieties of pregnant femininity. One (the most expensive) is a kind of bitch-goddess power dressing reminiscent of the costumes in Dynasty; the second is the stereotypical feminine look with lace and frills; and the third is a wholesome, domestic, Doris-Day-in-the-1950s version of the second. ... A fourth version of pregnant femininity – large overalls – is much closer to my image of my own style, but years of clambering in and out of such garments in public toilets has convinced me that they are essentially an impractical mode of dress, and as silly a fashion as any other.[7]

This means that I have basically made do with my existing wardrobe, supplemented by borrowing. ... Wearing familiar garments I feel rather more cumbersome and, hence, pregnant, than in the borrowed clothes – presumably due to their cut. But in the borrowed clothes I feel that the pregnancy is smaller than in my own. ... I have no real data on the impact that clothing has on how others perceive my size. Neither, given the fact that the only clothes I have bought for a pregnant me are underwear and a nondescript pair of maternity jeans can I speculate on what impact buying a maternity wardrobe would have had on my image of myself – whether I would have felt that enhanced sense of womanliness mentioned by some advice books.

A past or present state?

A point which connects with the last ... is the experience of the pregnant body as similar to or different from the body in the non-pregnant past. In my case, being pregnant has (intermittently) transformed my body so that it feels quite unlike the body I had 7 months ago, but in some ways rather like the body I had years before that. In this sense, pregnancy is a transition towards both the unknown and the familiar.

For example, medical writers and novelists alike have noted the impact that enlarged breasts may have on a newly pregnant woman. The medical writers tend to focus on the problem of no longer being able to sleep on one's stomach; writers like Nora Ephron (1983) are more sanguine: 'The beginning is glorious. ... Suddenly they begin to grow ... breasts fantastic tender apricot breasts, then charming plucky firm tangerines, and then, just as you were on the verge of peaches, oranges, grapefruits, cantaloupes, God knows what other blue-ribbon county-fair specimens, your stomach starts to grow and the other fruits are suddenly irrelevant because they're outdistanced by an honest-to God watermelon' (Ephron 1983: 43).

My own experience combined enjoyment of increase in what I have seen for years as a substantial pair of assets with difficulty lying on my stomach and squeezing into jumpers, but had an extra and unexpected dimension. A growing bosom recalled the experiences of early adolescence, where for some years I had a sensation of being a large bust followed around by a smaller person ... now the experience has been reinvoked in miniature and at a faster speed. As a teenager, I did not really feel that my breasts belonged to me until I left an all-girls' school and discovered that they were a social and sexual asset; all of a sudden my bosom and I were integrated. As a pregnant woman I again spent some time following around large breasts which had somehow got put in front of me, but with more enjoyment, and for a shorter time, since the watermelon did indeed fairly quickly eclipse them.

The sensation of a return to the body of adolescence was succeeded by a sensation of return to an even earlier body, based partly on clothing style (the borrowed dress which I like best is a black item cut more or less on the lines of a school tunic) and hence on the feel of clothes (now that my waistline has disappeared everything either hangs from my shoulders or is hitched up more or less under my armpits). The outcome of this is a body which feels rather like the one I had at about 7, including periodically the sensation (as I sit up straight for the sake of my back, with my hands folded comfortably supporting the watermelon which now hides my feet from view) that my legs are not quite long enough to reach the ground yet. For my mother, I think, my body has in one sense finally reached an adult state with which she can sympathetically identify, and has in another sense regressed even further than it has for me. My rounded cheeks, rotund body, slight clumsiness and the way I am now walking remind her, she says of the toddler Helen.

[...]

THEMES FOR FUTURE THEORISING?

I think that themes raised by Grosz (1987, 1994) and others can clearly be seen in this small account. This is partly to be expected because my ruminations on my experience were shaped by some of the same concerns and some of the same reading as the authors I have cited in the first part of this article. Many women and perhaps most feminists will have experienced the pregnant body as involved in a struggle for control. The theme of medical regimentation is common to sociology, feminism, and much of the consumer-oriented literature on childbirth. Other themes, however, such as that of shifting time, seem to me to have emerged directly out of the data. This serendipity leads me to suggest tentatively the points below might help refine the emerging framework of corporeal feminism.

Use the Möbius Strip to Cut Through Difficulties with the Concept of Body

The first and most obvious comment to make about my personal data set is that it shows how slippery the concept of the body can be, and how pervasive is mind/body dualism. The notes above were written as part of an attempt to rethink the concept of body. Yet, they depict a Helen who is somehow detached from the physical entity that goes under her name. The shift between being largely pregnant, a little bit pregnant, or not pregnant is a corporeal one with implications for how I stood and moved and was in the world; not merely a shift in some free-floating 'body image', but in my notes it is only understood in terms of the disembodied notion of 'what seems to be'. More dramatically, transition in labour is experienced as splitting and loss of self, which somehow slips toward culture being overcome by nature. Helen disappears; her body labours on, and eventually Helen (equated with a will that is proactive) reemerges triumphant.

The difficulty of moving beyond a dualist account of the body is precisely what motivated the writers I discussed in the first section of this paper. I propose that, at least for the moment, we simply accept this difficulty, and give up trying to find a concept that integrates the body and the self. Instead, we should resort to the metaphor of the Möbius strip, of external (biological) body and internal (social) self as distinct at a given moment and from a given perspective, but as seamlessly united overall.

Focus on People's Images of their Bodies

The awkwardness with language (that twee counterpoising of self and body) in my accounts is typical of the difficulty of giving and interpreting experiential accounts of the body. As Elaine Scarry (1985) says of pain, we cannot consistently comprehend what another person's experience is because we know but cannot readily speak our own sensations. The less bodily experience is discussed, the harder it is for us to grasp the reality of others' experiences; the pains of torture, the extraordinary sensation of a contraction, an orgasm, all these 'flicker before the mind then disappear' (Scarry 1985: 4). When those sensations occur in a social context that privileges and disembodies the self and the will, description is even harder, and the difficulty extends beyond the moment. Declaring oneself pregnant, or in labour, may be equally difficult – as it clearly was for me.

Feminists have recognised and responded to the political and practical need to name experience with a plethora of experiential accounts. Theorising the body can advance by looking, as Grosz (1994) suggests, at body images.[8] This should entail reflecting on the language in which people describe themselves and looking for divisions as a way of getting a better grasp of totalities. In my notes, as I have said, there is a clear division between mind, equated with self, experienced as proactive, and unthreatening and body, experienced as potentially troublesome, just as Ortner (1974) argued, is nature to culture. The only hint that mind could get out of control is the reference to dreams, which are depicted as an emanation of the body.

This may or may not be a typical image; the point is that it suggests further strategies for research. It might be useful to pursue the image of the body in various situations using questions such as 'Are you the same as your body here?' and 'are you the same as your mind here?' Had the dominant image of the body in my account been one of a piece of machinery, different questions would be implied.[9]

Focus on What Other Bodies are Present

The body described in my data is as social as it is individual. Throughout the personal account an embodied self is being continuously created and recreated by and in social interaction that takes place within and around corporeal action. That interaction includes, of course, interaction with the dualist conception of self. No wonder a woman whose livelihood depends on a particular kind of thinking should experience childbirth as an un-making of self rather than, say, as a temporary enlarging to a state of total embodiment. My labour was, in Adrienne Rich's (1977) terms created by both 'experience' and by 'institution'.

The methodological point emerging from the last paragraph is that when we try to name our bodily experiences, we are always involved in a dialogue. So when researchers try to describe women's experiences in particular situations, we need to ask about who else is there – literally or in imagination. For example, one of my key reference points for comprehending what was happening to me was clearly the expert literature about what women's bodies do during labour. Research on the expert 'inscribing' of the body alone is not enough, but it is important precisely because it is the descriptions and prescriptions of the experts that we often use to comprehend our sensations. It is not just that my account is clearly that of a middle-class woman who has read the medically-oriented literature on birthing; the literature and the advice from antenatal classes are encoded and embodied in the experience itself (as is the technology which accompanies the literature). Hence, I am not only still unsure whether a contraction resembles a period cramp or a tram running over one's stomach, but I also cannot say truthfully whether contractions are pain or not in spite of having used the word in my notes. My mother has no such difficulty; she bore her children in pain, at a time when both the expert and the everyday advice about childbirth focused on the pangs of labour.

The pregnancy notes depict that plurality of bodily presence that Rothfield (1992: 110) notes. They reveal a multiplicity of bodies shifting and shimmering like images in a kaleidoscope according to the focus applied to them: pregnant/not, hugely pregnant or tidily pregnant, in control/out of control.

[...]

Thus, a useful question for both researchers and theorists to ask might be 'how many bodies are here, and of what kinds?' or 'which players (in the immediate situation and the broader social context) are involved in the creation and recreation of this body?'

Bodies Operate In and On Time

The shifting multiplicity of bodies occurs in time and across time. Thinking about transitions to pregnancy and then maternity helps us to understand how time is not merely something we inhabit. My difficulty with knowing when I got pregnant and whether labour had started is not unique. Barbara Katz Rothman's discussions of pregnancy (which I did not read until after I wrote the account) raise the same question about starting point (Rothman 1982). Both she and Emily Martin (1987) point out that the medical clock is a tool of medical power. Rothman demonstrates that the use of medical statistics in the United States has set up a spiral in which 'normal' pregnancies and labours grow shorter, and the proportion of medically effected and controlled births grows larger. Martin's work shows women using their knowledge of medical clocks to maintain control over their birthing (e.g., trying to stay out of hospitals until they are well into labour so that they will not be faced with surgical interferences to medically-defined abnormally long labours; Martin 1987: chapter 8).

The notes on pregnancy show an orientation variously to the present, the future (in growth) and the past of adolescence and childhood. In birthing, there is movement both within clock time (while I try to work out how long contractions last) and outside it ('I'd have ... contractions for ever, then something worse for ever').

The outcome of pregnancy and birth, not covered by the notes, is another set of varied interactions with time. Day by day, interaction with my daughter – the transition to a maternal self which Rossi (1968) describes – ensures that after maternity, Helen will never be the woman she used to be. It also, however, ensures that, as old experiences are witnessed again, some of the child she used to be is periodically recreated,[10] as they were when the tunic maternity dress and the disappearance from sight of her feet recreated the child for the woman on the tram.

[...]

While timing is an element which has been explored in some ways with relation to women's bodies in childbirth, disciplines such as sociology and history are only beginning to reflect on the concept (see Donaldson 1990). I was not conscious of time as in any sense problematic when I wrote my account of pregnancy. It now seems to me that the notion of bodies simultaneously existing within a socially structured clock-time system, helping through rhythmic action to structure that system and structuring a multiplicity of other nonclock systems might be an exciting way to structure both interpretations of phenomenological data and theoretical reflections.

Timing and structuring time relate to control, as when a band leader sets the tempo for players, the rhythm of the belt sets the speed and rhythm of the assembly-line workers, the commercial calender of sales-linked festivals dominates older calenders with their seasonal rituals. The query 'who is in charge of this body' is one that is asked in many research contexts. Asking it in conjunction with questions about time and the body (for example in research on menopause) could enhance our understandings of what embodiment means.

Allied with the theme of body time and clock time is the idea that Grosz calls exploring non-Euclidean and non-Kantian notions of space (Grosz 1987: 11). The personal notes above contain hints of this in the references to shifts (rather than a simple progression of growth) in size during pregnancy and in walking into contractions during labour. While I was pregnant, and for a few days after my daughter was born, my commonsense understanding of how I occupy space was augmented by a sense of self as existing inside space but at the same time comprehending and enveloping space. At the time I had no way of naming this, but a diary note from my sixth month of pregnancy gives some sense of this 'fantasmatic' perception:

> Woke up briefly when E got into my bed quite late, and had sleepy 30 minutes of absolute physical joy about being pregnant. ... Him lying behind me and I could feel the rhythm of his breathing. The cat lying curled up and purring in front of me, and Ruth inside kicking in a sort of counterpoint.

In the days following the birth, this extraordinary geometry was recreated every time I handled my baby and smelled again on her my interior scent, first encountered when the waters broke.

The geometry of our embodied experience merits further exploration, and may usefully be linked to other work on spaces, for and about women. The operative question for both theorists and researchers is 'Where is the body?'

I have reflected on my idiosyncratic data in order to suggest ways in which we could further our understanding of corporeality. The empirical strategy I argue for and the questions I suggest be pursued appear to be distant from the sophisticated and difficult literature reviewed in the first section of this article. It is precisely because it is so difficult to name our bodies and what they do that corporeal feminist theory needs to pay more attention to naive accounts of experiences.

NOTES

1. This, of course, is not a simple matter of uncovering what is there to be found. Neither will an approach that basically sees the construction of meaning as the material of social research sit unproblematically beside the theoretical enterprise of rethinking the body.

2. Hence, techniques like taking the second-order constructs back to the group on whose experience they are based for some kind of validation.
3. None of these works rely on phenomenological theory, but all contain data on how aspects of women's bodies are experienced. Laws (1990) studies how a small group of men understand menstruation; the rest focus on women's understandings and experiences.
4. I take the point made by Elizabeth Grosz (personal communication, June 6, 1992) that statistically normal bodily occasions are as extraordinary as rarer ones seen away from taken-for-granted frameworks.
5. It is based on relatively systematic notes I had kept during the pregnancy.
6. For years I have thought of myself as more or less a little teapot (short and stout), and have had the impression that the less I am wearing the stouter I am. In fact, it is my general observation that female bodies seem bigger and somehow bolder unclothed while male bodies diminish and appear more vulnerable when stripped. I assume that this has something to do with the circumstances in which I normally encounter naked bodies. When I see other women without clothes, it's usually in changing rooms at swimming pools, or communal dressing rooms in clothing shops, etc. The intimacy in these situations is one of shared purpose and frequently there are quite a lot of bodies involved. In contrast, I see naked male adult bodies only in situations of sexual intimacy and one at a time.
7. On this, see also Wilson (1985).
8. Her suggestion is for research on 'body image' and 'body schema' in the psychological sense – the subjects' 'sense of its place in the world and in connection with others'. My suggestion includes reflection on the linguistic images that give indications of the subjects' varied body images and schema.
9. Apart from the notion of looking after my pregnant body, there is nothing in my account of body as machine, but my impression is that it is an image often seen in male-oriented fitness literature.
10. Another example: My daughter's first infant ear infection brought back to me a vivid image of my mother's anguished face during an early earache of my own. The episode was not a suppressed memory. The visual image of my mother's face, however, was not one I recall having had prior to being in my mother's position. So there was both movement back in time and a shift in perspective involved. Further testament to the complex possibilities of movement in time is the way in which, when I told her of her grand-daughter's illness, the same anguish flickered briefly in her ageing face. Will my daughter, at some future time briefly sense her grandmother's pain? Is it, perhaps, also the pain of a more remote maternal ancestor?

REFERENCES

Boston Women's Health Collective (1976/1984) *Our Bodies Ourselves*, Harmondsworth: Penguin.

Caddick, Alison (1986) 'Feminism and the Body', *Arena*, 74: 60–90.

Caddick, Alison (1992) 'Feminist and Postmodern', *Arena* 99/100: 112–28.

Donaldson, Mike (1990) *Time of Our Lives: Labour and Love in the Working Class*, Sydney: Allen and Unwin.

Eagleton, Terry (1993) 'It is not quite true that, I have a body, and not quite true that I am one either' [Review of the book *Body Work*], *London Review of Books* 27(5): 7–8.

Ephron, Nora (1983) *Heartburn*, London: Pavanne/Heinemann.

Grosz, Elizabeth (1987) 'Notes Towards a Corporeal Feminism', *Australian Feminist Studies*, 5: 1–16.

Grosz, Elizabeth (1994) *Volatile Bodies: Toward a Corporeal Feminism*, Sydney: Allen and Unwin.

Laws, Sophie (1990) *Issues of Blood: The Politics of Menstruation*, Basingstoke: Macmillan.

Martin, Emily (1987) *The Woman in the Body: A Cultural Analysis of Reproduction*, Milton Keynes: Open University Press.
Meckelburg, Patricia (1993) '"*Practising Breaststroking*": *Women's Breasted Bodies*', Paper presented at the conference of The Australian Sociological Association, Sydney.
Ortner, S. (1974) 'Is Female to Male as Nature is to Culture?' in M. Rosaldo and L. Lamphere (Eds), *Woman, Culture and Society*, pp. 67–87, Stanford, CA: Stanford University Press.
Rich, Adrienne (1977) *Of Woman Born: Motherhood as Experience and Institution*, London: Virago.
Rossi, Alice (1968) 'Transition to Parenthood', *Journal of Marriage and the Family* 30: 26–39.
Rothfield, Philipa (1992) 'Backstage in the Theatre of Representation' *Arena* 99/100: 98–111.
Rothman, Barbara (1982) *In Labour: Women and Power in the Birthplace*, London: Junction Books.
Scarry, Elaine (1985) *The Body in Pain: The Making and Unmaking of the World*, New York: Oxford University Press.
Schutz, Alfred (1962) 'Common-sense and Scientific Interpretations of Human Action', in *Collected papers. 1: The problem of social reality*, pp. 3–47, The Hague: Martinus Nijhoff.
Stanley, Liz and Wise, Sue (1983) *Breaking Out*, London: Routledge and Kegan Paul.
Turner, Bryan S. (1984) *The Body and Society*, Oxford: Basil Blackwell.
Turner, Bryan S. (1990) 'The Talking Disease: Hilda Bruch and Anorexia Nervosa', *The Australian and New Zealand Journal of Sociology* 26: 157–169.
Wilson, Elizabeth (1985) *Adorned in Dreams*, London: Virago.

SECTION 2
SEXY BODIES

INTRODUCTION

This section features a range of articles and extracts which look at the production of sexual identities and sexuate bodies. It is not intended to cover sexuality as a set of practices as such, nor to include such issues as sexual violence, AIDS, sadomasochism or the sex industry, all of which can be found in *Feminism and Sexuality: A Reader* (Jackson and Scott 1996). Our own concern with sexy bodies focuses primarily on the ways in which feminism has theorised the sexual as intrinsic to embodied subjectivity, as an element in becoming, rather than as a characteristic of a pre-existing person which is then channelled into diverse forms. Moreover, what is taken to constitute sexuality is not taken for granted but is open to exploration in all and any form of sexy – even if not strictly sexual – bodies.

For women, of course, sexuality has always been linked in the dominant discourse to a certain excessiveness that stands against the attribution of full subjecthood, and that marks the feminine as sexual in itself. It is part of the classic move that identifies the male with the mind and the female with the body. And as Foucault (1979) makes clear, sexuality, in postEnlightenment thought, is the overloaded focus of the discursive strategies of power and knowledge; and women's bodies, he asserts, are saturated with sex. From these decidely negative connotations, feminist theory has sought to recuperate sexy bodies, and to extol the *jouissance* – particularly in the case of 'French' feminism, of which Luce Irigaray here stands representative – of feminine excess. Irigaray's turn to the metaphor of the multiplicity and tactility of women's sexualities – the two lips are both self and other-touching – stands in contradistinction to the singular, rigid, phallic standard that characterises masculinity. The plasticity of Irigarayan bodies is decidedly sexy, and no one part serves as a focus – a point also clearly exemplified in Jeanette Winterson's erotic celebration of the body in all its anatomical and physiological specificity. The two short passages here speak to the nose and mouth, and to the senses of smell and taste.

The question that these writers and many other theorists pose is what are the forms of a specifically feminine desire that are uncontained by the relations of reproduction? All would agree that women's sexuality has been deeply constrained by the discursive and disciplinary practices of patriarchy that have sought to channel bodies and subjectivities into predetermined gendered models. The feminist goal of breaking out of the boundaries of the proper body, of overflowing the sexual categories assigned to women is supplemented in the discourse of black women by the need to comprehensibly rewrite the legacies of a colonial history. Particularly in the context of slavery, black women's bodies and sexuality were seen as animal-like, good only for a dehumanised form of reproduction, and as Evelynn Hammonds' piece points out, often pathologised in relation to white women's sexual bodies. In response to such damaging and violent representations, resistance has to a large extent taken the form of what Hammonds calls a 'politics of silence', a silence that is still evident in the relative sparsity of writing by black feminists that *directly* addresses issues of the sexual body. Nonetheless, there is a growing recognition that the intersection of race, gender and sexuality is one that should be theorised by all feminists.

The goal of thinking sexy bodies and sexuality outside of conventional discourse is common to all the selections, and is as evident in Lynne Segal's challenge to the relations of heterosexuality from the perspective of 'straight' desire, as it is in Barbara Creed's piece where the nature of desire itself is at stake. Creed's lesbian bodies are themselves contested, however, in the politics and practices of both transgender, about which Judith Halberstam writes, and in the growing interest in cybersexuality, taken up here by Juniper Wiley, and by Sue-Ellen Case. Both issues raise the question of how sex, gender and sexuality line up, the first in the presence of a body that falls into no one category, and the second in the absence of bodies altogether. And while transexuality and transgender are not self-evidently transgressive, insofar as they display a nostalgia for fit, the issue for cybersexuality is to what extent virtual bodies can be said to be sexy at all. The desire to queer existing models of embodied sexuality and embodied identity is very strong, but is it trangression for its own sake, or does all destabilisation of sexual boundaries offer something positive to feminist theory?

REFERENCES AND FURTHER READING

Bornstein, Kate (1994) *Gender Outlaw: On Men, Women, and the Rest of Us*, London: Routledge.

Brown, Terry (1994) 'The Butch-Femme Fatale' in Laura Doan (ed.), *The Lesbian Postmodern*, New York: Columbia University Press.

Foucault, Michel (1979) *History of Sexuality, Vol. 1*, London: Penguin Books.

Galler, Roberta (1984) 'The Myth of the Perfect Body' in Carole S. Vance (ed.), *Pleasure and Danger: Exploring Female Sexuality*, London: Pandora Press.
Holland, S. and Ramazanoglu, C. (1994) 'Desire, Risk and Control: The Body as Site of Contestation' in Lesley Doyle, Jennie Naidoo and Tamsin Wilton (eds), *AIDS: Setting a Feminist Agenda*, London: Taylor and Francis.
hooks, bell (1994) 'Talking Sex: Beyond the Patriarchal Phallic Imaginary' in *Outlaw Culture: Resisting Representations*, London: Routledge.
Jackson, Stevi and Scott, Sue (1996) *Feminism and Sexuality: A Reader*, Edinburgh: Edinburgh University Press.
Lorde, Audre (1984) 'Uses of the Erotic: The Erotic as Power' in *Sister Outsider*, Trumansberg, NY: Crossing Press.
Najmabadi, Afsaneh (1993) 'Veiled Discourse – Unveiled Bodies', *Feminist Studies* 19 (3): 487–518.
Shapiro, Judith (1991) 'Transsexualism: Reflections on the Persistence of Gender and the Mutability of Sex' in Julia Epstein and Kristina Straub (eds), *Body Guards*, London: Routledge.
Singer, Linda (1993) *Erotic Welfare: Sexual Theory and Politics in the Age of Epidemic*, London: Routledge.
Suleiman, Susan Rubin (1986) '(Re)writing the Body: The Politics and Poetics of Female Eroticism' in *The Female Body in Western Culture*, Cambridge, MA: Harvard University Press.
Thadani, Giti (1996) 'The Dual Feminine' in *Sakhiyani*, London: Cassell
Wittig, Monique (1975) *The Lesbian Body*, London: Peter Owen.
Zita, Jacqueline (1994) 'Male Lesbians and the Postmodern Body' in Claudia Card (ed.), *Adventures in Lesbian Ethics*, Bloomington: Indiana University Press.

2.1

WHEN OUR LIPS SPEAK TOGETHER

Luce Irigaray

If we keep on speaking the same language together, we're going to reproduce the same history. Begin the same old stories all over again. Don't you think so? Listen: all round us, men and women sound just the same. The same discussions, the same arguments, the same scenes. The same attractions and separations. The same difficulties, the same impossibility of making connections. The same ... Same ... Always the same.

If we keep on speaking sameness, if we speak to each other as men have been doing for centuries, as we have been taught to speak, we'll miss each other, fail ourselves. Again ... Words will pass through our bodies, above our heads. They'll vanish, and we'll be lost. Far off, up high. Absent from ourselves: we'll be spoken machines, speaking machines. Enveloped in proper skins, but not our own. Withdrawn into proper names, violated by them. Not yours, not mine. We don't have any. We change names as men exchange us, as they use us, use us up. It would be frivolous of us, exchanged by them, to be so changeable.

How can I touch you if you're not there? Your blood has become their meaning. They can speak to each other, and about us. But what about us? Come out of their language. Try to go back through the names they've given

This text was originally published as 'Quand nos lèvres se parlent', in *Cahiers du Grif*, no. 12. English translation: 'When Our Lips Speak Together', trans. Carolyn Burke, in *Signs: Journal of Women in Culture and Society*, 6 (1): 69–79, Fall 1980.

you. I'll wait for you, I'm waiting for myself. Come back. It's not so hard. You stay here, and you won't be absorbed into familiar scenes, worn-out phrases, routine gestures. Into bodies already encoded within a system. Try to pay attention to yourself. To me. Without letting convention, or habit, distract you.

For example: 'I love you' is addressed by convention or habit to an enigma – an other. An other body, an other sex. I love you: I don't quite know who, or what. 'I love' flows away, is buried, drowned, burned, lost in a void. We'll have to wait for the return of 'I love'. Perhaps a long time, perhaps forever. Where has 'I love'. gone? What has become of me? 'I love' lies in wait for the other. Has he swallowed me up? Spat me out? Taken me? Left me? Locked me up? Thrown me out? What's he like now? No longer (like) me? When he tells me 'I love you', is he giving me back? Or is he giving himself in that form? His? Mine? The same? Another? But then where am I, what have I become?

When you say I love you – staying right here, close to you, close to me – you're saying I love myself. You don't need to wait for it to be given back; neither do I. We don't owe each other anything. That 'I love you' is neither gift nor debt. You 'give' me nothing when you touch yourself, touch me, when you touch yourself again through me. You don't give yourself. What would I do with you, with myself, wrapped up like a gift? You keep our selves to the extent that you share us. You find our selves to the extent that you trust us. Alternatives, oppositions, choices, bargains like these have no business between us. Unless we restage their commerce, and remain within their order. Where 'we' has no place.

I love you: body shared, undivided. Neither you nor I severed. There is no need for blood shed, between us. No need for a wound to remind us that blood exists. It flows within us, from us. Blood is familiar, close. You are all red. And so very white. Both at once. You don't become red by losing your candid whiteness. You are white because you have remained close to blood. White and red at once, we give birth to all the colors: pinks, browns, blonds, greens, blues ... For this whiteness is no sham. It is not dead blood, black blood. Sham is black. It absorbs everything, closed in on itself, trying to come back to life. Trying in vain ... Whereas red's whiteness takes nothing away. Luminous, without autarchy, it gives back as much as it receives.

We are luminous. Neither one nor two. I've never known how to count. Up to you. In their calculations, we make two. Really, two? Doesn't that make you laugh? An odd sort of two. And yet not one. Especially not one. Let's leave *one* to them: their oneness, with its prerogatives, its domination, its solipsism: like the sun's. And the strange way they divide up their couples, with the other as the image of the one. Only an image. So any move toward the other means turning back to the attraction of one's own mirage. A (scarcely) living mirror, she/it is frozen, mute. More lifelike. The ebb and flow of our lives spent in the exhausting labor of copying, miming. Dedicated to reproducing – that sameness in which we have remained for centuries, as the other.

But how can I put 'I love you' differently? I love you, my indifferent one? That still means yielding to their language. They've left us only lacks, deficiencies, to designate ourselves. They've left us their negative(s). We ought to be – that's already going too far – indifferent.

Indifferent one, keep still. When you stir, you disturb their order. You upset everything. You break the circle of their habits, the circularity of their exchanges, their knowledge, their desire. Their world. Indifferent one, you mustn't move, or be moved, unless they call you. If they say 'come', then you may go ahead. Barely. Adapting yourself to whatever need they have, or don't have, for the presence of their own image. One step, or two. No more. No exuberance. No turbulence. Otherwise you'll smash everything. The ice, the mirror. Their earth, their mother. And what about your life? You must pretend to receive it from them. You're an indifferent, insignificant little receptacle, subject to their demands alone.

So they think we're indifferent. Doesn't that make you laugh? At least for a moment, here and now? *We* are *indifferent?* (If you keep on laughing that way, we'll never be able to talk to each other. We'll remain absorbed in their words, violated by them. So let's try to take back some part of our mouth to speak with.) Not different; that's right. Still . . . No, that would be too easy. And that 'not' still keeps us separate so we can be compared. Disconnected that way, no more 'us'? Are we alike? If you like. It's a little abstract. I don't quite understand 'alike'. Do you? Alike in whose eyes? in what terms? by what standard? with reference to what third? I'm touching you, that's quite enough to let me know that you are my body.

I love you: our two lips cannot separate to let just *one* word pass. A single word that would say 'you', or 'me'. Or 'equals'; she who loves, she who is loved. Closed and open, neither ever excluding the other, they say they both love each other. Together. To produce a single precise word, they would have to stay apart. Definitely parted. Kept at a distance, separated by *one word*.

But where would that word come from? Perfectly correct, closed up tight, wrapped around its meaning. Without any opening, any fault. 'You'. 'Me'. You may laugh . . . Closed and faultless, it is no longer you or me. Without lips, there is no more 'us'. The unity, the truth, the propriety of words comes from their lack of lips, their forgetting of lips. Words are mute, when they are uttered once and for all. Neatly wrapped up so that their meaning – their blood – won't escape. Like the children of men? Not ours. And besides, do we need, or want, children? What for? Here and now, we are close. Men and women have children to embody their closeness, their distance. But we?

I love you, childhood. I love you who are neither mother (forgive me, mother, I prefer a woman) nor sister. Neither daughter nor son. I love you – and where I love you, what do I care about the lineage of our fathers, or their desire for reproductions of men? Or their genealogical institutions? What need have I for husband or wife, for family, persona, role, function? Let's leave all those to men's reproductive laws. I love you, your body, here and now. I/you touch you/me, that's quite enough for us to feel alive.

Open your lips; don't open them simply. I don't open them simply. We – you/I – are neither open nor closed. We never separate simply: *a single word* cannot be pronounced, produced, uttered by our mouths. Between our lips, yours and mine, several voices, several ways of speaking resound endlessly, back and forth. One is never separable from the other. You/I: we are always several at once. And how could one dominate the other? impose her voice, her tone, her meaning? One cannot be distinguished from the other; which does not mean that they are indistinct. You don't understand a thing? No more than they understand you.

Speak, all the same. It's our good fortune that your language isn't formed of a single thread, a single strand or pattern. It comes from everywhere at once. You touch me all over at the same time. In all senses. Why only one song, one speech, one text at at time? To seduce, to satisfy, to fill one of my 'holes'? With you, I don't have any. We are not lacks, voids awaiting sustenance, plenitude, fulfilment from the other. By our lips we are women: this does not mean that we are focused on consuming, consummation, fulfilment.

Kiss me. Two lips kissing two lips: openness is ours again. Our 'world'. And the passage from the inside out, from the outside in, the passage between us, is limitless. Without end. No knot or loop, no mouth ever stops our exchanges. Between us the house has no wall, the clearing no enclosure, language no circularity. When you kiss me, the world grows so large that the horizon itself disappears. Are we unsatisfied? Yes, if that means we are never finished. If our pleasure consists in moving, being moved, endlessly. Always in motion: openness is never spent nor sated.

We haven't been taught, nor allowed, to express multiplicity. To do that is to speak improperly. Of course, we might – we were supposed to? – exhibit one 'truth' while sensing, withholding, muffling another. Truth's other side – its complement? its remainder? – stayed hidden. Secret. Inside and outside, we were not supposed to be the same. That doesn't suit their desires. Veiling and unveiling: isn't that what interests them? What keeps them busy? Always repeating the same operation, every time. On every woman.

You/I become two, then, for their pleasure. But thus divided in two, one outside, the other inside, you no longer embrace yourself, or me. Outside, you try to conform to an alien order. Exiled from yourself, you fuse with everything you meet. You imitate whatever comes close. You become whatever touches you. In your eagerness to find yourself again, you move indefinitely far from yourself. From me. Taking one model after another, passing from master to master, changing face, form, and language with each new power that dominates you. You/we are sundered; as you allow yourself to be abused, you become an impassive travesty. You no longer return indifferent; you return closed, impenetrable.

Speak to me. You can't? You no longer want to? You want to hold back? Remain silent? White? Virginal? Keep the inside self to yourself? But it doesn't exist without the other. Don't tear yourself apart like that with choices imposed on you. *Between us*, there's no rupture between virginal and non-virginal. No event that makes us women. Long before your birth, you touched yourself, innocently. Your/my body doesn't acquire its sex through an operation. Through the action of some power, function, or organ. Without any intervention or special manipulation, you are a woman already. There is no need for an outside; the other already affects you. It is inseparable from you. You are altered forever, through and through. That is your crime, which you didn't commit: you disturb their love of property.

How can I tell you that there is no possible evil in your sexual pleasure – you who are a stranger to good(s). That the fault only comes about when they strip you of your openness and close you up, marking you with signs of possession; then they can break in, commit infractions and transgressions and play other games with the law. Games in which they – and you? – speculate on your whiteness. If we play along, we let ourselves be abused, destroyed. We remain indefinitely distant from ourselves to support the pursuit of their ends. That would be our flaw. If we submit to their reasoning, we are guilty. Their strategy, intentional or not, is calculated to make us guilty.

You come back, divided: 'we' are no more. You are split into red and white, black and white: how can we find each other again? How can we touch each other once more? Cut up, dispatched, finished: our pleasure is trapped in their system, where a virgin is one as yet unmarked by them, for them. One who is not yet made woman by and for them. Not yet imprinted with their sex, their language. Not yet penetrated, possessed by them. Remaining in that candor that waits for them, that is nothing without them, a void without them. A virgin is the future of their exchanges, transactions, transports. A kind of reserve for their explorations, consummations, exploitations. The advent of their desire, Not of ours.

How can I say it? That we are women from the start. That we don't have to be turned into women by them, labelled by them, made holy and profaned by them. That that has always already happened, without their efforts. And that their history, their stories, constitute the locus of our displacement. It's not that we have a territory of our own; but their fatherland, family, home, discourse, imprison us in enclosed spaces where we cannot keep on moving, living, as ourselves. Their properties are our exile. Their enclosures, the death of our love. Their words, the gag upon our lips.

How can we speak so as to escape from their compartments, their schemas, their distinctions and oppositions: virginal/deflowered, pure/impure, innocent/experienced ... How can we shake off the chain of these terms, free ourselves from their categories, rid ourselves of their names? Disengage ourselves, *alive*, from their concepts? Without reserve, without the immaculate whiteness that shores up their systems. You know that we are never completed, but that we only embrace ourselves whole. That one after another, parts – of the body, of space, of time – interrupt the flow of our blood. Paralyze, petrify, immobilize us. Make us paler. Almost frigid.

Wait. My blood is coming back. From their senses. It's warm inside us again. Among us. Their words are emptying out, becoming bloodless, Dead skins. While our lips are growing red again. They're stirring, moving, they want to speak. You mean ... ? What? Nothing. Everything. Yes. Be patient. You'll say it all. Begin with what you feel, right here, right now. Our all will come.

But you can't anticipate it, foresee it, program it. Our all cannot be projected, or mastered. Our whole body is moved. No surface holds. No figure, line, or point remains. No ground subsists. But no abyss, either. Depth, for us, is not a chasm. Without a solid crust, there is no precipice. Our depth is the thickness of our body, our all touching itself. Where top and bottom, inside and outside, in front and behind, above and below are not separated, remote, out of touch. Our all intermingled. Without breaks or gaps.

If you/I hesitate to speak, isn't it because we are afraid of not speaking well? But what is 'well' or 'badly'? With what are we conforming if we speak 'well'? What hierarchy, what subordination lurks there, waiting to break our resistance? What claim to raise ourselves up in a worthier discourse? Erection is no business of ours: we are at home on the flatlands. We have so much space to share. Our horizon will never stop expanding; we are always open. Stretching out, never ceasing to unfold ourselves, we have so many voices to invent in order to express all of us everywhere, even in our gaps, that all the time there is will not be enough. We can never complete the circuit, explore our periphery: we have so many dimensions. If you want to speak 'well', you pull yourself in, you become narrower as you rise. Stretching upward, reaching higher, you pull yourself away from the limitless realm of your body. Don't make yourself erect, you'll leave us. The sky isn't up there: it's between us.

And don't worry about the 'right' word. There isn't any. No truth between our lips. There is room enough for everything to exist. Everything is worth exchanging, nothing is privileged, nothing is refused. Exchange? Everything is exchanged, yet there are no transactions. Between us, there are no proprietors, no purchasers, no determinable objects, no prices. Our bodies are nourished by our mutual pleasure. Our abundance is inexhaustible: it knows neither want nor plenty. Since we give each other (our) all, with nothing held back, nothing hoarded, our exchanges are without terms, without end. How can I say it? The language we know is so limited ...

Why speak? you'll ask me. We feel the same things at the same time. Aren't my hands, my eyes, my mouth, my lips, my body enough for you? Isn't what they are saying to you sufficient? I could answer 'yes', but that would be too easy. Too much a matter of reassuring you/us.

If we don't invent a language, if we don't find our body's language, it will have too few gestures to accompany our story. We shall tire of the same ones, and leave our desires unexpressed, unrealized. Asleep again, unsatisfied, we shall fall back upon the words of men – who, for their part, have 'known' for a long time. But *not our body*. Seduced, attracted, fascinated, ecstatic with our becoming, we shall remain paralyzed. Deprived of *our movements*. Rigid, whereas we are made for endless change. Without leaps or falls, and without repetition.

Keep on going, without getting out of breath. Your body is not the same today as yesterday. Your body remembers. There's no need for *you* to remember. No need to hold fast to yesterday, to store it up as capital in your head. Your memory? Your body expresses yesterday in what it wants today. If you think: yesterday I was, tomorrow I shall be, you are thinking: I have died a little. Be what you are becoming, without clinging to what you might have been, what you might yet be. Never settle. Let's leave definitiveness to the undecided; we don't need it. Our body, right here, right now, gives us a very different certainty. Truth is necessary for those who are so distanced from their body that they have forgotten it. But their 'truth' immobilizes us, turns us into statues, if we can't loose its hold on us. If we can't defuse its power by trying to say, right here and now, how we are moved.

You are moving. You never stay still. You never stay. You never 'are'. How can I say 'you', when you are always other? How can I speak to you? You remain in flux, never congealing or solidifying. What will make that current flow into words? It is multiple, devoid of causes, meanings, simple qualities. Yet it cannot be decomposed. These movements cannot be described as the passage from a beginning to an end. These rivers flow into no single, definitive sea. These streams are without fixed banks, this body without fixed boundaries. This unceasing mobility. This life – which will perhaps be called our restlessness, whims, pretenses, or lies. All this remains very strange to anyone claiming to stand on solid ground.

Speak, all the same. Between us, 'hardness' isn't necessary. We know the contours of our bodies well enough to love fluidity. Our density can do without trenchancy or rigidity. We are not drawn to dead bodies.

But how can we stay alive when we are far apart? There's the danger. How can I wait for you to return if when you're far away from me you cannot also be near? If I have nothing palpable to help me recall in the here and now the touch of our bodies. Open to the infinity of our separation, wrapped up in the intangible sensation of absence, how can we continue to live as ourselves? How can we keep ourselves from becoming absorbed once again in their violating language? From being embodied as mourning. We must learn to speak to each other so that we can embrace from afar. When I touch myself, I am surely remembering you. But so much has been said, and said of us, that separates us.

Let's hurry and invent our own phrases. So that everywhere and always we can continue to embrace. We are so subtle that nothing can stand in our way, nothing can stop us from reaching each other, even fleetingly, if we can find means of communication that have *our* density. We shall pass imperceptibly through every barrier, unharmed, to find each other. No one will see a thing. Our strength lies in the very weakness of our resistance. For a long time now they have appreciated what our suppleness is worth for their own embraces and impressions. Why not enjoy it ourselves? Rather than letting ourselves be subjected to their branding. Rather than being fixed, stabilized, immobilized. Separated.

Don't cry. One day we'll manage to say ourselves. And what we say will be even lovelier than our tears. Wholly fluent.

Already, I carry you with me everywhere. Not like a child, a burden, a weight, however beloved and precious. You are not *in me*. I do not contain you or retain you in my stomach, my arms, my head. Nor in my memory, my mind, my language. You are there, like my skin. With you I am certain of existing beyond all appearances, all disguises, all designations. I am assured of living because you are duplicating my life. Which doesn't mean that you give me yours, or subordinate it to mine. The fact that you live lets me know I am alive, so long as you are neither my counterpart nor my copy.

How can I say it differently? We exist only as two? We live by twos beyond all mirages, images, and mirrors. Between us, one is not the 'real' and the other her imitation; one is not the original and the other her copy. Although we can dissimulate perfectly within their economy, we relate to one another without simulacrum. Our resemblance does without semblances: for in our bodies, we are already the same. Touch yourself, touch me, you'll 'see'.

No need to fashion a mirror image to be 'doubled', to repeat ourselves – a second time. Prior to any representation, we are two. Let those two – made for you by your blood, evoked for you by my body – come together alive. You will always have the touching beauty of a first time, if you aren't congealed in reproductions. You will always be moved for the first time, if you aren't immobilized in any form of repetition.

We can do without models, standards, or examples. Let's never give ourselves orders, commands, or prohibitions. Let our imperatives be only appeals to move, to be moved, together. Let's never lay down the law to each other, or moralize, or make war. Let's not claim to be right, or claim the right to criticize one another. If one of us sits in judgment, our existence comes to an end. And what I love in you, in myself, in us no longer takes place: the birth that is never accomplished, the body never created once and for all, the form never definitively completed, the face always still to be formed. The lips never opened or closed on a truth.

Light, for us, is not violent. Not deadly. For us the sun does not simply rise or set. Day and night are mingled in our gazes. Our gestures. Our bodies. Strictly speaking, we cast no shadow. There is no danger that one or the other may be a darker double. I want to remain nocturnal, and find my night softly luminous, in you. And don't by any means imagine that I love you shining like a beacon, lording it over everything around you. If we divide light from night, we give up the lightness of our mixture, solidify those heterogeneities that make us so consistently whole. We put ourselves into watertight compartments, break ourselves up into parts, cut ourselves in two, and more. Whereas we are always one and the other, at the same time. If we separate ourselves that way, we 'all' stop being born. Without limits or borders, except those of our moving bodies.

And only the limiting effect of time can make us stop speaking to each other. Don't worry. I – continue. Under all these artificial constraints of time and space, I embrace you endlessly. Others may make fetishes of us to separate us: that's their business. Let's not immobilize ourselves in these borrowed notions.

And if I have so often insisted on negatives: *not*, *nor*, *without* . . . it has been to remind you, to remind us, that we only touch each other naked. And that, to find ourselves once again in that state, we have a lot to take off. So many representations, so many appearances separate us from each other. They have wrapped us for so long in their desires, we have adorned ourselves so often to please them, that we have come to forget the feel of our own skin. Removed from our skin, we remain distant. You and I, apart.

You? I? That's still saying too much. Dividing too sharply between us: all.

2.2

THE NOSE AND TASTE

Jeanette Winterson

THE NOSE: THE SENSE OF SMELL IN HUMAN BEINGS IS GENERALLY LESS ACUTE THAN IN OTHER ANIMALS

The smells of my lover's body are still strong in my nostrils. The yeast smell of her sex. The rich fermenting undertow of rising bread. My lover is a kitchen cooking partridge. I shall visit her gamey low-roofed den and feed from her. Three days without washing and she is well-hung and high. Her skirts reel back from her body, her scent is a hoop about her thighs.

From beyond the front door my nose is twitching, I can smell her coming down the hall towards me. She is a perfumier of sandalwood and hops. I want to uncork her. I want to push my head against the open wall of her loins. She is firm and ripe, a dark compound of sweet cattle straw and Madonna of the Incense. She is frankincense and myrrh, bitter cousin smells of death and faith.

When she bleeds the smells I know change colour. There is iron in her soul on those days. She smells like a gun.

My lover is cocked and ready to fire. She has the scent of her prey on her. She consumes me when she comes in thin white smoke smelling of saltpetre. Shot against her all I want are the last wreaths of her desire that carry from the base of her to what doctors like to call the olfactory nerves.

TASTE: THERE ARE FOUR FUNDAMENTAL SENSATIONS OF TASTE: SWEET, SOUR, BITTER AND SALT

My lover is an olive tree whose roots grow by the sea. Her fruit is pungent and

---

From: J. Winterson, *Written on the Body*, London: Jonathan Cape, 1992.

green. It is my joy to get at the stone of her. The little stone of her hard by the tongue. Her thick-fleshed salt-veined swaddle stone.

Who eats an olive without first puncturing the swaddle? The waited moment when the teeth shoot a strong burst of clear juice that has in it the weight of the land, the vicissitudes of the weather, even the first name of the olive keeper.

The sun is in your mouth. The burst of an olive is breaking of a bright sky. The hot days when the rains come. Eat the day where the sand burned the soles of your feet before the thunderstorm brought up your skin in bubbles of rain.

Our private grove is heavy with fruit. I shall worm you to the stone, the rough swaddle stone.

2.3

TOWARD A GENEALOGY OF BLACK FEMALE SEXUALITY: THE PROBLEMATIC OF SILENCE

Evelynn M. Hammonds

Sexuality has become one of the most visible, contentious and spectacular features of modern life in the United States during this century. Controversies over sexual politics and sexual behaviour reveal other tensions in US society, particularly those around changing patterns of work, family organization, disease control, and gender relations. In the wake of Anita Hill's allegations of sexual harassment by Supreme Court nominee Clarence Thomas and the more recent murder charges brought against football star O. J. Simpson, African Americans continue to be used as the terrain upon which contested notions about race, gender, and sexuality are worked out. Yet, while black men have increasingly been the focus of debates about sexuality in the academy and in the media, the specific ways in which black women figure in these discourses has remained largely unanalyzed and untheorized.

In this essay, I will argue that the construction of black women's sexuality, from the nineteenth century to the present, engages three sets of issues. First, there is the way black women's sexuality has been constructed in a binary opposition to that of white women: it is rendered simultaneously invisible, visible (exposed), hypervisible, and pathologized in dominant discourses. Secondly, I will describe how resistance to these dominant discourses has been coded and lived by various group of black women within black communities at different historical moments. Finally, I will discuss the limitations of these

---

From: J. Alexander and C. T. Mohanty (eds), *Feminist Genealogies, Colonial Legacies, Democratic Futures*, New York: Routledge, 1997.

strategies of resistance in disrupting dominant discourses about black women's sexuality and the implications of this for black women with AIDS.

In addressing these questions, I am specifically interested in interrogating the writing of black feminist theorists on black women's sexuality. As sociologist Patricia Hill Collins has noted, while black feminist theorists have written extensively on the impact of such issues as rape, forced sterilization, and homophobia on black women's sexuality, 'when it comes to other important issues concerning the sexual politics of Black womanhood, ... Black feminists have found it almost impossible to say what has happened to Black women'.[1] To date, there has been no full length historical study of African American women's sexuality in the United States. In this essay, I will examine some of the reasons why black feminists have failed to develop a complex, historically specific analysis of black women's sexuality.

Black feminist theorists have almost universally described black women's sexuality, when viewed from the vantage of the dominant discourses, as an absence. In one of the earliest and most compelling discussions of black women's sexuality, literary critic Hortense Spillers wrote, 'Black women are the beached whales of the sexual universe, unvoiced, misseen, not doing, awaiting *their* verb'.[2] For writer Toni Morrison, black women's sexuality is one of the 'unspeakable things unspoken' of the African American experience. Black women's sexuality is often described in metaphors of speechlessness, space, or vision; as a 'void' or empty space that is simultaneously ever-visible (exposed) and invisible, where black women's bodies are always already colonized. In addition, this always already colonized black female body has so much sexual potential it has none at all.[3] Historically, black women have reacted to the repressive force of the hegemonic discourses on race and sex that constructed this image with silence, secrecy, and a partially self-chosen invisibility.[4]

Black feminist theorists – historians, literary critics, sociologists, legal scholars, and cultural critics – have drawn upon a specific historical narrative which purportedly describes the factors that have produced and maintained perceptions of black women's sexuality (including their own). Three themes emerge in this history. First, the construction of the black female as the embodiment of sex and the attendant invisibility of black women as the unvoiced, unseen – everything that is not white. Secondly, the resistance of black women both to negative stereotypes of their sexuality and to the material effects of those stereotypes on black women's lives. And, finally, the evolution of a 'culture of dissemblance' and a 'politics of silence' by black women on the issue of their sexuality.

COLONIZING BLACK WOMEN'S BODIES

By all accounts, the history of discussions of black women's sexuality in Western thought begins with the Europeans' first contact with peoples on the African continent. As Sander Gilman argued in his widely cited essay 'Black Bodies, White Bodies: Toward an Iconography of Female Sexuality in Late

Nineteenth-Century Art, Medicine, and Literature',[5] the conventions of human diversity that were captured in the iconography of the period linked the image of the prostitute and the black female through the Hottentot female. The Hottentot female most vividly represented in this iconography was Sarah Bartmann, known as the 'Hottentot Venus'. This southern African black woman was crudely exhibited and objectified by European audiences and scientific experts because of what they regarded as unusual aspects of her physiognomy – her genitalia and buttocks. Gilman argued that Sarah Bartmann, along with other black females brought from southern Africa, became the central image for the black female in Europe through the nineteenth century. The 'primitive' genitalia of these women were defined by European commentators as the sign of their 'primitive' sexual appetites. Thus, the black female became the antithesis of European sexual mores and beauty and was relegated to the lowest position on the scale of human development. The image of the black female constructed in this period reflected everything the white female was not, or, as art historian Lorraine O'Grady has put it, 'White is what woman is [was]; not-white (and the stereotypes not-white gathers in) is what she had better not be'.[6] Gilman shows that by the end of the nineteenth century European experts in anthropology, public health, medicine, biology, and psychology had concluded, with ever-increasing 'scientific' evidence, that the black female embodied the notion of uncontrolled sexuality.

In addition, as white European elites' anxieties surfaced over the increasing incidence of sexually transmitted diseases, especially syphilis, high rates of these diseases among black women were used to define them further as a source of corruption and disease. It was the association of prostitutes with disease that provided the final link between the black female and the prostitute. Both were bearers of the stigmata of sexual difference and deviance. Gilman concluded that the construction of black female sexuality as inherently immoral and uncontrollable was a product of nineteenth-century biological sciences. Ideologically, these sciences reflected Europeans males' fear of difference in the period of colonialism, and their consequent need to control and regulate the sexuality of those rendered 'other'.

Paula Giddings, following Gilman, pointed out that the negative construction of black women's sexuality as revealed by the Bartmann case also occurred at a time when questions about the entitlement of nonenslaved blacks to citizenship was being debated in the United States. In part, the contradiction presented by slavery was resolved in the US by ascribing certain inherited characteristics to blacks, characteristics that made them unworthy of citizenship; foremost among these was the belief in the unbridled sexuality of black people and specifically that of black women.[7] Thus, racial difference was linked to sexual difference in order to maintain white male supremacy during the period of slavery.

During slavery, the range of ideological uses for the image of the always-already sexual black woman was extraordinarily broad and familiar. This stereotype was used to justify the enslavement, rape, and sexual abuse of black women by white men; the lynching of black men; and, not incidentally, the maintenance of a coherent biological theory of human difference based on fixed racial typologies. Because African American women were defined as property, their social, political, and legal rights barely exceeded those of farm animals – indeed, they were subjected to the same forms of control and abuse as animals. For black feminist scholars, the fact that black women emerged under slavery as speaking subjects at all is worthy of note.[8] And it is the fact that African American women of this period do speak to the fact of their sexual exploitation that counts as their contestation to the dominant discourses of the day. Indeed, as Hazel Carby has described, black women during slavery were faced with having to develop ways to be recognized within the category of woman by whites by asserting a positive value to their sexuality that could stand in both public and private.[9]

THE POLITICS OF 'RECONSTRUCTING WOMANHOOD'[10]

As the discussion of sex roles and sexuality began to shift among whites in the US by the end of the nineteenth century, the binary opposition which characterized black and white female sexuality was perpetuated by both Victorian sexual ideology and state practices of repression. White women were characterized as pure, passionless, and de-sexed, while black women were the epitome of immorality, pathology, impurity, and sex itself. 'Respectability' and 'sexual control' were set against 'promiscuity' in the discourse of middle-class whites, who viewed the lifestyles of black people and the new white immigrants in urban centers as undermining the moral values of the country.[11] Buttressed by the doctrine of the Cult of True Womanhood, this binary opposition seemed to lock black women forever outside the ideology of womanhood so celebrated in the Victorian era. As Beverly Guy Sheftall notes, black women were painfully aware that 'they were devalued no matter what their strengths might be, and that the Cult of True Womanhood was not intended to apply to them no matter how intensely they embraced its values'.[12]

In the late-nineteenth century, with increasing exploitation and abuse of black women despite the legal end of slavery, US black women reformers recognized the need to develop different strategies to counter negative stereotypes of their sexuality which had been used as justifications for the rape, lynching, and other abuses of black women by whites. More than a straightforward assertion of a normal female sexuality and a claim to the category of protected womanhood was called for in the volatile context of Reconstruction where, in the minds of whites, the political rights of black men were connected to notions of black male sexual agency.[13] Politics, sexuality, and race were already inextricably linked in the US, but the problematic established by this link reached new heights of visibility during the period of Reconstruction through the increased lynchings of black men and women by the early decades of the twentieth century.

[...]

THE POLITICS OF SILENCE

Although some of the strategies used by these black women reformers might have initially been characterized as resistance to dominant and increasingly hegemonic constructions of their sexuality, by the early twentieth century, they had begun to promote a public silence about sexuality which, it could be argued, continues to the present.[14] This 'politics of silence', as described by historian Evelyn Brooks Higginbotham, emerged as a political strategy by black women reformers who hoped by their silence and by the promotion of proper Victorian morality to demonstrate the lie of the image of the sexually immoral black woman.[15] Historian Darlene Clark Hine argues that the 'culture of dissemblance' which this politics engendered was seen as a way for black women to 'protect the sanctity of inner aspects of their lives'.[16] She defines this culture as 'the behavior and attitudes of Black women that created the appearance of openness and disclosure but actually shielded the truth of their inner lives and selves from their oppressors'. 'Only with secrecy', Hine argues, 'thus achieving a self-imposed invisibility, could ordinary Black women accrue the psychic space and harness the resources needed to hold their own'.[17] And by the projection of the image of a 'super moral' black woman, they hoped to garner greater respect, justice, and opportunity for all black Americans. Of course, as Higginbotham notes, there were problems with this strategy. First, it did not achieve its goal of ending the negative stereotyping of black women. And second, some middle-class black women engaged in policing the behavior of poor and working-class women and others who deviated from a Victorian norm, in the name of protecting the 'race'.[18] Black women reformers were responding to the ways in which any black woman could find herself 'exposed' and characterized in racist sexual terms no matter what the truth of her individual life; they saw any so-called deviant individual behavior as a threat to the race as a whole. But the most enduring and problematic aspect of this 'politics of silence' is that in choosing silence, black women have also lost the ability to articulate any conception of their sexuality.

Yet, this last statement is perhaps too general. Carby notes that during the 1920s, black women in the US risked having all representations of black female sexuality appropriated as primitive and exotic within a largely racist society.[19] She continues, 'Racist sexual ideologies proclaimed black women to be rampant sexual beings, and in response black women writers either focused on defending their morality or displaced sexuality onto another terrain'.[20] As many black feminist literary and cultural critics have noted, the other terrain on which black women's sexuality was displaced was music, notably the blues. The early blues singers – who were most decidedly not middle class – have been called 'pioneers who claimed their sexual subjectivity through their songs and produced a Black women's discourse on Black sexuality'.[21] At a moment when middle-class black women's sexuality was 'completely underwritten to avoid endorsing sexual stereotypes', the blues women defied and exploited those stereotypes.[22] Yet, ultimately, neither silence nor defiance was able to dethrone negative constructions of black female sexuality. Nor could these strategies allow for the unimpeded expression of self-defined black female sexualities. Such approaches did not allow African American women to gain control over their sexuality.

In previous eras, black women had articulated the ways in which active practices of the state – the definition of black women as property, the sanctioned rape and lynching of black men and women, the denial of the vote – had been supported by a specific ideology about black female sexuality (and black male sexuality). These state practices effaced any notion of differences among and between black women, including those of class, color, and educational and economic privilege; all black women were designated as the same. The assertion of a supermoral black female subject by black women activists did not completely efface such differences nor did it directly address them. For black women reformers of this period, grounded in particular religious traditions, to challenge the negative stereotyping of black women directly meant continuing to reveal the ways in which state power was complicit in the violence against black people. The appropriation of respectability and the denial of sexuality was, therefore, a nobler path to emphasizing that the story of black women's immorality was a lie.[23]

Without more detailed historical studies of black female sexuality in each period, we do not know the extent of this 'culture of dissemblance', and many questions remain unanswered.[24] Was it expressed differently in rural and in urban areas; in the north, west, or south? How was it maintained? Where and how was it resisted? How was it shaped by class, color, economic, and educational privilege? And furthermore, how did it change over time? How did something that was initially adopted as a political strategy in a specific historical period become so ingrained in black life as to be recognizable as a culture? Or was it? In the absence of detailed historical studies we can say little about the ways social constructions of sexuality change in tandem with changing social conditions in specific historical moments within black communities.

PERSISTENT LEGACIES: THE POLITICS OF COMMODIFICATION

[...]

It should not surprise us that black women are silent about sexuality. The imposed production of silence and the removal of any alternatives to the production of silence, reflect the deployment of power against racialized subjects 'wherein those who could speak did not want to and those who did want to speak were prevented from doing so'.[25] It is this deployment of power at the level of the social and the individual which has to be historicized. It seems clear that what is needed is a methodology that allows us to contest rather than reproduce the ideological system that has, up to now, defined the terrain of black women's sexuality. Hortense Spillers made this point over a decade ago when she wrote: 'Because black American women do not participate, as a category of social and cultural agents, in the legacies of symbolic power, they maintain no allegiances to a strategic formation of texts, or ways of talking about sexual experience, that even remotely resemble the paradigm of symbolic domination, except that such paradigm has been their concrete disaster'.[26] To date, largely through the work of black feminist literary critics, we know more about the elision of sexuality by black women than we do about the possible varieties of expression of sexual desire.[27] Thus, what we have is a very narrow view of black women's sexuality. Certainly it is true, as Crenshaw notes, that 'in feminist contexts, sexuality represents a central site of the oppression of women; rape and the rape trial are its dominant narrative trope. In antiracist discourse, sexuality is also a central site upon which the repression of blacks has been premised; the lynching narrative is embodied as its trope'.[28] Sexuality is also, as Carole Vance defines it, 'simultaneously a domain of restriction, repression, and danger as well as a domain of exploration, pleasure, and agency'.[29] In the past the restrictive, repressive, and dangerous aspects of black female sexuality have been emphasized by black feminists writers, while pleasure, exploration, and agency have gone underanalyzed.

I want to suggest that contemporary black feminist theorists have not taken up this project in part because of their own status in the academy. Reclaiming the body as well as subjectivity is a process that black feminist theorists in the academy must go through themselves while they are doing the work of producing theory. Black feminist theorists are themselves engaged in a process of fighting to reclaim the body – the maimed, immoral, black female body – which can be and is still being used by others to discredit them as producers of knowledge and as speaking subjects. Legal scholar Patricia J. Williams illuminates my point: 'No matter what degree of professional I am, people will greet and dismiss my black femaleness as unreliable, untrustworthy, hostile, angry, powerless, irrational, and probably destitute'.[30] When reading student evaluations, she finds comments about her teaching and her body: 'I marvel, in a moment of genuine bitterness, that anonymous student evaluations speculating on dimensions of my anatomy are nevertheless counted into the statistical measurement of my teaching proficiency'.[31] The hypervisibility of black women academics and the contemporary fascination with what bell hooks calls the 'commodification of Otherness'[32] means that black women today find themselves precariously perched in the academy. Ann du Cille notes:

> Mass culture, as hooks argues, produces, promotes, and perpetuates the commodification of Otherness through the exploitation of the black female body. In the 1990s, however, the principal sites of exploitation are not simply the cabaret, the speakeasy, the music video, the glamour magazine; they are also the academy, the publishing industry, the intellectual community.[33]

In tandem with the notion of silence, contemporary black women writers have repeatedly drawn on the notion of the 'invisible' to describe aspects of black women's lives in general and sexuality in particular. Audre Lorde writes that 'within this country where racial difference creates a constant, if unspoken distortion of vision, Black women have on the one hand always been highly visible, and on the other hand, have been rendered invisible through the depersonalization of racism'.[34] The hypervisibility of black women academics means that visibility, too, can be used to control the intellectual issues that black women can and cannot speak about. Already threatened with being sexualized and rendered inauthentic as knowledge producers in the academy by students and colleagues alike, this avoidance of theorizing about sexuality can be read as one contemporary manifestation of their structured silence. I want to stress here that the silence about sexuality on the part of black women academics is no more a 'choice' than was the silence practiced by early twentieth-century black women. This production of *silence* instead of *speech* is an effect of the institutions such as the academy which are engaged in the commodification of Otherness.

The 'politics of silence' and the 'commodification of Otherness' are not simply abstractions. These constructs have material effects on black women's lives. In shifting the site of theorizing about black female sexuality from the literary or legal terrain to that of medicine and the control of disease, we can see some of these effects. In the AIDS epidemic, the experiences and needs of black women have gone unrecognized. I have argued elsewhere that the set of controlling images of black women with AIDS has foregrounded stereotypes of these women that have prevented them from being embraced by the public as people in need of support and care. The AIDS epidemic is being used to 'inflect, condense and rearticulate the ideological meanings of race, sexuality, gender, childhood, privacy, morality, and nationalism'.[35] Black women with AIDS are largely poor and working-class; many are single mothers; they are constantly represented with regard to their drug use and abuse and uncontrolled sexuality. The supposedly 'uncontrolled sexuality' of black women is one of the key features in the representation of black women in the AIDS epidemic.

The position of black women in this epidemic was dire from the beginning and worsens with each passing day. Silence, erasure, and the use of images of immoral sexuality abound in narratives about the experiences of black women with AIDS. Their voices are not heard in discussions of AIDS, while intimate details of their lives are exposed to justify their victimization. In the 'war of representation' that is being waged through this epidemic, black women are the victims that are the 'other' of the 'other', the deviants of the deviants, irrespective of their sexual identities or practices. The representation of black women's sexuality in narratives about AIDS continues to demonstrate the disciplinary practices of the state against black women. The presence of disease is now used to justify denial of welfare benefits, treatment, and some of the basic rights of citizenship, such as privacy for black women and their children. Given the absence of black feminist analyses or a strong movement (such as the one Ida B. Wells led against lynching), the relationship between the treatment of black women in the AIDS epidemic and state practices has not been articulated. While white gay male activists are using the ideological space framed by this epidemic to contest the notion that homosexuality is 'abnormal' and to preserve the right to live out their homosexual desires, black women are rendered silent. The gains made by gay activists will do nothing for black women if the stigma continues to be attached to their sexuality. Black feminist critics must work to find ways to contest the historical construction of black female sexualities by illuminating how the dominant view was established and maintained and how it can be disrupted. This work might very well save some black women's lives.

[...]

The question remains: how can black feminists dislodge the negative stereotyping of their sexuality and the attendant denials of citizenship and protection?

Developing a complex analysis of black female sexuality is critical to this project. Black feminist theorizing about black female sexuality has, with a few exceptions (Cheryl Clarke, Jewelle Gomez, Barbara Smith, and Audre Lorde), been focused relentlessly on heterosexuality. The historical narrative that dominates discussions of black female sexuality does not address even the possibility of a black lesbian sexuality or of a lesbian or queer subject. Spillers confirms this point when she notes that 'the sexual realities of black American women across the spectrum of sexual preference and widened sexual styles tend to be a missing dialectical feature of the entire discussion'.[36] Discussions of black lesbian sexuality have most often focused on differences from or equivalencies with white lesbian sexualities, with 'black' added to delimit the fact that black lesbians share a history with other black women. However, this addition tends to obsfucate rather than illuminate the subject position of black lesbians. One obvious example of distortion is that black lesbians do not experience homophobia in the same way as white lesbians do. Here, as with other oppressions, the homophobia experienced by black women is always shaped by racism. What has to be explored and historicized is the specificity of black lesbian experience. I want to understand in what way black lesbians are 'outsiders' within black communities. This I think, would force us to examine the construction of the 'closet' by black lesbians. Although this is the topic for another essay, I want to suggest here that if we accept the existence of the 'politics of silence' as an historical legacy shared by all black women, then certain expressions of black female sexuality will be rendered as dangerous, for individuals and for the collectivity. It follows, then, that the culture of dissemblance makes it acceptable for some heterosexual black women to cast black lesbians as proverbial traitors to the race.[37] And this, in turn, explains why black lesbians – whose 'deviant' sexuality is framed within an already existing deviant sexuality – have been wary of embracing the status of 'traitor', and the potential loss of community such an embrace engenders.[38]

Of course, while some black lesbians have hidden the truth of their lives, others have developed forms of resistance to the formulation of lesbian as traitor within black communities. Audre Lorde is one obvious example. Lorde's claiming of her black and lesbian difference 'forced both her white and Black lesbian friends to contend with her historical agency in the face of [this] larger racial/sexual history that would reinvent her as dead'.[39] I would also argue that Lorde's writing, with its focus on the erotic, on passion and desire, suggests that black lesbian sexualities can be read as one expression of the reclamation of the despised black female body. Therefore, the works of Lorde and other black lesbian writers, because they foreground the very aspects of black female sexuality that are submerged – namely, female desire and agency – are critical to our theorizing of black female sexualities. Since silence about sexuality is being produced by black women and black feminist theorists, that silence itself suggests that black women do have some degree of agency. A focus on black lesbian sexualities implies that another discourse – other than silence – can be produced. Black lesbian sexualities are not simply identities. Rather they represent discursive and material terrains where there exists the possibility for the active production of speech, desire, and agency. Black lesbians theorizing sexuality is a site that disrupts silence and imagines a positive affirming sexuality. I am arguing here for a different level of engagement between black heterosexual and black lesbian women as the basis for the development of a black feminist praxis that articulates the ways in which invisibility, otherness, and stigma are produced and re-produced on black women's bodies. And ultimately my hope is that such an engagement will produce black feminist analyses which detail strategies for differently located black women to shape interventions that embody their separate and common interests and perspectives.

NOTES

1. Patricia Hill Collins, *Black Feminist Thought: Knowledge, Consciousness and the Politics of Empowerment* (Boston: Unwin Hyman, 1990), p. 164.
2. Hortense Spillers, 'Interstices: A Small Drama of Words', in Carole S. Vance, ed., *Pleasure and Danger*, pp. 73–100.
3. Ibid.
4. Darlene Clark Hine, 'Rape and the Inner Lives of Black Women in the Middle West: Preliminary Thoughts on the Culture of Dissemblance', *Signs* 14, no. 4 (1989): 915–20.
5. Sander L. Gilman, 'Black Bodies, White Bodies: Toward an Iconography of Female Sexuality in Late Nineteenth Century Art, Medicine, and Literature', *Critical Inquiry* 12, no. 1 (Autumn 1985): 204–42.
6. O'Grady, 'Olympia's Maid'.
7. Paula Giddings, 'The Last Taboo', in Toni Morrison, ed., *Race-ing Justice, Engendering Power: Essays on Anita Hill, Clarence Thomas and the Construction of Social Reality* (New York: Pantheon Books, 1992), p. 445.
8. See the studies of Harriet Jacobs and Linda Brent.

9. Hazel Carby, *Reconstructing Womanhood: The Emergence of the Black Female Novelist* (New York: Oxford University Press, 1987), pp. 40–61.
10. This heading is taken from the title of Carby's book cited above.
11. Giddings, 'The Last Taboo'.
12. Beverly Guy-Sheftall, *Daughters of Sorrow: Attitudes Toward Black Women, 1880–1920* Black Women in United States History, vol. 11 (Brooklyn: Carlson Publishing, 1990), p. 90.
13. Martha Hodes, 'The Sexualization of Reconstruction Politics: White Women and Black Men in the South after the Civil War', in John C. Fout and Maura S. Tantillo, eds, *American Sexual Politics: Sex Gender and Race since the Civil War* (Chicago: The University of Chicago Press, 1993), pp. 60–1.
14. See Evelyn Brooks Higginbotham, 'African-American Women's History and the Metalanguage of Race', *Signs* 17, no. 2 (1992): 251–74; Elsa Barkley Brown, 'Negotiating and Transforming the Public Sphere: African American Political Life in the Transition From Slavery to Freedom', *Public Culture* 7, no. 1 (Fall 1994): 107–46; as well as Hine, Giddings, and Carby.
15. Higginbotham, 'African American Women's History', p. 262.
16. Hine, 'Rape and the Inner Lives', p. 915.
17. Ibid.
18. See Carby, 'Policing the Black Woman's Body'. Elsa Barkley Brown argues that the desexualization of black women was not just a middle-class phenomenon imposed on working-class women. Though many working-class women resisted Victorian attitudes toward womanhood and developed their own notions of sexuality and respectability, some, also from their own experiences, embraced a desexualized image. Brown, 'Negotiating and Transforming the Public Sphere', p. 144.
19. Hazel Carby, *Reconstructing Womanhood*, p. 174.
20. Ibid.
21. Ann du Cille, 'Blues Notes on Black Sexuality: Sex and the Texts of Jessie Fauset and Nella Larsen', *Journal of the History of Sexuality* 3, no. 3 (1993): 419. See also Hazel Carby, '"It Just Be's Dat Way Sometime": The Sexual Politics of Black Women's Blues', in Ellen DuBois and Vicki Ruiz, eds, *Unequal Sisters: A Multicultural Reader in U.S. Women's History* (New York: Routledge, 1990).
22. Here, I am paraphrasing Ducille, 'Blues Notes', p. 443.
23. Evelyn Brooks Higginbotham, *Righteous Discontent: The Women's Movement in the Black Baptist Church, 1880–1920* (Cambridge, Mass.: Harvard University Press, 1993).
24. The historical narrative discussed here is very incomplete. To date, there are no detailed historical studies of black women's sexuality.
25. Abdul JanMohamed, 'Sexuality on/of the Racial Border: Foucault, Wright, and the Articulation of Racialized Sexuality', in Domna Stanton, ed., *Discourses of Sexuality: From Aristotle to AIDS* (Ann Arbor: University of Michigan Press, 1992), p. 105.
26. Spillers, 'Interstices', p. 80.
27. See analyses of novels by Nella Larsen and Jessie Fauset by Carby, McDowell, and others.
28. Crenshaw, 'Whose Story Is It Anyway?' p. 405.
29. Carole S. Vance, 'Pleasure and Danger: Towards a Politics of Sexuality', in Carole S. Vance (ed.) *Pleasure and Danger: Exploring Female Sexuality*, London: Pandora Press, 1989.
30. Patricia J. Williams, *The Alchemy of Race and Rights* (Cambridge: Harvard University Press, 1991), p. 95.
31. Ibid.
32. bell hooks, *Black Looks: Race, and Representation* (Boston: South End Press, 1992), p. 21.
33. Ann du Cille, 'The Occult of True Black Womanhood: Critical Demeanor and Black Feminist Studies', *Signs* 19, no. 3 (1994): 591–629.

34. Karla Scott, as quoted in Teresa De Lauretis, *The Practice of Love: Lesbian Sexuality and Perverse Desire* (Bloomington: Indiana University Press, 1994), p. 36.
35. Simon Watney, *Policing Desire: Pornography, AIDS and the Media* (Minneapolis: University of Minnesota Press, 1989), p. ix.
36. Spillers, *Ibid.*, 'Interstices'.
37. In a group discussion of two novels written by black women, Jill Nelson's *Volunteer Slavery* and Audre Lorde's *Zami*, a black woman remarked that while she thought Lorde's book was better written than Nelson's, she was disturbed that Lorde spoke so much about sex, and 'aired all of her dirty linen in public'. She held to this view even after it was pointed out to her that Nelson's book also included descriptions of her sexual encounters.
38. I am reminded of my mother's response when I 'came out' to her. She asked me why, given that I was already black and had a nontraditional profession for a woman, I would want to take on one more thing to make my life difficult. My mother's point, which is echoed by many black women, is that in announcing my homosexuality, I was choosing to alienate myself from the black community.
39. See Scott, quoted in Teresa De Lauretis, *The Practice of Love: Lesbian Sexuality and Perverse Desire* (Bloomington: Indiana University Press, 1994), p. 36.

2.4

BODY MATTERS: CULTURAL INSCRIPTIONS

Lynne Segal

Western science has defined not only sexual difference but the purpose and function of sexuality, primarily, in terms of reproductive biology: stressing the role of the penis and testes in male fertilization, and of the breast, ovaries, uterus and vagina in pregnancy and lactation. The reality, however, which twentieth-century sexology could not dismiss, is that while *some* biological narrative may provide a reproductive purpose for men's sexual pleasure in terms of the penis, no parallel narrative can centre women's sexual pleasure in pregnancy and lactation. The human clitoris, physiological site of female orgasm and without reproductive purpose, undermines all attempts to link sexual pleasure to reproductive outcome.[1] And it was Freud, of course, often against the grain of his own teleological thinking, who first suggested that human sexuality in adulthood is built out of its foundations in the autoeroticism and polymorphous perversities of childhood. Here bodily pleasures bear no relation to reproductive ends or acts; nor, significantly, to the supposed gender polarity of adult heterosexuality along lines of activity/passivity. Exactly which areas of the body surface would be eroticized, and with what fixity or fluidity of focus, his case studies suggested, would depend upon the psychic meanings they acquired in young children's interactions with the world around them. Indeed, Freud found the model for adult sexuality in infantile thumb-sucking, propped upon the biological mechanisms for feeding.

It is not so hard to establish that a scientific paradigm picturing the female body as functioning passively in the service of reproductive demands is primarily

From: L. Segal, *Straight Sex: The Politics of Pleasure*, London: Virago, 1994.

ideological. It is even easier to expose the ideological nature of the biological paradigm of male sexuality as active and initiating. The knowledge that human females do not have oestrous cycles like other primates (with their highly visible genital changes and 'sexual signalling' behaviour during fertile periods) is presented to us in the most familiar biological texts, like Donald Symons's *The Evolution of Human Sexuality*, through accounts of the 'year-round receptivity' of female sexuality, which is said to possess the passive feature of being 'continuously copulable'. Meanwhile human male sexuality is portrayed as perpetually ready to copulate: 'women inspire male sexual desire simply by existing.'[2] Proudly, if naively, Symonds draws our attention to the connection between his biological description of men's ever-ready sexual desire (accompanying women's ever-available sexual condition) and men's pornographic fantasies of 'basic male wishes' as 'easy, anonymous, impersonal, unencumbered sex with an endless succession of lustful, beautiful, orgasmic women.'[3] Quite. His description belongs to the world of male *fantasy*. Corporeal reality is different. Outside the fabulations and fantasmagoria of 'scientific' or pornographic texts the hominid penis is anything but permanently erect, anything but endlessly ready for unencumbered sex, anything but triggered by the nearest passing female – even when she happens to be his wife, mistress or lover, and eager for sex.

With chilling if humorous detail, Emily Martin illustrates how this same imagery of passive 'femininity' and active 'masculinity' is attributed even to the human ovum (or 'egg') and sperm. The ovum is variously depicted in contemporary medical texts as 'floating', 'drifting', 'transported' or 'swept along' the fallopian tube, and contrasted with the 'masculinity' of the 'streamlined', 'strong' sperm: 'lashing their tails' as they make their 'perilous journey' through the 'hostile environment' of the vagina, to 'penetrate', 'assault' or (as one illustration in *Science News* would have us envisage the physiological process) 'ferociously attack the egg' with jackhammer, pickaxe and sledgehammer. These stereotyped images persist, Martin comments, even as new research suggests the extreme fragility of the sperm, which swim 'blindly' in circles and mill around, unless 'captured' and 'held fast' by the adhesive molecules on the surface of the egg.[4]

There is no doubt that respected biological discourses have traditionally been used to prop up, while themselves feeding upon, a multitude of overlapping discourses and narratives from literature and pornography (although less uniformly in the latter). Tales of the all-powerful, ever-ready, male sex drive, located in the activities of the male sex organ – the penis (or sperm!) – function all too well to console and titillate (if also to intimidate) men, as well as to tease and to silence women. Exemplifying such discourses, Gay Talese reports from his journalistic survey of North American sexual mores in the 1970s on the male sexual member:

> Sensitive but resilient, equally available during the day or night with a minimum of coaxing, it has performed purposefully if not always skil-

> fully for an eternity of centuries, endlessly searching, sensing, expanding, probing, penetrating, throbbing, wilting, and wanting more ... it is men's most honest organ.[5]

A load of old codswallop, obviously. Its ideological function as mythic male fantasy is seemingly so blatant. Yet it remains entrenched as biological 'truth', however sexist and silly we know it to be from our personal experiences of sex, however familiar we may be with competing biological and clinical narratives of male impotence and sexual dysfunction. It is these myths of penile prowess which we need relentlessly to expose in our rethinking of heterosexuality.

There is a mass of medical and clinical data documenting men's chronic anxieties in relation to their penile performance. Partly as a result of feminist probing, one text after another in recent years has embarked upon the task of 'deconstructing the phallus': turning the medical and therapeutic gaze upon the penis and its persistent premature ejaculation, impotence, loss of desire; then widening the focus to include venereal disease, testicular cancer and infertility – contemplating all the while the painful remedies men seek for their sexual disorders. But knowledge of suffering and uncertainty in men's sex lives, supposed until recently to be one of men's best kept secrets, is not something which is new. On the contrary, Freud had written in 1908 that to many it was 'scarcely credible how seldom normal potency is to be found in the husband', and Wilhelm Stekel had added in 1927 that the 'percentage of relatively impotent people cannot be placed too high' (himself placing it near 50 per cent and citing premature ejaculation as modern man's most characteristic sexual practice).[6] However, men's routine inability to use the penis as they might wish was to remain a largely private torment, mainly receiving contempt if exposed to medical practitioners. Indeed, Lesley Hall concludes her research on the help and advice men sought and received in Britain on their sexual problems in the first half of this century with the thought: 'Whatever the social potency of men ... their actual sexual potency is always dubious and open to question.'[7]

The standard biological narrative of active penile prompting and passive vaginal receptivity as the paradigm for human sexual encounter thus serves above all to hide, as well as to create and sustain, the severe anxieties attaching to the penis, while also revealing men's fear of recognizing the existence of women's sexual agency – verbal, behavioural or physiological. As some women artists have recently been keen to affirm, producing art works of the male genitals as changeable, soft, vulnerable and comic, the more we display new bodily images and meanings exposing penile precariousness and mutability, the more we challenge traditional phallic narratives.[8] But there are pitfalls all the way. Seventies feminism was mistaken, according to current feminist rethinking, in its attempt to reclaim women's 'own' bodily experiences in some direct and unmediated way.

Feminists from the late sixties had stressed the distinction between biological 'sex' and socially constructed 'gender', in order to challenge and reject the ubiquitous mythologizing of women's 'nature' and place in the world. They wanted to contrast women's socialization into the constricting roles or performances of submissive femininity they called 'gender', with what their own collectively shared experiences of the body suggested about their biological 'sex': 'We refuse to accept anything as true that we can't confirm by our feelings and experience.'[9] But bodily experiences are themselves socially constructed, not only by the culturally specific ways we have of interpreting them, but by socially variable factors like diet, exercise, training and reactions to ageing, illness and so on. So while aware of the consequences of men's greater social and cultural power in controlling women's behaviour, feminists were – at least at first – less able to problematize the physiological reductionism of the biomedical sciences they wished to transcend. Sharing experiences, as the widely influential *Our Bodies, Ourselves* had promised in the 1970s, 'we rid ourselves of many fears and obstacles and can start to make better use of our untapped energies'.[10] Bodily responses were thus seen as speaking directly to us, rather than through our particular interpretative community. And this supposed access to the body, we know in retrospect, was all the more likely to create a collective prescriptiveness when seeking to bring sexual experiences into speech. Sexual experiences are so tied in with the most keenly felt but peculiarly inexpressible hopes and deprivations, promising either the confirmation of, or threats to, our identites as worthwhile or lovable people, that they can scarcely avoid invoking insecurities and anxieties. This is why, as Carole Vance has written, 'there is a very fine line between talking about sex and setting norms'.[11] And so there was.

When feminist-inspired research, like that of Shere Hite, reported that only 30 per cent of women reach orgasm during penetrative sex, this was quickly transformed, by Hite and by others, into the spurious announcement that most women did not like penetrative sex (against the grain of the complexity of feelings Hite herself uncovered). Before long the coercive message of much feminist sex-advice literature was that wise women, in touch with their 'authentic' needs, would avoid penetrative sex. ('*Hmn ... do I put it somewhere??*' a feminist cartoon muses, depicting a strong, naked woman, looking dubiously at a penis-shaped vibrator. She moves it around a bit, only to fling it down in horror, repeating in outrage the absurd suggestion, '*In my CUNT?!*')[12] Yet, any feminist preference for clitoral over vaginal, 'active' over 'passive', self-directed over self-shattering, sexual engagement not only ignores the unruliness of desire but reflects, more than transcends, the repudiation of 'femininity' in our misogynist culture.

The repetition of this repudiation is easy to understand: even the most recent feminist encyclopaedia or 'Companion' on sexuality, *The Sexual Imagination* (1993), has no entry under 'vagina', although the history and meaning of the 'clitoris' is boldly covered by its presiding editor as playing 'a disproportionately major role in women's sexual pleasure'.[13] It did not go unchallenged, but when affirmed, the reproductive resonance of vaginal iconography as 'birth canal' always threatened to over-ride or undermine any pleasure-encoding signification. It was the pioneer of post-war Western feminism, Simone de Beauvoir, who affirmed, with reference to the vagina that, 'the feminine sex organ is mysterious even to the woman herself. . . . Woman does not recognise its desires as hers.' Her own description of this 'sex organ', so often 'sullied with body fluids', tells us why:

> woman lies in wait like the carnivorous plant, the bog, in which insects and children are swallowed up. She is absorption, suction, humus, pitch and glue, a passive influx, insinuating and viscous: thus, at least, she vaguely feels herself to be.[14]

There is no vagueness in this description. It is a perfect illustration of the horror of what Kristeva has elaborated in her (currently much over-used) conception of the 'abject' object. Kristeva describes abjection as the process whereby the child takes up its own clearly defined ('clean and proper') body image through detaching itself from – expelling and excluding – the pre-Oedipal space and self-conception associated with its improper and unclean, 'impure', connection with the body of the mother. The mother's body, having been everything to the child, threatens its engulfment. On this view, entering the symbolic space of language brings with it a horror of (and fearful attraction to) everything without clear boundaries, everything which suggests a non-distinctiveness between inside and outside.[15] Elaborating Kristeva's thoughts, Elizabeth Grosz explains that in her notion of an 'unnamable, pre-oppositional, permeable barrier, the abject requires some mode of control or exclusion to keep it at a safe distance from the symbolic and its orderly proceedings'.[16] However culturally specific this psychoanalytic narrative of the child's entry into the symbolic may be (and Kristeva, with unconvincing but characteristic Lacanian grandiosity, takes it to be universal), it would seem to resonate with the place of vaginal iconography in our culture, and its absence from respectable discourses and contexts. The vagina has served as a condensed symbol of all that is secret, shameful and unspeakable in our culture.

The question which Grosz raises is whether it is discourse itself which confers the horror of 'abjection' onto female bodies, and whether there might thus be other ways of registering, or resignifying, the sexual specificity of female sexual bodies (which may include, but would not reduce to, reference to the mother's body – however conceived). Neither de Beauvoir nor Kristeva address this question. It is indeed a formidable task. That some interference and shift in standard perceptions and meaning are possible, when old images are repeated in contexts where they may be seen in new ways (always involving contention, and fears of recuperation), is evident from the battles which have already been fought around women's film and art works involving female genital anatomy.

[. . .]

NOTES

1. See, for example, Sarah Blaffer Hardy, *The Woman That Never Evolved*, Cambridge, Mass., Harvard University Press, 1981, p. 166.
2. Donald Symons, *The Evolution of Human Sexuality*, Oxford, Oxford University Press, 1979, p. 284.
3. Ibid., p. 177.
4. Emily Martin, 'Body Narratives, Body Boundaries', in Lawrence Grossberg *et al.* eds, *Cultural Studies*, Routledge, 1992, pp. 412–13.
5. Gay Talese, quoted in Louise Kaplan, *Female Perversions*, London, Pandora, 1991, p. 123.
6. These quotes are taken from Lesley Hall's excellent research into the history of male sexual anxieties and disorders in Lesley A. Hall, *Hidden Anxieties: Male Sexuality, 1900–1950*, Cambridge, Polity, 1991, pp. 117–18.
7. Ibid., p. 173.
8. See Robin Shaw, 'Open your Eyes for a Big Surprise: Erotic Photos of Men' and 'What She Wants: Clare Bayley talks to Naomi Salaman', both in *Body Politic: Feminism, Masculinity, Cultural Politics*, no. 1, Winter 1992; Jack Butler, 'Before Sexual Difference: Helen Chadwick's "Piss Flowers"', in Andrew Benjamin ed., *The Body: Journal of Philosophy and the Visual Arts*, London, The Academy Group, 1993.
9. These are the words of a feminist consciousness-raising group, as presented by Alix Kates Shulman, *Burning Questions*, London, Fontana, 1980, p. 229.
10. Boston Women's Health Book Collective, *Our Bodies Ourselves: A Health Book By and For Women*, 2nd edn, New York, Simon & Schuster, 1973, pp. 11, 13.
11. Carole S. Vance, 'Pleasure and Danger: Towards a Politics of Sexuality' in Carole S. Vance ed., *Pleasure and Danger: Exploring Female Sexuality*, Routledge & Kegan Paul, 1984, p. 21.
12. This cartoon appears in Anja Meulenbelt, Johanna's daughter, *For Ourselves: Our Bodies and Sexuality – from Women's Point of View*, London, Sheba, 1981, pp. 100–1.
13. Harriett Gilbert, *The Sexual Imagination: From Acker to Zola*, Harriett Gilbert ed., London, Jonathan Cape, 1993, p. 56.
14. Simone de Beauvoir, *The Second Sex* (1949), London, Picador, 1988, pp. 406–7.
15. Julia Kristeva, *Powers of Horror: An Essay on Abjection*, New York, Columbia University Press, 1982; see especially pp. 3–10, 61–71.
16. Elizabeth Grosz, 'The Body of Signification', in John Fletcher and Andrew Benjamin eds, *Abjection, Melancholia and Love: The Work of Julia Kristeva*, London, Routledge, 1990, p. 95.

2.5

LESBIAN BODIES: TRIBADES, TOMBOYS AND TARTS

Barbara Creed

Femme, vampiric, muscled, tattooed, pregnant, effete, foppish, amazonian – the lesbian body comes in a myriad of shapes and sizes. Images of the lesbian body in cultural discourse and the popular imagination abound. Various popular magazines and newspapers have announced that it is now chic to be a lesbian – 'in', fashionable, popular, desirable. The 'Saturday Extra' section of the Melbourne *Age* recently ran a cover story entitled 'Wicked Women', in which lesbians were described as 'glamorous gorgeous and glad to be gay'. On the front cover of the August 1993 edition of *Vanity Fair* we find an image of femme super-model Cindy Crawford shaving imaginary whiskers from the boyish, smiling lathered face of k.d. lang, the out-lesbian country and western singer. 'Oh, to be a lesbian, now that spring is here', seems to be the latest media hype. Fashion aside, is there a quintessential stereotype, or stereotypes, of the lesbian body?

[...]

Unlike man's body, the female body is frequently depicted within patriarchal cultural discourses as fluid, unstable, chameleon-like. Michèle Montrelay has argued that in western discourse, woman signifies 'the ruin of representation' (Montrelay 1978: 89). Julia Kristeva distinguishes between two kinds of bodies: the symbolic and the imaginary or abject body. In *Powers of Horror*, she argues that the female body is quintessentially the abject body because of

From: E. Grosz and E. Probyn (eds), *Sexy Bodies: The Strange Carnalities of Feminism*, London: Routledge, 1995.

its procreative functions. Unlike the male body, the proper female body is penetrable, changes shape, swells, gives birth, contracts, lactates, bleeds. Woman's body reminds man of his 'debt to nature' and as such threatens to collapse the boundary between human and animal, civilized and uncivilized (Kristeva 1982: 102). Bakhtin argued that the essentially grotesque body was that of the pregnant, birth-giving woman (1984: 339). When man is rendered grotesque, his body is usually feminized (Creed 1993: 122): it is penetrated, changes shape, swells, bleeds, is cut open, grows hair and fangs. Insofar as woman's body signifies the human potential to return to a more primitive state of being, her image is accordingly manipulated, shaped, altered, stereotyped to point to the dangers that threaten civilization from all sides. If it is the female body in general – rather than specifically the lesbian body – which signifies the other, how, then, does the lesbian body differ from the body of the so-called 'normal' woman?

There are at least three stereotypes of the lesbian body which are so threatening they cannot easily be applied to the body of the non-lesbian. These stereotypes are: the lesbian body as active and masculinized; the animalistic lesbian body; the narcissistic lesbian body. Born from a deep-seated fear of female sexuality, these stereotypes refer explicitly to the lesbian body, and arise from the nature of the threat lesbianism offers to patriarchal hetereosexual culture.

The central image used to control representations of the potentially lesbian body – to draw back the female body from entering the dark realm of lesbian desire – is that of the tomboy. The narrative of the tomboy functions as a liminal journey of discovery in which feminine sexuality is put into crisis and finally recuperated into the dominant patriarchal order – although not without first offering the female spectator a series of contradictory messages which may well work against their overtly ideological purpose of guiding the young girl into taking up her proper destiny. In other words, the well-known musical comedy, *Calamity Jane*, which starred Doris Day as the quintessential tomboy in love with another woman, could be recategorized most appropriately, in view of its subversive subtextual messages about the lure of lesbianism, as a 'lesbian western', that ground-breaking subgenre of films so ardently championed by Hollywood.

THE MASCULINIZED LESBIAN BODY

There is one popular stereotype about the nature of lesbianism which does posit a recognizable lesbian body. This view, which has been dominant in different historical periods and is still prevalent today, is that the lesbian is really a man trapped in a woman's body. The persistent desire to see the lesbian body as a pseudo male body certainly does not begin with Freud's theory of penis envy. We find evidence of the masculinized lesbian body in a number of pre-Freudian historical and cultural contexts: Amazonian society in which the Amazon is seen as a masculinized, single-breasted, man-hating warrior; cross-cultural woman-marriage (Cavin 1985: 129–37) whereby women don men's clothes and marry other women; female transvestism or cross-dressing; and the history of tribadism and female sodomy. It is the last category I wish to discuss in some detail.

In earlier centuries, prior to the invention (Katz 1990: 7–34) in the mid-nineteenth century of the homosexual and heterosexual as a person with a specific identity and lifestyle, women and men who engaged in same-sex relations – presumed to consist primarily of sodomy – were described as sodomites. Sodomites – heterosexual and homosexual – were 'guilty' of carrying out a specific act, not of being a certain kind of person with readily identifiable characteristics. Specifically, women were thought to take part in sodomy with other women in one of two ways: through clitoral penetration of the anus or with the use of diabolical instruments. In general terms, however, the term sodomite was used to refer to anyone engaged in unorthodox practices.

[. . .]

By the the time of the Renaissance it was believed that in some cases women with extremely large clitorises could commit acts of penetration – vaginal and anal – with another woman. One woman, accused of such acts, was said to possess a clitoris that 'equalled the length of half a finger and in its stiffness was not unlike a boy's member'. The woman was accused of 'exposing her clitoris outside the vulva and trying not only licentious sport with other women . . . but even stroking and rubbing them' (Laqueur 1990: 137).

In *Making Sex*, Laqueur traces the way in which our views of sex and sexual difference have changed, along with cultural and social changes, over the centuries. Prior to the eighteenth century, thinking about the body was dominated by the 'one-sex model':

> In the one-sex model, dominant in anatomical thinking for two thousand years, woman was understood as man inverted: the uterus was the female scrotum, the ovaries were testicles, the vulva was a foreskin, and *the vagina was a penis*. (Laqueur 1990: 236; emphasis in original)

There was only one archetypal body: on the male body the organs had descended, in the female body, due to a lack of bodily heat, they remained bottled up inside. By virtue of her coldness, woman was, at best, a potential man, at worst, a failed one. By the late eighteenth century the one-sex model had given way to a new model – the two-sex model in which men and women no longer correspond but are radically different. In the one-sex model, a number of vital female organs had no names of their own – the ovaries were female 'testicles' while the vagina did not have a name at all before 1700. The clitoris did not even appear in this model.

[. . .]

After the official discovery of the clitoris, the notion of the lesbian body as a pseudo male body becomes more credible because the clitoris is seen as a male penis which, it was also believed, ejaculated semen. What are the implications of this for the lesbian? In normal man – woman sex, the woman was the one rubbed against, that is, she assumed the passive role – despite her smaller 'penis'. But in cases where the clitoris is deemed too large (a complaint never directed at the penis) and the woman therefore is capable of adopting the active rubbing position in sex with another woman, she stands in violation of the sumptuary laws (Laqueur 1990: 136). Some women are therefore potentially capable of performing sodomitic acts on the bodies of other women. The woman who assumed a male role in sex with another woman was deemed a 'tribade'. The lesbian/tribade is a pseudo man, her body an inferior male body. In cases which were brought before the law, the offender, if found guilty, was usually burned as a tribade. In such cases the size of the female penis was crucial.

In 1601 Marie de Marcis was accused of sodomy. She declared publicly that she was a man, altered her name to Marlin and announced her intention to marry the woman she loved. At her trial (women could be tried for sodomy in French law) she was sentenced to be burned alive, but a sympathetic doctor intervened and demonstrated she was really a 'man' because when her genitals were rubbed a penis emerged which also ejaculated semen. She was decreed a man and escaped execution, although she was forbidden to have sex with women (and men) or dress as a man until she was 25 (Laqueur 1990: 136–7). In another case, in Holland in the early seventeenth century, Henrike Schuria, who donned men's clothes and joined the army, was caught having sex with another woman. Her clitoris was measured and found to equal the length of half a finger. She was found guilty of tribadism and sentenced to the stake, but the judge intervened and ordered she be sent into exile after enduring a clitoridectomy (Laqueur 1990: 137; van der Meer 1990: 191).

It is relevant to point out that laws against sodomy not only varied according to place and time, but that they were not universally applied. In England at the time there were no laws against sodomy or cross-dressing – only dressing outside one's class was illegal. Nonetheless there is still a strong stigma attached to lesbianism. In Renaissance England, Ben Jonson accused a female critic of literary tribadism in order to denigrate her. He accuses her of raping the Muse:

> What though with Tribade lust she force a Muse,
> And in an Epiceone fury can write newes
> Equall with that, which for the best newes goes ...
> (Jones and Stallybrass 1991: 103)

In eighteenth-century Holland, however, a wave of sodomy trials took place in which women figured prominently and were referred to as committing 'sodomitical filthiness' (van der Meer 1990: 190). In France, after the rediscovery of the clitoris, the hermaphrodite was classified as a woman with a large clitoris who could legally be tried for engaging in acts of sodomy with other women (Jones and Stallybrass 1991: 90). Women were, of course, also punished for committing other sexual acts with women, such as mutual masturbation, but these were not seen in the same light as sodomy which was a far more serious offence.

In the cases discussed above, the solution to the female body which threatens to confuse gender boundaries is either legal ('she really is a man') or surgical ('cut her back to size') or lethal ('burn the witch'). In all three instances the offending body challenges gender boundaries in terms of the active/passive dualism, a dichotomy which is crucial to the definition of gender in patriarchal culture. Marie de Marcis was not judged a woman with a large clitoris but a man. There is a clear distinction here between penis and clitoris in which the former grants its possessor the status of manhood and all of its attendant rights. Henrike Schuria was not lucky enough to be deemed a man; rather, she was judged a freakish woman and forced to have the offending organ cut out. As Laqueur points out: 'Getting a certifiable penis is getting a phallus, in Lacanian terms, but getting a large clitoris is not' (1990: 140–1).

The tribade is the woman who assumes a male role in sexual intercourse with another woman – either because she is the one 'on top' or because she has a large clitoris and can engage in penetration. She threatens because she is active, desiring, hot. Theo van der Meer argues that the tribade does not really fit into the world of romantic, but asexual female friendships, nor into the tradition of female transvestism. Van de Meer claims that perhaps the tribades, with their overtly sexual desires, 'may represent the more – if not the most – important and direct predecessors of the modern lesbian' (1990: 209). I have used the word tribade for the early modern period, because, not only did the term 'lesbian' not exist in the eighteenth century, but 'lesbian' also conveys the idea of a sexual identity which was not really invented for the female homosexual until the mid-nineteenth century. According to Barbara Walker (1983: 536), in Christian Europe, lesbianism was 'a crime without a name'. The sixteenth-century definition of the tribade as a pseudo male has much in common with Freud's later definition of the homosexual woman as one suffering from unresolved penis-envy. Both definitions adopt male anatomy as the defining norm. The difference is that Freud's model of sexual difference is based on the two-sex theory; in this, woman is not an ill-formed man, she is the 'other' – a creature who has already (in male eyes) been castrated. The lesbian body of Freudian theory is one that attempts to overcome its 'castration' by assuming a masculine role in life and/or masculine appearance through clothing, gesture, substitution.

In the one-sex model, the tribade is guilty of assuming the male role which she is seen as perfectly capable of doing because she is already potentially a man; in the two-sex model the lesbian is deemed ultimately as incapable of even assuming a pseudo-male position because, like all women, she signifies an irremediable lack. Her genitals are not in danger of falling through her body and transforming her from male to female, nor does she possess a clitoris that might be taken seriously enough by a judge or medical doctor to suggest she might adopt an active, rubbing role in sex – she signifies only castration and lack. Her lack, however, can be overcome artificially by the use of a dildo – a popular male fantasy about lesbian practices. It is worth noting that the phallic woman, the woman with a penis, who is central to the Freudian theory of fetishism, has much in common with the image of the sodomitic tribade. The phallic woman, who straps on a dildo and sodomizes the male, is a popular figure in pornography specifically designed for the burgeoning male masochist market. Perhaps the phallic woman of male fantasy is not just a Freudian fetish but also represents male desire for an active, virile woman – a lesbian!

Freud attributes lesbianism not to woman's own specifically female desires but to her desire to be a man. The lesbian is the woman who either has never relinquished, or seeks to recover, her repressed phallic sexuality. She refuses to relinquish her pre-Oedipal or phallic love for the mother and develops a masculinity complex. She may also become a lesbian out of a desire for revenge. In his single study of lesbianism, Freud (1920) argues that the woman becomes a lesbian to enact revenge on her father who she feels betrayed her because he made the mother, her rival, pregnant. He states that 'she changed into a man, and took her mother in place of her father as the object of her love' (1920: 384). He notes the 'masculine' physique of his client and states that only inverts assume the mental characteristics of the opposite sex. Freud likens the female homosexual to the male heterosexual – both desire the feminine woman. In his footnotes to the Dora case history Freud refers to Dora's 'homosexual (gynaecophilic) love for Frau K.' as the 'strongest unconscious current in her mental life' (1905a: 162) and to her aggressive identification with masculinity. Although Freud does not appear to see lesbianism as pathological (he does not prescribe any form of therapy) his emphasis on vaginal – not clitorial – orgasm as offering the only true source of sexual pleasure for women makes it clear that he regarded lesbian sexual practices as inferior and immature. On her journey into proper womanhood, the girl gives up the pleasures of clitoral orgasm for vaginal orgasm. In his discussion of proper femininity and masculinity, Freud writes: 'Maleness combines [the factors of] subject, activity and possession of the penis; femaleness takes [those of] object and passivity. The vagina is now valued as a place of shelter for the penis; it enters into the heritage of the womb' (Freud 1923: 312).

In the Freudian model of sexual difference, the vagina – no longer an inverted penis – is now 'a place of shelter for the penis'. It passively awaits the male member, husband, master of the house. The clitoris loses its earlier active prowess and becomes 'like pine shavings' waiting to be 'kindled' in order to make the home warm, friendly: 'to set a log of harder wood on fire' (Freud 1905b: 143). The woman who refuses to see her sexual organs as mere wood chips, designed to make the man's life more comfortable, is in danger of becoming a lesbian – an active, phallic woman, an intellectual virago with a fire of her own.

Freud's theory regarding the shift of pleasure from the clitoris to the vagina has no basis in fact. He seems bent on ascribing a specific role to the vagina as a means of convincing women that they should assume a passive position within the family and society. Not only Freud, however, feared the active woman. Bram Dijkstra points out that at the time there were a number of popular beliefs in circulation about the dangers of the active, masculine woman who threatened to destroy the fragile boundary which kept the sexes different and separate. He cites the work of Bernard Talmey who believed that masturbation made women into lesbians and led to abnormal conditions such as 'hypertrophy of the clitoris' which caused the clitoris to expand and become erect: 'The female masturbator becomes excessively prudish, despises and hates the opposite sex, and forms passionate attachments for other women' (Dijkstra 1986: 153). The lesbian body is a particularly pernicious and depraved version of the female body in general; it is susceptible to auto-eroticism, clitorial pleasure and self-actualization.

Freud's narrative of woman's sexual journey from clitorial pubescence into mature vaginal bliss is a bit like the transformation fairy tales in which the ugly duckling matures into a beautiful swan and marries the handsome cygnet. Literary and filmic narratives replay this scenario of female fulfilment through the figure of the tomboy. The tomboy's journey is astonishingly similar to that of the clitoris. During the early stage, the tomboy/clitoris behaves like a 'little man' enjoying boy's games, pursuing active sports, refusing to wear dresses or engage in feminine pursuits; on crossing into womanhood the youthful adventurer relinquishes her earlier tomfoolery, gives up boyish adventures, dons feminine clothes, grows her hair long and sets out to capture a man whose job it is to 'tame' her as if she were a wild animal.

We see this narrative played out in *Calamity Jane* where the heroine (Doris Day) relinquishes her men's clothing, foul language, guns and horse for a dress, feminine demeanour, sweet talk and a man. She also gives up the woman, Alice, with whom she has set up house and whom she clearly loves. Katherine Hepburn in *Sylvia Scarlett* adopts the name of Sylvester, dons boy's clothes and masquerades as a youth until she falls in love and exchanges her masculine appearance for a feminine one. *Queen Christina* depicts the lesbian queen (Greta Garbo) in the first part of the narrative wearing men's clothes and long riding boots, striding about the palace accompanied by two great danes and muttering to her manservant that all men are fools and she will never marry. Predictably, she falls in love, throws off her mannish trappings, gives up the Lady Ebba and redirects her erotic desires towards the Spanish ambassador, one of the 'fools' she vowed she would never marry. In *Marnie*, the journey into womanhood is presented in the context of a psychological crisis. Marnie, played by Tippi Hedren, is sexually frigid, a thief who steals from her male employers. She loves only her mother and her horse, Forio. Before she can begin her transformation into proper womanhood, and learn to desire the man she has been forced to marry, she has to shoot her horse, which has a broken leg, and give up her criminal activities. Her horse/virile ways are replaced by his. Passivity and propriety are essential preconditions for the transition from active, virile femininity into passive, feminine conformity.

The liminal journey of the tomboy – one of the few rites of passage stories available to women in the cinema – is a narrative about the forging of the proper female identity. It is paralleled by Freud's anatomical narrative about the journey of the clitoris which is, at base, a narrative about culture. The tomboy who refuses to travel Freud's path, who clings to her active, virile pleasures, who rejects the man and keeps her horse is stigmatized as the lesbian. She is a threatening figure on two counts. First, her image undermines patriarchal gender boundaries that separate the sexes. Second, she pushes to its extreme the definition of the active heterosexual woman – she represents the other side of the heterosexual woman, her lost phallic past, the autonomy she surrenders in order to enter the heritage of the Freudian womb. In this context, it is the lesbian – not woman in general – who signifies the 'ruin of representation'.

ANIMALISTIC LESBIAN BODY

The stereotype that associates lesbianism with bestiality also pushes representation to its limits. As discussed earlier, woman is, in the popular (male) imagination, associated more with the world of abject nature because of her procreative and birth-giving functions. In religious discourse, her sinful nature makes her a natural companion of the serpent. The embodiment of mother nature, woman represents the fertile womb, the Freudian hearth of domestic bliss. Whereas woman's function is to replicate that of the natural world, man's function is to control and cultivate that world for his own uses. Like the animal world, woman has an insatiable sexual appetite that must be controlled by man. Modern pornography depicts woman's link with nature in images of women posed in the 'doggy' position or engaged in sex with animals – particularly horses and dogs.

In the first part of the twentieth century, woman was particularly aligned with nature because of a widely held belief in a pseudo-scientific theory known as the theory of 'devolution'. According to this belief, while man was in general constantly evolving, some men and all women were in danger of devolving to lower animal forms. Dijkstra presents a fascinating study of the representation of devolution in *fin-de-siècle* art. He points out that whereas 'half-bestial creatures [such] as satyrs and centaurs' were used to depict such men, often caricatured as semitic or negroid, 'there was no need to find a symbolic form to represent [woman's] bestial nature' as 'women, being female, were, as a matter of course, already directly representative of degeneration' (Dijkstra 1986: 275). Hence many paintings of the period depicted women frolicking with satyrs and cavorting with animals in the dark recesses of the woods. If women in general were associated with the animal world, the lesbian was an animal. Dijkstra also refers to the work of Havelock Ellis to support his argument. Drawing on Darwin's view that animals could become sexually excited by the smell of women, Havelock Ellis argued that 'the animal is taught to give gratification by cunnilinctus. In some cases there is really sexual intercourse between the animal and the woman'. Apparently, Ellis drew connections between lesbianism in young girls and 'later predilection for encounters with animals' (Dijkstra 1986: 297). The association of homosexuality with bestiality, however, extends much further back than Victorian England. One of the most widely read books of the medieval period, said to be as popular as the Bible, was the *Physiologus*, also known as 'the medieval bestiary'. It consisted of a collection of stories, many without any accuracy whatsoever, about animal behaviour and its relationship to human behaviour. It was widely translated, and its influence felt for centuries. According to Boswell it was a 'manual of piety, a primer of zoology, and a form of entertainment' (1980: 141).

The *Physiologus*, which incorporated the *Epistle of Barnabas* from the first century AD, advanced various arguments about animal behaviour that were used to decry homosexual behaviour. It claimed that he who ate the meat of hare would become 'a boy-molester' because 'the hare grows a new anal opening each year, so that however many years he has lived, he has that many anuses' (Boswell 1980: 137–8). Those who ate the meat of the hyena would, like the hyena, change their gender from male to female every year. So women could develop male sexual organs and vice versa. Those who ate the weasel would become like those women who engage in oral sex and who conceive and give birth orally. The abject practices of the hare, weasel and hyena were associated with homosexual practices, abnormal birth and sex changes. In this context, homosexual acts were seen as unclean and animalistic.

Desire transforms the body; abject desire makes the body abject. This belief is similar to the view that women gave birth to monsters because of the kinds of desires they experienced during pregnancy (Huet 1993: 13). Desire can also affect the sexual organs. The story of the hyena was used to explain gender changes for both male and female. The image of a hare with multiple anuses constructs the body from the perspective of the feminized creature, the one being penetrated. It also suggests a fantasy about passivity and an excess of pleasure. No doubt the medieval story of the hare was applied primarily to the male sodomite, but given that the female homosexual was also seen as a sodomite, she would have been associated with the monstrous, transforming body of the hare.

A recent film, *Face Of A Hare* (Liliana Ginanneschi), which explored an unusual friendship between two women, draws on associations between woman, the hare and repressed lesbian desire. The narrative tells the story of two women who have lost their daughters. One woman, a derelict, who lives on the streets, takes up a maternal role in relation to the younger woman in that she appears to possess knowledge about the meaning of life that the younger woman needs. In this way, the film constructs three mother – daughter relationships. Men have no place in the story. In the pre-credit sequence, we are told the story of the 'Moon and the Hare' in which the moon, referred to as 'she', punishes the hare for delivering a false message to men about the meaning of life. The moon hit the hare on the snout with a stick and flattened its nose forever. We are then told that the younger woman, Elena, who visibly resembles a hare, also felt 'flattened like the hare'. The hare, with its flattened/castrated nose, is associated with woman. The two women are also symbolically castrated in that both have lost their daughters. They form an unusual and close friendship – brought together by their mutual experience of loss and their feelings of despair. But a growing bond of friendship helps to ease their pain. At one point the older woman announces she is Marlene Dietrich, a star whose screen persona has always signified lesbian desire, and shortly after seizes the younger woman in an embrace and begins to dance with her. The women form a couple but, as in virtually all male friendship films, one of the couple dies.

[. . .]

Another popular image of the lesbian as non-human creature appeared in stories of the female vampire. A seductive creature of the night, the lesbian vampire – still a popular monster of the horror film – not only attacked young girls but also men whose blood she drank in order to assume their masculine virility. Like an animal, the lesbian vampire was prey to her own sexual lusts and primitive desires.

The tomboy, the girl whose sexual identity is androgynous, is almost always associated with animals, particularly the horse and dog. The image of the lesbian as part of the natural world – as distinct from the civilized – might repel some, but it is also immensely appealing.

NARCISSISTIC LESBIAN BODY

A popular convention of *fin-de-siècle* painting, the cinema and fashion photography is the image of two women, posed in such a way as to suggest one is a mirror-image of the other. We see the image of the lesbian as narcissist in films about lesbianism. After the two women in *Les Biches* begin a relationship they start to imitate each other in dress and appearance; the women in *Persona* also wear identical clothes and beach hats, making it almost impossible to tell them apart; in *Single White Female* the mentally disturbed girl, in love with her flatmate, deliberately vampirizes her appearance and behaviour until they look like identical twins. In lesbian vampire horror films, such as *Vampyres*, the female friends are also depicted as identical, even the blood that smears their lips seems to trickle from identical mouths and fangs.

Contemporary fashion images in magazines and shop windows also exploit the idea of female narcissism, using models dressed in similar clothes and similar poses – sometimes caught together in an embrace – to sell their products. More overt forms of lesbian behaviour (butch–femme displays) are now also used, particularly as many younger lesbians, who have rejected the lesbian refusal of fashion associated with the 1970s, opt to explore fashion possibilities. Whether or not the general buying public reads lesbianism into these advertisements is another matter. In her discussion of lesbian consumerism, Danae Clark points out that advertisers, as a matter of conscious policy, now attempt to appeal to the gay community through what they describe as 'code behaviour' that only gays would understand: 'If heterosexual consumers do not notice these subtexts or subcultural codes, then advertisers are able to reach the homosexual market along with the heterosexual market without ever revealing their aim' (Clark 1991: 183). However, images that exploit the notion of the feminine/lesbian narcissism draw on a much older tradition than that represented in the contemporary fashion industry. If this tradition suggests that woman is, by her very nature, vain, the lesbian couple represents, by definition, feminine narcissism and auto-eroticism *par excellence*.

In a chapter entitled, 'The Lesbian Glass', Dijkstra (1986) discusses the popular belief, championed by Havelock Ellis, that women are vain, narcissists capable of completely losing themselves in self-admiration. Turn-of-the-century medical writers pointed to the supposed connection between masturbation in women, narcissism and lesbianism. Masturbation increased the size of the clitoris; the woman with a large clitoris was likely to become a lesbian and to engage in those 'excesses' called 'lesbian love' (Gilman 1985: 89). According to Dijkstra, women were painted kissing themselves in mirrors – vain, self-absorbed, completely uninterested in men: 'Woman's desire to embrace her own reflection, her "kiss in the glass", became the turn of the century's emblem of her enmity towards man' (ibid.: 150). Dijkstra cites the eponymous heroine of the film, *Lulu*, played by Louise Brooks, the notorious *femme fatale* whose beauty attracts both men and women, is depicted as a completely self-absorbed narcissist. At one point she says: 'When I look at myself in the mirror I wish I were a man ... my own husband.'

[...]

Like masturbation, lesbianism was seen as inextricably linked to self-absorption and narcissism. Men were shut out from this world – hence they understood the threat offered by the lesbian couple. (According to popular male mythology, what the lesbian really needs is a good fuck, that is, a phallic intrusion to break up the threatening duo.) The representation of the lesbian couple as mirror-images of each other constructs the lesbian body as a reflection or an echo. Such an image is dangerous to society and culture because it suggests there is no way forward – only regression and circularity are possible.

Representations of the lesbian as female narcissist in painting, film and fashion images almost always depict the lesbian as conventionally feminine. This is the key area in which popular fantasies about the nature of lesbianism do not draw on the cliché of the lesbian as a thwarted man. The narcissistic femme lesbian, however, almost always adopts an ambiguous position in relation to the gaze of the camera/spectator. She is on display, her pose actively designed to lure the gaze; the crucial difference is, however, that the spectator is shut out from her world. He may look but not enter. Images of the lesbian double are designed to appeal to the voyeuristic desires of the male spectator.

In the first two stereotypes discussed, the lesbian body is constructed in terms of the heterosexual model of sex which involves penetration; there was no attempt to define the nature of lesbian pleasure from the point of view of the feminine. The threat offered by the image of the lesbian-as-double is not specifically related to the notion of sexual penetration. Instead, the threat is associated more with auto-eroticism and exclusion.

Representations of the lesbian double – circulated in fashion magazines, film and pornography – draw attention to the nature of the image itself, its association with the feminine, and the technologies that enable duplication and repetition. The lesbian double threatens because it suggests a perfectly sealed world of female desire from which man is excluded, not simply because he is a man, but also because of the power of the technology to exclude the voyeuristic spectator. But exclusion is also part of the nature of voyeuristic pleasure which demands that a distance between the object and the subject who is looking should always be preserved. Photographic technology, with its powers of duplication, reinforces a fear that, like the image itself, the lesbian couple-as-double will reduplicate and multiply.

THE LESBIAN BODY/COMMUNITY

The body is both so important in itself and yet so clearly a sign or symbol referring to things outside itself in our culture. So far I have discussed the representation of the lesbian body in terms of male fantasies and patriarchal stereotypes. Historically and culturally, the lesbian body – although indistinguishable in reality from the female body itself – has been represented as a body in extreme: the pseudo-male, animalistic and narcissistic body. Although all of these deviant tendencies are present in the female body, it is the ideological function of the lesbian body to warn the 'normal' woman about the dangers of undoing or rejecting her own bodily socialization. This is why the culture points with most hypocritical concern at the mannish lesbian, the butch lesbian, while deliberately ignoring the femme lesbian, the woman whose body in no way presents itself to the straight world as different or deviant. To function properly as ideological litmus paper, the lesbian body must be instantly recognizable. In one sense, the femme lesbian is potentially as threatening – although not as immediately confronting – as the stereotyped butch because she signifies the possibility that all women are potential lesbians. Like the abject, the stereotyped mannish/animalistic/auto-erotic lesbian body hovers around the borders of gender socialization, luring other women to its side, tempting them with the promise of deviant pleasures.

Within the lesbian community itself, however, a different battle has taken place around the definition of the lesbian body. This battle has nothing to do with the size of the clitoris, animals or self-reflecting mirrors. Preoccupied with the construction of the properly socialized feminine body, lesbian–feminism of the 1970s became obsessed with appearance, arguing that the true lesbian should reject all forms of clothing that might associate her image with that of the heterosexual woman and ultimately with patriarchal capitalism. The proper lesbian had short hair, wore sandshoes, jeans or boiler suit, flannel shirt and rejected all forms of make-up. In appearance she hovered somewhere between the look of the butch lesbian, who wore men's clothes and parodied men's behaviour and gestures, and the tomboy. She was a dyke – not a butch – whose aim was to capture an androgynous uniformed look. Lesbians who rejected this model were given a difficult time. In debates that raged in Melbourne in the mid-1970s, some of us who refused the lesbian uniform were labelled 'heterosexual lesbians', an interesting concept that constructs a lesbian as an impossibility – a figure perhaps more in tune with the queer world of the 1990s.

From the 1970s onwards, the lesbian community has adopted a series of fashion styles ranging from flannel shirts to the leather and lipstick lesbians of the 1990s. A recent film, *Framing Lesbian Fashion* (Karen Everett, 1991), pays tribute to the flannel lesbians while celebrating the changing styles of recent years. The film is structured around a series of inter-titles which point to the key changes in style which have involved flannel, leather, corporate drag, tattooing and body piercing. There are a series of interviews with lesbians who have lived through these changes, as well as a lesbian fashion show. The opening credits are accompanied by the words 'I like to shop, shop, shop, shop – shop until I drop'. The film concludes with a tribute to the lesbians of the 1970s who set out to liberate themselves from the patriarchal stereotypes of feminine dress and appearance. The problem was that they also imposed a fairly rigid code of dress on themselves and anyone who wanted to join the lesbian community. There was certainly no place for femme or older style butch lesbians. Only with the butch–femme renaissance of the 1980s did butch and femme lesbians come out of the closet and begin to assert their own needs to express themselves without fear of retribution. Today, with the liberating influence of queer theory and practice (often quite separate entities), almost any form of dress is acceptable.

The film makes one thing very clear: most women enjoyed wearing the different 'uniforms' such as flannel, leather, lipstick because it gave them a sense of belonging to a community, the gang, the wider lesbian body. They speak of having a sense of family and shared identity via their common forms of dress. The need to construct a sense of community, through dress and appearance, suggests quite clearly that there is no such thing as an essential lesbian body – lesbians themselves have to create this body in order to feel they belong to the larger lesbian community, recognizable to its members not through essentialized bodily forms but through representation, gesture and play. The 1990s lesbian is most interested in playing with appearance and with sex roles. Women interviewed in *Framing Lesbian Fashion* were very clear about the element of parody in their dress styles. One woman who cross-dressed even wore a large dildo in her leather pants ('packing it') to simulate the penis – the male penis as well as the one that male fantasy has attributed throughout the centuries to the lesbian and her tribade forebears. Unlike

Calamity Jane, whose outfit would have caused a sensation at *Club Q*, the 1990s lesbian refuses to exchange her whip and leathers for home, hearth and the seal of social approval. She has a body that is going places.

ACKNOWLEDGEMENTS

Thanks to Lis Stoney for her perceptive comments and suggestions. This article is indebted to Thomas Laqueur's brilliant book *Making Sex: Body and Gender from the Greeks to Freud.*

REFERENCES

Bakhtin, M. (1984) *Rabelais and His World*, translated by Hélène Iswolsky, Bloomington, Ind.: Indiana University Press.

Boswell, J. (1980) *Christianity, Social Tolerance, and Homosexuality*, Chicago, Ill. and London: University of Chicago Press.

Cavin, S. (1985) *Lesbian Origins*, San Francisco, Calif.: Ism Press Inc.

Clark, D. (1991) 'Commodity Lesbianism', *Camera Obscura*, 25–7: 181–201.

Creed, B. (1993) 'Dark Desires: Male Masochism in the Horror Film', in S. Cohan and I. R. Hark (eds), *Screening The Male: Exploring Masculinities in Hollywood Cinema*, London and New York: Routledge.

Dijkstra, B. (1986) *Idols of Perversity: Fantasies of Feminine Evil in Fin-de-Siècle Culture*, New York: Oxford University Press.

Freud, S. (1905a) 'Fragment of an Analysis of a Case of Hysteria ("Dora")', *Case Histories 1* (Pelican Freud Library, vol. 8), Harmondsworth: Penguin, pp. 31–164.

Freud, S. (1905b) 'Three Essays on The Theory of Sexuality', in J. Strachey (trans. and ed.) *On Sexuality* (Pelican Freud Library, vol. 7), Harmondsworth: Penguin.

Freud, S. (1920) 'The Psychogenesis of a Case of Homosexuality in a Woman', *Case Histories 11* (Pelican Freud Library, vol. 9), Harmondsworth: Penguin.

Freud, S. (1923) 'Fragment of an Analysis of a Case of Hysteria ("Dora")', *Case Histories 1* (Pelican Freud Library, vol. 8), Harmondsworth: Penguin.

Gilman, S. L. (1985) *Difference and Pathology: Stereotypes of Sexuality, Race and Madness*, Ithaca, NY and London: Cornell University Press.

Huet, M. (1993) *Monstrous Imagination*, Cambridge, Mass. and London: Harvard University Press.

Jones, A. R. and Stallybrass, P. (1991) 'Fetishizing Gender: Constructing the Hermaphrodite in Renaissance Europe', in J. Epstein and K. Straub (eds), *Bodyguards: The Cultural Politics of Gender Ambiguity*, New York and London: Routledge.

Katz, J. N. (1990) 'The Invention of Heterosexuality', *Socialist Review* 20 (1): 7–34.

Kristeva, J. (1982) *Powers of Horror: An Essay on Abjection*, translated by L. S. Roudiez, New York: Columbia University Press.

Laqueur, T. (1990) *Making Sex: Body and Gender from the Greeks to Freud*, Cambridge, Mass. and London: Harvard University Press.

Montrelay, M. (1978) 'Inquiry into Femininity', *m/f*, 1: 83–102.

van der Meer, T. (1990) 'Tribades on Trial: Female Same-Sex Offenders in Late Eighteenth-Century Amsterdam', in J. C. Fout (ed.), *Forbidden History: The State, Society and the Regulation of Sexuality in Modern Europe*, Chicago, Ill. and London: University of Chicago Press.

Walker, B. (1983) *The Woman's Encyclopedia of Myths and Secrets*, San Francisco, Calif.: Harper & Row.

2.6

F2M: THE MAKING OF FEMALE MASCULINITY

Judith Halberstam

The postmodern lesbian body as visualized by recent film and video, as theorized by queer theory, and as constructed by state of the art cosmetic technology breaks with a homo-hetero sexual binary and remakes gender as not simply performance but also as fiction. Gender fictions are fictions of a body taking its own shape, a cut-up genre that mixes and matches body parts, sexual acts, and postmodern articulations of the impossibility of identity. Such fictions demand readers attuned to the variegated contours of desire. The end of identity in this gender fiction does not mean a limitless and boundless shifting of positions and forms, rather it indicates the futility of stretching terms like *lesbian* or *gay* or *straight* or *male* or *female* across vast fields of experience, behavior, and self-understanding. It further hints at the inevitable exclusivity of any claim for identity and refuses the respectability of being named, identified, known. This essay will call for new sexual vocabularies that acknowledge sexualities and genders as styles rather than life-styles, as fictions rather than facts of life, and as potentialities rather than as fixed identities.

Axiom 1 of Eve Kosofsky Sedgwick's *Epistemology of the Closet*: 'People Are Different From Each Other'. Sedgwick's genealogy of the unknown suggests the vast range of identities and events that remain unaccounted for by the 'coarse axes of categorization' that we have come to see as indispensable. Sedgwick claims that to attend to the 'reader relations' of texts can potentially access the 'nonce taxonomies' or 'the making and unmaking and remaking and redissolution of hundreds of old and new categorical meanings

From: L. Doan (ed.), *The Lesbian Postmodern*, New York: Columbia University Press, 1994.

concerning all the kinds it may take to make up a world'.[1] All kinds of people, all kinds of identities, in other words, are simply not accounted for in the taxonomies we live with. Nonce taxonomies indicate a not-knowing already embedded in recognition.

We live with difference even though we do not always have the conceptual tools to recognize it. One recent film, Jenny Livingston's *Paris Is Burning*, shocked white gay and straight audiences with its representations of an under-exposed subculture of the African-American and Latino gay world of New York. The shock value of the film lay in its ability to confront audiences with subcultural practices that the audience thought they knew already. People knew of voguing through Madonna, of drag shows through gay popular culture, but they did not know, in general, about Houses, about walking the Balls, about Realness. Livingston's film, which has been criticized in some circles for adopting a kind of pedagogical approach, was in fact quite sensitive to the fact that there were lessons to be learned from the Balls and the Houses, lessons about how to read gender and race, for example, as not only artificial but highly elaborate and ritualistic significations. *Paris Is Burning* focused questions of race, class, and gender and their intersections with the drag performances of poor, gay men of color.

How and in what ways does the disintegration and reconstitution of gender identities focus upon the postmodern lesbian body? What is postmodern about lesbian identity? In the 1990s lesbian communities have witnessed an unprecedented proliferation of sexual practices or at least of the open discussion of lesbian practices. Magazines like *Outlook* and *On Our Backs* have documented ongoing debates about gender, sexuality, and venues for sexual play, and even mainstream cinema has picked up on a new visibility of lesbian identities (*Basic Instinct* [1992], for example). Lesbians are particularly invested in proliferating their identities and practices because, as the sex debates of the 1980s demonstrated, policing activity within the community and commitment to a unitary conception of lesbianism has had some very negative and problematic repercussions.[2]

Some queer identities have appeared recently in lesbian zines and elsewhere: guys with pussies, dykes with dicks, queer butches, aggressive femmes, F2Ms, lesbians who like men, daddy boys, gender queens, drag kings, pomo afro homos, bulldaggers, women who fuck boys, women who fuck like boys, dyke mommies, transsexual lesbians, male lesbians. As the list suggests, gay/lesbian/straight simply cannot account for the range of sexual experience available. In this essay, I home in on the transsexual lesbian, in particular, the female to male transsexual or F2M, and I argue that within a more general fragmentation of the concept of sexual identity, the specificity of the transsexual disappears. In a way, I claim, we are all transsexuals.

We are all transsexuals except that the referent of the *trans* becomes less and less clear (and more and more queer). We are all cross-dressers but where are we crossing from and to what? There is no 'other' side, no 'opposite' sex, no natural divide to be spanned by surgery, by disguise, by passing. We all pass or we don't, we all wear our drag, and we all derive a different degree of pleasure – sexual or otherwise – from our costumes. It is just that for some of us our costumes are made of fabric or material, while for others they are made of skin; for some an outfit can be changed; for others skin must be resewn. There are no transsexuals.

Desire has a terrifying precision. Pleasure might be sex with a woman who looks like a boy; pleasure might be a woman going in disguise as a man to a gay bar in order to pick up a gay man. Pleasure might be two naked women; pleasure might be masturbation watched by a stranger; pleasure might be a man and a woman; but pleasure seems to be precise. In an interview with a pre-op female-to-male transsexual called Danny, Chris Martin asks Danny about his very particular desire to have sex with men as a man. 'What's the difference', she asks, 'between having sex with men now and having sex with men before?' Danny responds: 'I didn't really. If I did it was oral sex ... it was already gay sex ... umm ... that was a new area. It depends upon your partner's perception. If a man thought I was a woman, we didn't do it.'[3] Danny requires that his partners recognize that he is a man before he has 'gay' sex with them. He demands that they read his gender accurately according to his desire, in other words, though, he admits, there is room for the occasional misreading. On one occasion, for example, he recalls that a trick he had picked up discovered that Danny did not have a penis. Danny allowed his partner to penetrate him vaginally because, 'it was what he had been looking for all his life only he hadn't realized it. When he saw me it was like "Wow. I want a man with a vagina".'

Wanting a man with a vagina or wanting to be a woman transformed into a man having sex with other men are fairly precise and readable desires – precise and yet not at all represented by the categories for sexual identity we have settled for. And, as another pre-op female-to-male transsexual, Vern, makes clear, the so-called gender community is often excluded by or vilified by the gay community. Vern calls it genderphobia: '*Genderphobia* is my term. I made it up because there is a clone movement in the non-heterosexual community to make everybody look just like heterosexuals who sleep with each other. The fact is that there is a whole large section of the gay community who is going to vote Republican.'[4]

Genderphobia, as Vern suggests, indicates all kinds of gender trouble in the mainstream gay and lesbian community. Furthermore, the increasing numbers of female-to-male transsexuals (f to m's) appearing particularly in metropolitan or urban lesbian communities has given rise to interesting and sometimes volatile debates among lesbians about f to m's.[5]

Genderbending among lesbians is not limited to sex change operations. In New York, sex queen Annie Sprinkle has been running. 'Drag King For A Day' workshops with pre-op f to m Jack Armstrong, a longtime gender activist. The workshops instruct women in the art of passing and culminate in a night out on the town as men. Alisa Solomon wrote about her experience in the workshop for *The Village Voice*, reporting how eleven women flattened their breasts, donned strips of stage makeup facial hair, 'loosened our belts a notch to make our waistlines fall, pulled back hair, put on vests.'[6] Solomon felt inclined, however, to draw the line at putting a sock in her Jockeys because she 'was interested in gender, not sex. A penis has nothing to do with it.' She also notes in response to Jack Armstrong's discussion of his transsexuality: 'I could have done without his photo-aided descriptions of phalloplasties and other surgical procedures. After all I had no interest in how to *be* a man; I only wanted, for the day, to be *like* one.'

Solomon's problematic response to the issue of transsexualism is indicative of the way that many lesbians embrace the idea of gender performance, but they reduce it to just that, an act with no relation to biology, real or imagined. Solomon disavows the penis here as if that alone is the mark of gender – she is comfortable with the clothes and the false facial hair, but the suggestion of a constructed penis leads her to make an essential difference between feigning maleness for a day and being a man. In fact, as she wanders off into the Village in her drag, Alisa Solomon, inasmuch as she passes successfully, *is* a man, is male, is a man for a day. The insistence here that the penis alone signifies maleness, corresponds to a tendency within academic discussion of gender to continue to equate masculinity solely with men. Recent studies on masculinity[7] persist in making masculinity an extension or discursive effect of maleness. But what about female masculinity or lesbian masculinity?

In the introduction to her groundbreaking new study of transvestism, *Vested Interests*, Marjorie Garber discusses the ways in which transvestism and transsexualism provoke a 'category crisis'.[8] Garber elaborates this term suggesting that often the crisis occurs elsewhere but is displaced onto the ambiguity of gender. Solomon obviously confronts a 'category crisis' as she ponders the politics of stuffing her Jockeys, and presumably such a crisis is one of the intended by-products of Sprinkle/Armstrong's workshop. Solomon attempts to resolve her category crisis by assuring herself that she wants to look *like* a man, not *be* a man, and that therefore her desire has nothing to do with possession of the penis. But, in fact, what Solomon misunderstands is that penises as well as masculinity become artificial and constructible when we challenge the naturalness of gender. Socks in genetic girls' jockeys are part and parcel of creating fictitious genders; they are not reducible to sex.

But what then is the significance of the surgically constructed penis in this masquerade of sex and gender? In a chapter of her study called 'Spare Parts: The Surgical Construction of Gender', Garber discusses the way in which the phenomenon of transsexuality 'demonstrates that essentialism *is* cultural construction'.[9] She suggests that f to m surgery has been less common and less studied than male-to-female transsexual operations, partly because medical technology has not been able to construct a functional penis but also on account of 'a sneaking feeling that it should not be so easy to "construct" a "man" – which is to say, a male body' (Garber 1992: 102). Garber is absolutely right, I think, to draw attention to a kind of conscious or unconscious unwillingness within the medical establishment to explore the options for f to m surgery. After all, the construction of a functional penis for f to m transsexuals could alter inestimably the most cherished fictions of gender in the Western world.

If penises were purchasable, in other words – functional penises, that is – who exactly might want one? What might the effect of surgically produced penises be upon notions like – 'penis envy', 'castration complex', 'size queens'? If anyone could have one, who would want one? How would the power relations of gender be altered by a market for the penis? Who might want a bigger one? Who might want an artificial one rather than the 'natural' one they were born with? What if surgically constructed models 'work' better? Can the penis be improved upon? Certainly the folks at *Good Vibrations*, who have been in the business of selling silicone dildos for years now, could tell you about many models as good as, if not better than, the 'real' thing.

Obviously, the potential of medical technology to alter bodies makes natural gender and biological sex merely antiquated categories in the history of sexuality, that is, part of the inventedness of sex. Are we then, as Jan Morris claims in her autobiography *Conundrum: An Extraordinary Narrative of Transsexualism*, possibly entering a post-transsexual era?[10] I believe we are occupying the transition here and now, that we are experiencing a boundary change, a shifting of focus, that may have begun with the invention of homosexuality at the end of the nineteenth century but that will end with the invention of the sexual body at the end of the twentieth century. This does not mean that we will all in some way surgically alter our bodies; it means that we will begin to acknowledge the ways in which we have already surgically, technologically, and ideologically altered our bodies, our identities, ourselves.

One might expect, then, in these postmodern times that as we posit the artificiality of gender and sex with increasing awareness of how and why our bodies have been policed into gender identities, there might be a decrease in the incidence of such things as sex-change operations. On the contrary, however, especially in lesbian circles (and it is female to male transsexualism that I am concerned with here) there has been, as I suggested, a rise in discussions of, depictions of, and requests for f to m sex change operations. In a video documenting the first experience of sexual intercourse by a new f to m transsexual, Annie Sprinkle introduces the viewers to the world of f to m sex changes. The video, *Linda/Les and Annie*, is remarkable as a kind of post-op, postporn, postmodern artifact of what Sprinkle calls 'gender flexibility'. It is archaic, however, in its tendency to fundamentally realign sex and gender. In the video, Les Nichols, a post-op f to m transsexual sexually experiments with his new surgically constructed penis. The video records the failure of Les's first attempt at intercourse as a 'man', and yet it celebrates the success of his gender flexibility.

[...]

[Alongside] these fictions of gender, it is worth examining the so-called facts of gender – the facticities at least – that are perhaps best revealed by the medical discourse surrounding transsexual operations. While I want to avoid the inevitable binarism of a debate about whether transsexual operations are redundant, I do think that the terms we have inherited from medicine to think through transsexualism, sex changes and sexual surgery must change. Just as the idea of cross-dressing presumes an immutable line between two opposite sexes, so transsexualism, as a term, as an ideology, presumes that if you are not one you are the other. I propose that we call all elective body alterations for whatever reason (postcancer or postaccident reconstruction, physical disabilities, or gender dysphoria) *cosmetic* surgery and that we drop altogether the constrictive terminology of crossing.[11]

An example from a recent series on plastic surgery in the *Los Angeles Times* may illustrate my point. The series by Robert Scheer, entitled 'The Revolution in Cosmetic Surgery', covers the pros and cons of the plastic surgery industry. By way of making a point about the interdependence of the business of cosmetic surgery and the fashion industry, the writer states the obvious, namely, that very often media standards for beauty impose a 'world-wide standard of beauty' that leads non-Western, nonwhite women to desire the 'eyes, cheekbones or breasts of their favorite North American television star'.[12] By way of illustrating his point, Scheer suggests that 'turning a Japanese housewife ... into a typical product of the dominant white American genetic mix – for whatever that is worth – is now eminently doable'. He quotes from an Asian woman who says she wants to be like an American, 'You know. Big eyes. Everybody, all my girlfriends did their eyes deeper, so I did.' Scheer asks her what is next on her cosmetic surgery agenda: 'Nose and chin this time around.' Scheer comments:

> Eyelids are often redone too. Asian women don't have a crease in the middle. Why does one need an extra fold like two tracks running horizontally across the eyelid? Why is the smooth expanse of eyelid skin not perfect enough? The answer is that the desirable eye, the one extolled in the massive cosmetic industry blitz campaigns, is the Western eye, and the two lines provide the border for eye shadow and other make-up applications.

Scheer's rhetorical question as to why 'the smooth expanse of eyelid skin' is not acceptable is supposed to ironize the relationship between body politics and market demands. His answer to his own question is to resolve that the dictates of the marketplace govern seemingly aesthetic considerations. And, we might add, the racially marked face is not only marginalized by a kind of economy of beauty, it is also quite obviously the product of imperialist, sexist, and racist ideologies. The cosmetic production of occidental beauty in this scene of cosmetic intervention, then, certainly ups the ante on racist and imperialist notions of aesthetics, but it also has the possibly unforeseen effect of making race obviously artificial, another fiction of culture.

Cosmetic surgery, then, can, in a sometimes contradictory way, both bolster dominant ideologies of beauty and power, and it can undermine completely the fixedness of race, class, and gender by making each one surgically or sartorially reproducible. By commenting only upon the racist implications of such surgery in his article, Scheer has sidestepped the constructedness of race altogether. To all intents and purposes, if we are to employ the same rhetoric that pertains to transsexualism, the Japanese woman paying for the face job has had a race change (and here we might also think of the surgical contortions of Michael Jackson). She has altered her appearance until she appears to be white.

Why then do we not mark surgery that focuses on racial features in the same way that we positively pathologize surgery that alters the genitals? In 'Spare Parts', Marjorie Garber makes a similar point. She writes:

> Why does a 'nose job' or 'breast job' or 'eye job' pass as mere self-improvement, all – as the word 'job' implies – in a day's work for a surgeon (or an actress), while a sex change (could we imagine it called a 'penis job'?) represents the dislocation of everything we conventionally 'know' or believe about gender identities and gender roles, 'male' and 'female' subjectivities?[13]

The rhetoric of cosmetic surgery, in other words, reveals that identity is nowhere more obviously bound to gender and sexuality than in the case of transsexual surgery. And gender and sexuality are nowhere more obviously hemmed in by binary options.

Transsexual surgery, in other words, unlike any other kind of body-altering operations, requires that the medically produced body be resituated ontologically. All that was known about this body has now to be relearned; all that was recognizable about this body has to be renamed. But oppositions break down rather quickly in the area of body-altering surgery. Transsexual lesbian playwright Kate Bornstein perhaps phrases it best in her latest theater piece called 'The Opposite Sex Is Neither'. Describing herself as a 'gender outlaw', Bornstein writes: 'See, I'm told I must be a man or a woman. One or the other. Oh, it's OK to be a transsexual, say some – just don't talk about it. Don't question your gender any more, just be a woman now – you went to so much trouble – just be satisfied. I am not so satisfied.'[14] As a gender outlaw, Bornstein gives gender a new context, a new definition. She demands that her audience read her not as man or woman, or lesbian or heterosexual, but as some combination of presumably incompatible terms.

[...]

We are all transsexuals, I wrote earlier in this essay, and there are no transsexuals. I want both claims to stand and find a place in relation to the postmodern lesbian body, the body dressed up in its gender or surgically constructed in the image of its gender. What is the relationship between the transsexual body and the postmodern lesbian body? Both threaten the binarism of homo/hetero sexuality by performing and fictionalizing gender. The postmodern lesbian body is a body fragmented by representation and theory, overexposed and yet inarticulate, finding a voice finally in the underground culture of zines and sex clubs.

Creating gender as fiction demands that we learn how to read it. In order to find our way into a posttranssexual era, we must educate ourselves as readers of gender fiction, we must learn how to take pleasure in gender and how to become an audience for the multiple performances of gender we witness everyday. In a 'Posttranssexual Manifesto' entitled 'The Empire Strikes Back', Sandy Stone also emphasizes the fictionality or readability of gender. She proposes that we constitute transsexuals as a 'genre – a set of embodied texts whose potential for *productive* disruption of structured sexualities and spectra of desire has yet to be explored'.[15] The *post* in posttranssexual demands, however, that we examine the strangeness of all gendered bodies, not only the transsexualized ones and that we rewrite the cultural fiction that divides a sex from a transsex, a gender from a transgender. All gender should be transgender, all desire is transgendered, movement is all.

The reinvention of lesbian sex, indeed of sex in general, is an ongoing project, and it coincides, as I have tried to show, with the formation of, or surfacing of, many other sexualities. The transgender community, for example, people in various stages of gender transition, have perhaps revealed the extent to which lesbians and gay men are merely the tip of the iceberg when it comes to identifying sexualities that defy heterosexual definition or the label straight. The breakdown of genders and sexualities into identities is in many ways, therefore, an endless project, and it is perhaps preferable therefore to acknowledge that gender is defined by its transitivity, that sexuality manifests as multiple sexualities, and that therefore we are all transsexuals. There are no transsexuals.

NOTES

1. See Sedgwick, *Epistemology of the Closet*, 23.
2. See Alice Echols, 'The New Feminism of Yin Yang', in Snitow, Stansell, and Thompson, eds, *Powers of Desire*, 439–59; and 'The Taming of the Id: Feminist Sexual Politics, 1968–1983', in Vance, ed., *Pleasure and Danger: Exploring Female Sexuality*, 50–72.
3. Interview, 'Guys With Pussies' by Chris Martin with 'Vern and Danny'. Part of this interview was published in *Movement Research Performance Journal* 3 (Fall 1991): 6–7.
4. Interview with Chris Martin, 'World's Greatest Cocksucker', in *Movement Research Journal* 3 (Fall 1991): 6.
5. See, for example, Marcie Sheiner, 'Some Girls Will Be Boys', in *On Our Backs* (March/April 1991): 20.
6. Alisa Solomon, 'Drag Race: Rites of Passing', *Village Voice* (November 15, 1991): 46.
7. For example, see Kaja Silverman, *Masculinity in the Margins* (New York: Routledge, 1992) or Victor Seidler, *Rediscovering Masculinity: Reason, Language, and Sexuality* (London and New York: Routledge, 1989).

8. Garber, *Vested Interests*, 16.
9. Ibid., 109.
10. Morris, *Conundrum*.
11. As I was writing this piece, I read in a copy of *Seattle Gay News* (January 1992) that a transsexual group in Seattle was meeting to discuss how to maintain the definition of transsexual operations as medical rather than cosmetic, because if they are termed 'cosmetic', then insurance companies can refuse to pay for them. As always, discursive effects are altered by capitalist relations in ways that are unforeseeable. I do not think we should give up on the cosmeticization of transsexualism in order to appease insurance companies. Rather, we should argue that cosmetics are never separate from 'health', and insurance companies should not be the ones making such distinctions, anyway.
12. Robert Scheer, 'The Cosmetic Surgery Revolution: Risks and Rewards', *Los Angeles Times* (December 22, 1991): A1, A24, A42.
13. Garber, *Vested Interests*, 117.
14. Bornstein's play, *The Opposite Sex is Neither*, played in San Diego at the Sushi Performance Gallery, December 13–14, 1991. The quotation is from 'Transsexual Lesbian Playwright Tells All' in Scholder and Silverberg, eds, *High Risk*, 261.
15. Sandy Stone, 'The Empire Strikes Back: A Posttranssexual Manifesto', in Epstein and Straub, eds, *Body Guards*, 296.

References

Butler, Judith (1990) *Gender Trouble: Feminism and the Subversion of Identity*, New York: Routledge.

Epstein, Julia and Kristina Straub (eds) (1991) *Body Guards: The Cultural Politics of Gender Ambiguity*, New York: Routledge.

Garber, Marjorie (1992) *Vested Interests: Cross-Dressing and Cultural Anxiety*, New York: Routledge.

Morris, Jan (1974) *Conundrum*, New York: Harcourt, Brace, Jovanovich.

Scholder, Amy and Ira Silverberg (eds) (1991) *High Risk*, New York: Penguin.

Sedgwick, Eve Kosofsky (1990) *Epistemology of the Closet*, Berkeley and Los Angeles: University of California Press.

Snitow, Ann, Christine Stansell, and Sharon Thompson (eds) (1983) *Powers of Desire: The Politics of Sexuality*, New York: Monthly Review Press.

Vance, Carole (ed.) (1984) *Pleasure and Danger: Exploring Female Sexuality*, Boston and London: Routledge, Kegan Paul.

2.7

NO BODY IS 'DOING IT': CYBERSEXUALITY

Juniper Wiley

[Wiley considers what is 'revolutionary' about computer-mediated-communication]

Computer-mediated-communication provides no context for nonverbal communication leakage. All information is conveyed in verbal mode. The lack of vocalistic cues such as tone, pitch, quality of speech, removes an entire realm for interpreting social meanings. Verbal communication is always in written form, input line-by-line, and requiring some measure of typing skills for a reasonable rate of interaction. The 'asides' that are conveyed in face-to-face or telephone communications, like laughter or tone of sarcasm, must be included in a written text.

The feedback loop of backchannel work in face-to-face encounters and telephone transmissions that are continuous and immediate in communicating turn-taking, listening, understanding, interest, emotionality, on and on, is transformed from synchronistic feedback to asynchronistic as a transmission must be completed before a response is possible. This is true even if the Bulletin Board System (BBS) interaction is ongoing and not simply e-mail or postings. Each line of thought must be completed before the next proceeds rather than simultaneous feedback.

In face-to-face encounters, participants can pick up various social role information such as age, race, nationality, gender, demeanour, style of dress, etc., that are absent in computer-mediated-communication as any aspect of social role performance, presentation of self, and physical appearance are

From: *Body & Society* (1995) 1 (1): 151–60.

communicated within the written text. Face-to-face interactions occur within real time whereas in computer-mediated-communication the user may respond hours, weeks, or months later to a statement in e-mail or a posting. Even within a one-to-one chat, a pause is not easily interpreted as allowances for typing skills, transmission problems, differences in transmission systems enter the field of possibility.

Computer-mediated-communication extends interactional possibilities beyond an immediate geographical environment, with capacity to talk with vast numbers of others, together as a group or individually, at anytime or anyplace, and the social perception of sight, sound, activity within the capabilities of the text-producers. The distinctions between public and private, interpersonal and mass-mediated contexts are blurred and fused. Even the CB-radio craze of the 1970s cannot compare to the BBS world as it was measured in miles. Ruben (1985: 6) sums it up in the following manner:

> At the core of the information age is the progressive convergence of media that were once distinct. Print, broadcast, and common carrier technologies are increasingly indistinguishable from one another because of changes in the devices themselves and changes in the uses to which the technologies are put ... the information age is marked by the wholesale availability and application of electronic technologies in a wide range of personal and professional contexts. While some of these technologies are new, many others have resulted from the convergence and transformation of existing technological forms and uses.

Online Interactivity: Co-Mingling Through Simulation

On the BBS everyday life reality is transformed into a virtual reality. From the first moments of logging on, new users creatively craft ironically-intentioned or whimsically-concocted 'handles' that replace everyday names. Newly generated personas – faceless, voiceless, bodiless – displace history with a timeless present and multiple selves easily co-exist with the flick of a finger. Fantasy is freed.

Cybersexuality is a postmodern narrative, like 'having carnal knowledge' or 'doing the wild thing,' that works to mark an era, distinguish a generational attitude, and to transform the meaning of the act. Simulation and fantasy merge with 'reality' as BBS generated personas with disembodied anonymity join dreams, create mutual adventures, share secrets, lies, personal disclosures, and act on each other in cybersex. That which follows may be considered by some the bleak side of a postmodern era, an 'era of final solutions' (Baudrillard, 1990a: 2) that suggests:

> Nothing is less certain today than sex, behind the liberation of its discourse. And nothing today is less certain than desire, behind the proliferation of its images ... In matters of sex, the proliferation is approaching total loss. Here lies the secret of the ever increasing production of sex and its signs, and the hyperrealism of sexual pleasure ... No

more want, no more prohibitions, and no more limits: it is the loss of every referential principle ... It is the ghost of desire that haunts the defunct reality of sex. Sex is everywhere, except in sexuality [Barthes]. (Baudrillard, 1990a: 5)

Computer humming, modem clicking ... we are logging onto 'In Contact' to an 'Open Forum' (group chat) in the 'Hotel California' (where 1–99 'rooms' represent different 'places' for separate group chats). The lines of text, originally fast paced across my monitor screen, re-present the narrative spun by two bisexual women story-tellers as I joined them. 'The Nasty Club' scenario displays one form of the 'acting on' that is alluded to within the concept of 'interactivity'.

'Fieldnote-Capture File', Saturday, 29–06–91 at approx. 2:30 p.m.

***** Welcome to the Hotel California *****
-> Welcome to the Hotel California, there are 2 guests registered
-> Your room number please! (1–99)?
=> 77
//
\\\ Entering Open Forum \\\
//
<Carisa 405> well i'd rather have you
<Searcher 468> hello out there ...
<Rebecca 1109> hi
<Carisa 405> speak of the devil. ... hi Searcher
<Searcher 468> am i interrupting or may i stay?
<Carisa 405> please stay
<Rebecca 1109> stay
<Searcher 468> devil?
<Rebecca 1109> i have the devil in me today ... feeling very wicked ... he he
<Carisa 405> we can maybe seduce her, huh Rebecca
<Searcher 468> oh many have/do try/tried ...
<Carisa 405> we'll bend this STRAIGHT woman our way
<Rebecca 1109> hhmmm .. will she let us?
<Carisa 405> i hope so
<Carisa 405> we can always use a new member in the nasty club
<Searcher 468> is this called flirting or am i just to remain
<Searcher 468> silent?
<Rebecca 1109> we can start with soft sexy kisses..
<Searcher 468> 'nasty club'????
<Carisa 405> speak up
<Searcher 468> shall i play innocent here???!
<Rebecca 1109> the only way to be is totally nasty ... he he

<Carisa 405> yes slow wet ones, you start at her head and i'll start with her feet
<Carisa 405> Searcher you can just moan, and feel good
<Rebecca 1109> just close your eyes..feel my lips over yours ... you taste so good,
<Rebecca 1109> so wet..
<Carisa 405> well spread your legs and let me start there
<Carisa 405> well did you spread those legs for me?
<Searcher 468> you mean YOU don't know????
<Carisa 405> oh baby, let me lick that sweet wet box
<Rebecca 1109> maybe we should work on both sides at the same time ... feeling
<Rebecca 1109> tongues all over
<Carisa 405> i like where my tongue is, it tastes so sweet ... he he
<Rebecca 1109> let me taste some..please ...
[Some exchanges about my 'straight' status occur for a few minutes, then 'interactivity' resumes]

[...]

BBS realities, like everyday life, are woven in a tapestry of meaning. But of what sort? The ordinary sensory cues of touch, sight, sound, smell, and taste that we give meaning to in a taken-for-granted manner in everyday life are suddenly exposed as the narratives of story-tellers during online interactivity. The textually reproduced sexuality displays another arena for the playful and imaginative disregard for everyday life boundaries. According to Poster (1990: 117):

> For the first time individuals engage in telecommunications with other individuals, often on an enduring basis, without considerations that derive from the presence to the partner of their body, their voice, their sex, many of the markings of their personal history. Conversationalists are in the position of fiction writers who compose themselves as characters in the process of writing, inventing themselves from their feelings, their needs, their ideas, their desires, their social position, their political views, their economic circumstances, their family situation – their entire humanity.

[...]

Motion is (trans)formed as a cursor bounces across the monitor screen. Touch is (tran)scribed by (dis)embodied flying fingers on keyboards. Feeling the touch requires dematerializing the corporeality of 'the body' and relocating sexuality in a more evanescent (non)place. But, 'the body' is no more an objectively given reality than anything else. It is also an historical narrative and a social construct. 'The body' is bestowed with meaning (and different meanings for the male and female body), defined and redefined, by human beings through a history of interaction. Bodies are less objects and

> ... instead almost trace phenomena which are produced by the wheelings-about of great technologies and politics ... In a strong sense the body is a concept, and so hardly intelligible unless it is read in relation to whatever else supports it and surrounds it. (Riley 1988: 102)

Cybersexuality is mindful. It is inter-active. It is a communicated reality that (re)moves the physical world of things to somewhere other and encompasses 'a game of masks' ...

> The domain of freedom then retreats to the computer monitor and the invented identities that can be communicated through the modem ... Playfulness, spontaneity, imagination and desire all are absent or diminished from the public and private domains of career-building. Only the messagerie, with its fictional self-constitution and perfect anonymity, offers an apparent respite from what has become for many a treadmill of reason. (Poster, 1990: 120)

Paradoxically, in this new 'space' focus is upon the exchanges of signs representing the senses (essence?) of the other. In this universe of sign and simulation ...

> The real does not efface itself in favour of the imaginary; it effaces itself in favour of the more real than real: the hyperreal. The truer than true: this is simulation ... Sexuality does not fade into sublimation, repression and morality, but fades much more surely into the more sexual than sex: porn, the hyperreality contemporaneous with the hyperreal. (Baudrillard, 1990b: 11)

'K. J. Jeep' related something similar to this as her online experience:

> Each day brings a new surprise here for me. I have found romance and an invitation to open myself to experiences of both love and life that some will never find ... given the opportunity to be jestful, serious, or just stone cold nuts. Through the intensity of such a communication system I have gotten to know the essence of people ... the people here for the most part remain faceless. There is no other choice other than to leave the walls behind that people use for protection in a face to face, live, social situation. There is no need for them here. It is a fast pace Peyton Place also ... but, what I have found is a world of faceless, nameless strangers that (sic) connect on various levels.

[...]

Licentious dialogue weaves a warm cloak to wrap around the remote nights. Boundaries blur between fantasy and 'reality'. And, why not? 'Reality' warns that 'the body' is vulnerable. Sexual repression (AIDS and safe sex campaigns) and denial of pleasures (sins of smoking, eating meat, drinking coffee or alcohol, war on drugs campaign, on and on) – 'disavowing the (punished) body from an enjoyment' (Olalquiaga, 1992: 7) – run concurrent with the intensification of technological simulations and voyeurism that satiate the carnal appetite *vis-à-vis* imagination and sex without a body.

> ... pleasure is attained precisely where conventional reality and simulacra (reality and fantasy in pornography) become indistinguishable ... As the body disappears, it is imaginarily reconstructed from its leftover fragments and traces ... In the same contradictory manner that advertising promotes the reality/unreality of a referent, bodies have become the locus of a fierce battle between permanence and evanescence. (Olalquiaga, 1992: 6, 10)

Cyberspace and cybersexuality emerge and surge from the ambiguous and shifting ground of an everyday lifeworld and the indeterminacy of what is known and knowable. Simulacra and hyperreality brew in this space of NONPLACE. No other cues exist. The signs point to signs and fantasy explodes amid jostling confusions and lurking darkness beyond the glitter. But, many of the social actors behind the monitors make connections and move beyond the glitter.

Some BBSers wrestle to distinguish their fantasies and realities as they continue to articulate their text world or leap offline and meet face-to-face. Not all, but many, create strong alliances and develop reciprocal relationships that reconstitute the 'nonplace' and imbue it with a sense of community. And yet there are those who continue to find the mutual production of electronic pornography powerful, and even satisfying in a bounded way. A transformation of sexuality – its meanings and production – emphasizes the social and historical construct of 'the body' as pornographic imagery excites 'sex without a body' in an era of advanced technology and information implosion, heightened awareness of the diversity in sexual expression, and corporeal vulnerability.

[...]

References

Baudrillard, J. (1990a) *Seduction*, New York: St Martin's Press.

Baudrillard, J. (1990b) *Fatal Strategies*, New York: Semiotext(e).

Manning, P. K. (1991) 'Strands in the Postmodern Rope: Ethnographic Themes', pp. 3–27 in N. Denzin (ed.), *Studies in Symbolic Interaction*, 12, Greenwich, Conn: JAI Press.

Olalquiaga, Celeste (1992) *Megalopolis: Contemporary Cultural Sensibilities*, Minneapolis: University of Minnesota Press.

Poster, Mark (1990) *The Mode of Information: Poststructuralism and Social Context*, Chicago: Chicago University Press.

Riley, Denise (1988) *'Am I That Name?' Feminism and the Category of 'Women' in History*, Minneapolis, MN: University of Minnesota Press.

Ruben, Brent D. (1985) 'The Coming of the Information Age: Information, Technology, and the Study of Behavior', pp. 3–26 in B. D. Ruben (ed.), *Information and Behavior*, 1. New Brunswick: Transaction Books

2.8

THE HOT ROD BODIES OF CYBERSEX

Sue-Ellen Case

> In the chat room for new members on America Online, the conversation soon turned to the password for a certain porn bulletin board. Upon receipt of the password, half the people exited. I followed them, to find myself in a world of sexist remarks and the kind of talk that reminded me of dirty phone calls. I remembered reading about the first 'rape' in cyberspace, where a woman was accosted by a violent, sadistic scenario someone sent to her before asking her permission. I looked up at the menu, where I saw the button that would censor incoming data, in case I had children in my house with access to the computer.

While Sappho's body is being reconstituted on a bulletin board, and spatial/address encryption may be suggested indirectly in feminist/lesbian theories and modernist writings, the hot rod cybersex body is already in full production. A multitude of porn bulletin boards already exist. One called 'Chicago's Windy City Freedom Fortress' serves up pictorial displays of 'everything from a woman going down on a dog, to up close and personal action potty shots to fisting the next door neighbor'. 'Hot chatting', online flirtation, and sex cover more territory. Suddenly the bulletin boards begin to reveal the profiles of those who run them and those who subscribe. Some few are run by women, and many maintain an interesting, even high quality of 'chat'. For example, the book *The Joy of Cybersex* identifies Stacy Horn, the woman who runs ECHO, one of the large, arty boards, as 'a stylish avenger to the prototypical computer nerd, [who] commands ECHO with grace, and a distinct and welcoming

---

From: S-E. Case, *The Domain Matrix*, Bloomington: Indiana University Press, 1996.

literary bent.' Horn refers to her own board as an 'electronic salon'. Women are in the minority, however, online. According to *Boardwatch Magazine*, only 10 per cent of bulletin board callers are female.

Online activities often follow those of the first computer games. The games were composed of two genres: violence and sex. *The Joy of Cybersex* notes that the first 'dirty' computer game was *Softporn Adventure*, published in 1980 by Chuck Benton. Fifty thousand copies were sold at the time, when there were fewer than 400 000 Apple computers. A new version of the game is now available on Compuserve as donationware. In 1983, Leather Goddesses of Phobos appeared, which had three 'filth levels' and in which gender-switching was possible, but the book identifies 1984 as the turning point. In Leisure Suit Larry in the Land of Lounge Lizards, animated graphics, stories, a central character, and graphic anatomy are portrayed. The success of this game, reviewed positively in major newspapers and presses, encouraged the growth of the industry. Then came Macplaymate – eventually with a button that allowed the player to return to text immediately, in order to play at the work site and avoid surveillance. The electronic bunny, the Macplaymate, is available for the insertion of dildo icons, moved by the mouse into her orifice, exacting moans and groans. Penetration fantasies reached new depths with the release in CD-ROM of Virtual Valerie, in 1990. In full-color graphics, the player can enter her apartment, play her CDs, and enter her with a full regalia of sex toys. Macfoxes offers Misty, the cheerleader: 'Across the top of the screen is a selection of tools: a vibrator, a cucumber, an inanimate object, which the program calls a "dick", a telescoping dildo', and you get a score for how hot you make her on the meter.

Unfortunately, sex games do not seem to be produced by or for women. Nor do women seem to buy them. Larry Miller of Interotica stresses that 'regardless of the content, it's men who buy it [the software]. Men are more voyeuristic. . . . Generally, nothing is being done by and for women'. So cybersex reiterates the old sexist structures of the passive female and the active male, who literally 'scores' when he penetrates her. The Adult Reference Library on CD-ROM offers these possibilities: 'Do you like young love? You'll find it here. Do you like . . . Your women pregnant, or lactating? There are Oriental women and women who aren't women at all. . . . You'll see both-sex folks, same-sex folks disporting, contortionists contorting.' The 'you' emerges as pretty gender-specific, appealing to the traditional mix of racism and sexism.

In 1993, at the Comdex Fall Trade Show in Las Vegas, where leading hardware and software producers gather, CD-ROM sales hit new heights. The mobbed booths were selling porn. CD-ROM, as it turns out, is the best way to ship or carry porn into countries or states with strict laws, for the mirror-like disks offer no clue as to their contents and can be carried in an audio CD case. So what difference does it make? In one of the fastest-growing industries in the world, owned, operated, and produced primarily by men, the center of entertainment is in sexist porn. And the future for women online?

Certainly, the chat rooms on America Online are humming with lesbian encounters. The lesbian or queer dyke sex-radical project of the last decade or so has been to write sex. An interactive sex-writing site promises the pleasures of fantasy scenarios, roles, and appearances. Online sex seems to provide much of the pleasure that the sex-radical movement seeks. Group sex takes on a whole new dimension. A friend of mine reported having online sex with someone who was in her office in Hong Kong, before the work day began, and at the same time with another who was enjoying a break in Australia. Writing sex can happen anywhere the computer can be set up: the workplace, the private office at home, the jacked-in laptop in the lonely hotel room, etc. Sex may turn into relationships. One colleague has flown to several distant cities to meet her online sex partners in the flesh. The first divorce case has been filed against a wife who was having sexual relations online. The status of the virtual is tested by online sex. The regime of the flesh is tested as well. Writing sex finds a lively format.

[. . .]

SECTION 3
BODIES IN SCIENCE AND BIOMEDICINE

INTRODUCTION

The discourses of science and medicine are a powerful influence – some would say *the* most powerful influence – on constructions of the female body, and on what it is to be a woman. Within the fields of biology and biomedicine, the characteristic notion has been of the biological body as a universal, stable entity, outside of history, culture, geography and language, and of a belief in the givenness of the biological foundations of femininity. At the same time as claiming a neutral status, those perspectives are nevertheless culturally and ideologically situated, such that the body in biomedicine has long been a contested field. The following articles offer a range of positions on how feminism has engaged with such discourses around the (sexed) body, and they reflect upon key themes within feminist responses to post-Enlightenment scientific thought.

Early second wave feminists, such as Adrienne Rich (1976), Mary Daly (1979) and Barbara Ehrenreich and Deirdre English (1979), focus on the oftimes negative manifestations of conventional beliefs, as evidenced, for example, in the devaluation of women through the practices of biomedicine. They document the violence exercised against women, particularly in relation to their reproductive potential, and address the desire of women to take control of their own bodies. New knowledges of the body and of health and illness drawing powerfully upon women's own experiences, are developed in such approaches which chart how medicine has constructed the female body – as deficiency, as uncontrolled, as inherently diseased – in short, as failing to meet the norms of the male (white/middle-class/fit) body. *Our Bodies, Ourselves* (1976), produced by the Boston Women's Health Collective, was one of the first in a wide range of writing through which women have developed new ways of understanding their bodies, and of challenging the apparent hegemony of the masculinist medical profession. Whilst the selection here does not address directly the widely varying campaigns, self-help groups and writings of the women's health movement, they do explore the developments in feminist theories of the body that have worked alongside, fed into and drawn upon this political activism. The extracts from Audre Lorde and Eve Kosofsky Sedgwick, whilst drawing upon differing theoretical understandings, both offer examples of how feminists have used theories of the body to live through, and with, experiences of illness.

Concurrent with the growth of the women's health movement, feminist inspired studies of the body in science and biomedicine began to historicise the taken for granted categories of female/male; nature/culture; body/mind. The apparent immutability of such binary categories is displaced, in part through tracing their emergence within the context of European Enlightenment thought and by revealing the ways in which female corporeality and subjectivity were constituted through the practices of a science marked not only by its espoused values of objectivity and rationality, but also by androcentrism, racism, classism, and misogyny. Writers such as Ehrenreich and English (1979), Smith-Rosenberg (1972), Schiebinger (Section 1) and Jordanova (included here) have shown how the categories of race, gender and class have been central to the development of the practices and discourses of biomedicine, and through these, to specific contextualisations of scientific and biological bodies as always already marked.

The relationship of feminists to bioscience has been an ambivalent one. It reflects on the one hand the desire for a way of understanding the lived experiences of women, embodied as they are in flesh and blood and bones, and it strives to see in science a source of good in the world. Yet on the other hand, there exists a distrust of a set of ideas and practices that have seemed to function both historically and contemporaneously to devalue, damage and exclude women as figures of dense, unspeaking, gross corporeality. A number of different moves have evolved, many drawing on other contemporary radical critiques of science, but bringing to them an additional sensitivity to feminist concerns (see Rose 1994). That response has been given particular urgency in the 1970s and 80s by the growth of sociobiology, with all its racist and sexist implications, and its reinvigorated belief in biological determinism, the view that biology is destiny. Writing against such material, Anne Fausto-Sterling offers a way of rereading masculine science through her reinterpretation of the 'facts' of menopause.

One of the central themes of feminist activism within health has been the call for women to 'take control of our bodies'. Whilst serving as a powerful rallying point, it promotes, as Wendy Hadd points out, 'the notion of being able to control one's body . . . within a discourse which accepts as given the concept of mind/body dualism' (1991: 165). The body is positioned as an object apart, something upon which the forces of biomedicine act, and of which women must struggle to regain control. In contrast, the moves towards what can be broadly termed postmodernist critiques of science and biomedicine reject the possibility of universal truths and argue instead for localised and specifically contextual feminist knowledges in which bodies are understood as both the objects and effects of biomedical discourse. Central to these critiques is a view of the body not as something given, waiting to have its mysteries revealed by the gaze of science (however feminist in its practice), but as something that can only be accessed through language. In other words, our understanding of the body's form and structure, its very anatomy and physiology, are shaped by the *discourses* of science and biomedicine. Recent feminist theorists thus offer new readings of the body that allow us to see beyond Enlightenment claims to reveal the truth of the natural world, and of bodies in their gross materiality, to understand instead the body as representation, medicine as a political practice, and disease as a language (Treichler 1989). Emily Martin offers a cogent example of this in her discussion of the metaphors used in scientific descriptions of the process of conception.

Medicine, then, is re-evaluated as a discursive force that has shaped and constituted bodies as differentially masculine and feminine, and health is viewed not as an absolute, but as a set of norms held in place by the regulatory practices of both health care professionals and their clients. The evolution of such mechanisms, cogently analysed by Foucault (1973, 1979), marks biomedicine as a discourse that serves to constitute the self by maintaining the integrity of bodily boundaries. In focusing on an area of ongoing feminist concern, Jana Sawicki's piece on the new reproductive technologies offers a feminist take on a Foucauldian reading of the disciplinary practices that operate at the level both of the population and the individual.

The enduring power and authority of biomedicine to reproduce bodies as predictable collections of matter, fixed and held in place by ever more detailed empirical scientific analyses, is challenged by the insight that such grasping for certainty itself produces new bodies that function, as Donna Haraway outlines, in ever more dynamic, elusive and uncertain ways. New influences on biomedicine, in the shape of genetic engineering, and biofeedback technology for example, have introduced ways of thinking the body which disrupt the idea of a holistic, discrete body and draw on notions of loop systems, responsiveness to the environment, adaptation to change: in short, a cybernetics which suggests that the body functions not with corporeal integrity, but in terms of boundary constraints and rates of flow. Haraway's contribution both addresses these new theorisations of the body, and explores their political implications. While the take on postmodern bioscience engages hitherto unexplored areas, it continues, nonetheless, to ask questions of recognisably feminist concern.

REFERENCES AND FURTHER READING

Arditti, Rita, Klein, Renate and Minden, Shelley (eds) (1984) *Test-Tube Women. What Future for Motherhood*, London: Pandora Press.

Balsamo, Anne (1996) 'Public Pregnancies and Cultural Narratives of Surveillance' in *Technologies of the Gendered Body. Reading Cyborg Women*, Durham: Duke University Press.

Boston Women's Health Collective (1976) *Our Bodies, Ourselves*, Harmondsworth: Penguin.

Braidotti, Rosi (1989) 'Organs without Bodies' in *differences* 1 (1): 147–61.

Braidotti, Rosi (1994) 'Body-images and the Pornography of Representation' in Kathleen Lennon and Margaret Whitford (eds), *Knowing the Difference. Feminist Perspectives in Epistomology*, London: Routledge.

Chhachhi, Sheba (1998) 'Raktpushp (Blood Flower)' in Margrit Shildrick and Janet Price (eds), *Vital Signs. Feminist Reconfiguration of the Bio/logical Body*, Edinburgh: Edinburgh University Press.

Daly, Mary (1979) *Gyn/Ecology. The Metaethics of Radical Feminism*, London: The Women's Press.

Ehrenreich, Barbara and English, Deirdre (1979) *For Her Own Good. 150 Years of the Experts' Advice to Women*, London: Pluto Press.

Foucault, Michel (1973) *The Birth of the Clinic: An Archaeology of Medical Perception*, trans. A. M. Sheridan, London: Tavistock Publications

Foucault, Michel (1979) *History of Sexuality*, Volume 1: An Introduction, trans. Robert Hurley, London: Allen Lane.

Hadd, Wendy (1991) 'A Womb with a View: Women as Mothers and the Discourse of the Body' in *Berkeley Journal of Sociology* 36: 165–75.

Jacobus, Mary, Fox-Keller, Evelyn and Shuttleworth, Sally (eds) (1989) *Body/Politics: Women and the Discourses of Science*, New York: Routledge.

Lorde, Audre (1980) *The Cancer Journals*, San Fransisco: Spinsters Ink.

Lupton, Deborah (1994) 'The Body in Medicine' in *Medicine as Culture. Illness, Disease and the Body in Western Society*, London: Sage Publications.

Martin, Emily (1987) *The Woman In the Body: A Cultural Analysis of Reproduction*, Boston: Beacon Press.

Petchesky, Rosalind P. (1987) 'Foetal Images: The Power of Visual Culture in the Politics of Reproduction' in Michelle Stanworth (ed.), *Reproductive Technologies: Gender, Motherhood and Medicine*, Cambridge: Polity Press

Rich, Adrienne (1976) *Of Woman Born. Motherhood as Experience and Institution*, New York: W. W. Norton.

Rose, Hilary (1994) *Love, Power and Knowledge. Towards a Feminist Transformation of the Sciences*, Cambridge: Polity Press.

Smith-Rosenberg, Carroll (1972) 'The Hysterical Woman: Sex Roles and Role Conflict in Nineteenth Century America' in *Social Research* 39: 652–78.

Shildrick, Margrit (1997) 'Fabrica(tions): On the Construction of the Human Body' in *Leaky Bodies and Boundaries. Feminism, Postmodernism and (Bio)ethics*, London: Routledge.

Singer, Linda (1993) 'Regulating Women in the Age of Sexual Epidemic' in *Erotic Welfare. Sexual Theory and Poitics in the Age of Epidemic*, New York: Routledge.

Smyth, Ailbhe (1997) 'Taking Pain. A Particular History' in *f/m* 2: 2–9.

Stacey, Jackie (1997) 'Bodies' in *Teratologies. A Cultural Study of Cancer*, London: Routledge.

Treichler, Paula (1989) 'Aids, Homophobia and Biomedical Discourse. An Epidemic of Signification' in Douglas Crimp (ed.), *AIDS: Cultural Analysis/Cultural Activism*, Cambridge, MA: MIT Press.

Tuana, Nancy (1993) 'The Hysteria of Women' in *The Less Noble Sex*, Bloomington: Indiana University Press.

3.1

A BURST OF LIGHT: LIVING WITH CANCER

Audre Lorde

November 8, 1986
New York City

If I am to put this all down in a way that is useful, I should start with the beginning of the story.

Sizable tumor in the right lobe of the liver, the doctors said. Lots of blood vessels in it means it's most likely malignant. Let's cut you open right now and see what we can do about it. Wait a minute, I said. I need to feel this thing out and see what's going on inside myself first, I said, needing some time to absorb the shock, time to assay the situation and not act out of panic. Not one of them said, I can respect that, but don't take too long about it.

Instead, that simple claim to my body's own processes elicited such an attack response from a reputable Specialist In Liver Tumors that my deepest – if not necessarily most useful – suspicions were totally aroused.

What that doctor could have said to me that I would have heard was, 'You have a serious condition going on in your body and whatever you do about it you must not ignore it or delay deciding how you are going to deal with it because it will not go away no matter what you think it is.' Acknowledging my responsibility for my own body. Instead, what he said to me was, 'If you do not do exactly what I tell you to do right now without questions you are going to die a horrible death.' In exactly those words.

I felt the battle lines being drawn up within my own body.

From: A. Lorde, *A Burst of Light*, New York: Firebrand Books, 1988.

I saw this specialist in liver tumors at a leading cancer hospital in New York City, where I had been referred as an outpatient by my own doctor.

The first people who interviewed me in white coats from behind a computer were only interested in my health-care benefits and proposed method of payment. Those crucial facts determined what kind of plastic ID card I would be given, and without a plastic ID card, no one at all was allowed upstairs to see any doctor, as I was told by the uniformed, pistoled guards at all the stairwells.

From the moment I was ushered into the doctor's office and he saw my x-rays, he proceeded to infantalize me with an obviously well-practiced technique. When I told him I was having second thoughts about a liver biopsy, he glanced at my chart. Racism and Sexism joined hands across his table as he saw I taught at a university. 'Well, you look like an *intelligent girl*', he said, staring at my one breast all the time he was speaking. 'Not to have this biopsy immediately is like sticking your head in the sand.' Then he went on to say that he would not be responsible when I wound up one day screaming in agony in the corner of his office!

I asked this specialist in liver tumors about the dangers of a liver biopsy spreading an existing malignancy, or even encouraging it in a borderline tumor. He dismissed my concerns with a wave of his hand, saying, instead of answering, that I really did not have any other sensible choice.

I would like to think that this doctor was sincerely motivated by a desire for me to seek what he truly believed to be the only remedy for my sickening body, but my faith in that scenario is considerably diminished by his $250 consultation fee and his subsequent medical report to my own doctor containing numerous supposedly clinical observations of *obese abdomen* and *remaining pendulous breast.*

In any event, I can thank him for the fierce shard lancing through my terror that shrieked there must be some other way, this doesn't feel right to me. If this is cancer and they cut me open to find out, what is stopping that intrusive action from spreading the cancer, or turning a questionable mass into an active malignancy? All I was asking for was the reassurance of a realistic answer to my real questions, and that was not forthcoming. I made up my mind that if I was going to die in agony on somebody's office floor, it certainly wasn't going to be his! I needed information, and pored over books on the liver in Barnes & Noble's Medical Textbook Section on Fifth Avenue for hours. I learned, among other things, that the liver is the largest, most complex, and most generous organ in the human body. But that did not help me very much.

In this period of physical weakness and psychic turmoil, I found myself going through an intricate inventory of rage. First of all at my breast surgeon – had he perhaps done something wrong? How could such a small breast tumor have metastasized? Hadn't he assured me he'd gotten it all, and what was this now anyway about micro-metastases? Could this tumour in my liver have been seeded at the same time as my breast cancer? There were so many unanswered questions, and too much that I just did not understand.

But my worst rage was the rage at myself. For a brief time I felt like a total failure. What had I been busting my ass doing these past six years if it wasn't living and loving and working to my utmost potential? And wasn't that all a guarantee supposed to keep exactly this kind of thing from ever happening again? So what had I done wrong and what was I going to have to pay for it and WHY ME?

But finally a little voice inside me said sharply, 'Now really, is there any other way you would have preferred living the past six years that would have been more satisfying? And be that as it may, *should* or *shouldn't* isn't even the question. How do you want to live the rest of your life from now on and what are you going to do about it?' Time's awasting!

Gradually, in those hours in the stacks of Barnes & Noble, I felt myself shifting into another gear. My resolve strengthened as my panic lessened. Deep breathing, regularly. I'm not going to let them cut into my body again until I'm convinced there is no other alternative. And this time, the burden of proof rests with the doctors because their record of success with liver cancer is not so good that it would make me jump at a surgical solution. And scare tactics are not going to work. I have been scared now for six years and that hasn't stopped me. I've given myself plenty of practice in doing whatever I need to do, scared or not, so scare tactics are just not going to work. Or I hoped they were not going to work. At any rate, thank the goddess, they were not working yet. One step at a time.

But some of my nightmares were pure hell, and I started having trouble sleeping.

In writing this I have discovered how important some things are that I thought were unimportant. I discovered this by the high price they exact for scrutiny. At first I did not want to look again at how I slowly came to terms with my own mortality on a level deeper than before, nor with the inevitable strength that gave me as I started to get on with my life in actual time. Medical textbooks on the liver were fine, but there were appointments to be kept, and bills to pay, and decisions about my upcoming trip to Europe to be made. And what do I say to my children? Honesty has always been the bottom line between us, but did I really need them going through this with me during their final difficult years at college? On the other hand, how could I shut them out of this most important decision of my life?

I made a visit to my breast surgeon, a doctor with whom I have always been able to talk frankly, and it was from him that I got my first trustworthy and objective sense of timing. It was from him that I learned that the conventional forms of treatment for liver metastases made little more than one year's difference in the survival rate. I heard my old friend Clem's voice coming back to me through the dimness of thirty years: 'I see you coming here trying to make sense where there is no sense. Try just living in it. Respond, alter, see what happens.' I thought of the African way of perceiving life, as experience to be lived rather than as problem to be solved.

Homeopathic medicine calls cancer the cold disease. I understand that down to my bones that quake sometimes in their need for heat, for the sun, even for just a hot bath. Part of the way in which I am saving my own life is to refuse to submit my body to cold whenever possible.

In general, I fight hard to keep my treatment scene together in some coherent and serviceable way, integrated into my daily living and absolute. Forgetting is no excuse. It's as simple as one missed shot could make the difference between a quiescent malignancy and one that is growing again. This not only keeps me in an intimate, positive relationship to my own health, but it also underlines the fact that I have the responsibility for attending my own health. I cannot simply hand over that responsibility to anybody else.

Which does not mean I give in to the belief, arrogant or naive, that I know everything I need to know in order to make informed decisions about my body. But attending my own health, gaining enough information to help me understand and participate in the decisions made about my body by people who know more medicine than I do, are all crucial strategies in my battle for living. They also provide me with important prototypes for doing battle in all other arenas of my life.

Battling racism and battling heterosexism and battling apartheid share the same urgency inside me as battling cancer. None of these struggles are ever easy, and even the smallest victory is never to be taken for granted. Each victory must be applauded, because it is so easy not to battle at all, to just accept and call that acceptance inevitable.

And all power is relative. Recognizing the existence as well as the limitations of my own power, and accepting the responsibility for using it in my own behalf, involve me in direct and daily actions that preclude denial as a possible refuge. Simone de Beauvoir's words echo in my head: 'It is in the recognition of the genuine conditions of our lives that we gain the strength to act and our motivation for change.'

3.2

BREAST CANCER: AN ADVENTURE IN APPLIED DECONSTRUCTION

Eve Kosofsky Sedgwick

PROJECT 3

This project involves thinking and writing about something that's actually structured a lot of my daily life over the past year. Early in 1991 I was diagnosed, quite unexpectedly, with a breast cancer that had already spread to my lymph system, and the experiences of diagnosis, surgery, chemotherapy, and so forth, while draining and scary, have also proven just sheerly *interesting* with respect to exactly the issues of gender, sexuality, and identity formation that were already on my docket. (Forget the literal-mindedness of mastectomy, chemically induced menopause, etc.: I would warmly encourage anyone interested in the social construction of gender to find some way of spending half a year or so as a totally bald woman.) As a general principle, I don't like the idea of 'applying' theoretical models to particular situations or texts – it's always more interesting when the pressure of application goes in both directions – but all the same it's hard not to think of this continuing experience as, among other things, an adventure in applied deconstruction.[1] How could I have arrived at a more efficient demonstration of the instability of the supposed oppositions that structure an experience of the 'self'? – the part and the whole (when cancer so dramatically corrodes that distinction); safety and danger (when fewer than half of the women diagnosed with breast cancer display any of the statistically defined 'risk factors' for the disease); fear and hope (when I feel – I've got a quarterly physical coming up – so much less

From: E. Kosofsky Sedgwick, *Tendencies*, Durham: Duke University Press, 1994.

prepared to deal with the news that a lump or rash *isn't* a metastasis than that it is); past and future (when a person anticipating the possibility of death, and the people who care for her, occupy temporalities that more and more radically diverge); thought and act (the words in my head are aswirl with fatalism, but at the gym I'm striding treadmills and lifting weights); or the natural and the technological (what with the exoskeleton of the bone-scan machine, the uncanny appendage of the IV drip, the bionic implant of the Port-a-cath, all in the service of imaging and recovering my 'natural' healthy body in the face of its spontaneous and endogenous threat against itself). Problematics of undecidability present themselves in a new, unfacile way with a disease whose very *best* outcome – since breast cancer doesn't respect the five-year statute of limitations that constitutes cure for some other cancers – will be decades and decades of free-fall interpretive panic.

Part of what I want to see, though, is what's to be learned from turning this experience of dealing with cancer, in all its (and my) marked historical specificity, and with all the uncircumscribableness of the turbulence and threat involved, back toward a confrontation with the theoretical models that have helped me make sense of the world so far. The phenomenology of life-threatening illness; the performativity of a life threatened, relatively early on, by illness; the recent crystallization of a politics explicitly oriented around grave illness: exploring these connections *has* (at least for me it has) to mean hurling my energies outward to inhabit the very farthest of the loose ends where representation, identity, gender, sexuality, and the body can't be made to line up neatly together.

It's probably not surprising that gender is so strongly, so multiply valenced in the experience of breast cancer today. Received wisdom has it that being a breast cancer patient, even while it is supposed to pose unique challenges to one's sense of 'femininity', nonetheless plunges one into an experience of almost archetypal Femaleness. Judith Frank is the friend whom I like to think of as Betty Ford to my Happy Rockefeller – the friend, that is, whose decision to be public about her own breast cancer diagnosis impelled me to the doctor with my worrisome lump; she and her lover, Sasha Torres, are only two of many women who have made this experience survivable for me: compañeras, friends, advisors, visitors, students, lovers, correspondents, relatives, caregivers (these being anything but discrete categories). Some of these are indeed people I have come to love in feminist- and/or lesbian-defined contexts; beyond that, a lot of the knowledge and skills that keep making these women's support so beautifully apropos derive from distinctive feminist, lesbian, and women's histories. (I'd single out, in this connection, the contributions of the women's health movement of the 70s – its trenchant analyses, its grass-roots and antiracist politics, its publications,[2] the attitudes and institutions it built and some or the careers it seems to have inspired.)

At the same time, though, another kind of identification was plaited inextricably across this one – not just for me, but for others of the women I have been close to as well. Probably my own most formative influence from a quite early age has been a viscerally intense, highly speculative (not to say inventive) cross-identification with gay men and gay male cultures as I inferred, imagined, and later came to know them. It wouldn't have required quite so overdetermined a trajectory, though, for almost any forty year old facing a protracted, life-threatening illness in 1991 to realize that the people with whom she had perhaps most in common, and from whom she might well have most to learn, are people living with AIDS, AIDS activists, and others whose lives had been profoundly reorganized by AIDS in the course of the 1980s.

As, indeed, had been my own life and those of most of the people closest to me. 'Why me?' is the cri de coeur that is popularly supposed to represent Everywoman's deepest response to a breast cancer diagnosis – so much so that not only does a popular book on the subject have that title, but the national breast cancer information and support hotline is called Y-ME! Yet 'Why me?' was not something it could have occurred to me to ask in a world where so many companions of my own age were already dealing with fear, debilitation, and death. I wonder, too, whether it characterizes the responses of the urban women of color forced by violence, by drugs, by state indifference or hostility, by AIDS and other illnesses, into familiarity with the rhythms of early death. At the time of my diagnosis the most immediate things that were going on in my life were, first, that I was coteaching (with Michael Moon) a graduate course in queer theory, including such AIDS-related material as Cindy Patton's stunning *Inventing AIDS*. Second, that we and many of the students in the class, students who indeed provided the preponderance of the group's leadership and energy at that time, were intensely wrapped up in the work (demonstrating, organizing, lobbying) of a very new local chapter of the AIDS activist organization ACT UP. And third, that at the distance of far too many miles I was struggling to communicate some comfort or vitality to a beloved friend, Michael Lynch, a pioneer in gay studies and AIDS activism, who seemed to be within days of death from an AIDS-related infection in Toronto.

'White Glasses', the final essay in *Tendencies*, tells more about what it was like to be intimate with this particular friend at this particular time. More generally though, the framework in which I largely experienced my diagnosis – and the framework in which my friends, students, house sharers, life companion, and others made available to me almost over whelming supplies of emotional, logistical, and cognitive sustenance[3] – was very much shaped by AIDS and the critical politics surrounding it, including the politics of homophobia and of queer assertiveness. The AIDS activist movement, in turn, owes much to the women's health movement of the 70s; and in another turn, an activist politics of breast cancer, spearheaded by lesbians, seems in the last year or two to have been emerging based on the model of AIDS activism.[4] The dialectical epistemology of the two diseases, too – the kinds of secret each has constituted; the kinds of *out*ness each has required and inspired – has made an intimate motive for me. As 'White Glasses' says,

It's as though there were transformative political work to be done just by being available to be identified with in the very grain of one's illness (which is to say, the grain of one's own intellectual, emotional, bodily self as refracted through illness and as resistant to it) – being available for identification to friends, but as well to people who don't love one; even to people who may not like one at all nor even wish one well.

NOTES

1. That deconstruction can offer crucial resources of thought for survival under duress will sound astonishing, I know, to anyone who knows it mostly from the journalism on the subject – journalism that always depicts 'deconstructionism', not as a group of usable intellectual tools, but as a set of beliefs involving a patently absurd dogma ('nothing really exists'), loopy as Christian Science but as exotically aggressive as (American journalism would also have us find) Islam. I came to my encounter with breast cancer not as a member of a credal sect of 'deconstructionists' but as someone who needed all the cognitive skills she could get. I found, as often before, that I had some good and relevant ones from my deconstructive training.
2. The work of this movement is most available today through books like the Boston Women's Health Book Collective's *The New Our Bodies, Ourselves: Updated and Expanded for the Nineties* (New York: Simon and Schuster, 1992). An immensely important account of dealing with breast cancer in the context of feminist, antiracist, and lesbian activism is Audre Lorde, *The Cancer Journals*, 2d ed. (San Francisco: Spinsters Ink, 1988) and *A Burst of Light* (Ithaca, NY: Firebrand Books, 1988).
3. And physical: I can't resist mentioning the infallibly appetite-provoking meals that Jonathan Goldberg, on sabbatical in Durham, planned and cooked every night during many queasy months of my chemotherapy.
4. On this, see Alisa Solomon, 'The politics of Breast Cancer', *Village Voice* 14 May 1991, pp. 22–27; Judy Brandy, ed., *4 in 3: Women with Cancer Confront an Epidemic* (Pittsburgh and San Francisco: Cleis Press, 1991); Midge Stocker, ed., *Cancer as a Women's Issue: Scratching the Surface* (Chicago: Third Side Press, 1991); and Sandra Butler and Barbara Rosenblum, *Cancer in Two Voices* (San Francisco: Spinster, 1991.).

3.3

NATURAL FACTS: A HISTORICAL PERSPECTIVE ON SCIENCE AND SEXUALITY

Ludmilla Jordanova

Introduction

The distinction between women as natural and men as cultural appeals to a set of ideas about the biological foundations of womanhood. Understanding the historical dimensions of these two inter-related pairs of dichotomies in European thought entails revealing the connections between science and sexuality. Sex roles were constituted in a scientific and medical language, and, conversely, the natural sciences and medicine were suffused with sexual imagery. This paper explores the links between nature/culture, woman/man through a historical study of the biomedical sciences and the metaphors and symbols they employed. I draw my examples principally from eighteenth- and nineteenth-century France and Britain.

Since the eighteenth century the polarities seem to have hardened, yet the lived experience to which they supposedly relate was extremely complex. Recent feminist history has shown the diversity of women's social and occupational roles despite the inflexibility of contemporary ideas about them.[1] The lack of fit between ideas and experience clearly points to the ideological function of the nature/culture dichotomy as applied to gender. This ideological message was increasingly conveyed in the language of medicine.

[...]

The oppositions between women as nature and men as culture were pressed concretely through distinctions commonly made, such as that between women's

From: C. MacCormack and M. Strathern (eds), *Nature, Culture and Gender*, Cambridge: Cambridge University Press, 1980. Ludmilla Jordanova's most recent book is *Nature Displayed. Gender, Science and Medicine 1760–1820*, Longman, 1999 work and men's work. The ideological dimension to these oppositions is discernable in the dichotomy constructed by the élite of the medical profession between male strength and female vulnerability. Social and conceptual changes take place slowly and in piecemeal and fragmented ways. Our project is not to search for neat consistent ideological structures, but through the contradictions, tensions and paradoxes to find patterns we can understand.

There are strong reasons for beginning with the Enlightenment. In this period the shifts in meaning and usage of words such as culture, civil, civilize, nature and life, provide indicators of deep changes in the way human society and its relations with the natural world were conceived. Ultimately, the Enlightenment is no easier to define than notions of nature and culture are, but, in the term itself, we can see an appeal to light as a symbol of a certain form of knowledge which had the potential for improving human existence. Rational knowledge based on empirical information derived from the senses was deemed the best foundation for secure knowledge. Starting with a sensualist epistemology, and a number of assumptions about the potential social application of an understanding of natural laws, many Enlightenment writers critically examined forms of social organization. In so doing, they employed a language fraught with sexual metaphor, and systematically examined the natural facts of sexuality.

Science and medicine were fundamental to this endeavour in three different ways. First, natural philosophers and medical writers addressed themselves to phenomena in the natural world, such as reproduction and generation, sexual behaviour, and sex-related diseases. Second, science and medicine held a privileged position because their methods appeared to be the only ones that would lead away from religious orthodoxy and towards a secular, empirically based knowledge of the natural and social worlds. Finally, as I hope to show, science and medicine as activities were associated with sexual metaphors, which were clearly expressed in designating nature as a woman to be unveiled, unclothed and penetrated by masculine science. The relationship between women and nature, and men and culture must therefore be examined through the mediations of science and medicine.

ENLIGHTENED ENVIRONMENTALISM

In the self-conscious scientism of the Enlightenment, the capacity of the human mind to delve into the secrets of nature was celebrated. Increasingly this capacity for scientific prowess was conceptualized as a male gift, just as nature was the fertile woman, and sometimes the archetypal mother (Kolodny 1975). People had explored their capacity to master and manipulate nature for many centuries (Glacken 1967), but the powerful analytical tools of the natural sciences and the techniques of engineering and technology enormously enhanced their confidence that human power over the environment was boundless. As Bacon expressed it in the early seventeenth century, 'My only earthly wish is ... to stretch the deplorably narrow limits of man's dominion over the universe to their promised bounds' (Farrington 1964: 62). And the process by which Bacon thought this would be achieved was a casting off of 'the darkness of antiquity' in favour of the detailed study of nature (Farrington 1964: 69). 'I am come in very truth leading to you Nature with all her children to bind her to your service and make her your slave' (Farrington 1964: 62).

In discussions of human domination over nature, the concept of environment comes to hold an important, and complex, place from the late eighteenth century onwards (Jordanova 1979). Above all, the environment was that cluster of variables which acted upon organisms and was responsible for many of their characteristics. An understanding of human beings in sickness and in health was to be based on a large number of powerful environmental factors: climate, diet, housing, work, family situation, geography and atmosphere. This notion of environment could be split into two. First, there were variables such as custom and government which were human creations and were, at least in principle, amenable to change. Second, there were parameters such as climate, meteorology in general, and geographical features such as rivers and mountains, which were in the province of immutable natural laws and proved more challenging to human power. In the first case environment denoted culture, in the second, nature.

Taking environment in the sense of culture, it was clear to people at the end of the eighteenth century that living things and their environment were continually interacting and changing each other in the process. This was also true of sexuality, for, although sex roles were seen as being in some sense 'in nature' because of their relationship to physical characteristics, it was also acknowledged that they were mutable, just as physiology and anatomy in general were taken to be. The customs and habits of day-to-day life such as diet, exercise and occupation, and more general social forces, such as modes of government, were taken to have profound effects on all aspects of people's lives; their sexuality was no exception. The foundation for these beliefs was a complex conceptual framework that spoke naturalistically about the physiological, mental and social aspects of human beings. An understanding of this framework is therefore an essential background for any account of the relations between nature, culture and gender in the period.

In the bio-medical sciences of the late eighteenth century, mind and body were not seen as incommensurable, absolutely distinct categories. Mental events, such as anger, fear or grief, were known to have physical effects, while illnesses such as fevers produced emotional and intellectual changes. I would argue that at the end of the eighteenth century a model of health and illness became dominant in which lifestyle and social roles were closely related to health. This model was applied to both men and women, but with different implications. A tight linkage was assumed between jobs performed in the social arena (for women, the production, suckling and care of children, the creation of a natural morality through family life) and health and disease. Women thus became a distinct class of persons, not by virtue of their reproductive organs, but through their social lives. The total physiology of women could, it was argued, only be understood in terms of lifestyle and the social roles they ought to fulfil, if they were not doing so already.

I want to stress that the model applied to both sexes. For example, people who lived in certain climates, such as men who worked in mines or factories, were known to be susceptible to particular diseases. Physicians therefore advocated that they take precautions to preserve their health: appropriate diet, exercise, housing, clothing, behaviour, regimen. The same argument applied to women, and in fact each way of life held its own particular dangers for the health of men and women, which could be held at bay by the appropriate preventive measures. In the case of women, permissible occupation was tightly defined according to putatively natural criteria. There was thus a reflexive relationship between physiology and lifestyle; each affected the other. Through habit and custom, physiological changes took place which had been socially induced.

The emphasis on occupation and lifestyle as determinants of health, which led to a radical boundary being drawn between the sexes, had as its explicit theoretical basis a physiology that recognized few basic boundaries. It conflated moral and physical, mind and body; it created a language capable of containing biological, psychological and social considerations. This is clearly revealed in the use of bridging concepts such as 'temperament', 'habit', 'constitution' and 'sensibility' as technical terms in medicine. These concepts alluded to aspects of human physiology which were not just physical or mental, but contained something of both while being also closely bound to social change. As a result, the temperament and constitution of an individual were seen as products of biological, psychological and social interactions.

Because health was determined to a large extent by variables outside the human body, each person had a distinct physiological make-up which corresponded to his or her unique experience. Groups of people living under the same environmental conditions displayed similar biological and social characteristics. The systematic understanding of these conditions, on which appropriate therapy could be based, was derived from the analysis of a number of distinct variables. The factors affecting groups and individuals had to be clearly delineated. Yet although it was seldom made explicit, there were considered to be limits to the extent to which people could be changed. It was widely acknowledged that there was much variation among women which derived from different climates, patterns of work and so on, but that, nevertheless, all women had in common certain physiological features, not directly a matter of their reproductive organs. For it was a basic premise of physicians in late-eighteenth-century France that women were quite distinct from men by virtue of their whole anatomy and physiology. As Cabanis put it at the end of the eighteenth century: 'Nature has not simply distinguished the sexes by a single set of organs, the direct instruments of reproduction: between men and women there exist other differences of structure which relate more to the role which has been assigned to them' (1956, I: 275). The teleological argument was made more explicit by his contemporary, Roussel: 'The soft parts which are part of the female constitution ... also manifest differences which enable one to catch a glimpse of the functions to which a woman is called, and of the passive state to which nature has destined her' (1803: 11–12).

The ways in which gender differences were conceptualized can be illustrated by referring to the medical notion of sensibility. This was a physiological property which, although present in all parts of the body, was most clearly expressed through the state of the nervous system (Figlio 1975). The nervous system was taken by many to be that physiological system which, because it brought together physical and mental dimensions of human beings, expressed most precisely the total state of the individual, especially with respect to the impact of social changes. Thus it was said that increases in hysterical illnesses in women during the eighteenth century were evidence of the growing use of luxuries such as tea and coffee, and of other changes (Pomme 1782: 578–82). By virtue of their sex, women had a distinct sensibility, which could be further modified during their lifetimes. Women, it was said, are highly *sensible* (in the sense of sensitive, or even sensitized) like children, and more passionate than men. This is because of 'the great mobility of their fibres, especially those in the uterus; hence their irritability, and suffering from vapours' (Macquart 1799, II: 511). The peculiar sensibility of women could also be used to explain their greater life expectancy in a way which associated lifestyle with the physical consistency of the constituent fibres of their bodies. Barthez, a prominent eighteenth-century French physician explained:

> Probably women enjoy this increase in their average age because of the softness and flexibility of the tissue of their fibres, and particularly because of their periodic evacuations which rejuvenate them, so to speak, each month, renew their blood, and re-establish their usual freshness ... Another important cause of women living longer than men is that they are usually more accustomed to suffering infirmities, or to experiencing miseries in life. This habit gives their vital sensibility more moderation, and can only render them less susceptible to illness. (Barthez 1806, II: 298)

However, he went on to say that because of their 'delicate and feeble constitution', women feel things more deeply than men. This aptly portrays the ambivalence which we have already noted in the association of woman and nature. Women are tougher *and* softer, more vulnerable *and* more tenacious of life than men. However, more often than not, the softness of women was returned to again and again, and it was a metaphor that was imaginatively built on to construct a whole image of the dependent nature of woman:

> This muscular feebleness inspires in women an instinctive disgust of strenuous exercise; it draws them towards amusements and sedentary occupations. One could add that the separation of their hips makes walking more painful for women ... This habitual feeling of weakness

> inspires less confidence . . . and as a woman finds herself less able to exist on her own, the more she needs to attract the attention of others, to strengthen herself using those around her whom she judges most capable of protecting her. (Cabanis 1956, I: 278)

Eighteenth-century physiology was based upon necessary links between biological, psychological and social phenomena, not on the anatomical organs of reproduction alone. Although the physiological presuppositions on which Cabanis' views were based applied to both sexes, there was an important asymmetry in that women's occupations were taken to be rooted in and a necessary consequence of their reproductive functions, whereas men's jobs were unrestricted. Women's destiny to bear and suckle children was taken to define their whole body and mind, and therefore their psychological capacities and social tasks. Men were thereby potential members of the broadest social and cultural groups, while women's sphere of action, it was constantly insisted, was the private arena of home and family. As a result, women became a central part of contemporary social debates which focused on the family as the natural, i.e. biological, element in the social fabric, and on women, who through motherhood were the central figures in the family.

The links between women, motherhood, the family and natural morality may help to explain the emphasis on the breast in much medical literature. There is a danger in our seeing the uterus as the constant object of attention in the search for the biological roots of womanhood. It seems likely that different parts of the body were emphasized at different periods, and from different points of view. While the uterus and ovaries interested nineteenth-century gynaecologists, the breast caught the attention of eighteenth-century medical practitioners, who were concerned with moral philosophy and ethics. The breast symbolized women's role in the family through its association with the suckling of babies. It appeared to define the occupational status of females in private work in the family, not in public life. The breast was visible – it was the sign of femininity that men recognized. It could thus be said to be a social law that sexual attraction was founded on the breast, and a natural law that women should breast feed their own children. Based on the natural goodness of the breast it was easy to create a moral injunction on women to feed their own children. It was, it was claimed, an undeniable law which, if thwarted, resulted in suffering for the child and in punishment for the rest of the mother's life, including the miscarriage of subsequent children. 'It is thus that one exposes oneself to cries of pain, for having been unfeeling about those of nature' (Macquart 1799, I: 77). The breasts of women not only symbolized the most fundamental social bond, that between mother and child, but they were also the means by which families were made since their beauty elicited the desires of the male for the female. An excellent example of this fusion of aesthetic, medical and social arguments is Roussel's book on women, *Système Physique et Moral de la Femme* which was an instant success when it first appeared in

1775 and at once became part of literary culture (Alibert 1803: 7). It is significant that in praising Roussel, Alibert employed the metaphor of science unclothing woman: 'I would like to see the author ... portrayed receiving ... homage from the enchanting sex whose organism he has unveiled with so much delicacy and so much insight [pénétration]' (Alibert 1803: 7).

There was a strong aesthetic component in medical writings on women in this period. Discussing the beauty of the breast in the same breath as its vital nutritive function was not undisciplined confusion but indicative of the conflation of social and physiological functions. The breast was good, both morally and biologically, hence its attractiveness and the resultant sociability between the sexes. Indeed the family and thus society were predicated on natural sociability, a quality which Roussel characterized as a major universal law. In these ways the physiological, the social and the aesthetic aspects of human existence were brought together.

So far we have noted a number of overlapping sets of dichotomies and the extent to which the two members of each pair were blurred:

nature	:	culture
woman	:	man
physical	:	mental
mothering	:	thinking
feeling and superstition	:	abstract knowledge and thought
country	:	city
darkness	:	light
nature	:	science and civilization

I have also stressed that these associations worked in two ways so that the association of women with nature had a positive and a negative side. Their sentiment and simple, pure morality constituted the first side, their ignorance and lack of intellectual powers, the second. It was common in the eighteenth century to emphasize the second, negative aspects of female naturalness in attacks on superstition and credulity. *Philosophes* in the vanguard of the Enlightenment believed that they had to fight against the superstition and ignorance of the mass of the people because these were impediments to social progress, and one of the vehicles for their polemic was a form of sex-role stereotyping. The classic example of the problem was the uneducated woman under the thumb of her priest who fed her a diet of religious dogma, urging her to believe things which served his interests alone. This situation was the antithesis of that the savants were trying to promote, where people, free from the influence of the entrenched powers of the aristocracy and clergy, lived according to simple moral precepts derived from the direct study of nature. Women were seen as a major impediment in this process of enlightenment, because they repeated hearsay and tittle tattle and were more prone than men to religious enthusiasm. It was therefore in the interests of savants to polarize women and men, reaction and progress.

[...]

IMAGES OF WOMAN – NATURE DISROBES BEFORE REASON

An important eighteenth-century example of images of woman in visual representations is the wax models of human figures used for making anatomical drawings and for display in popular museums (Haviland and Parish 1970; Thompson 1925; Deer 1977). These were intended both for teaching, both popular and technical, and for decoration. Although male and female anatomical organs, especially the female abdomen, were commonly depicted in anatomy texts from the sixteenth century onwards (Choulant 1962), these models are distinctly different. In the wax series, many of which were made in Florence at the end of the eighteenth century (Azzaroli 1975), the female figures are recumbent, frequently adorned with pearl necklaces. They have long hair, and occasionally they have hair in the pubic area also. These 'Venuses' as they were significantly called lie on velvet or silk cushions, in a passive, almost sexually inviting pose. Comparable male figures are usually upright, and often in a position of motion. The female models can be opened to display the removable viscera, and most often contain a foetus, while the male ones are made in a variety of forms to display the different physiological systems.[2] The figures of recumbent women seem to convey, for the first time, the sexual potential of medical anatomy. Until this time it was usual in engravings for the actual genitals to be covered by a cloth, but in the waxes, as in some contemporaneous medical illustrations, they are not just present, but drawn to the attention. Not only is the literal naturalness of women portrayed, in their total nakedness and by the presence of a foetus, but their symbolic naturalness is implied in the whole conception of such figures. Female nature had been unclothed by male science, making her understandable under general scrutiny. The image was made explicit in the late nineteenth-century statue in the Paris medical faculty of a young woman, her breasts bare, her head slightly bowed beneath the veil she is taking off, which bears the inscription 'Nature unveils herself before Science'.

Women's bodies as objects of medical enquiry as well as of sexual desire became the focus for a physiological literature which expressed a refined aesthetic of women's natural beauty, and found in their bodies an expression of their social condition. To understand women was thus a scientific and medical task which involved revealing the manner of physiological functioning, both normal and pathological, that was peculiar to them. It was for this reason that when Jules Michelet wished to comprehend the condition of women in mid-nineteenth-century France, his first port of call was the dissecting room, and his reading was anatomy texts. In the cadavers of women, Michelet saw their lives revealed and explained before his very eyes. Once again a dual meaning of woman as natural was evoked: she was taken as a creature defined by her biology and as the feminine natural object of masculine science. But perhaps we should add a new third sense. In her pregnant state woman evoked nature yet again through her capacity to reproduce the species, to pass on life. With the definition of life as a new guiding concept at the end of the eighteenth century (Figlio 1976: 25ff.), the mechanism whereby life was transmitted took on fresh significance. The capacity to engender life seemed a special elusive force, made concrete through the female reproductive system. This sacred function went hand in hand with female anatomy. One expression of this was the concern among anatomists to discover ideal female beauty. During the eighteenth century medical writers placed great emphasis on the aesthetics of the human body, and on the natural beauty of women, which, they argued, should remain undeformed by clothing, and especially by corsets (Choulant 1962: 304).

The peak of the sexualized female anatomy was a German painting and lithograph of a beautiful young woman, who had been drowned, being dissected by an anatomist, Professor Lucae, who was interested in the physical basis of female attractiveness. A group of men stand around the table on which a female corpse is lying. She has long hair and well-defined breasts. One of the men has begun the dissection and is working on her thorax. He is holding up a sheet of skin, the part which covers her breast, as if it were a thin article of clothing so delicate and fine is its texture. The corpse is being undressed scientifically, the constituent parts of the body are being displayed for scrutiny and analysis. The powerful sexual image is integral to the whole pictorial effect.[3]

By the 1860s when this engraving was produced, the image it contained might be associated with others that were relatively new to the general public. I am thinking in particular of the fierce public debates about vivisection, which in Britain was opposed, interestingly enough, by a number of women's groups and early feminists (French 1975: 239–50). In vivisectional experiments, pictures of which were prominently displayed in the propaganda put out by critics, there was the same contrast between the utter passivity of the living material used and the active intrusion and manipulation of the experimenters. Despite the long history of anatomical dissections, and the fact that the victim was dead, the anatomizing of the corpse, especially as portrayed in Hasselhorst's picture of Lucae, seems to have similar qualities to the vivisectional experiment. And the exaggerated femininity of the corpse reinforced its passivity. It is almost as if women in their sexually stereotyped roles were made kin to all living objects brought under the penetrating enquiry of male reason.

[...]

CONCLUSION: GENDER LOGIC

Perhaps returning to the Enlightenment can teach us two lessons. First, our ways of thought have a long history, despite the belief that the social sciences are a relatively new field. A more textured understanding of our historical inheritance helps us to analyse current ideas and problems more adequately. Second, recent notions of nature and culture have taken them to be infinitely more simple, reduced categories than they were in the eighteenth and nineteenth centuries. I have stressed the complexity and ambiguity with which ideas of nature, culture and gender were endowed. But I have also pointed to another layer – that of science itself as a sexual activity in its relationship to nature.

While it is important to realize that nature was endowed with a remarkable range of meanings during the period of the Enlightenment (Lovejoy 1960), there was also one common theme. Nature was taken to be that realm on which mankind acts, not just to intervene in or manipulate directly, but also to understand and render it intelligible. This perception of nature includes people and the societies they construct. Such an interpretation of nature led to two distinct positions: nature could be taken to be that part of the world which human beings have understood, mastered and made their own. Here, through the unravelling of laws of motion for example, the inner recesses of nature were revealed to the human mind. But secondly, nature was also that which has not yet been penetrated (either literally or metaphorically), the wilderness and deserts, unmediated and dangerous nature. To these two positions correspond two senses in which women are nature. According to the first, they, as repositories of natural laws, can be revealed and understood. This was Michelet's point in denying that women are unpredictable. On the contrary, he claimed, they are so clearly subsumed under nature's laws, expressed for example in the menstrual cycle, that their states of mind and body can be read by the trained person. For this reason, a systematic study of the anatomy and physiology of women was of great importance. According to the second position it was woman's emotions and uncontrolled passions which gave her special qualities. Women, being endowed with less reason than men, indeed with less need for reason since their social lives required of them feeling and not thought, were more easily dominated by extreme emotions. Women were therefore conceptualized as dangerous because less amenable to the guiding light of reason. According to this second perspective, moves to contain women's dangerous potential are more appropriate than attempts to subject them to scientific scrutiny. Their potential for disorder can be minimized by drawing and maintaining strong social boundaries around them. To these two positions corresponded these positive and negative moral evaluations of the female sex discussed earlier. Ultimately we might say that nature, culture and gender in the history of our own society were and are concepts which express the desire for clarity in areas of life which appear constantly subject to change. Their historical interrelatedness does teach us important lessons about the ways in which apparently distinct areas of life are linked through sets of symbols and metaphors. Furthermore, the links between these cognitive structures and the behavioural level of sexuality are immensely complex, and still largely unexplored.

I have stressed the role of science and medicine as mediators of our ideas of nature, culture and gender, and argued for the rootedness of these ways of thought in recent Western history. One of the most powerful aspects of scientific and medical constructions of sexuality is the way in which apparently universal categories were set up which implied the profound similarities of all women, and to a lesser extent, of all men. Perhaps one of the problems with the current promiscuous use of the nature/culture dichotomy in relation to gender is that it has taken the claims of Western science at face value, and so lapsed into a biologism which it is the responsibility of the social sciences, including history and anthropology, to combat.[4]

NOTES

1. See for example, Hufton (1971: 1975–6).
2. Here I have alluded briefly to what is in fact a very complex issue. There were many different traditions of anatomical models but little has been written about them. See however Thompson (1925), on early ivory manikins, and on anatomical illustration in general, Wolf-Heidegger and Cetto (1967: especially pp. 434, 438, 504, 505, 546–7). Collections which include these or similar figures are: Wellcome Collections, Science Museum, London; Institut für Geschichte der Medizin der Universität, Vienna; Museo 'La Specola', Florence.
3. The original painting is in the Historisches Museum, Frankfurt, the lithograph is in the Wellcome Institute for the History of Medicine, London. An illustration of the former is in Wolf-Heidegger and Cetto (1967: 546).
4. Haraway (1979), is a good starting point in this enterprise.

REFERENCES

Alibert, J. L. (1803), 'Eloge historique de Pierre Roussel', in *Système Physique et Moral de la Femme*, P. Roussel, pp. 1–52, Paris: Crapart, Caille et Ravier.

Azzaroli, M. L. (1975), 'La Specola. The Zoological Museum of Florence University', *Atti del 1°* Congresso Internazionale sulla Ceroplastica nella Scienza e Nell'Arte, pp. 5–31 + 9 plates.

Babbage, C. (1830), *Reflections on the Decline of Science in England, and on some of its causes*, London: B. Fellowes and J. Booth.

Barthes, R. (1954) *Michelet*, Paris: Editions du Seuil.

Barthez, P. J. (1806) *Nouveaux Eléments de la Science de l'Homme*, 2 vols, Paris: Goujor et Brunot.

Brett, R. L. (1968) *George Crabbe*, revised edition, London: Longmans.

Brown, P. (1980) 'Women as "nature", men as "culture": an anthropological debate as object lesson', mimeo.

Cabanis, J. (1978) *Michelet, le Prêtre et la Femme*, Paris: Gallimard.

Cabanis, P. J. G. (1956) *Oeuvres Philosophiques*, 2 vols, Paris: Presses Universitaires de France.

Choulant, J. L. (1962) *History and Bibliography of Anatomic Illustration*, New York and London: Hafner.

Coser, R. L. (1978) 'The Principle of Patriarchy', *Signs. Journal of Women in Culture and Society*, 4 (2); 337–48.

Crabbe, G. (1823) *The Works of the Rev. George Crabbe*, 8 vols, London: John Murray.

de Beauvoir, S. (1972) *The Second Sex*, Harmondsworth: Penguin.

Deer, L. (1977) 'Italian Anatomical Waxes in the Wellcome Collection: the Missing Link', *Rivista di Storia delle Scienze mediche e naturali*, 20: 281–98.

Donnison, J. (1977) *Midwives and Medical Men. A History of Inter-Professional Rivalries and Women's Rights*, London: Heinemann.

Durkheim, E. (1952) *Suicide. A Study in Sociology*, London: Routledge and Kegan Paul.

Farrington, B. (1964) *The Philosophy of Francis Bacon. An Essay on its Development from 1603 to 1609 with new Translations of Fundamental Texts*, Liverpool: Liverpool University Press.

Figlio, K. (1975) 'Theories of Perception and the Physiology of Mind in the Late Eighteenth Century', *History of Science* 12: 177–212.

Figlio, K. (1976) 'The Metaphor of Organisation: a Historiographical Perspective on the Bio-medical Sciences of the Early Nineteenth Century', *History of Science* 14: 17–53.

Foucault, M. (1973) *Moi, Pierre Rivière. Ayant Egorgé Ma Mère, Ma Soeur et Mon Frère ... Un Cas de Parricide au XIXe Siècle*, Paris: Gallimard/Julliard. (Available in English, 1978, Harmondsworth: Penguin.)
French, R. D. (1975) *Antivivisection and Medical Science in Victorian Society*, Princeton and London: Princeton University Press.
Gélis, J. (1977) 'Sages-femmes et accoucheurs: l'obstétrique populaire aux XVIIe et XVIIIe siècles', *Annales: Économies, Sociétés, Civilisations* 32 (part 5): 927–57.
Glacken, C. (1967) *Traces on the Rhodian Shore. Nature and Culture in Western Thought from ancient times to the end of the eighteenth century*, Berkeley, Los Angeles and London: University of California Press.
Haraway, D. (1979) 'The Biological Enterprise: Sex, Mind, and Profit from Human Engineering to Sociobiology', *Radical History Review*, no. 20: 206–37.
Haviland, T. N. and Parish, L. C. (1970) 'A Brief Account of the Use of Wax Models in the Study of Medicine', *Journal of the History of Medicine* 25: 52–75.
Hufton, O. (1971) 'Women in Revolution 1789–1796', *Past and Present*, no. 53: pp. 90–108.
Hufton, O. (1975–6) 'Women and the Family Economy in Eighteenth-Century France', *French Historical Studies*, 9: 1–22.
Jordanova, L. J. (1979) 'Earth Science and Environmental Medicine: the synthesis of the late enlightenment', in *Images of the Earth: Essays in the History of the Environmental Sciences*, ed. L. J. Jordanova and R. Porter, pp. 119–46, Chalfont St Giles: British Society for the History of Science.
Knibiehler, Y. (1976) 'Les médecins et la "nature feminine" au temps du code civil', *Annales: Économies, Sociétés, Civilisations* 31 (part 4): 824–45.
Kolodny, A. (1975) *The Lay of the Land*, Chapel Hill: University of North Carolina Press.
Lovejoy, A. O. (1960) '"Nature" as an Aesthetic Norm', in *Essays in the History of Ideas*, pp. 69–77, New York: Putnam's.
Macquart, L. C. H. (1799) *Dictionnaire de la Conservation de l'Homme*, 2 vols, Paris: Bidault.
Michelet, J. (1860) *La Femme*, Paris: Hachette.
Morel, M.-F. (1977) 'Ville et compagne dans le discours medical sur la petite enfance au XVIIIe siècle', *Annales: Économies, Sociétés, Civilisations* 32 (part 5): 1007–24.
Ortner, S. A. (1974) 'Is female to male as nature is to culture?', in *Woman, Culture and Society*, ed. M. Z. Rosaldo and L. Lamphere, pp. 67–87, Stanford: Stanford University Press.
Ploss, H. H., Bartels, M. and Bartels, P. (1935) *Woman. An Historical Gynaecological and Anthropological Compendium*, 3 vols, London: Heinemann.
Pomme, P. (1782) *Traité des Affections Vaporeuses des Deux Sexes*, Paris: L'Imprimerie Royale.
Roussel, P. (1803) *Système Physique et Moral de la Femme*, 2nd edition, Paris: Crapart, Caille et Ravier.
Rush, B. (1947) *The Selected Writings of Benjamin Rush*, New York: Philosophical Library.
Sterne, L. (1967) *The Life and Opinions of Tristram Shandy Gentleman*, Harmondsworth: Penguin.
Thompson, C. J. S. (1925) 'Anatomical manikins', *Journal of Anatomy* 59 (part 4): 442–5 + 2 plates.
Williams, R. (1975) *The Country and the City*, St Albans: Paladin.
Wolf-Heidegger, G. and Cetto, A. M. (1967), *Die Anatomische Sektion in Bildlicher Darstellung*, Basle and New York: S. Karger.

3.4

MENOPAUSE: THE STORM BEFORE THE CALM

Anne Fausto-Sterling

An unlikely specter haunts the world. It is the ghost of former womanhood, 'unfortunate women abounding in the streets walking stiffly in twos and threes, seeing little and observing less. ... The world appears [to them] as through a grey veil, and they live as docile, harmless creatures missing most of life's values.' According to Dr Robert Wilson and Thelma Wilson, though, one should not be fooled by their 'vapid cow-like negative state' because 'There is ample evidence that the course of history has been changed not only by the presence of estrogen, but by its absence. The untold misery of alcoholism, drug addiction, divorce and broken homes caused by these unstable estrogen-starved women cannot be presented in statistical form'.[1]

Rather than releasing women from their monthly emotional slavery to the sex hormones, menopause involves them in new horrors. At the individual level one encounters the specter of sexual degeneration, described so vividly by Dr David Reuben: 'The vagina begins to shrivel, the breasts atrophy, sexual desire disappears. ... Increased facial hair, deepening voice, obesity ... coarsened features, enlargement of the clitoris, and gradual baldness complete the tragic picture. Not really a man but no longer a functional woman, these individuals live in the world of intersex.'[2] At the demographic level writers express foreboding about women of the baby-boom generation, whose life span has increased from an average forty-eight years at the turn of the century to a

From: A. Fausto-Sterling, 'Hormonal Hurricanes: Menstruation, Menopause, and Female Behaviour' in *Myths of Gender: Biological Theories about Women and Men*, New York: Basic Books, 1986.

projected eighty years in the year 2000.[3] Modern medicine, it seems, has played a cruel trick on women. One hundred years ago they didn't live long enough to face the hardships of menopause but today their increased longevity means they will live for twenty-five to thirty years beyond the time when they lose all possibility of reproducing. To quote Dr Wilson again: 'The unpalatable truth must be faced that all postmenopausal women are castrates.'[4]

But what medicine has wrought, it can also rend asunder. Few publications have had so great an effect on the lives of so many women as have those of Dr Robert A. Wilson who pronounced menopause to be a disease of estrogen deficiency. At the same time in an influential popular form, in his book *Feminine Forever*, he offered a treatment: estrogen replacement therapy.[5] During the first seven months following publication in 1966, Wilson's book sold one hundred thousand copies and was excerpted in *Vogue* and *Look* magazines. It influenced thousands of physicians to prescribe estrogen to millions of women, many of whom had no clinical 'symptoms' other than cessation of the menses. As one of his credentials Wilson lists himself as head of the Wilson Research Foundation, an outfit funded by Ayerst Labs, Searle, and Upjohn, all pharmaceutical giants interested in the large potential market for estrogen. (After all, no woman who lives long enough can avoid menopause.) As late as 1976 Ayerst also supported the Information Center on the Mature Woman, a public relations firm that promoted estrogen replacement therapy. By 1975 some six million women had started long-term treatment with Premarin (Ayerst Labs' brand name for estrogen), making it the fourth or fifth most popular drug in the United States. Even today, two million of the forty million postmenopausal women in the United States contribute to the $70 million grossed each year from the sale of Premarin-brand estrogen.[6] The 'disease of menopause' is not only a social problem: it's big business.[7]

The high sales of Premarin continue despite the publication in 1975 of an article linking estrogen treatment to uterine cancer.[8] Although in the wake of that publication many women stopped taking estrogen and many physicians became more cautious about prescribing it, the idea of hormone replacement therapy remains with us. At least three recent publications in medical journals seriously consider whether the benefits of estrogen might not outweigh the dangers.[9] The continuing flap over treatment for this so-called deficiency disease of the aging female forces one to ask just what *is* this terrible state called menopause? Are its effects so unbearable that one might prefer to increase, even ever-so-slightly, the risk of cancer rather than suffer the daily discomforts encountered during 'the change of life'?

Ours is a culture that fears the elderly. Rather than venerate their years and listen to their wisdom, we segregate them in housing built for 'their special needs' separated from the younger generations from which we draw hope for the future. At the same time we allow millions of old people to live on inadequate incomes, in fear that serious illness will leave them destitute. The happy, productive elderly remain invisible in our midst. (One must look to feminist publications such as *Our Bodies, Ourselves* to find women who express pleasure in their postmenopausal state.) Television ads portray only the arthritic, the toothless, the wrinkled, and the constipated. If estrogen really is the hormone of youth and its decline suggests the coming of old age, then its loss is a part of biology that our culture ill equips us to handle.

There is, of course, a history to our cultural attitudes toward the elderly woman and our views about menopause. In the nineteenth century physicians believed that at menopause a woman entered a period of depression and increased susceptibility to disease. The postmenopausal body might be racked with 'dyspepsia, diarrhea ... rheumatic pains, paralysis, apoplexy ... hemorrhaging ... tuberculosis ... and diabetes', while emotionally the aging female risked becoming irritable, depressed, hysterical, melancholic, or even insane. The more a woman violated social laws (such as using birth control or promoting female suffrage), the more likely she would be to suffer a disease-ridden menopause.[10] In the twentieth century, psychologist Helene Deutsch wrote that at menopause 'woman has ended her existence as a bearer of future life and has reached her natural end – her partial death – as a servant of the species'.[11] Deutsch believed that during the postmenopausal years a woman's main psychological task was to accept the progressive biological withering she experienced. Other well-known psychologists have also accepted the idea that a woman's life purpose is mainly reproductive and that her postreproductive years are ones of inevitable decline. Even in recent times postmenopausal women have been 'treated' with tranquilizers, hormones, electroshock, and lithium.[12]

But should women accept what many see as an inevitable emotional and biological decline? Should they believe, as Wilson does, that 'from a practical point of view a man remains a man until the end', but that after menopause 'we no longer have the "whole woman" – only the "part woman"'?[13] What is the real story of menopause?

THE CHANGE: ITS DEFINITION AND PHYSIOLOGY

In 1976 under the auspices of the American Geriatric Society and the medical faculty of the University of Montpellier, the First International Congress on the Menopause convened in the south of France. In the volume that emerged from that conference, scientists and clinicians from around the world agreed on a standard definition of the words *menopause* and *climacteric*. 'Menopause', they wrote, 'indicates the final menstrual period and occurs during the climacteric. The climacteric is that phase in the aging process of women marking the transition from the reproductive stage of life to the non-reproductive stage.'[14] By consensus, then, the word *menopause* has come to mean a specific event, the last menstruation, while *climacteric* implies a process occurring over a period of years. (There is also a male climacteric, which entails a gradual reduction in production of the hormone testosterone over the years as part of the male aging process. What part it plays in that process is poorly understood and seems frequently to be ignored by researchers, who prefer to contrast continuing male reproductive potency with the loss of childbearing ability in women.[15])

[. . .]

What happens to the intricately balanced hormone cycle during the several years preceding menopause is little understood, although it seems likely that gradual changes occur in the balance between pituitary activity (FSH and LH production) and estrogen synthesis.[16] One thing, however, is clear: menopause does not mean the *absence* of estrogen, but rather a gradual lowering in the availability of *ovarian* estrogen.

[. . .]

The other estrogenic hormones, as well as progesterone and testosterone, drop off to some extent but continue to be synthesized at a level comparable to that observed during the early phases of the menstrual cycle. Instead of concentrating on the notion of estrogen deficiency, however, it is more important to point out that: (1) postmenopausally the body makes different kinds of estrogen; (2) the ovaries synthesise less and the adrenals more of these hormones; and (3) the monthly ups and downs of these hormones even out following menopause.

While estrogen levels begin to decline, the levels of FSH and LH start to increase. Changes in these hormones appear as early as eight years before menopause.[17] At the time of menopause and for several years afterward these two hormones are found in very high concentrations compared to menstrual levels (FSH as many as fourteen times more concentrated than premenopausally, and LH more than three times more). Over a period of years such high levels are reduced to about half their peak value, leaving the postmenopausal woman with one-and-one-half times more LH and seven times more FSH circulating in her blood than when she menstruated regularly.

It is to all of these changes in hormone levels that the words such as *climacteric* and *menopause* refer. From these alterations Wilson and others have chosen to blame estrogen for the emotional deterioration they believe appears in postmenopausal women. Why they have focused on only one hormone from a complex system of hormonal changes is anybody's guess. I suspect, however, that the reasons are (at least) twofold. First, the normative biomedical disease model of female physiology [. . .] looks for simple cause and effect. Most researchers, then, have simply assumed estrogen to be a 'cause' and set out to measure its 'effect'. The model or framework out of which such investigators work precludes an interrelated analysis of all the different (and closely connected) hormonal changes going on during the climacteric. But why single out estrogen? Possibly because this hormone plays an important role in the menstrual cycle as well as in the development of 'feminine' characteristics such as breasts and overall body contours. It is seen as the quintessential female hormone. So where could one better direct one's attention if, to begin with, one views menopause as the loss of true womanhood?

Physical changes do occur following menopause. Which, if any, of these are caused by changing hormone levels is another question. Menopause research comes equipped with its own unique experimental traps.[18] The most obvious is that a postmenopausal population is also an aging population. Do physical and emotional differences found in groups of postmenopausal women have to do with hormonal changes or with other aspects of aging? It is a difficult matter to sort out. Furthermore, many of the studies on menopause have been done on preselected populations, using women who volunteer because they experience classic menopausal 'symptoms' such as the hot flash. Such investigations tell us nothing about average changes within the population as a whole. In the language of the social scientist, we have no baseline data, nothing to which we can compare menopausal women, no way to tell whether the complaint of a particular woman is typical, a cause for medical concern, or simply idiosyncratic.

Since the late 1970s feminist researchers have begun to provide us with much-needed information. Although their results confirm some beliefs long held by physicians, these newer investigators present them in a more sophisticated context. Dr Madeleine Goodman and her colleagues designed a study in which they drew information from a large population of women ranging in age from thirty-five to sixty. All had undergone routine multiphasic screening at a health maintenance clinic, but none had come for problems concerning menopause. From the complete clinic records they selected a population of women who had not menstruated for at least one year and compared their health records with those who still menstruated, looking at thirty-five different variables, such as cramps, blood glucose levels, blood calcium, and hot flashes, to see if any of these symptoms correlated with those seen in postmenopausal women. The results are startling. They found that only 28 per cent of Caucasian women and 24 per cent of Japanese women identified as postmenopausal 'reported traditional menopausal symptoms such as hot flashes, sweats, etc., while in non-menopausal controls, 16 per cent in Caucasians and 10 per cent in Japanese also reported these same symptoms'.[19] In other words, 75 per cent of menopausal women in their sample reported no remarkable menopausal symptoms, a result in sharp contrast to earlier studies using women who identified themselves as menopausal.

In a similar exploration, researcher Karen Frey found evidence to support Goodman's results. She wrote that menopausal women 'did not report significantly greater frequency of physical symptoms or concern about these symptoms than did pre- or post-menopausal women.'[20] The studies of Goodman, Frey, and others[21] draw into serious question the notion that menopause is generally or necessarily associated with a set of disease symptoms. Yet at least three physical changes – hot flashes, vaginal dryness and irritation, and osteoporosis – and one emotional one – depression – remain associated in the minds of many with the decreased estrogen levels of the climacteric. Goodman's work indicates that such changes may be far less widespread than previously believed, but if they are troublesome to 26 per cent of all menopausal women they remain an appropriate subject for analysis.

We know only the immediate cause of hot flashes: a sudden expansion of the blood flow into the skin. The technical term to describe them, *vasomotor instability*, means only that nerve cells signal the widening of blood vessels allowing more blood into the body's periphery. A consensus has emerged on two things: (1) the high concentration of FSH and LH in the blood probably causes hot flashes, although exactly how this happens remains unknown; and (2) estrogen treatment is the only currently available way to suppress the hot flashes. One hypothesis is that by means of a feedback mechanism, artificially raised blood levels of estrogen signal the brain to tell the pituitary to call off the FSH and LH. Although estrogen does stop the hot flashes, its effects are only temporary; remove the estrogen and the flashes return. Left alone, the body eventually adjusts to the changing levels of FSH and LH. Thus a premenopausal woman has two choices in dealing with hot flashes: she can either take estrogen as a permanent medication, a course Wilson refers to as embarking 'on the great adventure of preserving or regaining your full femininity',[22] or suffer some discomfort while nature takes its course. Since the longer one takes estrogen the greater the danger of estrogen-linked cancer, many health-care workers recommend the latter.[23]

Some women experience postmenopausal vaginal dryness and irritation that can make sexual intercourse painful. Since the cells of the vaginal wall contain estrogen receptors it is not surprising that estrogen applied locally or taken in pill form helps with this difficulty. Even locally applied, however, the estrogen enters into the bloodstream, presenting the same dangers as when taken in pill form. There are alternative treatments, though, for vaginal dryness. The Boston Women's Health Collective, for example, recommends the use of nonestrogen vaginal creams or jellies, which seem to be effective and are certainly safer. Continued sexual activity also helps – yet another example of the interaction between behavior and physiology.

Hot flashes and vaginal dryness are the *only* climacteric-associated changes for which estrogen unambiguously offers relief. Since significant numbers of women do not experience these changes and since for many of those that do the effects are relatively mild, the wisdom of estrogen replacement therapy must be examined carefully and on an individual basis. Both men and women undergo certain changes as they age, but Wilson's catastrophic vision of postmenopausal women – those ghosts gliding by 'unnoticed and, in turn, notic[ing] little'[24] – is such a far cry from reality that it is a source of amazement that serious medical writers continue to quote his work.

In contrast to hot flashes and vaginal dryness, osteoporosis, a brittleness of the bone which can in severe cases cripple, has a complex origin. Since this potentially life-threatening condition appears more frequently in older women than in older men, the hypothesis of a relationship with estrogen levels seemed plausible to many. But as one medical worker has said, a unified theory of the disease 'is still non-existent, although sedentary life styles, genetic predisposition, hormonal imbalance, vitamin deficiencies, high-protein diets, and cigarette smoking all have been implicated.'[25] Estrogen treatment seems to arrest the disease for a while, but may lose effectiveness after a few years.[26]

Even more so than in connection with any physical changes, women have hit up against a medical double bind whenever they have complained of emotional problems during the years of climacteric. On the one hand physicians dismissed these complaints as the imagined ills of a hormone-deficient brain, while on the other they generalized the problem, arguing that middle-aged women are emotionally unreliable, unfit for positions of leadership and responsibility. Women had two choices: to complain and experience ridicule and/or improper medical treatment, or to suffer in silence. Hormonal changes during menopause were presumed to be the cause of psychiatric symptoms ranging from fatigue, dizziness, irritability, apprehension, and insomnia to severe headaches and psychotic depression. In recent years, however, these earlier accounts have been supplanted by a rather different consensus now emerging among responsible medical researchers.

To begin with, there are no data to support the idea that menopause has any relationship to serious depression in women. Postmenopausal women who experience psychosis have almost always had similar episodes premenopausally.[27] The notion of the hormonally depressed woman is a shibboleth that must be laid permanently to rest. Some studies have related irritability and insomnia to loss of sleep from nighttime hot flashes. Thus, for women who experience hot flashes, these emotional difficulties might, indirectly, relate to menopause. But the social, life history, and family contexts in which middle-aged women find themselves are more important links to emotional changes occurring during the years of the climacteric. And these, of course, have nothing whatsoever to do with hormones. Quite a number of studies suggest that the majority of women do not consider menopause a time of crisis. Nor do most women suffer from the so-called 'empty nest syndrome' supposedly experienced when children leave home. On the contrary, investigation suggests that women without small children are less depressed and have higher incomes and an increased sense of well-being.[28] Such positive reactions depend upon work histories, individual upbringing, cultural background, and general state of health, among other things.

In a survey conducted for *Our Bodies, Ourselves*, one which in no sense represents a balanced cross section of US women, the Boston Women's Health Collective recorded the reactions of more than two hundred menopausal or postmenopausal women, most of whom were suburban, married, and employed, to a series of questions about menopause. About two-thirds of them felt either positively or neutrally about a variety of changes they had undergone, while a whopping 90 per cent felt okay or happy about the loss of childbearing ability![29] This result probably comes as no surprise to most women, but it flies in the face of the long-standing belief that women's lives and emotions are driven in greater part by their reproductive systems.

No good account of adult female development in the middle years exists. Levinson,[30] who studied adult men, presents a linear model of male development designed primarily around work experiences. In his analysis, the male climacteric plays only a secondary role. Feminist scholars Rosalind Barnett and Grace Baruch have described the difficulty of fitting women into Levinson's scheme: 'It is hard to know how to think of women within this theory – a woman may not enter the world of work until her late thirties, she seldom has a mentor, and even women with life-long career commitments rarely are in a position to reassess their commitment pattern by age 40', as do the men in Levinson's study.[31]

Baruch and Barnett call for the development of a theory of women in their middle years, pointing out that an adequate one can emerge only when researchers set aside preconceived ideas about the central role of biology in adult female development and listen to what women themselves say. Paradoxically, in some sense we will remain unable to understand more about the role of biology in women's middle years until we have a more realistic *social* analysis of women's postadolescent psychological development. Such an analysis must, of course, take into account ethnic, racial, regional, and class differences among women, since once biology is jettisoned as a universal cause of female behavior it no longer makes sense to lump all women into a single category.

Much remains to be understood about menopause. Which biological changes, for instance, result from ovarian degeneration and which from other aspects of aging? How does the aging process compare in men and women? What causes hot flashes and can we find safe ways to alleviate the discomfort they cause? Do other aspects of a woman's life affect the number and severity of menopausally related physical symptoms? What can we learn from studying the experience of menopause in other, especially non-Western cultures? A number of researchers have proposed effective ways of finding answers to these questions.[32] We need only time, research dollars, and an open mind to move forward.

CONCLUSION

The premise that women are by nature abnormal and inherently diseased dominates past research on menstruation and menopause. While appointing the male reproductive system as normal, this viewpoint calls abnormal any aspect of the female reproductive life cycle that deviates from the male's. At the same time such an analytical framework places the essence of a woman's existence in her reproductive system. Caught in her hormonal windstorm, she strives to attain normality but can only do so by rejecting her biological uniqueness, for that too is essentially deformed: a double bind indeed. Within such an intellectual structure no medical research of any worth to women's health can be done, for it is the blueprint itself that leads investigators to ask the wrong questions, look in the wrong places for answers, and then distort the interpretation of their results.

Reading through the morass of poorly done studies on menstruation and menopause, many of which express deep hatred and fear of women, can be a discouraging experience. One begins to wonder how it can be that within so vast a quantity of material so little quality exists. But at this very moment the field of menstrual-cycle research (including menopause) offers a powerful antidote to that disheartenment in the form of feminist researchers (both male and female) with excellent training and skills, working within a new analytical framework. Rejecting a strict medical model of female development, they understand that men and women have different reproductive cycles, *both* of which are normal. Not binary opposites, male and female physiologies have differences *and* similarities. These research pioneers know too that the human body functions in a social milieu and that it changes in response to that context. Biology is not a one-way determinant but a dynamic component of our existence. And, equally important, these new investigators have learned not only to *listen* to what women say about themselves but to *hear* as well. By and large, these researchers are not in the mainstream of medical and psychological research, but we can look forward to a time when the impact of their work will affect the field of menstrual-cycle research for the better and for many years to come.

If women are seen as emotional slaves of their reproductive physiologies, men do not get off scot-free: *their* 'problem hormone' is testosterone. In normal amounts it is seen as the source of many positive traits – drive, ambition, success. But too high a quantity, so the theory goes, can cause antisocial behaviour – crime, violence, and even war. In contrast, women's lower testosterone levels mean they are less likely to make it in the world of the hard-driving professional, although they can derive pride from their more peaceable manner. This tale of testosterone and aggression forms yet another subplot in the collection of biological stories about male and female behaviour.

NOTES

1. Robert A. Wilson and Thelma A. Wilson, 'The Fate of the Nontreated Postmenopausal Woman: A Plea for the Maintenance of Adequate Estrogen from Puberty to the Grave', *Journal of the American Geriatric Society* 11(1963):352–56.
2. David Reuben, *Everything You Always Wanted to Know about Sex but Were Afraid to Ask* (New York: McKay, 1969), 292.
3. Wulf H. Utian, *Menopause in Modern Perspectives* (New York: Appleton-Century-Crofts, 1980).
4. Wilson and Wilson, 'The Fate of the Nontreated Postmenopausal Woman', 347.
5. Robert A. Wilson, *Feminine Forever* (New York: M. Evans, 1966).
6. Marilyn Grossman and Pauline Bart, 'The Politics of Menopause', in *The Menstrual Cycle*, vol. 1, ed. Dan, Graham, and Beecher.
7. Kathleen MacPherson, 'Menopause as Disease: The Social Construction of a Metaphor', *Advances in Nursing Science* 3(1981):95–113; A. Johnson, 'The Risks of Sex Hormones as Drugs', *Women and Health* 2(1977):8–11.
8. D. Smith et al., 'Association of Exogenous Estrogen and Endometrial Cancer', *New England Journal of Medicine* 293(1975):1164–7.
9. H. Judd et al., 'Estrogen Replacement Therapy', *Obstetrics and Gynecology* 58(1981):267–75; M. Quigley, 'Postmenopausal Hormone Replacement Therapy:

Back to Estrogen Forever?' *Geriatric Medicine Today* 1(1982):78–85; and Thomas Skillman, 'Estrogen Replacement: Its Risks and Benefits', *Consultant* (1982):115–27.

10. C. Smith-Rosenberg, 'Puberty to Menopause: The Cycle of Femininity in 19th Century America', *Feminist Studies* 1(1973):65.
11. Helene Deutsch, *The Psychology of Women* (New York: Grune and Stratton, 1945), 458.
12. J. H. Osofsky and R. Seidenberg, 'Is Female Menopausal Depression Inevitable?', *Obstetrics and Gynecology* 36(1970):611.
13. Wilson and Wilson, 'The Fate of the Nontreated Postmenopausal Woman', 348.
14. P. A. vanKeep, R. B. Greenblatt and M. Albeaux-Fernet, eds, *Consensus on Menopause Research* (Baltimore: University Park Press, 1976), 134.
15. Marcha Flint, 'Male and Female Menopause: A Cultural Put-on', in *Changing Perspectives on Menopause*, ed. A. M. Voda, M. Dinnerstein, and S. O'Donnell (Austin: University of Texas Press, 1982).
16. Utian, *Menopause in Modern Perspectives.*
17. *Ibid.*
18. *Madeleine Goodman, 'Toward a Biology of Menopause', Signs* 5 (1980):739–53.
19. Madeleine Goodman, C. J. Stewart and F. Gilbert, 'Patterns of Menopause: A Study of Certain Medical and Physiological Variables among Caucasian and Japanese Women Living in Hawaii', *Journal of Gerontology* 32(1977):297.
20. Karen Frey, 'Middle-Aged Women's Experience and Perceptions of Menopause', *Women and Health* 6(1981):31.
21. Eve Kahana, A. Kiyak and J. Liang, 'Menopause in the Context of Other Life Events', in *The Menstrual Cycle*, vol. 1, ed. Dan, Graham, and Beecher, 167–78.
22. Wilson, *Feminine Forever*, 134.
23. A. Voda and M. Eliasson, 'Menopause: The Closure of Menstrual Life', *Women and Health* 8(1983): 137–56.
24. Wilson and Wilson, 'The Fate of the Nontreated Postmenopausal Woman', 356.
25. Louis Avioli, 'Postmenopausal Osteoporosis: Prevention vs. Cure', *Federation Proceedings* 40(1981): 2418.
26. Voda and Eliasson, 'Menopause: The Closure of Menstrual Life'.
27. G. Winokur and R. Cadoret, 'The Irrelevance of the Menopause to Depressive Disease', in *Topics in Psychoendocrinology*, ed. E. J. Sachar (New York: Grune and Stratton, 1975).
28. Rosalind Barnett and Grace Baruch, 'Women in the Middle Years: A Critique of Research and Theory', *Psychology of Women Quarterly* 3(1978): 187–97.
29. Boston Women's Health Collective, *Our Bodies, Ourselves.*
30. D. Levinson et al., 'Periods in the Adult Development of Men: Ages 18–45', *The Counseling Psychologist* 6(1976): 21–25.
31. Barnett and Baruch, 'Women in the Middle Years', 189.
32. Ibid.; Goodman, 'Toward a Biology of Menopause', and Voda, Dinnerstein, and O'Donnell, eds, *Changing Perspectives on Menopause.*

3.5

THE EGG AND THE SPERM: HOW SCIENCE HAS CONSTRUCTED A ROMANCE BASED ON STEREOTYPICAL MALE–FEMALE ROLES

Emily Martin

> The theory of the human body is always a part of a world-picture. . . . The theory of the human body is always a part of a *fantasy*. (James Hillman, *The Myth of Analysis*)[1]

As an anthropologist, I am intrigued by the possibility that culture shapes how biological scientists describe what they discover about the natural world. If this were so, we would be learning about more than the natural world in high school biology class; we would be learning about cultural beliefs and practices as if they were part of nature. In the course of my research I realized that the picture of egg and sperm drawn in popular as well as scientific accounts of reproductive biology relies on stereotypes central to our cultural definitions of male and female. The stereotypes imply not only that female biological processes are less worthy than their male counterparts but also that women are less worthy than men. Part of my goal in writing this article is to shine a bright light on the gender stereotypes hidden within the scientific language of biology. Exposed in such a light, I hope they will lose much of their power to harm us.

Egg and Sperm: A Scientific Fairy Tale

At a fundamental level, all major scientific textbooks depict male and female reproductive organs as systems for the production of valuable substances, such as eggs and sperm.[2] In the case of women, the monthly cycle is described as

From: *Signs: Journal of Women in Culture and Society* 16 (31): 485–501, 1991.

being designed to produce eggs and prepare a suitable place for them to be fertilized and grown – all to the end of making babies. But the enthusiasm ends there. By extolling the female cycle as a productive enterprise, menstruation must necessarily be viewed as a failure. Medical texts describe menstruation as the 'debris' of the uterine lining, the result of necrosis, or death of tissue. The descriptions imply that a system has gone awry, making products of no use, not to specification, unsalable, wasted, scrap. An illustration in a widely used medical text shows menstruation as a chaotic disintegration of form, complementing the many texts that describe it as 'ceasing', 'dying', 'losing', 'denuding', 'expelling'.[3]

Male reproductive physiology is evaluated quite differently. One of the texts that sees menstruation as failed production employs a sort of breathless prose when it describes the maturation of sperm: 'The mechanisms which guide the remarkable cellular transformation from spermatid to mature sperm remain uncertain. . . . Perhaps the most amazing characteristic of spermatogenesis is its sheer magnitude: the normal human male may manufacture several hundred million sperm per day.'[4] In the classic text *Medical Physiology*, edited by Vernon Mountcastle, the male/female, productive/destructive comparison is more explicit: 'Whereas the female *sheds* only a single gamete each month, the seminiferous tubules *produce* hundreds of millions of sperm each day' (emphasis mine).[5] The female author of another text marvels at the length of the microscopic seminiferous tubules, which, if uncoiled and placed end to end, 'would span almost one-third of a mile!' She writes, 'In an adult male these structures produce millions of sperm cells each day.' Later she asks, 'How is this feat accomplished?'[6] None of these texts expresses such intense enthusiasm for any female processes. It is surely no accident that the 'remarkable' process of making sperm involves precisely what, in the medical view, menstruation does not: production of something deemed valuable.[7]

One could argue that menstruation and spermatogenesis are not analogous processes and, therefore, should not be expected to elicit the same kind of response. The proper female analogy to spermatogenesis, biologically, is ovulation. Yet ovulation does not merit enthusiasm in these texts either. Textbook descriptions stress that all of the ovarian follicles containing ova are already present at birth. Far from being *produced*, as sperm are, they merely sit on the shelf, slowly degenerating and aging like overstocked inventory: 'At birth, normal human ovaries contain an estimated one million follicles [each], and no new ones appear after birth. Thus, in marked contrast to the male, the newborn female already has all the germ cells she will ever have. Only a few, perhaps 400, are destined to reach full maturity during her active productive life. All the others degenerate at some point in their development so that few, if any, remain by the time she reaches menopause at approximately 50 years of age.'[8] Note the 'marked contrast' that this description sets up between male and female: the male, who continuously produces fresh germ cells, and the female, who has stockpiled germ cells by birth and is faced with their degeneration.

Nor are the female organs spared such vivid descriptions. One scientist writes in a newspaper article that a woman's ovaries become old and worn out from ripening eggs every month, even though the woman herself is still relatively young: 'When you look through a laparoscope ... at an ovary that has been through hundreds of cycles, even in a superbly healthy American female, you see a scarred, battered organ.'[9]

To avoid the negative connotations that some people associate with the female reproductive system, scientists could begin to describe male and female processes as homologous. They might credit females with 'producing' mature ova one at a time, as they're needed each month, and describe males as having to face problems of degenerating germ cells. This degeneration would occur throughout life among spermatogonia, the undifferentiated germ cells in the testes that are the long-lived, dormant precursors of sperm.

But the texts have an almost dogged insistence on casting female processes in a negative light. The texts celebrate sperm production because it is continuous from puberty to senescence, while they portray egg production as inferior because it is finished at birth. This makes the female seem unproductive, but some texts will also insist that it is she who is wasteful.[10] In a section heading for *Molecular Biology of the Cell*, a best-selling text, we are told that 'Oogenesis is wasteful.' The text goes on to emphasize that of the seven million oogonia, or egg germ cells, in the female embryo, most degenerate in the ovary. Of those that do go on to become oocytes, or eggs, many also degenerate, so that at birth only two million eggs remain in the ovaries. Degeneration continues throughout a woman's life: by puberty 300 000 eggs remain, and only a few are present by menopause. 'During the 40 or so years of a woman's reproductive life, only 400 to 500 eggs will have been released', the authors write. 'All the rest will have degenerated. It is still a mystery why so many eggs are formed only to die in the ovaries.'[11]

The real mystery is why the male's vast production of sperm is not seen as wasteful.[12] Assuming that a man 'produces' 100 million (10^8) sperm per day (a conservative estimate) during an average reproductive life of sixty years, he would produce well over two trillion sperm in his lifetime. Assuming that a woman 'ripens' one egg per lunar month, or thirteen per year, over the course of her forty-year reproductive life, she would total five hundred eggs in her lifetime. But the word 'waste' implies an excess, too much produced. Assuming two or three offspring, for every baby a woman produces, she wastes only around two hundred eggs. For every baby a man produces, he wastes more than one trillion (10^{12}) sperm.

How is it that positive images are denied to the bodies of women? A look at language – in this case, scientific language – provides the first clue. Take the egg and the sperm.[13] It is remarkable how 'femininely' the egg behaves and how 'masculinely' the sperm.[14] The egg is seen as large and passive.[15] It does not *move* or *journey*, but passively 'is transported', 'is swept',[16] or even 'drifts'[17] along the fallopian tube. In utter contrast, sperm are small, 'streamlined',[18] and invariably active. They 'deliver' their genes to the egg, 'activate the developmental program of the egg',[19] and have a 'velocity' that is often remarked upon.[20] Their tails are 'strong' and efficiently powered.[21] Together with the forces of ejaculation, they can 'propel the semen into the deepest recesses of the vagina'.[22] For this they need 'energy', 'fuel',[23] so that with a 'whiplashlike motion and strong lurches'[24] they can 'burrow through the egg coat'[25] and 'penetrate' it.[26]

At its extreme, the age-old relationship of the egg and the sperm takes on a royal or religious patina. The egg coat, its protective barrier, is sometimes called its 'vestments', a term usually reserved for sacred, religious dress. The egg is said to have a 'corona',[27] a crown, and to be accompanied by 'attendant cells'.[28] It is holy, set apart and above, the queen to the sperm's king. The egg is also passive, which means it must depend on sperm for rescue. Gerald Schatten and Helen Schatten liken the egg's role to that of Sleeping Beauty: 'a dormant bride awaiting her mate's magic kiss, which instills the spirit that brings her to life'.[29] Sperm, by contrast, have a 'mission',[30] which is to 'move through the female genital tract in quest of the ovum'.[31] One popular account has it that the sperm carry out a 'perilous journey' into the 'warm darkness', where some fall away 'exhausted'. 'Survivors' 'assault' the egg, the successful candidates 'surrounding the prize'.[32] Part of the urgency of this journey, in more scientific terms, is that 'once released from the supportive environment of the ovary, an egg will die within hours unless rescued by a sperm'.[33] The wording stresses the fragility and dependency of the egg, even though the same text acknowledges elsewhere that sperm also live for only a few hours.[34]

In 1948, in a book remarkable for its early insights into these matters, Ruth Herschberger argued that female reproductive organs are seen as biologically interdependent, while male organs are viewed as autonomous, operating independently and in isolation:

> At present the functional is stressed only in connection with women: it is in them that ovaries, tubes, uterus, and vagina have endless interdependence. In the male, reproduction would seem to involve 'organs' only.
> Yet the sperm, just as much as the egg, is dependent on a great many related processes. There are secretions which mitigate the urine in the urethra before ejaculation, to protect the sperm. There is the reflex shutting off of the bladder connection, the provision of prostatic secretions, and various types of muscular propulsion. The sperm is no more independent of its milieu than the egg, and yet from a wish that it were, biologists have lent their support to the notion that the human female, beginning with the egg, is congenitally more dependent than the male.[35]

Bringing out another aspect of the sperm's autonomy, an article in the journal *Cell* has the sperm making an 'existential decision' to penetrate the egg: 'Sperm are cells with a limited behavioral repertoire, one that is directed toward fertilizing eggs. To execute the decision to abandon the haploid state, sperm swim to an egg and there acquire the ability to effect membrane fusion.'[36] Is this a corporate manager's version of the sperm's activities – 'executing decisions' while fraught with dismay over difficult options that bring with them very high risk?

There is another way that sperm, despite their small size, can be made to loom in importance over the egg. In a collection of scientific papers, an electron micrograph of an enormous egg and tiny sperm is titled 'A Portrait of the Sperm'.[37] This is a little like showing a photo of a dog and calling it a picture of the fleas. Granted, microscopic sperm are harder to photograph than eggs, which are just large enough to see with the naked eye. But surely the use of the term 'portrait', a word associated with the powerful and wealthy, is significant. Eggs have only micrographs or pictures, not portraits.

One depiction of sperm as weak and timid, instead of strong and powerful – the only such representation in western civilization, so far as I know – occurs in Woody Allen's movie *Everything You Always Wanted To Know About Sex* *But Were Afraid to Ask*. Allen, playing the part of an apprehensive sperm inside a man's testicles, is scared of the man's approaching orgasm. He is reluctant to launch himself into the darkness, afraid of contraceptive devices, afraid of winding up on the ceiling if the man masturbates.

The more common picture – egg as damsel in distress, shielded only by her sacred garments; sperm as heroic warrior to the rescue – cannot be proved to be dictated by the biology of these events. While the 'facts' of biology may not *always* be constructed in cultural terms, I would argue that in this case they are. The degree of metaphorical content in these descriptions, the extent to which differences between egg and sperm are emphasized, and the parallels between cultural stereotypes of male and female behavior, and the character of egg and sperm all point to this conclusion.

NEW RESEARCH, OLD IMAGERY

As new understandings of egg and sperm emerge, textbook gender imagery is being revised. But the new research, far from escaping the stereotypical representations of egg and sperm, simply replicates elements of textbook gender imagery in a different form. The persistence of this imagery calls to mind what Ludwik Fleck termed 'the self-contained' nature of scientific thought. As he described it, 'the interaction between what is already known, what remains to be learned, and those who are to apprehend it, go to ensure harmony within the system. But at the same time they also preserve the harmony of illusions, which is quite secure within the confines of a given thought style.'[38] We need to understand the way in which the cultural content in scientific descriptions changes as biological discoveries unfold, and whether that cultural content is solidly entrenched or easily changed.

In all of the texts quoted above, sperm are described as penetrating the egg, and specific substances on a sperm's head are described as binding to the egg. Recently, this description of events was rewritten in a biophysics lab at Johns

Hopkins University – transforming the egg from the passive to the active party.[39]

Prior to this research, it was thought that the zona, the inner vestments of the egg, formed an impenetrable barrier. Sperm overcame the barrier by mechanically burrowing through, thrashing their tails and slowly working their way along. Later research showed that the sperm released digestive enzymes that chemically broke down the zona; thus, scientists presumed that the sperm used mechanical *and* chemical means to get through to the egg.

In this recent investigation, the researchers began to ask questions about the mechanical force of the sperm's tail. (The lab's goal was to develop a contraceptive that worked topically on sperm.) They discovered, to their great surprise, that the forward thrust of sperm is extremely weak, which contradicts the assumption that sperm are forceful penetrators.[40] Rather than thrusting forward, the sperm's head was now seen to move mostly back and forth. The sideways motion of the sperm's tail makes the head move sideways with a force that is ten times stronger than its forward movement. So even if the overall force of the sperm were strong enough to mechanically break the zona, most of its force would be directed sideways rather than forward. In fact, its strongest tendency, by tenfold, is to escape by attempting to pry itself off the egg. Sperm, then, must be exceptionally efficient at *escaping* from any cell surface they contact. And the surface of the egg must be designed to trap the sperm and prevent their escape. Otherwise, few if any sperm would reach the egg.

The researchers at Johns Hopkins concluded that the sperm and egg stick together because of adhesive molecules on the surfaces of each. The egg traps the sperm and adheres to it so tightly that the sperm's head is forced to lie flat against the surface of the zona, a little bit, they told me, 'like Br'er Rabbit getting more and more stuck to tar baby the more he wriggles.' The trapped sperm continues to wiggle ineffectually side to side. The mechanical force of its tail is so weak that a sperm cannot break even one chemical bond. This is where the digestive enzymes released by the sperm come in. If they start to soften the zona just at the tip of the sperm and the sides remain stuck, then the weak, flailing sperm can get oriented in the right direction and make it through the zona – provided that its bonds to the zona dissolve as it moves in.

Although this new version of the saga of the egg and the sperm broke through cultural expectations, the researchers who made the discovery continued to write papers and abstracts as if the sperm were the active party who attacks, binds, penetrates, and enters the egg. The only difference was that sperm were now seen as performing these actions weakly.[41] Not until August 1987, more than three years after the findings described above, did these researchers reconceptualize the process to give the egg a more active role. They began to describe the zona as an aggressive sperm catcher, covered with adhesive molecules that can capture a sperm with a single bond and clasp it to the zona's surface.[42] In the words of their published account: 'The innermost vestment, the *zona pellucida*, is a glycoprotein shell, which captures and tethers the sperm before they penetrate it. ... The sperm is captured at the initial contact between the sperm tip and the *zona*. ... Since the thrust [of the sperm] is much smaller than the force needed to break a single affinity bond, the first bond made upon the tip-first meeting of the sperm and *zona* can result in the capture of the sperm.'[43]

[...]

SOCIAL IMPLICATIONS: THINKING BEYOND

These revisionist accounts of egg and sperm cannot seem to escape the hierarchical imagery of older accounts. Even though each new account gives the egg a larger and more active role, taken together they bring into play another cultural stereotype: woman as a dangerous and aggressive threat. In the Johns Hopkins lab's revised model, the egg ends up as the female aggressor who 'captures and tethers' the sperm with her sticky zona, rather like a spider lying in wait in her web.[44] The Schatten lab has the egg's nucleus 'interrupt' the sperm's dive with a 'sudden and swift' rush by which she 'clasps the sperm and guides its nucleus to the center.'[45] Wassarman's description of the surface of the egg 'covered with thousands of plasma membrane-bound projections, called microvilli' that reach out and clasp the sperm adds to the spiderlike imagery.[46]

These images grant the egg an active role but at the cost of appearing disturbingly aggressive. Images of woman as dangerous and aggressive, the femme fatale who victimizes men, are widespread in Western literature and culture.[47] More specific is the connection of spider imagery with the idea of an engulfing, devouring mother.[48] New data did not lead scientists to eliminate gender stereotypes in their descriptions of egg and sperm. Instead, scientists simply began to describe egg and sperm in different, but no less damaging, terms.

Can we envision a less stereotypical view? Biology itself provides another model that could be applied to the egg and the sperm. The cybernetic model – with its feedback loops, flexible adaptation to change, coordination of the parts within a whole, evolution over time, and changing response to the environment – is common in genetics, endocrinology, and ecology and has a growing influence in medicine in general.[49] This model has the potential to shift our imagery from the negative, in which the female reproductive system is castigated both for not producing eggs after birth and for producing (and thus wasting) too many eggs overall, to something more positive. The female reproductive system could be seen as responding to the environment (pregnancy or menopause), adjusting to monthly changes (menstruation), and flexibly changing from reproductivity after puberty to nonreproductivity later in life. The sperm and egg's interaction could also be described in cybernetic terms. J. F. Hartman's research in reproductive biology demonstrated fifteen years ago that if an egg is killed by being pricked with a needle, live sperm cannot get through the zona.[50] Clearly, this evidence shows that the egg and sperm *do* interact on more mutual terms, making biology's refusal to portray them that way all the more disturbing.

We would do well to be aware, however, that cybernetic imagery is hardly neutral. In the past, cybernetic models have played an important part in the imposition of social control. These models inherently provide a way of thinking about a 'field' of interacting components. Once the field can be seen, it can become the object of new forms of knowledge, which in turn can allow new forms of social control to be exerted over the components of the field. During the 1950s, for example, medicine began to recognize the psychosocial *environment* of the patient: the patient's family and its psychodynamics. Professions such as social work began to focus on this new environment, and the resulting knowledge became one way to further control the patient. Patients began to be seen not as isolated, individual bodies, but as psychosocial entities located in an 'ecological' system: management of 'the patient's psychology was a new entrée to patient control.'[51]

The models that biologists use to describe their data can have important social effects. During the nineteenth century, the social and natural sciences strongly influenced each other: the social ideas of Malthus about how to avoid the natural increase of the poor inspired Darwin's *Origin of Species*.[52] Once the *Origin* stood as a description of the natural world, complete with competition and market struggles, it could be reimported into social science as social Darwinism, in order to justify the social order of the time. What we are seeing now is similar: the importation of cultural ideas about passive females and heroic males into the 'personalities' of gametes. This amounts to the 'implanting of social imagery on representations of nature so as to lay a firm basis for reimporting exactly that same imagery as natural explanations of social phenomena.'[53]

Further research would show us exactly what social effects are being wrought from the biological imagery of egg and sperm. At the very least, the imagery keeps alive some of the hoariest old stereotypes about weak damsels in distress and their strong male rescuers. That these stereotypes are now being written in at the level of the *cell* constitutes a powerful move to make them seem so natural as to be beyond alteration.

The stereotypical imagery might also encourage people to imagine that what results from the interaction of egg and sperm – a fertilized egg – is the result of deliberate 'human' action at the cellular level. Whatever the intentions of the human couple, in this microscopic 'culture' a cellular 'bride' (or femme fatale) and a cellular 'groom' (her victim) make a cellular baby. Rosalind Petchesky points out that through visual representations such as sonograms, we are given '*images* of younger and younger, and tinier and tinier, fetuses being "saved".' This leads to 'the point of visibility being "pushed back" *indefinitely*.'[54] Endowing egg and sperm with intentional action, a key aspect of personhood in our culture, lays the foundation for the point of viability being pushed back to the moment of fertilization. This will likely lead to greater acceptance of technological developments and new forms of scrutiny and manipulation, for the benefit of these inner 'persons': court-ordered restrictions on a pregnant woman's activities in order to protect her fetus, fetal surgery, amniocentesis, and rescinding of abortion rights, to name but a few examples.[55]

Even if we succeed in substituting more egalitarian, interactive metaphors to describe the activities of egg and sperm, and manage to avoid the pitfalls of cybernetic models, we would still be guilty of endowing cellular entities with personhood. More crucial, then, than what *kinds* of personalities we bestow on cells is the very fact that we are doing it at all. This process could ultimately have the most disturbing social consequences.

One clear feminist challenge is to wake up sleeping metaphors in science, particularly those involved in descriptions of the egg and the sperm. Although the literary convention is to call such metaphors 'dead', they are not so much dead as sleeping, hidden within the scientific content of texts – and all the more powerful for it.[56] Waking up such metaphors, by becoming aware of when we are projecting cultural imagery onto what we study, will improve our ability to investigate and understand nature. Waking up such metaphors, by becoming aware of their implications, will rob them of their power to naturalize our social conventions about gender.

NOTES

1. James Hillman, *The Myth of Analysis* (Evanston, Ill.: Northwestern University Press, 1972), 220.
2. The textbooks I consulted are the main ones used in classes for undergraduate premedical students or medical students (or those held on reserve in the library for these classes) during the past few years at Johns Hopkins University. These texts are widely used at other universities in the country as well.
3. Arthur C. Guyton, *Physiology of the Human Body*, 6th ed. (Philadelphia: Saunders College Publishing, 1984), 624.
4. Arthur J. Vander, James H. Sherman and Dorothy S. Luciano, *Human Physiology: The Mechanisms of Body Function*, 3rd ed. (New York: McGraw Hill, 1980), 483–4.
5. Vernon B. Mountcastle, *Medical Physiology*, 14th ed. (London: Mosby, 1980), 2: 1624.
6. Eldra Pearl Solomon, *Human Anatomy and Physiology* (New York: CBS College Publishing, 1983), 678.
7. For elaboration, see Emily Martin, *The Woman in the Body: A Cultural Analysis of Reproduction* (Boston: Beacon, 1987), 27–53.
8. Vander, Sherman, and Luciano, 568.
9. Melvin Konner, 'Childbearing and Age', *New York Times Magazine* (December 27, 1987), 22–23, esp. 22.
10. I have found but one exception to the opinion that the female is wasteful: 'Smallpox being the nasty disease it is, one might expect nature to have designed antibody molecules with combining sites that specifically recognize the epitopes on smallpox virus. Nature differs from technology, however: it thinks nothing of wastefulness. (For example, rather than improving the chance that a spermatozoon will meet an egg cell, nature finds it easier to produce millions of spermatozoa.' (Niels Kaj Jerne, 'The Immune System', *Scientific American* 229, no. 1 [July 1973]: 53). Thanks to a *Signs* reviewer for bringing this reference to my attention.
11. Bruce Alberts et al., *Molecular Biology of the Cell* (New York: Garland, 1983), 795.
12. In her essay 'Have Only Men Evolved?' (in *Discovering Reality: Feminist Perspectives on Epistemology, Metaphysics, Methodology, and Philosophy of Science*, ed. Sandra

Harding and Merrill B. Hintikka [Dordrecht: Reidel, 1983], 45–69, esp. 60–61), Ruth Hubbard points out that sociobiologists have said the female invests more energy than the male in the production of her large gametes, claiming that this explains why the female provides parental care. Hubbard questions whether it 'really takes more "energy" to generate the one or relatively few eggs than the large excess of sperms required to achieve fertilization'. For further critique of how the greater size of eggs is interpreted in sociobiology, see Donna Haraway, 'Investment Strategies for the Evolving Portfolio of Primate Females', in *Body/Politics*, ed. Mary Jacobus, Evelyn Fox Keller, and Sally Shuttleworth (New York: Routledge, 1990), 155–6.

13. The sources I used for this article provide compelling information on interactions among sperm. Lack of space prevents me from taking up this theme here, but the elements include competition, hierarchy, and sacrifice. For a newspaper report, see Malcolm W. Browne, 'Some Thoughts on Self Sacrifice', *New York Times* (July 5, 1988), C6. For a literary rendition, see John Barth, 'Night-Sea Journey', in his *Lost in the Funhouse* (Garden City, NY: Doubleday, 1968), 3–13.
14. See Carol Delaney, 'The Meaning of Paternity and the Virgin Birth Debate', *Man* 21, no. 3 (September 1986): 494–513. She discusses the difference between this scientific view that women contribute genetic material to the fetus and the claim of long-standing Western folk theories that the origin and identity of the fetus comes from the male, as in the metaphor of planting a seed in soil.
15. For a suggested direct link between human behavior and purportedly passive eggs and active sperm, see Erik H. Erikson, 'Inner and Outer Space: Reflections on Womanhood', *Daedalus* 93, no. 2 (Spring 1964): 582–606, esp. 591.
16. Guyton (n. 3 above), 619; and Mountcastle (n. 5 above), 1609.
17. Jonathan Miller and David Pelham, *The Facts of Life* (New York: Viking Penguin, 1984), 5.
18. Alberts et al., 796.
19. Ibid., 796.
20. See, e.g., William F. Ganong, *Review of Medical Physiology*, 7th ed. (Los Altos, Calif.: Lange Medical Publications, 1975), 322.
21. Alberts et al. (n. 11 above), 796.
22. Guyton, 615.
23. Solomon (n. 6 above), 683.
24. Vander, Sherman, and Luciano (n. 4 above), 4th ed. (1985), 580.
25. Alberts et al., 796.
26. All biology texts quoted above use the word 'penetrate'.
27. Solomon, 700.
28. A. Beldecos et al., 'The Importance of Feminist Critique for Contemporary Cell Biology', *Hypatia* 3, no. 1 (Spring 1988): 61–76.
29. Gerald Schatten and Helen Schatten, 'The Energetic Egg', *Medical World News* 23 (January 23, 1984): 51–53, esp. 51.
30. Alberts et al., 796.
31. Guyton (n. 3 above), 613.
32. Miller and Pelham (n. 17 above), 7.
33. Alberts et al. (n. 11 above), 804.
34. Ibid., 801.
35. Ruth Herschberger, *Adam's Rib* (New York: Pelligrini & Cudaby, 1948), esp. 84. I am indebted to Ruth Hubbard for telling me about Herschberger's work, although at a point when this paper was already in draft form.
36. Bennett M. Shapiro. 'The Existential Decision of a Sperm', *Cell* 49, no. 3 (May 1987): 293–94, esp. 293.
37. Lennart Nilsson, 'A Portrait of the Sperm', in *The Functional Anatomy of the Spermatozoan*, ed. Bjorn A. Afzelius (New York: Pergamon, 1975), 79–82.
38. Ludwik Fleck, *Genesis and Development of a Scientific Fact*, ed. Thaddeus J. Trenn and Robert K. Merton (Chicago: University of Chicago Press, 1979), 38.

39. Jay M. Baltz carried out the research I describe when he was a graduate student in the Thomas C. Jenkins Department of Biophysics at Johns Hopkins University.
40. Far less is known about the physiology of sperm than comparable female substances, which some feminists claim is no accident. Greater scientific scrutiny of female reproduction has long enabled the burden of birth control to be placed on women. In this case, the researchers' discovery did not depend on development of any new technology. The experiments made use of glass pipettes, a manometer, and a simple microscope, all of which have been available for more than one hundred years.
41. Jay Baltz and Richard A. Cone, 'What Force Is Needed to Tether a Sperm?' (abstract for Society for the Study of Reproduction, 1985), and 'Flagellar Torque on the Head Determines the Force Needed to Tether a Sperm' (abstract for Biophysical Society, 1986).
42. Jay M. Baltz, David F. Katz and Richard A. Cone, 'The Mechanics of the Sperm-Egg Interaction at the Zona Pellucida', *Biophysical Journal* 54, no. 4 (October 1988): 643–54. Lab members were somewhat familiar with work on metaphors in the biology of female reproduction. Richard Cone, who runs the lab, is my husband, and he talked with them about my earlier research on the subject from time to time. Even though my current research focuses on biological imagery and I heard about the lab's work from my husband every day, I myself did not recognize the role of imagery in the sperm research until many weeks after the period of research and writing I describe. Therefore, I assume that any awareness the lab members may have had about how underlying metaphor might be guiding this particular research was fairly inchoate.
43. Ibid., 643, 650.
44. Baltz, Katz, and Cone (n. 42 above), 643, 650.
45. Schatten and Schatten, (n. 29 above), 53.
46. Paul M. Wassarman, 'The Biology and Chemistry of Fertilization', *Science* 235, no. 4788 (January 30, 1987): 553–60, esp. 554.
47. Mary Ellman, *Thinking about Women* (New York: Harcourt Brace Jovanovich, 1968), 140; Nina Auerbach, *Woman and the Demon* (Cambridge, Mass.: Harvard University Press, 1982), esp. 186.
48. Kenneth Alan Adams, 'Arachnophobia: Love American Style', *Journal of Psychoanalytic Anthropology* 4, no. 2 (1981): 157–97.
49. William Ray Arney and Bernard Bergen, *Medicine and the Management of Living* (Chicago: University of Chicago Press, 1984).
50. J. F. Hartman, R. B. Gwatkin and C. F. Hutchison, 'Early Contact Interactions between Mammalian Gametes *In Vitro*', *Proceedings of the National Academy of Sciences (US)* 69, no. 10 (1972): 2767–69.
51. Arney and Bergen, 68.
52. Ruth Hubbard, 'Have Only Men Evolved?' (n. 12 above), 51–52.
53. David Harvey, personal communication, November 1989.
54. Rosalind Petchesky, 'Fetal Images: The Power of Visual Culture in the Politics of 'Reproduction', *Feminist Studies* 13, no. 2 (Summer 1987): 263–92, esp. 272.
55. Rita Arditti, Renate Klein and Shelley Minden, *Test-Tube Women* (London: Pandora, 1984); Ellen Goodman, 'Whose Right to Life?' *Baltimore Sun* (November 17, 1987); Tamar Lewin, 'Courts Acting to Force Care of the Unborn', *New York Times* (November 23, 1987), A1 and B10; Susan Irwin and Brigitte Jordan, 'Knowedge, Practice, and Power: Court Ordered Cesarean Sections', *Medical Anthropology Quarterly* 1, no. 3 (September 1987): 319–34.
56. Thanks to Elizabeth Fee and David Spain, who in February 1989 and April 1989, respectively, made points related to this.

3.6

DISCIPLINING MOTHERS: FEMINISM AND THE NEW REPRODUCTIVE TECHNOLOGIES

Jana Sawicki

Although he intended to, Foucault never wrote a history of women's bodies. Yet, had he proceeded according to his original plan, he would have written a volume in the *History of Sexuality* entitled *Woman, Mother and Hysteric*. It was to be a study of the sexualization of women's bodies and of concepts of pathology related to it such as hysteria, neurasthenia, and frigidity. Foucault intended to locate the processes through which women's bodies were controlled through a set of discourses and practices governing both the individual's body and the health, education and welfare of the population, namely, the discourses and practices of 'biopower'.[1]

In contrast to the often sporadic, violent power over a relatively anonymous social body exercised under older, monarchical forms of power, biopower emerges as an apparently benevolent, but peculiarly invasive and effective form of social control. It evolved in two basic and inter-related forms. One of these, disciplinary power, is a knowledge of and power over the individual body – its capacities, gestures, movements, location, and behaviors. Disciplinary practices represent the body as a machine. They aim to render the individual both more powerful, productive, useful *and* docile. They are located within institutions such as hospitals, schools, and prisons, but also at the microlevel of society in the everyday activities and habits of individuals. They secure their hold not through the threat of violence or force, but rather by creating desires, attaching individuals to specific identities, and establishing norms against

From: J. Sawicki, *Disciplining Foucault. Feminism, Power and the Body*, New York: Routledge, 1991.

which individuals and their behaviors and bodies are judged and against which they police themselves.

The other form of biopower is a regulatory power inscribed in policies and interventions governing the population. This so-called 'biopolitics of the population' is focused on the 'species body', the body that serves as the basis of biological processes affecting birth, death, the level of health and longevity. It is the target of state interventions and the object of study in demography, public health agencies, health economics and so forth.

If, as Foucault claimed, biopower was an indispensable element in the development of capitalism insofar as it made possible a 'controlled insertion of bodies into the machinery of production', then it must also have been indispensable to patriarchal power insofar as it provided instruments for the insertion of women's bodies into the machinery of reproduction.[2] And if claiming a right to one's body only makes sense against the background of these new life-administering forms of power and knowledge, then the history of modern feminist struggles for reproductive freedom is a key dimension of the history of biopower.

[...]

THE 'ORIGINS' OF NEW REPRODUCTIVE TECHNOLOGIES: A FOUCAULDIAN ACCOUNT

Told from a Foucauldian perspective, the history of women's procreative bodies is a history with multiple origins, that is, a history of multiple centers of power, multiple innovations, with no discrete or unified origin. It is a history marked by resistance and struggle. Thinking specifically about the history of childbirth in America, a Foucauldian feminist does not assume a priori that the new reproductive technologies are the product of a long standing male 'desire' to control women's bodies or to usurp procreation. This does not mean that such motives do not play a role in this history of medicalization, but it does deny that they direct the historical process overall.[3]

Foucault described the social field as a network of intersecting practices and discourses, an interplay of non-egalitarian, shifting power relations. Individuals and groups do not possess power but rather occupy various and shifting positions in this network of relations – positions of power and resistance. Thus, although policies governing reproductive medicine and new reproductive technologies in the United States today are indeed largely controlled by non-feminist and anti-feminist forces, it is plausible to assume that women and feminists have played a role in defining past and current practices, for better or worse. It is also the case that these non- and anti-feminist forces are not unified or monolithic. Their control is neither total nor centrally orchestrated.

Employing a bottom-up analysis, a Foucauldian feminist would describe the present situation as the outcome of a myriad of micro-practices, struggles, tactics and counter-tactics among such agencies. Consequently, in describing the history of childbirth practices, she would focus not only on the dominant discourses and practices, namely, those of medical experts and the so-called 'technodocs', but also on the moments of resistance that have resulted in transforming these practices over the years.[4] Indeed, there are many discourses and practices in the contexts of medicine, law, religion, family planning agencies, consumer protection agencies, the insurance and pharmaceutical industries, the women's health movement, and social welfare agencies that struggle to influence reproductive politics and the social construction of motherhood. As Paula Treichler states:

> The position of modern medicine was not monolithic but emerged gradually in the course of key debates, federal initiatives, strains between private practitioners and academic physicians, and debates within medicine over what its professional hierarchy was to be ... Physicians did not uniformly declare a war on nature, nor decide that they should adopt an ideology of intervention and subordination.[5]

Medicalized childbirth has come under attack from many camps since the birth of modern medicine. Individual men and women as well as organized groups representing scientific, economic and feminist interests have consistently challenged the Western medical model of childbirth. In the second half of this century there have been continual demands for alternatives to the medical model of childbirth. For example, after World War II, Grantly Read, a British obstetrician criticized physicians for ignoring women's subjective experience of childbirth and for interfering with the natural birthing process. Lamaze was introduced and became increasingly popular. Natural childbirth was reintroduced as an option and a home-birth movement emerged. La Leche League encouraged women to disregard medical advice that favored bottle feeding and encouraged them to return to breastfeeding. In addition, there were proposals to admit fathers into delivery rooms, to eradicate the routine use of the lithotomy position, and to stop separating mothers and babies at birth. Both feminists and non-feminist critics have challenged the routine use of episiotomy, and of drugs for pain and the induction of labor. Furthermore, individual women attempt to control the terms of their own hospital childbirths by staying home longer before going into the hospital and thereby avoiding unnecessary C-sections due to prolonged labor, by demanding that they have an advocate present during the birthing process, by finding physicians who support their desire to minimize medical intervention, and so forth. Such resistance has served as the basis of forms of client resistance and has worked to counter tendencies toward depoliticizing motherhood and childbirth.

Highlighting the struggles surrounding the definition of childbirth and motherhood does not suggest that medicine has not had a monopoly over childbirth during this century, but rather that this control was not simply imposed from the top down. It had to be won and continually faces resistance. Nor do I mean to imply that challenges to medicine such as those advocating 'natural' childbirth have not been coopted to some extent. After all, sometimes resistance is wholly neutralized by the counter strategies of the hegemonic, white, upper middle-class, heterosexual scientific and medical establishments. For example, many natural childbirth classes offered in hospitals continue to train women to expect and accept medical interventions such as fetal monitors, intravenous drugs, labor induction, forceps, pain medication, and even Caesarean sections.[6] The ideology of the 'natural' has been used in the service of both feminist and anti-feminist struggles. Regardless, identifying such struggles undercuts assumptions that the medical control of childbirth is one-dimensional or total.

DISCIPLINING MOTHERS

As I have indicated, Foucault identified the history of women's bodies as a key dimension of the history of biopower. Understood as part of this history, new reproductive technologies represent the most recent of a set of discourses (systems of knowledge, classification, measurement, testing, treatment and so forth) that constitute a disciplinary technology of sex that was developed and implemented by the bourgeoisie at the end of the eighteenth century as a means of consolidating its power, improving itself, 'maximizing life'. Disciplinary technologies are not primarily repressive mechanisms. In other words, they do not operate primarily through violence against or seizure of women's bodies or bodily processes, but rather by producing new objects and subjects of knowledge, by inciting and channeling desires, generating and focusing individual and group energies, and establishing bodily norms and techniques for observing, monitoring, and controlling bodily movements, processes, and capacities. Disciplinary technologies control the body through techniques that simultaneously render it more useful, more powerful and more docile.

New reproductive technologies represent one of a series of types of body management that have emerged over the past two decades rendering women's bodies more mobilizable in the service of changing utilities of dominant agencies.[7] Their aim is less to eliminate the need for women than to make their bodies even more useful. They enhance the utility of women's bodies for multiple shifting needs. As Linda Singer aptly noted:

> The well managed body of the 80s is constructed so as to be even more multifunctional than its predecessors. It is a body that can be used for wage, labor, sex, reproduction, mothering, spectacle, exercise, or even invisibility, as the situation demands.[8]

Singer points out that fertility technologies can be used either for purposes of consolidating race and class privilege or for eliminating competition in the labor market from white, upper middle-class women who have delayed pregnancy for careers.

New reproductive technologies clearly fit the model of disciplinary power. They involve sophisticated techniques of surveillance and examination (for instance, ultrasound, fetal monitors, amniocentisis, antenatal testing procedures) that make both female bodies and fetuses visible to anonymous agents in ways that facilitate the creation of new objects and subjects of medical as well as legal and state intervention. Among the individuals created by these new technologies are infertile, surrogate and genetically impaired mothers, mothers whose bodies are not fit for pregnancy (either biologically or socially), mothers who are psychologically unfit for fertility treatments, mothers whose wombs are hostile environments to fetuses, mothers who are deemed 'negligent' for not choosing to undergo tests, abort genetically 'deficient' fetuses, or consent to Caesarean sections. As these medical disciplines isolate specific types of abnormality or deviancy, they contruct new norms of healthy and responsible motherhood. Additionally, insofar as the new technologies locate the problem of infertility within individuals, they deflect attention and energy that could be used to address the environmental causes of infertility. Hence, they tend to depoliticize infertility. They link up with the logic of consumerism and commodification by inciting the desire for 'better babies' and by creating a market in reproductive body parts, namely, eggs, wombs, and embryos. Finally, they make women's bodies useful to agencies that regulate and coordinate populations.

At the same time that these new technologies create new subjects – that is, fit mothers, unfit mothers, infertile women, and so forth – they create the possibility of new sites of resistance. Lesbians and single women can challenge these norms by demanding access to infertility treatments. Women who have undergone infertility treatment can share their experiences and demand improvements or expose inadequacies in the model of treatment. The question is not whether these women are victims of false consciousness insofar as they desire to be biological mothers, as much as it is one of devising feminist strategies in struggles over who defines women's needs and how they are satisfied.

To suggest that the new reproductive technologies 'produce' problems and desires and thereby contribute to the further medicalization of mothers' bodies is not to suggest that these problems (for instance, infertility) are not real, that the experts are charlatans, and that those who seek their advice are blinded by the ideology of medical science. It does not imply that things were better before these technologies appeared. It does suggest, however, that part of the attraction of the new technologies is that many women perceive them as enabling. Of course, referring to them as disciplinary technologies does highlight their controlling functions. Yet, this control is not secured primarily through violence or coercion, but rather by producing new norms of motherhood, by attaching women to their identities as mothers, and by offering women specific kinds of solutions to problems they face. In fact, there may be better solutions; and there may be better ways of defining the problems. There is the danger that medical solutions will become the only ones and that other ways of defining them will be eclipsed.

This emphasis on normalization as opposed to violence represents a major advantage of the disciplinary model of power. If patriarchal power operated primarily through violence, objectification and repression, why would women subject themselves to it willingly? On the other hand, if it also operates by inciting desire, attaching individuals to specific identities, and addressing real needs, then it is easier to understand how it has been so effective at getting a grip on us.

Moreover, although the model of patriarchy as violence against women is appropriate in many contexts, I question its use in the context of critiques of medicine. Are all forms of objectification, even those that involve inequalities of power, inevitably violent? While many forms of surgical intervention are experienced as traumatic, are they best described as a form of 'violence against the body'? Emily Martin has written that Foucault was wrong to claim that the violent tactics of juridical or monarchical power have been replaced by the more subtle tactics of disciplinary power. She states: 'dismemberment is with us still, and the "hold on the body" has not so much slackened as it has moved from the law to science.'[9] Without suggesting that there have not been many incidences of unjust or callous use of power over women in medical contexts, it is important to avoid reducing all of Western medical science and technology to another example of violence against women. Many of its practices are clearly distinct from sadistic or coercive violence. At the very least the rhetoric of violence is likely to be politically ineffective since it does not resonate with so many of the women who must rely on medical institutions for health care.

This is only one of several political disadvantages of the repressive model of power with which radical feminists have operated. Another related difficulty is that it employs a binary model of alternatives, either repressive technology or a liberatory one, either a masculinist science or a feminist one, either mechanistic materialism or naturalism, either a technological approach or a natural one. This politically and cognitively restrictive binary logic stems in part from the tendency to portray patriarchal power in monolithic, essentialist and totalistic terms. It is limiting because it detemporalizes the process of social change by conceiving of it as a negation of the present rather than as emerging from possibilities in the present. In so doing, it restricts our political imaginations and keeps us from looking for the ambiguities, contradictions and liberatory possibilities in the technological tranformations of conception, pregnancy and childbirth.

The repressive model of power assumes that all women and men occupy essentially the same position in relation to patriarchy, namely, that of victims who are blinded by the ideology of science or perpetrators of violence, respectively. Like the discourses and practices they criticize, radical feminist discourses often position women as passive subjects not potential activists, as causally conditioned not self-determining, as morally or politically corrupted. Thus, they fail to take some women's expressed 'needs' (for fertility treatment, for genetic screening) seriously.[10] They provide inadequate explanations of how some women's interests appear to be bound up with the system of male domination. They also ignore the fact that some men, even physicians, are potential allies in struggles against domination.

In contrast, there are significant political advantages to adopting Foucault's disciplinary model of power for a feminist critique of new reproductive technologies. Operating with a model of the social field as a field of struggle consisting of multiple centers of power confronting multiple centers of resistance prompts us to look for the diverse relationships that women occupy in relation to these technologies, and for the many intersecting subject positions constituting the social field. We become focused on the pregnant or infertile woman in all of her social relationships, not simply her relationship to the physician. She may be a working woman, a welfare mother, a woman of color, a drug addict, a physician, a lawyer, a feminist. By directing our attention to the differences among women and to the intersecting social relations in which women are situated, we are more likely to locate the conflicting meanings and contradictions associated with the technological transformation of pregnancy.

Although it is crucial to continue to identify the ways in which new reproductive technologies threaten to erode women's power over their reproductive lives, it is also important to locate the potential for resistance in the current social field, that is, what Foucault refers to as 'subjugated knowledges' – forms of experience and knowledge that 'have been disqualified as inadequate ... or insufficiently elaborated: naive knowledges, located low down in the hierarchy beneath the required level of cognition or scientificity.'[11] This means looking not only at the discourses of the men who develop and implement the technologies, but also at the different ways in which women are being affected by them, that is, the material conditions of their lives, their own descriptions of their needs, and of their experiences of pregnancy and childbirth. Not all women have equal access to the most advanced medical technologies. For the majority of women, withdrawal from medical institutions is not an option. Indeed, inadequate access to health care and to information is the key issue for a majority of women in the United States today.

Despite her rejection of Foucault's position on violence, Emily Martin provides an example of such an approach in her book *The Woman in the Body*. She explores differences between middle-class and working-class women's experiences of pregnancy and childbirth and juxtaposes them with dominant mechanistic scientific and medical accounts of these same processes. She concludes that the dominant ideology is partial, that middle-class women are more inclined toward the medical view of themselves, and that working-class women described their experiences more in terms of non-medical aspects of their lives. She suggests that there is liberatory potential in resurrecting these subjugated discourses of bodily experience – that they might serve as critical feminist standpoints on medical discourse and practice.

Martin's observations are consistent with those of Rayna Rapp in her writings on amniocentesis and genetic counselling.[12] Rapp observes that the group best served by advances in reproductive medicine – white middle-class women – are also the most vulnerable to its powerful definitions of motherhood. This is the group of women who are most likely to become agents of fetal quality control. In her studies, Rapp found that the interpretations given of positive test results for Down's syndrome varies with differences in class, race or ethnicity. For example to a Hispanic woman in the urban ghetto whose 'normal' children confront serious obstacles to self-actualization, the meaning of being 'disabled' may be different. She may be more likely to opt to have a disabled child than her white middle-class counterpart.

Thus, although new reproductive technologies certainly threaten to reproduce and enhance existing power relations, they also introduce new possibilities for disruption and resistance.[13] Using Foucault's model of power as a shifting and unstable set of relations, and his understanding of discourses as ambiguous and polyvalent, we are encouraged to look for such possibilities in the present and to mobilize them as a means of challenging hegemonic reproductive relations on a variety of political fronts. As reproductive issues are increasingly taken up as cultural, not simply biological issues, more space is opened up for politicizing them. For example, Rapp suggests that the idea of quality control of fetuses can be used to support demands for adequate prenatal care for all women. And, ultimately, as these technologies destabilize current conceptions of motherhood, opportunities for identifying and legitimating alternative forms of motherhood are presented.

'Motherhood' and 'technology' are highly contested concepts in contemporary America. As an identity, an ideology and an institution, motherhood has been both a source of power and enslavement for women. But it is important for feminists not to assume that any one aspect of female practice is central to partriarchal control. We must not conflate motherhood and femininity, or motherhood and childbirth. Motherhood has biological meanings, but also many social ones. In arguing that new reproductive technologies may lead to the elimination of women altogether, Corea and others ignore the many, often conflicting, roles and positions that women occupy in contemporary society, the many services they provide, the many other ways in which their bodies are disciplined as mothers, workers, housewives, sexual beings, and so forth. Sandra Bartky suggests that normative femininity – a set of disciplinary practices regulating the body, its gestures, appetite, shape, size, movement, appearance and so forth – has come to center more on sexuality and appearance than on the maternal body.[14] Whether this is the case or not, it highlights the fact that women's bodies have many uses. Feminists must resist those forces that aim to enlist such practices in the service of docility and gender normalization and struggle to define them differently. But this is not tantamount to rejecting them entirely.

Similarly, there are many possibilities for a technological transformation of pregnancy that might benefit women. Of course, we cannot overlook the fact that scientific and technological practices are largely controlled by men. Nevertheless, there are good reasons to avoid reducing patriarchal domination to its technologies. As history reveals, technological developments are many edged. Who, in retrospect, would deny women many of the contraceptive technologies that were developed and introduced in this century? Both feminists and anti-feminists resisted the legalization of birth control. Feminists saw birth control as a means for men to escape their responsibility to women. Anti-feminists feared that if women had more control over their biological reproductive processes, they would reject their social roles as mothers and wives. (Like some radical feminists today, they tended to conflate control over the biological process of motherhood with control over motherhood itself.) As Linda Gordon points out, birth control has been a progressive development for women only to the extent that the women's movement has continually struggled to define policies regulating their development and use.[15] But rather than reject newly emerging technologies outright, feminists can meet multiple-edged developments with multiple-edged responses. We can resist the dangerous trends, the tendencies toward depoliticization, privatization, decreased autonomy, and the elision of women's experiences and interests in the process of developing and implementing reproductive technologies and the laws and policies regulating them. We can also allow ourselves to envision utopian possibilities for technologically transforming reproduction.[16]

Finally, at the same time that people in developed countries embrace new technologies with little resistance, there is also evidence in film, literature and the media of increasing cultural anxiety about the pace of technological change. The prospects of nuclear accidents or nuclear warfare loom large in the collective psyche. This ambivalence about technology can be tapped in efforts to democratize the process of technological innovation, design and implementation.

In an intriguing essay, Donna Haraway argues that images of the cyborg, of couplings between organism and machine, provide an imaginative resource for contemporary feminist politics.[17] Without either wholly natural or technological origins, cyborgs

> are not afraid of their joint kinship with animals and machines, not afraid of permanently partial identities and contradictory standpoints. The political struggle is to see from both perspectives at once because each reveals both dominations and possibilities unimaginable from the other vantage point.[18]

Similarly, Foucault's understanding of power as decentralized, as a myriad of shifting relations, enables us to avoid the extremes of dystopian or utopian political critique in favor of locating many positive and negative political and strategic possibilities presented in the present. From a Foucauldian perspective, every strategy or counter-strategy is potentially dangerous. Appeals to a more holistic, unified, natural', 'maternal', or 'feminine' experience of childbirth become merely one of several strategies that we might deploy in efforts to resist the medical takeover of women's bodies. In themselves, they are no less cooptable than high technology approaches. Indeed, holism and naturalism can and have been used for patriarchal as well as feminist ends.

De-medicalization is also not sufficient as a strategy for resisting the hegemonic forces that govern our bodies under patriarchal capitalism. De-medicalizing childbirth does remove it from this authoritarian context and open up more possibilities for contesting its meaning. Nevertheless, as Paula Treichler notes, de-medicalization also brings risks. It places childbirth in the public sphere where it 'can more easily be represented as a commodity, not only in the economic marketplace but in the ideological and social marketplace as well.'[19] Moreover, an open marketplace provides opportunities for exploitation and abuse. After all, infertility, pregnancy, and childbirth *are* partly medical issues. Rather than remove them from medical control, we can support efforts to build health care institutions that enable women and those whom they love to structure childbearing around their own needs. This will surely include expanding access to a variety of reproductive services that are currently made available only to privileged women, but will also require that we question current uses and modes of implementation.

What makes new reproductive technologies especially dangerous to women is not so much that they objectify and fragment bodily processes, but that they are designed and implemented by experts in contexts where scientific and medical authority is wielded with insufficient attention to the prerequisites for democratic or shared decision-making. The often unchallenged authority of experts makes possible an imposition of treatments and regimes that is in fact dangerous to women. Physicians and health care practictioners must be exhorted to further efforts to ensure that women are not treated solely as bodies, but also as subjects with desires, fears, special needs, and so forth. Of course, as I have attempted to show, many of the new reproductive techologies do operate through modes of 'subjectification', that is, by classifying and identifying subjects in efforts to further control them. So, attending to subjectivity is not a sufficient condition for ensuring that the pernicious effects of Western models of medicine are combatted. But it is necessary if individuals are to have more control over how their medical needs are satisfied. The authority of Western medicine has been challenged increasingly during the past three decades. Memories of thalidomide babies and the scandal of DES can be redeployed. Many individuals and groups resist medical authority and treatment everyday, for better or worse. Such resistance can be mobilized in efforts to eradicate the non-reciprocal relations of power so often still characteristic of the physician–patient relationship.

As I have already suggested, feminists are not the only ones resisting developments in reproductive medicine. We can also coalesce with other movements challenging current forms of organization and authority within health care institutions. We can build political unities not on the basis of some naturalized identity as women, or mothers, but on the basis of common political opposition and affinities with other political struggles. We can also continue to make demands for equal access to health care, for better information, and for more democratic processes of developing, designing, implementing and regulating new technologies.

Analyses that simply reject new reproductive technologies do not assist women in making choices. Nor do they lead to creative political strategies. We must provide analyses that enable women to assess risks and benefits – both individual and social, and that facilitate feminist and other oppositional struggles that are already ongoing in the context of health care institutions. We must provide analyses that bolster feminist political struggles for economic resources, information, access to health care, shared decision-making. We must build alliances across race, class, sexual differences and differences in ability.

On the basis of the Foucauldian analysis suggested here, one might conclude that some of the budding biotechnological developments should be resisted altogether at the present time. Up to this point I have treated new reproductive technologies as a whole, primarily for the sake of a more general metatheoretical analysis. Still, any adequate feminist analysis of new reproductive technologies must treat them separately. Of course, one must also pay attention to the ways in which they overlap and reinforce one another. Use of one technology often leads to or even requires use of another. Accordingly, if one opts for *in vitro* fertilization, one is also likely undergo the whole range of antenatal testing procedures as well as Caesarean section. But this does not mean that each of the procedures carries the same dangers or offers the same possibilities to women. We may ultimately want to preserve some and eliminate others.

In vitro fertilization is especially suspect. Success rates are exceedingly low, the procedures are physically and psychologically gruelling, and the health risks they pose to women have not been adequately measured. Most importantly, because *in vitro* techniques require significant expertise and are very expensive, much of the control over how they are implemented and who is eligible to receive them is monopolized by the scientists, technicians and administrators who offer them. At present, *in vitro* fertilization is available primarily to married, white, upper middle-class women who perceive biological motherhood as very desirable and who convince practitioners that they desire a child enough to withstand the treatment. The criteria for eligibility reinforce a traditional classist, racist, and heterosexist ideology of fit motherhood. Moreover, these women have little control over the process once they agree to undergo treatment. Finally, given that *in vitro* fertilization could be described as a failure due to such low success rates, we might ask what function this failure is serving. Scientists interested in research on human embryos have a vested interest in promoting the benefits of *in vitro* fertilization partly because the techniques employed yield 'surplus' embryos. As I am writing this, there is pressure on Congress to lift the ban on human embryo research. In short, there are reasons to believe that *in vitro* clinics are not really serving women at all.

But, in general, our conclusions about the value of these technologies will be reached on the basis of a different analysis – one that is more inclusive, pluralistic, and complex, one that looks for their ambiguous implications and identifies the many agencies struggling to define motherhood and childbirth in contemporary America in an effort to mobilize oppositional forces.

New reproductive technologies clearly threaten to make women's procreative bodies more effective targets for the intervention of hegemonic patriarchal and capitalist forces in contemporary America. As disciplinary technologies they represent a potentially insidious form of social control since they operate by inciting the desires of those who seek them out. But the question whether these techniques also offer more liberatory possibilities will depend on the extent to which mechanisms for resisting their pernicious disciplinary implications are devised. Feminist efforts to identify strategies of resistance will be aided by analyses that move beyond moralism, or nostalgia for a more female-centered era of 'natural' childbirth, and begin to look for the complex connections between the power/knowledge relations of biopower and other factors influencing sexual struggle. I have argued that a Foucauldian feminist approach is more likely to produce such analyses.[20]

NOTES

1. See Michel Foucault, *The History of Sexuality, Vol. 1: An Introduction*, trans. Robert Hurley (New York: Pantheon, 1978), pp. 140ff.
2. Ibid., p. 141.
3. As Susan Bordo points out in her feminist appropriation of Foucault, denying that the history of women's bodies is the product of patriarchal conspiracies

   > does not mean that individuals do not consciously pursue goals that advance their own positions, and advance certain power positions in the process. But it does deny that in doing so they are directing the overall movement of relations, or engineering their shape. They may not ever know what the shape is. Nor does the fact that power relations involve the domination of particular groups – say, prisoners by guards, females by males, amateurs by experts – entail that the dominators are in control of the situation, or that the dominated do not sometimes advance and extend the situation themselves.

   See her 'Anorexia Nervosa: Psychopathology as the Crystallization of Culture', in *Feminism and Foucault: Reflections on Resistance*, eds Irene Diamond and Lee Quinby (Boston: Northeastern University Press, 1988), p. 91.
4. See especially the work of Teresa de Lauretis.
5. Paula A. Treichler, 'Feminism, Medicine and the Meaning of Childbirth', in *Body/Politics: Women and the Discourses of Science*, eds Mary Jacobus, Evelyn Fox Keller, and Sally Shuttleworth (New York: Routledge, 1990), p. 118.
6. See D. A. Sullivan and R. Weitz, *Labor Pains: Modern Midwives and Homebirths* (New Haven: Yale University Press, 1988), p. 38.
7. See Susan Bordo's 'Anorexia Nervosa' and Sandra Bartky's 'Foucault, Femininity and Patriarchal Power', in *Foucault and Feminism* for two examples of analyses of how disciplinary practices produce specifically feminine forms of embodiment through the development of dietary and fitness regimens, pathologies related to them, and expert advice on how to walk, talk, dress, wear make-up, and so forth.
8. Linda Singer, 'Bodies, Pleasures, Powers', *differences*, Vol. 1 (Winter 1989), p. 57.
9. Emily Martin, *The Woman in the Body: A Cultural Analysis of Reproduction* (Boston: Beacon Press, 1987), p. 21.
10. For a fascinating account of a specific set of struggles over the interpretation of needs, see Nancy Fraser, 'Women, Welfare and the Politics of Needs Interpretation', in *Unruly Practices: Power, Discourse and Gender in Contemporary Social Theory* (Minneapolis: University of Minnesota Press, 1989), pp. 144–60.

11. Michel Foucault, 'Two Lectures', in *Power/Knowledge: Selected Interviews and Other Writings, 1972–1977*, ed. Colin Gordon (New York: Pantheon Books, 1980), p. 82.
12. 'Reproduction and Gender Hierarchy: Amniocentesis in Contemporary America', paper presented at an International Symposium of the Wenner-Gren Foundation for Anthropological Research, January 10–18, 1987, Mijas, Spain.
13. Ibid., p. 24.
14. See Sandra Lee Bartky, 'Foucault, Femininity and the Modernization of Patriarchal Power'.
15. Linda Gordon, 'The Struggle for Reproductive Freedom: Three Stages of Feminism', in *Capitalist Patriarchy and the Case for Socialist Feminism*, ed. Zillah Eisenstein (New York: Monthly Review Press, 1979), pp. 107–32.
16. See Marge Piercy, *Woman on the Edge of Time* (New York: Knopf, 1976), for a futuristic model of high technology childbirth that embodies feminist principles.
17. Donna Haraway, 'A Manifesto for Cyborgs: Science, Technology and Socialist Feminism in the 1980s', *Socialist Review, Vol.* 80 (1985), p. 66.
18. Ibid., p. 72.
19. Treichler, 'Feminism, Medicine and the Meaning of Childbirth', p. 131.
20. Earlier versions of this paper were presented to audiences at Vassar College, MIT, Northwestern University, and Notre Dame University. I wish to thank them for their helpful questions and comments. Special thanks go to Iris Young, Linda Singer, Nancy Fraser, Linda Nicholson, Sharon Barker and members of Sofphia (Socialist Feminist Philosophers) for helpful conversations and comments on previous drafts.

3.7

THE BIOPOLITICS OF POSTMODERN BODIES: DETERMINATIONS OF SELF IN IMMUNE SYSTEM DISCOURSE

Donna Haraway

For Robert Filomeno, who loved peace and died of AIDS, and with thanks to Scott Gilbert, Rusten Hogness, Rayna Rapp, and Joan Scott.

> If Koch's postulates must be fulfilled to identify a given microbe with a given disease, perhaps it would be helpful, in rewriting the AIDS text, to take 'Turner's postulates' into account (1984: 209): 1) disease is a language; 2) the body is a representation; and 3) medicine is a political practice. (Treichler 1987: 27)

> Non-self: A term covering everything which is detectably different from an animal's own constituents. (Playfair 1984: 1)

> . . . immune system must recognize self in some manner in order to react to something foreign. (Golub 1987: 482)

Lumpy Discourses and the Denatured Bodies of Biology and Medicine

It has become commonplace to emphasize the multiple and specific cultural dialects interlaced in any social negotiation of disease, illness, and sickness in the contemporary worlds marked by biological research, biotechnology, and scientific medicine. The language of biomedicine is never alone in the field of empowering meanings, and its power does not flow from a consensus about symbols and actions in the face of suffering. Paula Treichler's excellent phrase in the title of her essay on the constantly contested meanings of AIDS as an 'epidemic of signification' could be applied widely to the social text of sickness. The power of biomedical language – with its stunning artifacts, images, archi-

From: *differences: A Journal of Feminist Cultural Studies* 1 (1): 3–43, 1989.

tectures, social forms, and technologies – for shaping the unequal experience of sickness and death for millions is a social fact deriving from ongoing heterogeneous social processes. The power of biomedicine and biotechnology is constantly reproduced, or it would cease. This power is not a thing fixed and permanent, embedded in plastic and ready to section for microscopic observation by the historian or critic. The cultural and material authority of biomedicine's productions of bodies and selves is more vulnerable, more dynamic, more elusive, and more powerful than that.

But if there has been recognition of the many non-, para-, anti-, or extra-scientific languages in company with biomedicine that structure the embodied semiosis of mortality in the industrialized world, it is much less common to find emphasis on the multiple languages *within* the territory that is often so glibly marked scientific. 'Science says' is represented as a univocal language. Yet even the spliced character of the potent words in 'science' hints at a barely contained and inharmonious heterogeneity. The words for the overlapping discourses and their objects of knowledge, and for the abstract corporate names for the concrete places where the discourse-building work is done, suggest both the blunt foreshortening of technicist approaches to communication and the uncontainable pressures and confusions at the boundaries of meanings within 'science' – biotechnology, biomedicine, psychoneuroimmunology, immunogenetics, immunoendocrinology, neuroendocrinology, monoclonal antibodies, hybridomas, interleukines, Genentech, Embrex, Immunetech, Biogen.

This paper explores some of the contending popular and technical languages constructing biomedical, biotechnical bodies and selves in postmodern scientific culture in the United States in the 1980s. Scientific discourses are 'lumpy'; they contain and enact condensed contestations for meanings and practices. The chief object of my attention will be the potent and polymorphous object of belief, knowledge, and practice called the immune system. My thesis is that the immune system is an elaborate icon for principal systems of symbolic and material 'difference' in late capitalism. Pre-eminently a twentieth-century object, the immune system is a map drawn to guide recognition and misrecognition of self and other in the dialectics of western biopolitics. That is, the immune system is a plan for meaningful action to construct and maintain the boundaries for what may count as self and other in the crucial realms of the normal and the pathological. The immune system is an historically-specific terrain, where global and local politics; Nobel prize-winning research; heteroglossic cultural productions, from popular dietary practices, feminist science fiction, religious imagery, and children's games, to photographic techniques and military strategic theory; clinical medical practice; venture capital investment strategies; world-changing developments in business and technology; and the deepest personal and collective experiences of embodiment, vulnerability, power, and mortality, interact with an intensity matched perhaps only in the biopolitics of sex and reproduction.[2]

The immune system is both an iconic mythic object in high-technology culture and a subject of research and clinical practice of the first importance.

Myth, laboratory, and clinic are intimately interwoven. This mundane point was fortuitously captured in the title listings in the 1986–87 *Books in Print*, where I was searching for a particular undergraduate textbook on immunology. The several pages of entries beginning with the prefix 'immuno-' were bounded, according to the English rules of alphabetical listing, by a volume called *Immortals of Science Fiction*, near one end, and by *The Immutability of God*, at the other. Examining the last section of the textbook to which *Books in Print* led me, *Immunology: A Synthesis* (Golub 1987), I found what I was looking for: an historical progression of diagrams of theories of immunological regulation and an obituary for their draftsman, an important immunologist, Richard K. Gershon, who 'discovered' the suppressor T cell. The standard obituary tropes for the scientist, who 'must have had what the earliest explorers had, an insatiable desire to be the first person to see something, to know that you are where no man has been before', set the tone. The hero-scientist 'gloried in the layer upon layer of [the immune response's] complexity. He thrilled at seeing a layer of that complexity which no one had seen before' (Golub 1987: 531–2). It is reasonable to suppose that all the likely readers of this textbook have been reared within hearing range of the ringing tones of the introduction to the voyages of the federation starship 'Enterprise' in *Star Trek* – to go where no man has gone before. Science remains an important genre of western exploration and travel literature. Similarly, no reader, no matter how literal-minded, could be innocent of the gendered erotic trope that figures the hero's probing nature's laminated secrets, glorying simultaneously in the layered complexity and in his own techno-erotic touch that goes ever deeper. Science as heroic quest and as erotic technique applied to the body of nature are utterly conventional figures. They take on a particular edge in late twentieth-century immune system discourse, where themes of nuclear exterminism, space adventure, extraterrestrialism, exotic invaders, and military high-technology are pervasive.

But Golub's and Gershon's intended and explicit text is not about space invaders and the immune system as a Star Wars prototype. Their theme is the love of complexity and the intimate natural bodily technologies for generating the harmonies of organic life. In four diagrams – dated 1968, 1974, 1977, and 1982 – Gershon sketched his conception of 'the immunological orchestra'. This orchestra is a wonderful picture of the mythic and technical dimensions of the immune system. All of the diagrams are about cooperation and control, the major themes of organismic biology since the late eighteenth century. From his commanding position in the root of a lymph node, G.O.D. conducts the orchestra of T and B cells and macrophages as they march about the body and play their specific parts. The lymphocytes all look like Casper the Ghost, with the appropriate distinguishing nuclear morphologies drawn in the center of their shapeless bodies. Baton in hand, G.O.D.'s arms are raised in quotation of a symphonic conductor. G.O.D. recalls the other 1960s bioreligious, Nobel prize-winning 'joke' about the coded bodily text of post-DNA biology and medicine – the Central Dogma of molecular biology that specifies that 'information' flows only from DNA to RNA to protein. These three were called the Blessed Trinity of the secularized sacred body, and histories of the great adventures of molecular biology could be titled *The Eighth Day of Creation* (Judson 1979), an image that takes on a certain irony in the venture capital and political environments of current biotechnology companies, like Genentech. In the technical-mythic systems of molecular biology, code rules embodied structure and function, never the reverse. Genesis is a serious joke, when the body is theorized as a coded text whose secrets yield only to the proper reading conventions, and when the laboratory seems best characterized as a vast assemblage of technological and organic inscription devices. The Central Dogma was about a master control system for information flow in the codes that determine meaning in the great technological communication systems that organisms progressively have become after World War II. The body is an artificial intelligence system, and the relation of copy and original is reversed and then exploded.

G.O.D. is the Generator of Diversity, the source of the awe-inspiring multiple specificities of the polymorphous system of recognition and misrecognition we call the immune system. By the second diagram (1974), G.O.D. is no longer in front of the immune orchestra but is standing, arms folded, looking authoritative but not very busy, at the top of the lymph node, surrounded by the musical lymphocytes. A special cell, the T suppressor cell, has taken over the role of conductor. By 1977, the diagram no longer has a single conductor, but is 'led' by three rather mysterious subsets of T cells, who hold a total of twelve batons signifying their direction-giving surface identity markers; and G.O.D. scratches his head in patent confusion. But the immune band plays on. In the final diagram from 1982, 'the generator of diversity seems resigned to the conflicting calls of the angels of help and suppression', who perch above his left and right shoulders. Besides G.O.D. and the two angels, there is a T cell conductor and two conflicting prompters, 'each urging its own interpretation'. The joke of single masterly control of organismic harmony in the symphonic system responsible for the integrity of 'self' has become a kind of postmodern pastiche of multiple centers and peripheries, where the immune music that the page suggests would surely sound like nursery school space music. All the actors that used to be on the stage set for the unambiguous and coherent biopolitical subject are still present, but their harmonies are definitely a bit problematic.

By the 1980s, the immune system is unambiguously a postmodern object – symbolically, technically, and politically. Katherine Hayles characterizes postmodernism in terms of 'three waves of developments occurring at multiple sites within the culture, including literature and science' ('Denaturalizing'). Her archaeology begins with Saussurrean linguistics, through which symbol systems were 'denaturalized'. Internally generated relational difference, rather than mimesis, ruled signification. Hayles sees the culmination of this approach in Claude Shannon's mid-century statistical theory of information, developed for packing the largest number of signals on a transmission line for the Bell

Telephone Company and extended to cover communication acts in general, including those directed by the codes of bodily semiosis in ethology or molecular biology. 'Information' generating and processing systems, therefore, are postmodern objects, embedded in a theory of internally differentiated signifiers and remote from doctrines of representation as mimesis. A history-changing artifact, 'information' exists only in very specific kinds of universes.[3] Progressively, the world and the sign seemed to exist in incommensurable universes – there was literally no measure linking them, and the reading conventions for all texts came to resemble those required for science fiction. What emerged was a global technology that 'made the separation of text from context an everyday experience'. Hayles's second wave, 'energized by the rapid development of information technology, made the disappearance of stable, reproducible context an international phenomenon. . . . Context was no longer a natural part of every experience, but an artifact that could be altered at will.' Hayles's third wave of denaturalization concerned time. 'Beginning with the Special Theory of Relativity, time increasingly came to be seen not as an inevitable progression along a linear scale to which all humans were subject, but as a construct that could be conceived in different ways.'

Language is no longer an echo of the *verbum dei*, but a technical construct working on principles of internally generated difference. If the early modern natural philosopher or Renaissance physician conducted an exegesis of the text of nature written in the language of geometry or of cosmic correspondences, the postmodern scientist still reads for a living, but has as a text the coded systems of recognition – prone to the pathologies of mis-recognition – embodied in objects like computer networks and immune systems. The extraordinarily close tie of language and technology could hardly be overstressed in postmodernism. The 'construct' is at the center of attention; making, reading, writing, and meaning seem to be very close to the same thing. This near identity between technology, body, and semiosis suggests a particular edge to the mutually constitutive relations of political economy, symbol, and science that 'inform' contemporary research trends in medical anthropology.

THE APPARATUS OF BODILY PRODUCTION: THE TECHNO-BIOPOLITICS OF ENGAGEMENT

Bodies, then, are not born; they are made. Bodies have been as thoroughly denaturalized as sign, context, and time. Late twentieth-century bodies do not grow from internal harmonic principles theorized within Romanticism. Neither are they discovered in the domains of realism and modernism. One is not born a woman, Simone de Beauvoir correctly insisted. It took the political-epistemological terrain of postmodernism to be able to insist on a co-text to de Beauvoir's: one is not born an organism. Organisms are made; they are constructs of a world-changing kind. The constructions of an organism's boundaries, the job of the discourses of immunology, are particularly potent mediators of the experiences of sickness and death for industrial and post-industrial people.

In this overdetermined context, I will ironically – and inescapably – invoke a constructionist concept as an analytic device to pursue an understanding of what kinds of units, selves, and individuals inhabit the universe structured by immune system discourse: I call the conceptual tool 'the apparatus of bodily production'. In her analysis of the production of the poem as an object of literary value, Katie King offers tools that clarify the particular historicity of scientific bodies. Partially recalling Althusser's formulation, King suggests the term 'apparatus of literary production' to highlight the emergence of what is embodied as literature at the intersection of art, business, and technology. She applies this analytic frame to the relation of women and writing technologies.

I would like to adapt her work to the articulation of bodies and other objects of value in scientific productions of knowledge. At first glance, there is a limitation to using King's scheme, and that limitation is inherent in the 'facticity' of biological discourse, which is absent from literary discourse and its knowledge claims. But, by means of a social/epistemological/ethical operator, I wish to translate the ideological dimension of 'facticity' into an entity called a 'material-semiotic actor'. Scientific bodies are not ideological constructions. Always radically historically specific, bodies have a different kind of specificity and effectivity, and so they invite a different kind of engagement and intervention. 'Material-semiotic actor' is intended to highlight the object of knowledge as an active part of the apparatus of bodily production, without *ever* implying the immediate presence of such objects or, what is the same thing, their final or unique determination of what can count as objective knowledge of a biomedical body at a particular historical juncture.

Like King's objects called 'poems', sites of literary production where language also is an actor, bodies as objects of knowledge are material-semiotic generative nodes. Their boundaries materialize in social interaction; 'objects' like bodies do not pre-exist as such. Scientific objectivity (the siting/sighting of objects) is not about dis-engaged discovery, but about mutual and usually unequal structuring, about taking risks. The various contending biological bodies emerge at the intersection of biological research, writing, and publishing; medical and other business practices; cultural productions of all kinds, including available metaphors and narratives; and technology, such as the visualization technologies that bring color-enhanced killer T cells and intimate photographs of the developing fetus into high-gloss art books for every middle-class home (Nilsson 1977, 1987). But also invited into that node of intersection is the analogue to the lively languages that actively intertwine in the production of literary value: the coyote and protean embodiments of a world as witty agent and actor. Perhaps our hopes for accountability in the techno-biopolitics within postmodern frames turn on revisioning the world as coding trickster with whom we must learn to converse. Like a protein subjected to stress, the world for us may be thoroughly denatured, but it is not any less consequential. So while the late twentieth-century immune system is a construct of an elaborate apparatus of bodily production, neither the immune system nor any other of bio-medicine's world-changing bodies – like a virus – is a ghostly fantasy. Coyote is not a ghost, merely a protean trickster.

The following chart abstracts and dichotomizes two historical moments in the biomedical production of bodies from the late nineteenth century to the 1980s. The chart highlights epistemological, cultural, and political aspects of possible contestations for constructions of scientific bodies in this century. The chart itself is a traditional little machine for making particular meanings. Not a description, it must be read as an argument, and one which relies on a suspect technology for the production of meanings – binary dichotomization.

Representation	Simulation
Bourgeois novel	Science fiction
Realism and modernism	Postmodernism
Organism	Biotic component, code
Work	Text
Mimesis	Play of signifiers
Depth, integrity	Surface, boundary
Heat	Noise
Biology as clinical practice	Biology as inscription
Physiology	Communications engineering
Microbiology, tuberculosis	Immunology, AIDS
Magic Bullet	Immunomodulation
Small group	Subsystem
Perfection	Optimization
Eugenics	Genetic engineering
Decadence	Obsolescence
Hygiene	Stress management
Organic division of labor	Ergonomics, cybernetics
Functional specialization	Modular construction
Biological determinism	System constraints
Reproduction	Replication
Individual	Replicon
Community ecology	Ecosystem
Racial chain of being	United Nations humanism
Colonialism	Multinational capitalism
Nature/Culture	Fields of difference
Cooperation	Communications enhancement
Freud	Lacan
Sex	Surrogacy
Labor	Robotic
Mind	Artificial intelligence
World War II	Star Wars
White Capitalist Patriarchy	Informatics of Domination

It is impossible to see the entries in the right-hand column as 'natural' a realization that subverts naturalistic status for the left-hand column as well. From the eighteenth to the mid-twentieth centuries, the great historical constructions of gender, race, and class were embedded in the organically marked bodies of woman, the colonized or enslaved, and the worker. Those inhabiting these marked bodies have been symbolically other to the rational self of universal, and so unmarked, species man, a coherent subject. The marked organic body has been a critical locus of cultural and political contestation, crucial both to the languages of liberatory politics of identity and to systems of domination drawing on widely shared languages of nature constructed as resource for the appropriations of culture. For example, the sexualized bodies of nineteenth-century middle-class medical advice literature in England and the United States, in their female form organized around the maternal function and the physical site of the uterus and in their male form ordered by the spermatic economy tied closely to the nervous system, were part of an elaborate discourse of organic economy. The narrative field in which these bodies moved generated accounts of rational citizenship, bourgeois family life, and prophylaxis against sexual pollution and inefficiency, such as prostitution, criminality, or race suicide. Some feminist politics argued for the full inclusion of women in the body politic on grounds of maternal functions in the domestic economy extended to a public world. Late into the twentieth century, gay and lesbian politics have ironically and critically embraced the marked bodies constructed in nineteenth- and twentieth-century sexologies and gender-identity medicines to create a complex humanist discourse of sexual liberation. Negritude, feminine writing, various separatisms, and other recent cultural movements have both drawn on and subverted the logics of naturalization central to biomedical discourse on race and gender in the histories of colonization and male supremacy. In all of these various, oppositionally interlinked, political and biomedical accounts, the body remained a relatively unambiguous locus of identity, agency, labor, and hierarchicalized function. Both scientific humanisms and biological determinisms could be authorized and contested in terms of the biological organism crafted in post-eighteenth century life sciences.

But how do narratives of the normal and the pathological work when the biological and medical body is symbolized and operated upon, not as a system of work, organized by the hierarchical division of labor, ordered by a privileged dialectic between highly localized nervous and reproductive functions, but instead as a coded text, organized as an engineered communications system, ordered by a fluid and dispersed command-control-intelligence network? From the mid-twentieth century, biomedical discourses have been progressively organized around a very different set of technologies and practices, which have destabilized the symbolic privilege of the hierarchical, localized, organic body. Concurrently – and out of some of the same historical matrices of decolonization, multinational capitalism, worldwide hi-tech militarization, and the emergence, in local and global politics, of new collective political actors from among those persons previously consigned to labor in silence – the question of 'differences' has destabilized humanist discourses of liberation based on a politics of identity and substantive unity. Feminist theory as a self-conscious discursive practice has been generated in this post-World War II period characterized by the translation of western scientific and political languages of nature from those based on work, localization, and the marked body to those based on codes, dispersal and networking, and the fragmented post-modern subject. An account of the biomedical, biotechnical body must start from the multiple molecular interfacings of genetic, nervous, endocrine, and immune systems. Biology is about recognition and misrecognition, coding errors, the body's reading practices (e.g., frameshift mutations), and billion dollar projects to sequence the human genome to be published and stored in a national genetic 'library'. The body is conceived as a strategic system, highly militarized in key arenas of imagery and practice. Sex, sexuality, and reproduction are theorized in terms of local investment strategies; the body ceases to be a stable spatial map of normalized functions and instead emerges as a highly mobile field of strategic differences. The biomedical–biotechnical body is a semiotic system, a complex meaning-producing field, for which the discourse of immunology, i.e., the central biomedical discourse on recognition/misrecognition, has become a high-stakes practice in many senses.

In relation to objects like biotic components and codes, one must think not in terms of laws of growth and essential properties, but rather in terms of strategies of design, boundary constraints, rates of flow, system logics, and costs of lowering constraints. Sexual reproduction becomes one possible strategy among many, with costs and benefits theorized as a function of the system environment. Disease is a subspecies of information malfunction or communications pathology; disease is a process of misrecognition or transgression of the boundaries of a strategic assemblage called self. Ideologies of sexual reproduction can no longer easily call upon the notions of unproblematic sex and sex role as organic aspects of 'healthy' natural objects like organisms and families. Likewise for race, ideologies of human diversity have to be developed in terms of frequencies of parameters and fields of power-charged differences, not essences and natural origins or homes. Race and sex, like individuals, are artifacts sustained or undermined by the discursive nexus of knowledge and power. Any objects or persons can be reasonably thought of in terms of disassembly and reassembly; no 'natural' architectures constrain system design. Design is nonetheless highly constrained. What counts as a 'unit', a one, is highly problematic, not a permanent given. Individuality is a strategic defense problem.

One should expect control strategies to concentrate on boundary conditions and interfaces, on rates of flow across boundaries, not on the integrity of natural objects. 'Integrity' or 'sincerity' of the western self gives way to decision procedures, expert systems, and resource investment strategies. 'Degrees of freedom' becomes a very powerful metaphor for politics. Human beings, like any other component or subsystem, must be localized in a system architecture whose basic modes of operation are probabilistic. No objects, spaces, or bodies are sacred in themselves; any component can be interfaced with any other if the proper standard, the proper code, can be constructed for processing signals in a common language. In particular, there is no ground for ontologically opposing the organic, the technical, and the textual.[4] But neither is there any ground for opposing the *mythical* to the organic, textual, and technical. Their convergences are more important than their residual oppositions. The privileged pathology affecting all kinds of components in this universe is stress – communications breakdown. In the body, stress is theorized to operate by 'depressing' the immune system. Bodies have become cyborgs – cybernetic organisms – compounds of hybrid techno-organic embodiment and textuality (Haraway 1985). The cyborg is text, machine, body, and metaphor – all theorized and enaged in practice in terms of communications.

[. . .]

NOTES

1. Research and writing for this project were supported by the Alpha Fund and the Institute for Advanced Study, Princeton, NJ; Academic Senate Faculty Research Grants of the University of California, Santa Cruz; and the Silicon Valley Research Project, UCSC. Crystal Gray was an excellent research assistant. Benefiting from many people's comments, this paper was first presented at the Wenner Gren Foundation's Conference on Medical Anthropology, Lisbon, Portugal, March 5–13, 1988.
2. Even without taking much account of questions of consciousness and culture, the extensive importance of immunological discourse and artifacts has many diagnostic signs: 1) The first Nobel prize in medicine in 1901 was given for an originary development, namely, the use of diphtheria antitoxin. With many intervening awards, the pace of Nobel awards in immunology since 1970 is stunning, covering work on the generation of antibody diversity, the histocompatibility system, monoclonal antibodies and hybridomas, the network hypothesis of immune regulation, and development of the radioimmunoassay system. 2) The products and processes of immunology enter into present and projected medical, pharmaceutical, and other industrial practices. This situation is exemplified by monoclonal antibodies, which can be used as extremely specific tools to identify, isolate, and manipulate components of production at a molecular scale and then gear up to an industrial scale with unheard of specificity and purity for a wide array of enterprises – from food-flavoring technology, to the design and manufacture of industrial chemicals, to delivery systems in chemotherapy (see figure on 'Applications of monoclonal antibodies in immunology and related disciplines', Nicholas 1985: 12). The *Research Briefings* for 1983 for the federal Office of Science and Technology Policy and various other federal departments and agencies identified immunology, along with artificial intelligence and cognitive science, solid earth sciences, computer design and manufacture; and regions of chemistry, as research areas 'that were likely to return the highest scientific dividends as a result of incremental federal investment' (National Academy of Sciences et al.). The dividends in such fields are hardly expected to be simply 'scientific'. 'In these terms the major money spinner undoubtedly is hybridoma technology, and its chief product the monoclonal antibody' (Nicholas 1985: Preface). 3). The field of immunology is itself an international growth industry. The First International Congress of Immunology was held in 1971 in Washington,

DC attended by most of the world's leading researchers in the field, about 3500 people from 45 countries. Over 8000 people attended the Fourth International Congress in 1980 (Klein 1982: 663). The number of journals in the field has been expanding since 1970 from around 12 to over 80 by 1984. The total of books and monographs on the subject reached well over 1000 by 1980. The industrial–university collaborations characteristic of the new biotechnology pervade research arrangements in immunology, as in molecular biology (with which it crossreacts extensively), e.g., the Basel Institute for Immunology, entirely financed by Hoffman-La Roche but featuring all the benefits of academic practice, including publishing freedoms. The International Union of Immunological Societies began in 1969 with 10 national societies and increased to 33 by 1984 (Nicholas 1985). Immunology will be at the heart of global biotechnological inequality and 'technology transfer' struggles. Its importance approaches that of information technologies in global science politics. 4) Ways of writing about the immune system are also ways of determining which diseases – and which interpretations of them – will prevail in courts, hospitals, international funding agencies, national policies, memories and treatment of war veterans and civilian populations, and so on. See for example the efforts of oppositional people, like labor and consumer advocates, to establish a category called 'chemical AIDS' to call attention to widespread and unnamed ('amorphous') sickness in late industrial societies putatively associated with its products and environments, and to link this sickness with infectious AIDS as a political strategy (Hayes 1987, Marshall 1986). Discourse on infectious AIDS is part of mechanisms that determine what counts as 'the general population', such that over a million infected people in the US alone, not to mention the global dimensions of infection, can be named in terms that make them not part of the general population, with important national medical, insurance, and legal policy implications. Many leading textbooks of immunology in the United States give considerably more space to allergies or auto-immune diseases than to parasitic diseases, an allocation that might lead future Nobel Prize winners into some areas of research rather than others and that certainly does nothing to lead undergraduates or medical students to take responsibility for the differences and inequalities of sickness globally. (Contrast Golub with Desowitz for the sensitivities of a cellular immunology researcher and a parasitologist.) Who counts as an individual is not unrelated to who counts as the general population.

3. Like the universe inhabited by readers and writers of this essay.
4. This ontological continuity enables the discussion of the growing practical problem of 'virus' programs infecting computer software (McLellan 1988). The infective, invading information fragments that parasitize their host code in favor of their own replication and their own program commands are more than metaphorically like biological viruses. And like the body's unwelcome invaders, the software viruses are discussed in terms of pathology as communications terrorism, requiring therapy in the form of strategic security measures. There is a kind of epidemiology of virus infections of artificial intelligence systems, and neither the large corporate or military systems nor the personal computers have good immune defenses. Both are extremely vulnerable to terrorism and the rapid proliferation of foreign code that multiplies silently and subverts their normal functions. Immunity programs to kill the viruses, like Data Physician sold by Digital Dispatch, Inc., are being marketed. More than half the buyers of Data Physician in 1985 were military. I start up, my Macintosh shows the icon for its vaccine program – a hypodermic needle.

REFERENCES

Barnes, Deborah M. (1986) 'Nervous and Immune System Disorders Linked in a Variety of Diseases' *Science* 232: 160–1.

Barnes, Deborah M. (1987) 'Neuroimmunology Sits on Broad Research Base' *Science* 237: 1568–9.

Barthes, Roland (1982) 'The Photographic Message' *A Barthes Reader*, Ed. Susan Sontag, New York: Hill.
Berger, Steward (1985) *Dr Berger's Immune Power Diet*, New York: NAI.
Blalock, J. Edwin (1984) 'The Immune System as a Sensory Organ' *Journal of Immunology* 132 (3): 1067–70.
Brewer, Maria Minich (1987) 'Surviving Fictions: Gender and Difference in Postmodern and Postnuclear Narrative' *Discourse* 9: 37–52.
Bryan, C. D. B. (1987) *The National Geographic Society: 100 Years of Adventure and Discovery*, New York: Abrams.
Buss, Leo (1987) *The Evolution of Individuality*, Princeton: Princeton University Press.
Butler, Octavia (1984) *Clay's Ark*. New York: St Martin's.
Butler, Octavia (1987) *Dawn: Xenogenesis*, New York: Warner.
Clyne, N. and M. Klynes (1961) *Drugs, Space and Cybernetics: Evolution to Cyborg*, New York: Columbia University Press.
Dawkins, Richard (1982) *The Extended Phenotype: The Gene as the Unit of Selection*, Oxford: Oxford University Press.
Dawkins, Richard (1976) *The Selfish Gene*, Oxford: Oxford University Press.
Desowitz, Robert S. (1987) *The Immune System and How It Works*, New York: Norton.
Foucault, Michel (1972) *The Archaeology of Knowledge*, New York: Harper.
Foucault, Michel (1971) *The Order of Things*, New York: Pantheon.
Goleman, Daniel (1987) 'The Mind Over the Body' *New York Times Sunday Magazine*, 27 Sept. 1987: 36 +.
Golub, Edward S. (1987) *Immunology: A Synthesis*, Sunderland, MA: Sinauer.
Haraway, Donna (1985) 'A Manifesto for Cyborgs: Science, Technology and Socialist Feminism in the 1980s'. *Socialist Review* 80: 65–107.
Haraway, Donna (1985) 'Teddy Bear Patriarchy: Taxidermy in the Garden of Eden, New York City, 1908–36 *Social Text* 11: 20–64.
Hayes, Dennis (1987) 'Making Chips with Dust-Free Poison' *Science as Culture* 1: 89–104.
Hayles, N. Katherine (1987) 'Denaturalizing Experience: Postmodern Literature and Science' *Literature and Science as Modes of Expression*, Unpublished abstracts from a Conference, 8–11 Oct. 1987, Worcester: Society for Literature and Science, Worcester Polytechnic.
Judson, Horace Freeland (1979) *The Eighth Day of Creation*, New York: Simon.
King, Katie (1987) Unpublished book prospectus for research on feminism and writing technologies. College Park, MD: University of Maryland, 1987.
Klein, Jan (1982) *Immunology: The Science of Non-Self Discrimination*, New York: Wiley-Interscience.
Marshall, Eliot (1986) 'Immune System Theories on Trial' *Science* 234: 1490–2.
McLellan, Vin (1988) 'Computer Systems under Siege' *New York Times* 31 Jan. 1988, sec. 3: 1 +.
National Academy of Sciences, the National Academy of Medicine. Committee on Science, Engineering, and Public Policy and the Institute of Medicine (1983) *Research Briefings* Washington, DC: National Academy P.
Nicholas, Robin (1985) *Immunology: An Information Profile*, London: Mansell.
Nilsson, Lennart (1987) *The Body Victorious: The Illustrated Story of Our Immune System and Other Defenses of the Human Body*, New York: Delacorte P.
Nilsson, Lennart (1977) *A Child Is Born*, New York: Dell.
Playfair, J. H. L. (1984) *Immunology at a Glance*, 3rd ed., Oxford: Blackwell Scientific.
Treichler, Paula (1987) 'AIDS, Homophobia, and Biomedical Discourse: an Epidemic of Signification' *October* 43 1987: 31–70.
Turner, Bryan S. (1984) *The Body and Society*, New York: Basil Blackwell.

SECTION 4
AFTER THE BINARY

INTRODUCTION

This selection of pieces drawn mainly from the last decade stands as a companion to Section 1 in tracing the differential ways in which feminist theory has incorporated the body as an abstract concept. Where mainstream second wave feminism, which works within a modernist framework, has struggled with the heritage of a cultural devaluation of corporeality in general, and the female body in particular, this section charts the impact of poststructuralism and postmodernism on feminist theory. Although that influence is extremely complex and in many ways highly problematic for an agenda that seeks the revaluation of women, there are several clear features that have been incorporated into contemporary theory. The most obvious development, as the section title indicates, is the deconstruction of the familiar binaries that have shaped the structure of western thought. What postconventional analysis suggests is that the hierarchical oppositions which underpin our understanding of ourselves and of our environment, and which seem to guarantee the superiority of the male over the female, and mind over body, are by no means stable or enduring over time. Far from describing a given reality, they constitute an unstable and provisional mechanism of discursive power.

In uncovering the constructed nature of both our epistemological and ontological categories, postmodernism offers feminism the insight that things could be otherwise. One paradox is that binaries inscribe (not describe, note) not balanced oppositions, but a single standard whereby the devalued term – most often that associated with the feminine or indeed with the body as such – is measured against the primary norm and found wanting. At the same time, the very boundaries that apparently secure the either/or structure of binary difference – the body *or* the mind, black *or* white, male *or* female – are opened up to slippage and uncertainty such that the reliance on sameness and difference is lost. What falls to postmodernist feminism, then, is the task of reclaiming the marginalised female/feminine body without reinstating it as a unified, closed and given category. All the extracts concern themselves with a body in process, a body that is specific rather than universal, and yet which can never be pinned down. Both Denise Riley and Moira Gatens offer a valuable overview of the existing terrain and an indication of the difficulties that feminist theory faces in letting go of the category of the natural body. As Gatens makes clear, the concept of difference must be pluralised, and rewritten no longer as a crude opposition. And the fluidity and unpredictability of the body which both writers refer to problematises not simply sexual difference in its binary sense, but for Riley, the very category of woman.

Alongside the deconstruction of binary thinking, another widely influential aspect of postmodernist thought is the emphasis given to the inescapable relationship between embodiment, power and knowledge. In this respect, the name of Foucault is commonly invoked, not because he was particularly sympathetic to the feminist cause, or even sensitive to gender concerns, but because his understanding of the discursive construction of bodies offers the grounds for a fully politicised analysis. Gatens and Riley both acknowledge the debt, and Butler too was originally known as a Foucauldian, though the piece selected here is from her later work. Nonetheless, most of those who use his work also offer a critique of Foucault. Susan Bordo, for example, takes up the notion of a body in the grip of power, but is less enchanted with the postmodernist idea that all power is dispersed into decentralised sites. The implications for feminist resistance are not wholly acceptable to her, and she reworks the scenario from a more woman-identified perspective. One of the attractions of Foucault's approach is that it seems to counter the immateriality of which postmodernism is so often accused. It offers an analysis in which the very processes by which bodies are made become apparent. Rather than a pre-exisitng and completed entity, the body is 'the inscribed surface of events'. For many feminists, Foucauldian or not, that opens up the possibility of writing the female body differently, against the grain of custom and expectation. The metaphor underlines Trinh T. Minh-ha's self-consciously literary, but nonetheless highly politicised strategy, as well as much of what is known as high theory, which tends to take a somewhat abstract approach, less concerned with everyday practicalities.

Not surprisingly, given the activist background to the second wave, some feminists have felt uncomfortable with what they see as an over-reliance on theory that loses touch with the flesh and blood body. It is in our view, however, a misreading of the nature of the feminist use of postmodernism. Both Judith Butler and Elizabeth Grosz, as unashamedly complex theorists, have at times been accused of reducing bodies to mere inscription. We include here a piece from each that marks precisely why their work is valuable and theoretically challenging, but continues to speak directly to the lives of women. Butler's book *Bodies that Matter* is directly concerned, not with corporeality as an abstract category, but with the ways in which bodies are materialised. And moreover she looks very closely at the political consequences of the forms that the process of materialisation takes in terms of differential sexuality and race. For Grosz, the feminist appropriation of the terms of psychoanalysis allows for a thoroughgoing rejection of the biologistic body, not as a move toward the immaterial, but as an opportunity to reclaim women's corporeality beyond masculinist and essentialising categories. As she insists, the reworking of psychoanalysis is a ground on which '(t)here are only concrete bodies, bodies in the plural, bodies with a specific sex and colour'.

What is evident throughout these selections is that the considerable impact of poststructuralism and postmodernism, and its take-up within feminism, makes the body as central to feminist theory as it has been to activism.

FURTHER READING

Bordo, Susan (1991) 'Postmodern Subjects, Postmodern Bodies: A Review Essay', *Feminist Studies*, 18, 1: 159–76.

Bordo, Susan (1993) *Unbearable Weight: Feminism, Western Culture and the Body*, Berkeley: University of California Press.

Brown, Beverley and Adams, Parveen (1979) 'The Feminine Body and Feminist Politics', *M/F* 3: 35–50.

Butler, Judith (1990) 'Subversive Bodily Acts' (Chap. 3) in *Gender Trouble: Feminism and the Subversion of Identity*, London: Routledge.

Bynum, Caroline (1995) 'Why All the Fuss about the Body? A Medievalist's Perspective', *Critical Inquiry* 22.

Gross, Elizabeth (1986) 'Philosophy, Subjectivity and the Body: Kristeva and Irigaray' in Carole Pateman and Elizabeth Gross (eds), *Feminist Challenges*, Sydney: Allen and Unwin.

Grosz, Elizabeth (1993) 'Bodies and Knowledges: Feminism and the Crisis of Reason' in Linda Alcoff and Elizabeth Potter (eds), *Feminist Epistemologies*, London: Routledge.

Grosz, Elizabeth (1994) *Volatile Bodies: Towards A Corporeal Feminism*, Bloomington: Indiana University Press.

Irigaray, Luce (1985) *Speculum of the Other Woman*, trans. Gillian C. Gill, Ithaca, NY: Cornell University Press.

Kristeva, Julia (1982) *Powers of Horror: An Essay on Abjection*, New York: Columbia University Press.

McNay, Lois (1992) *Foucault and Feminism: Power, Gender and the Self*, Cambridge: Polity Press.

Probyn, Elspeth (1991) 'This Body Which Is Not One' *Hypatia* 6 (3): 111–24.

Shildrick, Margrit (1997) 'The Disciplinary Body' (Chap. 1); and 'Feminist Theory and Postmodernism' (Chap. 5) in *Leaky Bodies and Boundaries*, London: Routledge.

Singer, Linda (1993) 'Bodies – Pleasures – Powers' in *Erotic Welfare*, London: Routledge.

Suleri, Sara (1992) 'Women Skin Deep: Feminism and the Postcolonial Condition', *Critical Inquiry* 18: 756–69.

4.1

BODIES, IDENTITIES, FEMINISMS

Denise Riley

[Riley notes how 'women's experience' is underpinned by a core of identification, purity, and mothering – i.e. by the concept of the female body]

Here we are on notoriously difficult ground. Hard, indeed, to speak against the body. Even if it is allowed that the collective 'women' may be an effect of history, what about biology, materiality? Surely, it is argued, those cannot be evaporated into time. And from the standpoint of feminism, what has always been lacking is a due recognition of the specificity of women's bodies, sexual difference as lived. Indeed, Simone de Beauvoir – she who, ironically, has been so often upbraided for paying no attention to precisely what she does name here – wrote in *The Second Sex*:

> In the sexual act and in maternity not only time and strength but also essential values are involved for woman. Rationalist materialism tries in vain to disregard this dramatic aspect of sexuality.[1]

Several contemporary feminisms also set themselves against what they believe to be a damaging indifference to the powerful distinct realities of the body. Here Elizabeth Gross sets out her understanding of the Irigarayan conception:

> All bodies must be male or female, and the particularities, specificities and differences of each need to be recognised and represented in specific terms. The social and patriarchal disavowal of the specificity of women's bodies is a function, not only of discriminatory social practices, but, more

From: D. Riley, *'Am I That Name?' Feminism and the Category of 'Women' in History*, London: Macmillan, 1988.

> insidiously, of the phallocentrism invested in the régimes of knowledge – science, philosophy, the arts – which function only because and with the effect of the submersion of women under male categories, values and norms. For Irigaray, the reinscription, through discourses, of a positive, autonomous body for women is to render disfunctional all forms of knowledge that have hitherto presented themselves as neutral, objective or perspective-less.[2]

If, for the moment, we take up this conviction about the political-analytic force of women's bodies and lead it towards history, then our question becomes – In what ways have these social and patriarchal 'disavowals' functioned, and how could the subdued bodies of women be restored in a true form? Do the existing social histories of the female body answer that? They do not. We may leaf through voyeuristic and sensational catalogues of revulsion. That is not to deny that, could they escape being charmed by the morbid, the histories of trained, exploited, or distorted flesh – of bodies raped, circumcised, infected, ignorantly treated in childbirth or subjected to constant pregnancies – would carry some moral force.[3] In respect of the developed countries at least, such accounts would suggest that women are less relentlessly caught in physical toils than they were, as pregnancy can be restricted and gynaecological hazards are far less catastrophic – that in this sense, women can spend less of their lives awkwardly *in* their bodies. But even this fragile assumption of progress can be qualified if we recall that contraception was rarely a complete mystery even when the physiology of reproduction was not deciphered, and that medical Whiggishness must be shaken by many examples – the exhaustion of over-used antibiotics, the ascent of new viral strains, the deeply undemocratic distribution of resources, the advancing technologies of international genocide.

So to the history of the body as a narrative of morbidity and its defeats, we could contrast a historical sociology of the body. This would worry about the management of populations, about social policies drawing on demography or eugenics, about malnutrition caused by economic policy in another hemisphere, the epidemiology of industrial and nuclear pollution, and so forth. Yet in all this, both 'the body' and 'women's bodies' will have slipped away as objects, and become instead almost trace phenomena which are produced by the wheelings-about of great technologies and politics. Is this simply the predictable end of that peculiar hypostatisation, 'the body'? Perhaps it must always be transmuted into bodies in the plural, which are not only marked and marred by famine, or gluttony, destitution or plenty, hazard or planning, but are also shaped and created by them. 'The body' is not, for all its corporeality, an originating point nor yet a terminus; it is a result or an effect.

And yet this train of thought doesn't satisfy our original question of the bodies of women in history. Even a gender-specific historical sociology would somehow miss the point. For instance, we could consider what an account of *men's* bodies would look like; it would include historical descriptions of sex-related illnesses, heart disease, lung cancer and the statistical challenge here from women; the history of soldiery, war slaughter, conscription; of virility as a concept, of Sparta; of the greater vulnerability of the male foetus; of narcissism and its failures; of disabling conditions of work, of mining accidents; of the invention of the male homosexual as a species-being. A history of prostitution but this time written from the side of the clients, of contraception written from the side of the fathers – to add to the histories of bodily endurance, triumphant musculature, or the humiliations of the feebler of frame. All this and more could count up the male body in history, its frailties and its enjoyments, analogously to women's. Yet the sum of the two parts, men and women, would still not produce a satisfying total of 'the body', now democratically analysed with a proper regard to sexual difference.

What would have gone wrong, then, in the search? A chain of unease remains: that anyone's body is – the classifications of anatomy apart – only periodically either lived or treated as sexed, therefore the gendered division of human life into bodily life cannot be adequate or absolute. Only at times will the body impose itself or be arranged as that of a woman or a man. So that if we set out to track the bodies of women in history, we would assume in advance that which really we needed to catch, instead, on the wing of its formulation. Neither the body marked with time, nor the sexed body marked with time, are the right concepts here. For the impress of history as well as of individual temporality is to establish the body itself as lightly or as heavily gendered, or as indifferent, and for that to run in and out of the eye of 'the social'. It's more of a question of tracing the (always anatomically gendered) body as it is differently established and interpreted as sexed within different periods. If female bodies are thought of as perenially such, as constant and even embodiments of sexed being, that is a misconception which carries risks. If it leads to feminist celebrations of the body as female, which intoxicatingly forget the temporality and malleability of gendered existence, at the same time it makes the feminist critique of, say, the instrumental positioning of women's bodies all the harder to develop coherently, because this critique needs some notion of temporality too. It could be claimed that a characteristic of the sadist's gaze is to fix and freeze its object, to insist on absolute difference, to forbid movement.

There is a further reason for unease with the sufficiency of a historical sociology of the body, sexed or not. In a strong sense the body is a concept, and so is hardly intelligible unless it is read in relation to whatever else supports it and surrounds it. Indeed the queer neutrality of the phrase 'the body' in its strenuous colourlessness suggests that something is up. We could speculate that some of the persistent draw of this 'the body' lies in the tacit promise to ground the sexual, to make intimacy more readily decipherable, less evanescent. But then this enticement is undercut by the fact that the very location of 'the sexual' in the body is itself historically mutable. And 'the body' is never above – or below – history.

This is visible in the degree, for instance, to which it is held co-extensive with the person; to which the mind – body distinction is in play, if at all; to which the soul is held to have the capacity to dominate the flesh. If the contemporary body is usually considered as sexed, exactly what this means now is in part the residue remaining after a long historical dethronement of the soul's powers, which in turn has swayed the balance of sexed nature. The modern western body is what the soul has thoroughly vacated (in favour, for some, of the unconscious). But here there is no symmetry between the sexes, because we can show that 'sex' expanded differently into the old fields of soul and body in a different way for 'women'. I suggested earlier that the eighteenth-century remnants of the soul were flooded with the womanly body, preparing the way for the nineteenth-century naturalising of the species Woman. Any history of how far 'the body' has been read as the measure of the human being would have to include this – how far 'the body' has been read as co-extensive with the gender of its bearer.

Some philosophical writings now hint that 'the body' does have the status of a realm of underlying truth, and try to rescue it from medicine or sociology by making it vivid again. Sebastiano Timpanaro attempts this for socialism in his *On Materialism*.[4] And Michel Foucault, at points in his *History of Sexuality*, treats 'bodies and their pleasures' as touchstones of an anarchic truth, innocent brute clarities which are then scored through with the strategies of bio-technical management from on high.[5] But elsewhere in his work there is nothing of a last court of appeal in the body. On the contrary, in the essay, 'Nietzsche, Genealogy, History', he writes:

> The body is the inscribed surface of events (traced by language and dissolved by ideas), the locus of a dissociated self (adopting the illusion of a substantial unity), and a volume in perpetual disintegration. Genealogy, as an analysis of descent, is thus situated within the articulation of the body and history. Its task is to expose a body totally imprinted by history and the process of history's destruction of the body.[6]

The integrity of the body's claim to afford a starting-point for analysis is refused:

> 'Effective' history differs from traditional history in being without constraints. Nothing in man – not even his body – is sufficiently stable to serve as the basis for self-recognition or for understanding other men.[7]

This Foucauldian body is a deliquescing effect; composed but constantly falling away from itself. What if the 'man' attached to it is erased, and 'woman' set there instead? Has history 'totally imprinted' the bodies of women in different ways?

One train of thought must answer yes. That women's bodies become women's bodies only as they are caught up in the tyrannies, the overwhelming incursions of both nature and man – or, more optimistically, that there are also vehement pleasures and delights to offset a history of unbridled and violent subjection. But to be faithful to the suggestion that 'the body' is really constantly altering as a concept means that we must back off from the supposition that women's bodies are systematically and exhaustively different, that they are unified in an integral otherness. Instead we would need to maintain that women only sometimes live in the flesh distinctively of women, as it were, and this is a function of historical categorisations as well as of an individual daily phenomenology. To say that is by no means to deny that because of the cyclical aspects of female physiology, there may be a greater overall degree of slipping in and out of the consciousness of the body for many women. But even this will always be subject to different interpretations, and nothing more radical than the facts of intermittent physiology really holds the bodies of women together.

Where they are dragged together, a sort of miserable sexual democracy may obtain – of malnutrition, for instance – although then they may well move from being starved bodies to being starved sexed bodies as amenorrhoea sets in. But these are rare constructions which do produce 'women's bodies' as the victims of shared sufferings. Conditions of deprivation, of sex-specific hard labour, do also pull together the bent backs of women, but then it is the sexual division of labour which has made the partition – not a natural bodily unity. Another kind of massing of potentially maternal bodies belongs to demographic policies, although even here 'nature' is remote. Of course, if women did not have the capacity of childbearing they could not be arrayed by natalist or anti-natalist plans into populations to be cajoled or managed. But the point is that irrespective of natural capacities, only some prior lens which intends to focus on 'women's bodies' is going to set them in such a light. The body becomes visible *as* a body, and *as* a female body, only under some particular gaze – including that of politics.

So the sexed body is not something reliably constant, which can afford a good underpinning for the complications of the thousand discourses on 'women'. How and when even the body will be understood and lived as gendered, or indeed as a body at all, is not fully predictable. Again this isn't only a function of an individual phenomenology but of a historical and political phenomenology. There is no deep natural collectivity of women's bodies which precedes some subsequent arrangement of them through history or biopolitics. If the body is an unsteady mark, scarred in its long decay, then the sexed body too undergoes a similar radical temporality, and more transitory states.

Then what is the attraction of the category of the body at all? For those feminist philosophies which espouse it, it promises a means of destabilising the tyranny of systematic blindness of sexual difference. It has to be conceded that such philosophies do not have to assume any naturally bestowed identity of women; the female body can be harshly characterised from above. This is demonstrated through Elizabeth Gross's exposition of the Irigarayan schema, a clear, sympathetic account of the feminist reception of that work:

> Psychical, social and interpersonal meanings thus mark the body, and through it, the identities or interiority of sexed objects. The female body is inscribed socially, and most often, individually experienced as a

> lacking, incomplete or inadequate body ... Women's oppression is generated in part by these systems of patriarchal morphological inscription – that is, by a patriarchal symbolic order – or part by internalised, psychic representations of this inscribed body, and in part as a result of the different behaviours, values and norms that result from these different morphologies and psychologies. Irigaray's aim ... is to speak about a positive model or series of representations of femininity by which the female body may be positively marked, which in its turn may help establish the conditions necessary for the production of new kinds of discourse, new forms of knowledge and new modes of practice.[8]

It is the conclusion here which worries me – that the goal is a fresh and autonomous femininity, voiced in a revolutionary new language, to speak a non-alienated being of woman. Indeed the 'woman' we have available is severely damaged. But for myself – in common with many other feminists, but unlike many others again – I would not seek the freshly conceived creature, the revelatory Woman we have not yet heard. She is an old enough project, whose repeated failures testify to the impossibility of carving out a truly radical space; the damage flows from the very categorisation 'woman' which is and has always been circumscribed in advance from some quarter or other, rendering the ideal of a purely self-representing 'femininity' implausible. A true independence here would only be possible when all existing ideas of sexual difference had been laid to rest; but then 'woman', too, would be buried.

Such reflections undo the ambition to retrieve women's bodies from their immersions beneath 'male categories, values and norms'. The body circulates inexorably among the other categories which sometimes arrange it in sexed ranks, sometimes not. For the concept 'women's bodies' is opaque, and like 'women' it is always in some juxtaposition to 'human' and to 'men'. If this is envisaged as a triangle of identifications, then it is rarely an equilateral triangle in which both sexes are pitched at matching distances from the apex of the human. And the figure is further skewed by the asymmetries of the histories of the sexes as concepts as well as their present disjointedness. If 'women' after the late seventeenth century undergoes intensified feminising, this change does not occur as a linear shift alone, as if we had moved from mercifully less of 'women' through a later excess of them. Other notions which redefine understandings of the person have their influential upheavals: Reason, Nature, the Unconscious, among many. The periodic hardenings of 'women' don't happen alone or in any necessary continuum (as any history of individualism would need to take into account).

[...]

NOTES

1. Simone de Beauvoir, *Le Deuxième Sexe, 1949*: *The Second Sex*, transl. by H. Parshley, London: Jonathan Cape, 1953, p. 84.

2. Elizabeth Gross, 'Philosophy, subjectivity, and the body: Kristeva and Irigaray', in Carole Pateman and Elizabeth Gross (eds), *Feminist Challenges: Social and Political Theory*, Sydney: Allen & Unwin, 1986, p. 139.
3. See Edward Shorter, *A History of Women's Bodies*, New York: Basic Books, 1982.
4. Sebastiano Timpanaro, *On Materialism*, London: New Left Books, 1975.
5. Michel Foucault, *The History of Sexuality* – Volume 1: An Introduction, trans. R. Hurley, London: Allen Lane, 1979.
6. Michel Foucault, 'Nietzsche, Genealogy, History', in *Language, Counter-Memory, Practice; Selected Essays and Interviews*, Donald F. Bouchard and Sherry Simon (eds and transl.), Ithaca: Cornell University Press, 1977, p. 148.
7. Ibid, p. 153.
8. Elizabeth Gross, 'Philosophy, Subjectivity, and the Body: Kristeva and Irigaray' (as above), p. 142.

4.2

POWER, BODIES AND DIFFERENCE

Moira Gatens

The Body

There is probably no simple explanation for the recent proliferation of writings concerning the body. Clearly, Foucault's work has been influential in making the body a favoured subject for analysis in contemporary philosophy, sociology and anthropology. However, the impact of feminist theory on the social sciences has no less a claim to credit for bringing the body into the limelight. The difficulties encountered by primarily middle-class women, who have had the greatest access to 'equality' in the public sphere, may well have served as a catalyst for feminist reflections on the body.

One response to the differential powers and capacities of women and men in the context of public life is to claim that women just are biologically disadvantaged relative to men. From this perspective it seems crucial to call for the further erosion of the reproductive differences between the sexes by way of advances in medical science. On this view, social reform can only achieve so much, leaving the rectification of the remaining determinations of women's situation to the increase in control over nature; that is, biology. Simone de Beauvoir retains the doubtful privilege of being the clearest exponent of this view. In the 1970s, Shulamith Firestone's *The Dialectic of Sex* was influential in perpetuating the view that science could fulfil a liberatory role for women.[1] Both theorists assumed that the specificity of the reproductive body must be overcome if sexual equality is to be realized.

An alternative response to questions of corporeal specificity is to claim that

From: Moira Gatens, 'Power, Bodies and Difference', in M. Barrett and A. Phillips (eds). *Destabilising Theory*, Cambridge: Polity Press, 1992.

women should not aspire to be 'like men'. Interestingly, this response comes from both feminists and anti-feminists alike.[2] Recent feminist research suggests that the history of western thought shows a deep hatred and fear of the body.[3] This somatophobia is understood by some feminists to be specifically masculine and intimately related to gynophobia and misogyny.[4] In response to this negative attitude towards the body and women, some feminists advocate the affirmation and celebration of women's bodies and their capacity to recreate and nurture. In its strongest form this view argues that the specific capacities and powers of women's bodies imply an essential difference between men and women, where women may be presented as essentially peace-loving, 'biophilic' or caring, and men as essentially aggressive, 'necrophilic' or selfish.[5] These theorists argue that there is an essential sexual difference which should be retained, not eroded by scientific intervention.

These two responses to women's corporeal specificity are often taken to exhaust what has been termed the 'sexual equality versus sexual difference debate'. Yet both responses are caught up within the same paradigm. Both understand the body as a given biological entity which either has or does not have certain ahistorical characteristics and capacities. To this extent, the sexual difference versus sexual equality debate is located within a framework which assumes a body/mind, nature/culture dualism. The different responses are both in answer to the question of which should be given priority: the mind or the body, nature or culture.

An alternative view of the body and power might refuse this dualistic manner of articulating the issue of sexual difference. Specifically, to claim a history for the body involves taking seriously the ways in which diet, environment and the typical activities of a body may vary historically and create its capacities, its desires and its actual material form.[6] The body of a woman confined to the role of wife/mother/domestic worker, for example, is invested with particular desires, capacities and forms that have little in common with the body of a female Olympic athlete. In this case biological commonality fails to account for the specificity of these two bodies. Indeed, the female Olympic athlete may have more in common with a male Olympic athlete than with a wife/mother. This commonality is not simply at the level of interests or desires but at the level of the actual form and capacities of the body. By drawing attention to the context in which bodies move and recreate themselves, we also draw attention to the complex dialectic between bodies and their environments. If the body is granted a history then traditional associations between the female body and the domestic sphere and the male body and the public sphere can be acknowledged as historical realities, which have historical effects, without resorting to biological essentialism. The present capacities of female bodies are, by and large, very different to the present capacities of male bodies. It is important to create the means of articulating the historical realities of sexual difference without thereby reifying these differences. Rather, what is required is an account of the ways in which the typical spheres of movement of men and women and their respective activities construct and recreate particular kinds of body to perform particular kinds of task. This sort of analysis is necessary if the historical effects of the ways in which power constructs bodies are to be understood and challenged.[7]

This would involve not simply a study of how men and women become masculine and feminine subjects but how bodies become marked as male and female. Again, Foucault made this point well, arguing that what is needed is:

> an analysis in which the biological and the historical are not consecutive to one another, as in the evolutionism of the first sociologists, but are bound together in an increasingly complex fashion in accordance with the development of the modern technologies of power that take life as their objective. Hence, I do not envisage a 'history of mentalities' that would take account of bodies only through the manner in which they have been perceived and given meaning and value; but a 'history of bodies' and the manner in which what is most material and most vital in them has been invested.[8]

Foucault's studies tend to concentrate on the history of the construction of male bodies and are not forthcoming on the question of sexual difference.[9] However, a critical use of psychoanalytic theory, in particular the theory of the body image, in conjunction with Foucault's analysis of power can provide some very useful insights in this context.

The works of Jacques Lacan, Maurice Merleau-Ponty and Paul Schilder offer an account of the body image which posits that a body is not properly a human body, that is, a human subject or individual, unless it has an image of itself as a discrete entity, or as a *gestalt*.[10] It is this orientation of one's body in space, and in relation to other bodies, that provides a perspective on the world and that is assumed in the constitution of the signifying subject. Lacan, in particular, presents the emergence of this *gestalt* as, in some sense, genetic. His famous 'Mirror Stage' paper, for example, offers ethological evidence for the identificatory effect produced by images and movements of others of the same species and even images and movements which merely *simulate* those of the species in question.[11] Lacan takes this 'homeomorphic identification' to be at the origin of an organism's orientation toward its own species. It would seem that it is this genetic basis to his account of the mirror stage that allows him, even while stressing the cultural specificity of body images, to assert the 'natural' dominance of the penis in the shaping of the *gestalt*.[12]

Foucault's historically dynamic account of the manner in which the micropolitical operations of power produce socially appropriate bodies offers an alternative to Lacan's ethological account. Using Foucault's approach, the imaginary body can be posited as an effect of socially and historically specific practices: an effect, that is, not of genetics but of relations of power. It would be beside the point to insist that, none the less, this imaginary body is in fact the anatomical body overlaid by culture, since the anatomical body is itself a theoretical object for the discourse of anatomy which is produced by human beings in culture. There is a regress involved in positing the anatomical body as the touchstone for cultural bodies since it is a particular culture which chooses to represent bodies anatomically. Another culture might take the clan totem as the essence or truth of particular bodies. The human body is always a signified body and as such cannot be understood as a 'neutral object' upon which science may construct 'true' discourses. The *human* body and its history presuppose each other.

This conception of the imaginary body may provide the framework in which we can give an account of how power, domination and sexual difference intersect in the lived experience of men and women. Gender itself may be understood on this model not as the effect of ideology or cultural values but as the way in which power takes hold of and constructs bodies in particular ways. Significantly, the sexed body can no longer be conceived as the unproblematic biological and factual base upon which gender is inscribed, but must itself be recognized as constructed by discourses and practices that take the body both as their target and as their vehicle of expression. Power is not then reducible to what is imposed, from above, on naturally differentiated male and female bodies, but is also constitutive of those bodies, in so far as they are constituted as male and female.

Shifting the analysis of the operations of power to this micro-level of bodies and their powers and capacities has an interesting effect when one turns to a consideration of the political body. If we understand the masculinity or maleness of the political body and the public sphere as an *arbitrary* historical fact about the genesis of states, then sexual equality should be achievable provided we ensure that women have equal access to the political body and the public sphere. However, the relation between the public sphere and male bodies is not an arbitrary one. The political body was conceived historically as the organization of many bodies into one body which would itself enhance and intensify the powers and capacities of specifically male bodies.[13]

Female embodiment as it is currently lived is itself a barrier to women's 'equal' participation in socio-political life. Suppose our body politic were one which was created for the enhancement and intensification of women's historical and present capacities. The primary aim of such a body politic might be to foster conditions for the healthy reproduction of its members. If this were the case, then presumably some men would now be demanding that medical science provide ways for them to overcome their 'natural' or biological disadvantages, by inventing, for example, means by which they could lactate. This may seem a far-fetched suggestion, but it nevertheless makes the point that a biological disadvantage can be posited as such only in a cultural context.

DIFFERENCE

The crux of the issue of difference as it is understood here is that difference does not have to do with biological 'facts' so much as with the manner in which culture marks bodies and creates specific conditions in which they live and recreate themselves. It is beside the point to 'grant' equal access to women and others excluded from the traditional body politic, since this amounts to 'granting' access to the body politic and the public sphere in terms of an individual's ability to emulate those powers and capacities that have, in a context of male/masculine privilege, been deemed valuable by that sphere. The present and future enhancement of the powers and capacities of women must take account of the ways in which their bodies are presently constituted.

Clearly, the sketch of power and bodies that has been offered here is not one which would lend itself to an understanding of sexual difference in terms of essentialism or biologism. The female body cannot provide the ontological foundation required by those who assert an essential sexual difference. On the contrary, it is the construction of biological discourse as being able to provide this status that is in need of analysis. The cluster of terms 'the female body', 'femininity' and 'woman' need to be analysed in terms of their historical and discursive associations. If discourses cannot be deemed as 'outside', or apart from, power relations then their analysis becomes crucial to an analysis of power. This is why language, signifying practices and discourses have become central stakes in feminist struggles.

Writing itself is a political issue and a political practice for many contemporary feminists. For this reason it is inappropriate to reduce the project of *écriture féminine* to an essentialist strategy. The 'difference' which this form of writing seeks to promote is a difference rooted not in biology but rather in discourse – including biological discourses. It is unhelpful to quibble over whether this writing is an attempt to 'write the female body' or to 'write femininity', since it is no longer clear what this distinction amounts to.[14] What is clear is that discourses, such as Lacanian psychoanalysis, and social practices, such as marriage, construct female and male bodies in ways that constitute and validate the power relations between men and women.

The account of female sexuality offered by Lacanian psychoanalysis constructs female bodies as lacking or castrated and male bodies as full or phallic. This construction tells of a power relation where the actual understanding of sexual difference implies a passive/active relation. Writing of a sexuality that is not simply the inverse or the complement of male sexuality presents a discursive challenge to the traditional psychoanalytic understanding of sexual difference, where difference is exhausted by phallic presence or absence. Irigaray's writing of the 'two lips' of feminine morphology is an active engagement with the construction of what here has been called the imaginary body. It is not an attempt to construct a 'true' theory of sexual difference, starting from the foundation of female biology. Rather, it is a challenge to the traditional construction of feminine morphology where the bodies of women are seen as receptacles for masculine completeness. At the same time as Irigaray's writing offers a challenge to traditional conceptions of women, it introduces the possibility of *dialogue* between men and women in place of the monological pronouncements made by men over the mute body of the (female) hysteric.[15]

Legal practices and discourses surrounding marriage also assume this conception of sexual difference by allotting conjugal rights to the (active) male over the body of the (passive) female. Significantly, the act which is taken to consummate marriage is legally defined as an act performed by a man on a woman. Needless to say, these legal, psychoanalytic and social understandings of the female body have been articulated from the perspective of male writers, who take it upon themselves to represent women, femaleness and femininity. From this perspective, it is not surprising that women are represented as pale shadows and incomplete complements to the more excellent type: 'man'. The project of *écriture féminine* involves challenging the masculine monopoly on the construction of femininity, the female body and woman. It also involves a rejection of the notion that there can be *a* theory of woman, for this would be to accept that woman *is* some (*one*) thing.

The works of Luce Irigaray, Hélène Cixous and Adrienne Rich are each in their own ways involved in investigating the manner in which women's bodies are constructed and lived in culture.[16] Each could be seen to be writing from an embodied perspective about the female body, femininity and women. Yet none of these writers claims to *represent* (all) women or the multiplicity of women's experiences. This would be for them to take up a masculine attitude in relation to other women. Significantly, all three writers critically address the dualisms which have dominated western thought. Addressing constructions of the feminine in history necessarily involves addressing those terms which have been associated with femininity: the body, emotion and so on. When Irigaray, for example, writes of the 'repression of the feminine', she is also alluding to the repression of the body and passion in western thought. To attempt to 'write' the repressed side of these dualisms is not, necessarily, to be working for the reversal of the traditional values associated with each but rather to unbalance or disarrange the discourses in which these dualisms operate. It is to create new conditions for the articulation of difference.

To understand 'difference feminism' as the obverse of 'equality feminism' would be to miss entirely the point of this essay. Difference, as it has been presented here, is not concerned with privileging an essentially biological difference between the sexes. Rather, it is concerned with the mechanisms by which bodies are recognized as different only in so far as they are constructed as possessing or lacking some socially privileged quality or qualities. What is crucial in our current context is the thorough interrogation of the means by which bodies become invested with differences which are then taken to be fundamental ontological differences. Differences as well as commonality must be respected among those who have historically been excluded from speech/writing and are now struggling for expression. If bodies and their powers and capacities are invested in multiple ways, then accordingly their struggles will be multiple.

The conception of difference offered here is not one which seeks to construct a dualistic theory of an essential sexual difference. Rather, it entertains a multiplicity of differences. To insist on sexual difference as *the* fundamental and eternally immutable difference would be to take for granted the intricate and pervasive ways in which patriarchal culture has made that difference its insignia.

NOTES

1. S. Firestone, *The Dialectic of Sex* (Bantam Books, New York, 1970).
2. For an example of the former, see M. Daly, *Gyn/Ecology: The Metaethics of Radical Feminism* (Beacon Press, Boston, Mass., 1978); and for one of the latter, see C. McMillan, *Women, Reason and Nature* (Basil Blackwell, Oxford, 1982).
3. See E. Spelman, 'Woman as Body: Ancient and Contemporary Views', *Feminist Studies*, 8 (1982), pp. 109–31.
4. See Daly, *Gyn/Ecology*, pp. 109–12.
5. Ibid., pp. 61–2.
6. See M. Foucault, 'Nietzsche, Genealogy, History', in his *Language, Counter-Memory, Practice*, ed. D. Bouchard (Cornell University Press, Ithaca, NY, 1977), pp. 139–64.
7. See C. Gallagher and T. Laqueur (eds), *The Making of the Modern Body* (University of California Press, Berkeley, Cal., 1987).
8. Foucault, *History of Sexuality*, p. 152.
9. For a sympathetic feminist reading of Foucault's work, see J. Sawicki, 'Foucault and Feminism: Toward a Politics of Difference', in Shanley and Pateman, *Feminist Interpretations and Political Theory*, pp. 217–31.
10. See J. Lacan, 'Some Reflections on the Ego', *International Journal of Psycho-analysis*, 34 (1953), pp. 11–17; Lacan, 'The Mirror Stage' in *Ecrits* (Tavistock, London, 1977), pp. 1–7; M. Merleau-Ponty, 'The Child and his Relation to Others', in Merleau-Ponty, *The Primacy of Perception* (North-western University Press, Evanston, Ill., 1964), pp. 96–155; P. Schilder, *The Image and Appearance of the Human Body* (International University Press, New York, 1978).
11. Lacan writes:

    > it is a necessary condition for the maturation of the gonad of the female pigeon that it should see another member of its species, of either sex; so sufficient in itself is this condition that the desired effect may be obtained merely by placing the individual within reach of the field of reflection of a mirror. Similarly, in the case of the migratory locust, the transition within a generation from the solitary to the gregarious form can be obtained by exposing the individual, at a certain stage, to the exclusively visual action of a similar image, provided it is animated by movements of a style sufficiently close to that characteristic of the species. ('Mirror Stage', p. 3)

12. Lacan, 'Some Reflections', p. 13.
13. For a recent feminist account of the aims of the masculine political body, see Pateman, *Sexual Contract*, ch. 4; Gatens, 'Representation in/and the Body Politic', in R. Diprose and R. Ferrel (eds), *Cartographies: The Mapping of Bodies and Spaces* (Allen and Unwin, Sydney, 1991), pp. 79–87.
14. See, for example, Toril Moi's arguments in *Sexual/Textual Politics* (Methuen, London, 1985), pp. 102–26, which misunderstand the conception of difference being employed by Cixous.
15. See, for example, the writings of Freud and Breuer on hysteria and femininity in volume 2 of *The Standard Edition of the Complete Psychological Works of Freud*, ed. J. Strachey (Hogarth Press, London, 1974).

16. See L. Irigaray, *This Sex Which is Not One* (Cornell University Press, Ithaca, NY, 1985) and *Speculum of the Other Woman* (Cornell University Press, Ithaca, NY, 1985); H. Cixous, 'Castration or Decapitation?', *Signs*, 7 (1981), pp. 41–55; A. Rich, *Blood, Bread and Poetry* (Virago, London, 1987).

4.3

BODIES THAT MATTER

Judith Butler

> Why should our bodies end at the skin, or include at best other beings encapsulated by skin? (Donna Haraway, *A Manifesto for Cyborgs*)
>
> If one really thinks about the body as such, there is no possible outline of the body as such. There are thinkings of the systematicity of the body, there are value codings of the body. The body, as such, cannot be thought, and I certainly cannot approach it. (Gayatri Chakravorty Spivak, 'In a Word', interview with Ellen Rooney)
>
> There is no nature, only the effects of nature: denaturalization or naturalization. (Jacques Derrida, *Donner le Temps*)

Is there a way to link the question of the materiality of the body to the performativity of gender? And how does the category of 'sex' figure within such a relationship? Consider first that sexual difference is often invoked as an issue of material differences. Sexual difference, however, is never simply a function of material differences which are not in some way both marked and formed by discursive practices. Further, to claim that sexual differences are indissociable from discursive demarcations is not the same as claiming that discourse causes sexual difference. The category of 'sex' is, from the start, normative; it is what Foucault has called a 'regulatory ideal'. In this sense, then, 'sex' not only functions as a norm, but is part of a regulatory practice that produces the bodies it governs, that is, whose regulatory force is made clear as a kind of productive power, the power to produce – demarcate, circulate, differentiate – the bodies it controls. Thus, 'sex' is a regulatory ideal whose materialization is compelled, and this materialization takes place (or fails to

From: J. Butler, *Bodies that Matter*, New York: Routledge, 1993.

take place) through certain highly regulated practices. In other words, 'sex' is an ideal construct which is forcibly materialized through time. It is not a simple fact or static condition of a body, but a process whereby regulatory norms materialize 'sex' and achieve this materialization through a forcible reiteration of those norms. That this reiteration is necessary is a sign that materialization is never quite complete, that bodies never quite comply with the norms by which their materialization is impelled. Indeed, it is the instabilities, the possibilities for rematerialization, opened up by this process that mark one domain in which the force of the regulatory law can be turned against itself to spawn rearticulations that call into question the hegemonic force of that very regulatory law.

But how, then, does the notion of gender performativity relate to this conception of materialization? In the first instance, performativity must be understood not as a singular or deliberate 'act', but, rather, as the reiterative and citational practice by which discourse produces the effects that it names. What will, I hope, become clear in what follows is that the regulatory norms of 'sex' work in a performative fashion to constitute the materiality of bodies and, more specifically, to materialize the body's sex, to materialize sexual difference in the service of the consolidation of the heterosexual imperative.

In this sense, what constitutes the fixity of the body, its contours, its movements, will be fully material, but materiality will be rethought as the effect of power, as power's most productive effect. And there will be no way to understand 'gender' as a cultural construct which is imposed upon the surface of matter, understood either as 'the body' or its given sex. Rather, once 'sex' itself is understood in its normativity, the materiality of the body will not be thinkable apart from the materialization of that regulatory norm. 'Sex' is, thus, not simply what one has, or a static description of what one is: it will be one of the norms by which the 'one' becomes viable at all, that which qualifies a body for life within the domain of cultural intelligibility.[1]

At stake in such a reformulation of the materiality of bodies will be the following: (1) the recasting of the matter of bodies as the effect of a dynamic of power, such that the matter of bodies will be indissociable from the regulatory norms that govern their materialization and the signification of those material effects; (2) the understanding of performativity not as the act by which a subject brings into being what she/he names, but, rather, as that reiterative power of discourse to produce the phenomena that it regulates and constrains; (3) the construal of 'sex' no longer as a bodily given on which the construct of gender is artificially imposed, but as a cultural norm which governs the materialization of bodies; (4) a rethinking of the process by which a bodily norm is assumed, appropriated, taken on as not, strictly speaking, undergone *by a subject*, but rather that the subject, the speaking 'I', is formed by virtue of having gone through such a process of assuming a sex; and (5) a linking of this process of 'assuming' a sex with the question of *identification*, and with the discursive means by which the heterosexual imperative enables certain sexed identifications and forecloses and/or disavows other identifications. This exclusionary matrix by which subjects are formed thus requires the simultaneous production of a domain of abject beings, those who are not yet 'subjects', but who form the constitutive outside to the domain of the subject. The abject[2] designates here precisely those 'unlivable' and 'uninhabitable' zones of social life which are nevertheless densely populated by those who do not enjoy the status of the subject, but whose living under the sign of the 'unlivable' is required to circumscribe the domain of the subject. This zone of uninhabitability will constitute the defining limit of the subject's domain; it will constitute that site of dreaded identification against which – and by virtue of which – the domain of the subject will circumscribe its own claim to autonomy and to life. In this sense, then, the subject is constituted through the force of exclusion and abjection, one which produces a constitutive outside to the subject, an abjected outside, which is, after all, 'inside' the subject as its own founding repudiation.

The forming of a subject requires an identification with the normative phantasm of 'sex', and this identification takes place through a repudiation which produces a domain of abjection, a repudiation without which the subject cannot emerge. This is a repudiation which creates the valence of 'abjection' and its status for the subject as a threatening spectre. Further, the materialization of a given sex will centrally concern *the regulation of identificatory practices* such that the identification with the abjection of sex will be persistently disavowed. And yet, this disavowed abjection will threaten to expose the self-grounding presumptions of the sexed subject, grounded as that subject is in a repudiation whose consequences it cannot fully control. The task will be to consider this threat and disruption not as a permanent contestation of social norms condemned to the pathos of perpetual failure, but rather as a critical resource in the struggle to rearticulate the very terms of symbolic legitimacy and intelligibility.

Lastly, the mobilization of the categories of sex within political discourse will be haunted in some ways by the very instabilities that the categories effectively produce and foreclose. Although the political discourses that mobilize identity categories tend to cultivate identifications in the service of a political goal, it may be that the persistence of *dis*identification is equally crucial to the rearticulation of democratic contestation. Indeed, it may be precisely through practices which underscore disidentification with those regulatory norms by which sexual difference is materialized that both feminist and queer politics are mobilized. Such collective disidentifications can facilitate a reconceptualization of which bodies matter, and which bodies are yet to emerge as critical matters of concern.

FROM CONSTRUCTION TO MATERIALIZATION

The relation between culture and nature presupposed by some models of gender 'construction' implies a culture or an agency of the social which acts upon a nature, which is itself presupposed as a passive surface, outside the social and yet its necessary counterpart. One question that feminists have raised, then, is whether the discourse which figures the action of construction as a kind of imprinting or imposition is not tacitly masculinist, whereas the figure of the passive surface, awaiting that penetrating act whereby meaning is endowed, is not tacitly or – perhaps – quite obviously feminine. Is sex to gender as feminine is to masculine?[3]

Other feminist scholars have argued that the very concept of nature needs to rethought, for the concept of nature has a history, and the figuring of nature as the blank and lifeless page, as that which is, as it were, always already dead, is decidedly modern, linked perhaps to the emergence of technological means of domination. Indeed, some have argued that a rethinking of 'nature' as a set of dynamic interrelations suits both feminist and ecological aims (and has for some produced an otherwise unlikely alliance with the work of Gilles Deleuze). This rethinking also calls into question the model of construction whereby the social unilaterally acts on the natural and invests it with its parameters and its meanings. Indeed, as much as the radical distinction between sex and gender has been crucial to the de Beauvoirian version of feminism, it has come under criticism in more recent years for degrading the natural as that which is 'before' intelligibility, in need of the mark, if not the mar, of the social to signify, to be known, to acquire value. This misses the point that nature has a history, and not merely a social one, but, also, that sex is positioned ambiguously in relation to that concept and its history. The concept of 'sex' is itself troubled terrain, formed through a series of contestations over what ought to be decisive criterion for distinguishing between the two sexes; the concept of sex has a history that is covered over by the figure of the site or surface of inscription. Figured as such a site or surface, however, the natural is construed as that which is also without value; moreover, it assumes its value at the same time that it assumes its social character, that is, at the same time that nature relinquishes itself as the natural. According to this view, then, the social construction of the natural presupposes the cancellation of the natural by the social. Insofar as it relies on this construal, the sex/gender distinction founders along parallel lines; if gender is the social significance that sex assumes within a given culture – and for the sake of argument we will let 'social' and 'cultural' stand in an uneasy interchangeability – then what, if anything, is left of 'sex' once it has assumed its social character as gender? At issue is the meaning of 'assumption' where to be 'assumed' is to be taken up into a more elevated sphere, as in 'the Assumption of the Virgin'. If gender consists of the social meanings that sex assumes, then sex does not *accrue* social meanings as additive properties but, rather, *is replaced by* the social meanings it takes on; sex is relinquished in the course of that assumption, and gender emerges, not as a term in a continued relationship of opposition to sex, but as the term which absorbs and displaces 'sex', the mark of its full substantiation into gender or what, from a materialist point of view, might constitute a full *de*substantiation.

When the sex/gender distinction is joined with a notion of radical linguistic constructivism, the problem becomes even worse, for the 'sex' which is referred to as prior to gender will itself be a postulation, a construction, offered within language, as that which is prior to language, prior to construction. But this sex posited as prior to construction will, by virtue of being posited, become the effect of that very positing, the construction of construction. If gender is the social construction of sex, and if there is no access to this 'sex' except by means of its construction, then it appears not only that sex is absorbed by gender, but that 'sex' becomes something like a fiction, perhaps a fantasy, retroactively installed at a prelinguistic site to which there is no direct access.

But it is right to claim that 'sex' vanishes altogether, that it is a fiction over and against what is true, that it is a fantasy over and against what is reality? Or do these very oppositions need to be rethought such that if 'sex' is a fiction, it is one within whose necessities we live, without which life itself would be unthinkable? And if 'sex' is a fantasy, is it perhaps a phantasmatic field that constitutes the very terrain of cultural intelligibility? Would such a rethinking of such conventional oppositions entail a rethinking of 'constructivism' in its usual sense?

[. . .]

What I would propose in place of these conceptions of construction is a return to the notion of matter, not as site or surface, but as *a process of materialization that stabilizes over time to produce the effect of boundary, fixity, and surface we call matter*. That matter is always materialized has, I think, to be thought in relation to the productive and, indeed, materializing effects of regulatory power in the Foucaultian sense.[4] Thus, the question is no longer, How is gender constituted as and through a certain interpretation of sex? (a question that leaves the 'matter' of sex untheorized), but rather, Through what regulatory norms is sex itself materialized? And how is it that treating the materiality of sex as a given presupposes and consolidates the normative conditions of its own emergence?

Crucially, then, construction is neither a single act nor a causal process initiated by a subject and culminating in a set of fixed effects. Construction not only takes place *in* time, but is itself a temporal process which operates through the reiteration of norms; sex is both produced and destabilized in the course of this reiteration.[5] As a sedimented effect of a reiterative or ritual practice, sex acquires its naturalized effect, and, yet, it is also by virtue of this reiteration that gaps and fissures are opened up as the constitutive instabilities in such constructions, as that which escapes or exceeds the norm, as that which cannot be wholly defined or fixed by the repetitive labor of that norm. This instability is the *de*constituting possibility in the very process of repetition, the power that undoes the very effects by which 'sex' is stabilized, the possibility to put the consolidation of the norms of 'sex' into a potentially productive crisis.[6]

Certain formulations of the radical constructivist position appear almost compulsively to produce a moment of recurrent exasperation, for it seems that when the constructivist is construed as a linguistic idealist, the constructivist refutes the reality of bodies, the relevance of science, the alleged facts of birth, aging, illness, and death. The critic might also suspect the constructivist of a certain somatophobia and seek assurances that this abstracted theorist will admit that there are, minimally, sexually differentiated parts, activities, capacities, hormonal and chromosomal differences that can be conceded without reference to 'construction'. Although at this moment I want to offer an absolute reassurance to my interlocutor, some anxiety prevails. To 'concede' the undeniability of 'sex' or its 'materiality' is always to concede some version of 'sex', some formation of 'materiality'. Is the discourse in and through which that concession occurs – and, yes, that concession invariably does occur – not itself formative of the very phenomenon that it concedes? To claim that discourse is formative is not to claim that it originates, causes, or exhaustively composes that which it concedes; rather, it is to claim that there is no reference to a pure body which is not at the same time a further formation of that body. In this sense, the linguistic capacity to refer to sexed bodies is not denied, but the very meaning of 'referentiality' is altered. In philosophical terms, the constative claim is always to some degree performative.

In relation to sex, then, if one concedes the materiality of sex or of the body, does that very conceding operate – performatively – to materialize that sex? And further, how is it that the reiterated concession of that sex – one which need not take place in speech or writing but might be 'signalled' in a much more inchoate way – constitutes the sedimentation and production of that material effect?

The moderate critic might concede that *some part* of 'sex' is constructed, but some other is certainly not, and then, of course, find him or herself not only under some obligation to draw the line between what is and is not constructed, but to explain how it is that 'sex' comes in parts whose differentiation is not a matter of construction. But as that line of demarcation between such ostensible parts gets drawn, the 'unconstructed' becomes bounded once again through a signifying practice, and the very boundary which is meant to protect some part of sex from the taint of constructivism is now defined by the anti-constructivist's own construction. Is construction something which happens to a ready-made object, a pregiven thing, and does it happen *in degrees*? Or are we perhaps referring on both sides of the debate to an inevitable practice of signification, of demarcating and delimiting that to which we then 'refer', such that our 'references' always presuppose – and often conceal – this prior delimitation? Indeed, to 'refer' naively or directly to such an extra-discursive object will always require the prior delimitation of the extra-discursive. And insofar as the extra-discursive is delimited, it is formed by the very discourse from which it seeks to free itself. This delimitation, which often is enacted as an untheorized presupposition in any act of description, marks a boundary that includes and excludes, that decides, as it were, what will and will not be the stuff of the object to which we then refer. This marking off will have some normative force and, indeed, some violence, for it can construct only through erasing; it can bound a thing only through enforcing a certain criterion, a principle of selectivity.

What will and will not be included within the boundaries of 'sex' will be set by a more or less tacit operation of exclusion. If we call into question the fixity of the structuralist law that divides and bounds the 'sexes' by virtue of their dyadic differentiation within the heterosexual matrix, it will be from the exterior regions of that boundary (not from a 'position', but from the discursive possibilities opened up by the constitutive outside of hegemonic positions), and it will constitute the disruptive return of the excluded from within the very logic of the heterosexual symbolic.

[...]

PERFORMATIVITY AS CITATIONALITY

When, in Lacanian parlance, one is said to assume a 'sex', the grammar of the phrase creates the expectation that there is a 'one' who, upon waking, looks up and deliberates on which 'sex' it will assume today, a grammar in which 'assumption' is quickly assimilated to the notion of a highly reflective choice. But if this 'assumption' is *compelled* by a regulatory apparatus of heterosexuality, one which reiterates itself through the forcible production of 'sex', then the 'assumption' of sex is constrained from the start. And if there is *agency*, it is to be found, paradoxically, in the possibilities opened up in and by that constrained appropriation of the regulatory law, by the materialization of that law, the compulsory appropriation and identification with those normative demands. The forming, crafting, bearing, circulation, signification of that sexed body will not be a set of actions performed in compliance with the law; on the contrary, they will be a set of actions mobilized by the law, the citational accumulation and dissimulation of the law that produces material effects, the lived necessity of those effects as well as the lived contestation of that necessity.

Performativity is thus not a singular 'act' for it is always a reiteration of a norm or set of norms, and to the extent that it acquires an act-like status in the present, it conceals or dissimulates the conventions of which it is a repetition. Moreover, this act is not primarily theatrical; indeed, its apparent theatricality is produced to the extent that its historicity remains dissimulated (and, conversely, its theatricality gains a certain inevitability given the impossibility of a full disclosure of its historicity). Within speech act theory, a performative is that discursive practice that enacts or produces that which it names.[7] According to the biblical rendition of the performative, i.e., 'Let there be light', it appears that it is by virtue of *the power of a subject or its will* that a phenomenon is named into being. In a critical reformulation of the performative, Derrida makes clear that this power is not the function of an originating will, but is always derivative:

> Could a performative utterance succeed if its formulation did not repeat a 'coded' or iterable utterance, or in other words, if the formula I pro-

> nounce in order to open a meeting, launch a ship or a marriage were not identifiable as conforming with an iterable model, if it were not then identifiable in some way as a 'citation'? ... in such a typology, the category of intention will not disappear; it will have its place, but from that place it will no longer be able to govern the entire scene and system of utterance (*l'énonciation*).[8]

To what extent does discourse gain the authority to bring about what it names through citing the conventions of authority? And does a subject appear as the author of its discursive effects to the extent that the citational practice by which he/she is conditioned and mobilized remains unmarked? Indeed, could it be that the production of the subject as originator of his/her effects is precisely a consequence of this dissimulated citationality? Further, if a subject comes to be through a subjection to the norms of sex, a subjection which requires an assumption of the norms of sex, can we read that 'assumption' as precisely a modality of this kind of citationality? In other words, the norm of sex takes hold to the extent that it is 'cited' as such a norm, but it also derives its power through the citations that it compels. And how it is that we might read the 'citing' of the norms of sex as the process of approximating or 'identifying with' such norms?

Further, to what extent within psychoanalysis is the sexed body secured through identificatory practices governed by regulatory schemas? Identification is used here not as an imitative activity by which a conscious being models itself after another; on the contrary, identification is the assimilating passion by which an ego first emerges.[9] Freud argues that 'the ego is first and foremost a bodily ego', that this ego is, further, 'a projection of a surface',[10] what we might redescribe as an imaginary morphology. Moreover, I would argue, this imaginary morphology is not a presocial or presymbolic operation, but is itself orchestrated through regulatory schemas that produce intelligible morphological possibilities. These regulatory schemas are not timeless structures, but historically revisable criteria of intelligibility which produce and vanquish bodies that matter.

If the formulation of a bodily ego, a sense of stable contour, and the fixing of spatial boundary is achieved through identificatory practices, and if psychoanalysis documents the hegemonic workings of those identifications, can we then read psychoanalysis for the inculcation of the heterosexual matrix at the level of bodily morphogenesis?

[...]

As a result of the reformulation of performativity, (a) gender performativity cannot be theorized apart from the forcible and reiterative practice of regulatory sexual regimes; (b) the account of agency conditioned by those very regimes of discourse power cannot be conflated with voluntarism or individualism, much less with consumerism, and in no way presupposes a choosing subject; (c) the regime of heterosexuality operates to circumscribe and contour the 'materiality' of sex, and that 'materiality' is formed and sustained through and as a materialization of regulatory norms that are in part those of heterosexual hegemony; (d) the materialization of norms requires those identificatory processes by which norms are assumed or appropriated, and these identifications precede and enable the formation of a subject, but are not, strictly speaking, performed by a subject; and (e) the limits of constructivism are exposed at those boundaries of bodily life where abjected or delegitimated bodies fail to count as 'bodies'. If the materiality of sex is demarcated in discourse, then this demarcation will produce a domain of excluded and delegitimated 'sex'. Hence, it will be as important to think about how and to what end bodies are constructed as is it will be to think about how and to what end bodies are *not* constructed and, further, to ask after how bodies which fail to materialize provide the necessary 'outside', if not the necessary support, for the bodies which, in materializing the norm, qualify as bodies that matter.

How, then, can one think through the matter of bodies as a kind of materialization governed by regulatory norms in order to ascertain the workings of heterosexual hegemony in the formation of what qualifies as a viable body? How does that materialization of the norm in bodily formation produce a domain of abjected bodies, a field of deformation, which, in failing to qualify as the fully human, fortifies those regulatory norms? What challenge does that excluded and abjected realm produce to a symbolic hegemony that might force a radical rearticulation of what qualifies as bodies that matter, ways of living that count as 'life', lives worth protecting, lives worth saving, lives worth grieving?

NOTES

1. Clearly, sex is not the only such norm by which bodies become materialized, and it is unclear whether 'sex' can operate as a norm apart from other normative requirements on bodies. This will become clear in later sections of this text.
2. Abjection (in latin, *ab-jicere*) literally means to cast off, away, or out and, hence, presupposes and produces a domain of agency from which it is differentiated. Here the casting away resonates with the psychoanalytic notion of *Verwerfung*, implying a foreclosure which founds the subject and which, accordingly, establishes that foundation as tenuous. Whereas the psychoanalytic notion of *Verwerfung* translated as 'foreclosure', produces sociality through a repudiation of a primary signifier which produces an unconscious or, in Lacan's theory, the register of the real, the notion of *abjection* designates a degraded or cast out status within the terms of sociality. Indeed, what is foreclosed or repudiated *within* psychoanalytic terms is precisely what may not reenter the field of the social without threatening psychosis, that is, the dissolution of the subject itself. I want to propose that certain abject zones within sociality also deliver this threat, constituting zones of uninhabitability which a subject fantasizes as threatening its own integrity with the prospect of a psychotic dissolution ('I would rather die than do or be that!'). See the entry under 'Forclusion' in Jean Laplanche and J.-B. Pontalis, *Vocabulaire de la psychanalyse* (Paris: Presses Universitaires de France, 1967) pp. 163–7.
3. See Sherry Ortner, 'Is Female to Male as Nature is to Culture?', in *Woman, Culture, and Society*, Michele Rosaldo and Louise Lamphere (Stanford: Stanford University Press, 1974) pp. 67–88.

4. Although Foucault distinguishes between juridical and productive models of power in *The History of Sexuality, Volume One*, tr. Robert Hurley (New York: Vintage, 1978), I have argued that the two models presuppose each other. The production of a subject – its subjection (*assujetissement*) – is one means of its regulation. See my 'Sexual Inversions', in Domna Stanton, ed., *Discourses of Sexuality* (Ann Arbor: University of Michigan Press, 1992), pp. 344–61.
5. It is not simply a matter of construing performativity as a repetition of acts, as if 'acts' remain intact and self-identical as they are repeated in time, and where 'time' is understood as external to the 'acts' themselves. On the contrary, an act is itself a repetition, a sedimentation, and congealment of the past which is precisely foreclosed in its act-like status. In this sense an 'act' is always a provisional failure of memory. In what follows, I make use of the Lacanian notion that every act is to be construed as a repetition, the repetition of what cannot be recollected, of the irrecoverable, and is thus the haunting spectre of the subject's deconstitution. The Derridean notion of iterability, formulated in response to the theorization of speech acts by John Searle and J. L. Austin, also implies that every act is itself a recitation, the citing of a prior chain of acts which are implied in a present act and which perpetually drain any 'present' act of its presentness. See note 7 below for the difference between a repetition in the service of the fantasy of mastery (i.e., a repetition of acts which build the subject, and which are said to be the constructive or constituting acts of a subject) and a notion of repetition-compulsion, taken from Freud, which breaks apart that fantasy of mastery and sets its limits.
6. The notion of temporality ought not to be construed as a simple succession of distinct 'moments' all of which are equally distant from one another. Such a spatialized mapping of time substitutes a certain mathematical model for the kind of duration which resists such spatializing metaphors. Efforts to describe or name this temporal span tend to engage spatial mapping, as philosophers from Bergson through Heidegger have argued. Hence, it is important to underscore the effect of *sedimentation* that the temporality of construction implies. Here what are called 'moments' are not distinct and equivalent units of time, for the 'past' will be the accumulation and congealing of such 'moments' to the point of their indistinguishability. But it will also consist of that which is refused from construction, the domains of the repressed, forgotten and the irrecoverably foreclosed. That which is not included – exteriorized by boundary – as a phenomenal constituent of the sedimented effect called 'construction' will be as crucial to its definition as that which is included; this exteriority is not distinguishable as a 'moment'. Indeed, the notion of the 'moment' may well be nothing other than a retrospective fantasy of mathematical mastery imposed upon the interrupted durations of the past.

   To argue that construction is fundamentally a matter of iteration is to make the temporal modality of 'construction' into a priority. To the extent that such a theory requires a spatialization of time through the postulation of discrete and bounded moments, this temporal account of construction presupposes a spatialization of temporality itself, what one might, following Heidegger, understand as the reduction of temporality to time.

   The Foucaultian emphasis on *convergent* relations of power (which might in a tentative way be contrasted with the Derridean emphasis on iterability) implies a mapping of power relations that in the course of a genealogical process form a constructed effect. The notion of convergence presupposes both motion and space; as a result, it appears to elude the paradox noted above in which the very account of temporality requires the spatialization of the 'moment'. On the other hand, Foucault's account of convergence does not fully theorize what is at work in the 'movement' by which power and discourse are said to converge. In a sense, the 'mapping' of power does not fully theorize temporality.

   Significantly, the Derridean analysis of iterability is to be distinguished from simple repetition in which the distances between temporal 'moments' are treated as uniform in their spatial extension. The 'betweenness' that differentiates 'moments' of time is not one that can, within Derridean terms, be spatialized or bounded as an identifiable object. It is the nonthematizable différance which erodes and contests any and all claims to discrete identity, including the discrete identity of the 'moment'. What differentiates moments is not a spatially extended duration, for if it were, it would also count as a 'moment', and so fail to account for what falls between moments. This 'entre', that which is at once 'between' and 'outside', is something like non-thematizable space and non-thematizable time as they converge.

Foucault's language of construction includes terms like 'augmentation', 'proliferation', and 'convergence', all of which presume a temporal domain not explicitly theorized. Part of the problem here is that whereas Foucault appears to want his account of genealogical effects to be historically specific, he would favour an account of genealogy over a philosophical account of temporality. In 'The Subject and Power' (Hubert Dreyfus and Paul Rabinow, eds, *Michel Foucault: Beyond Structuralism and Hermeneutics*, Chicago: Northwestern University Press, 1983), Foucault refers to 'the diversity of ... logical sequence' that characterises power relations. He would doubtless reject the apparent linearity implied by models of iterability which link them with the linearity of older models of historical sequence. And yet, we do not receive a specification of 'sequence': Is it the very notion of 'sequence' that varies historically, or are there configurations of sequence that vary, with sequence itself remaining invariant? The specific social formation and figuration of temporality is in some ways unattended by both positions. Here one might consult the work of Pierre Bourdieu to understand the temporality of social construction.

7. See J. L. Austin, *How to Do Things With Words*, J. O. Urmson and Marina Sbisà, eds (Cambridge, Mass.: Harvard University Press, 1955), and *Philosophical Papers* (Oxford: Oxford University Press, 1961), especially pp. 233–52; Shoshana Felman, *The Literary Speech-Act: Don Juan with J. L. Austin, or Seduction in Two Languages*, tr. Catherine Porter (Ithaca: Cornell University Press, 1983); Barbara Johnson, 'Poetry and Performative Language: Mallarmé and Austin', in *The Critical Difference: Essays in the Contemporary Rhetoric of Reading* (Baltimore: Johns Hopkins University Press, 1980), pp. 52–66; Mary Louise Pratt, *A Speech Act Theory of Literary Discourse* (Bloomington: Indiana University Press, 1977); and Ludwig Wittgenstein, *Philosophical Investigations*, tr. G. E. M. Anscombe (New York: Macmillan, 1958), part 1.
8. Jacques Derrida, 'Signature, Event, Context', in *Limited, Inc.*, Gerald Graff, ed.; tr. Samuel Weber and Jeffrey Mehlman (Evanston: Northwestern University Press, 1988), p. 18.
9. See Michel Borch-Jacobsen, *The Freudian Subject*, tr. Catherine Porter (Stanford: Stanford University Press, 1988). Whereas Borch-Jacobsen offers an interesting theory of how identification precedes and forms the ego, he tends to assert the priority of identification to any libidinal experience, where I would insist that identification is itself a passionate or libidinal assimilation. See also the useful distinction between an imitative model and a mimetic model of identification in Ruth Leys, 'The Real Miss Beauchamp: Gender and the Subject of Imitation' in Judith Butler and Joan Scott, eds, *Feminists Theorize the Political* (New York: Routledge, 1992), pp. 167–214; Kaja Silverman, *Male Subjectivity at the Margins* (New York: Routledge, 1992), pp. 262–70; Mary Ann Doane, 'Misrecognition and Identity,' in Ron Burnett, ed., *Explorations in Film Theory: Selected Essays from Ciné-Tracts* (Bloomington: Indiana University Press, 1991), pp. 15–25; and Diana Fuss, 'Freud's Fallen Women: Identification, Desire, and "A Case of Homosexuality in a Women"', in *The Yale Journal of Criticism*, vol. 6, no. 1, (1993): pp. 1–23.
10. Sigmund Freud, *The Ego and the Id*, James Strachey, ed.; tr. Joan Riviere (New York: Norton, 1960), p. 16.

4.4

FEMINISM, FOUCAULT AND THE POLITICS OF THE BODY[1]

Susan Bordo

[...]

One of my goals in this chapter is to help restore feminism's rightful parentage of the 'politics of the body'. My point here is not only 'to set the record straight' out of some feminist chauvinism (although I admit frustration at the continual misunderstandings and caricatures of Anglo-American feminism, both from within and outside feminist scholarly circles). Rather, I think that we can learn something here from history and from the ways that we have re-membered and re-presented that history to ourselves; reflecting on my own participation in such representations, I certainly learned a great deal. In the next section I discuss the original feminist construction of the politics of the body. I then go on to describe what I view as the two key Foucauldian contributions to the further development of that construction, contributions which have significantly deepened, and (rightly) complicated, our understandings of both social 'normalisation' and social resistance.[2] But despite the fact that I view both these contributions as valuable, I am concerned about the recent theoretical *over*-appropriation (as it seems to me) of some of Foucault's more 'postmodern' ideas about resistance.[3] These ideas have been argued to represent more adequately the fragmented and unstable nature of contemporary power relations; my argument in the final section of this chapter is that 'normalisation' is still the *dominant* order of the day, even in a postmodern context, and especially with regard to the politics of *women's* bodies. Looking at contemporary commercials

From: C. Ramazanoglu (ed.), *Up Against Foucault*, London: Routledge, 1993.

and advertisements, I will also show how the rhetoric of resistance has itself been pressed into the service of such normalisation.[4]

FEMINISM AND THE POLITICS OF THE BODY

In my review of *The History of Sexuality* (Bordo 1980), I acknowledged what I felt to be truly innovative about Foucault's critique of the scientisation of sexuality. But I also pointed out that his notion of a power that works not through negative prohibition and restraint of impulse but proliferatively, at the level of the *production* of 'bodies and their materiality, their forces, energies, sensations and pleasures' was not itself new. I had in mind here Marcuse's notion, in *One-Dimensional Man* of the 'mobilisation and administration of libido', whose similarities and differences from Foucault's notion of the 'deployment of sexuality' I discussed in some detail in the review. Not for a moment did I consider the relevance of the extensive feminist literature (from the 1960s and 1970s) on the social construction and 'deployment' of female sexuality, beauty and 'femininity'. I was thoroughly familiar with that literature; I simply did not credit it with a *theoretical* perspective on power and the body. How could this have been? How could I have read Andrea Dworkin, for example, and failed to recognise the 'theory' in the following passage?

> Standards of beauty describe in precise terms the relationship that an individual will have to her own body. They prescribe her motility, spontaneity, posture, gait, the uses to which she can put her body. *They define precisely the dimensions of her physical freedom.* And of course, the relationship between physical freedom and psychological development, intellectual possibility, and creative potential is an umbilical one.
>
> In our culture, not one part of a woman's body is left untouched, unaltered. No feature or extremity is spared the art, or pain, of improvement. . . . From head to toe, every feature of a woman's face, every section of her body, is subject to modification, alteration. This alteration is an ongoing, repetitive process. It is vital to the economy, the major substance of male–female differentiation, the most immediate physical and psychological reality of being a woman. From the age of 11 or 12 until she dies, a woman will spend a large part of her time, money, and energy on binding, plucking, painting and deodorising herself. It is commonly and wrongly said that male transvestites through the use of makeup and costuming caricature the women they would become, but any real knowledge of the romantic ethos makes clear that these men have penetrated to the core experience of being a woman, a romanticised construct. (Dworkin 1974: 113–14; emphasis Dworkin's)

The answer to my question is complex. My failure to recognise the theoretical insight and authority of such work, as I suggested earlier, is in part attributable to the paucity of philosophical scaffolding and scholarly discussion in the works themselves. For the most part, these were not politically motivated academics (at least, not at that point in their lives), but writer/activists; their driving concern was *exposing* oppression, not elaborating the *ideas* most adequate to exposing that oppression (as was the case with Marcuse and Foucault and is arguably the case with much academic feminism today). Moreover, the way 'political writing' was conceived by feminists in those days was aimed at actually effecting *change* in readers' lives. This put a priority on clarity and immediacy, on startling and convincing argument and example, a shunning of obscurity and jargon. And yet: I cannot let myself entirely off the hook here (and of course I am hardly alone on that hook). In 1980, despite the fact that I was writing a dissertation historically critiquing the duality of male mind/female body, I still expected 'theory' only from men. Moreover – and here my inability to 'transcend' these dualisms reveals itself more subtly – I was unable to recognise *embodied* theory when it was staring me in the face. For it is hardly the case that these early feminist works were not theoretical, but rather that their theory never drew attention to *itself*, never made an appearance except as it shaped the 'matter' of their argument. That is, theory was rarely abstracted, objectified and elaborated as of interest in itself. Works that perform such abstraction and elaboration get taken much more seriously than works which do not. This is as true or truer in 1992 as it was in 1980.

Let me clarify here that I am *not* denying the value of such abstraction, or claiming that Foucault's complex theoretical contribution to the 'politics of the body' is contained or even anticipated in the work of Andrea Dworkin or any other feminist writer. Indeed, the next generation of feminist writers on the body often were drawn to Foucault precisely because his theoretical apparatus highlighted the inadequacies of the prevailing feminist discourse and was useful in reconstructing it. I will discuss these issues in more detail in the next section of this chapter. For now I only wish to point out, contrary to current narratives, that neither Foucault nor any other poststructuralist thinker discovered or invented the 'seminal' idea (to refer back to Johnson's account) that the 'definition and shaping' of the body is 'the focal point for struggles over the shape of power'. *That* was discovered by feminism, and long before it entered into its recent marriage with poststructuralist thought – as far back, indeed, as Mary Wollstonecraft's 1792 description of the production of the 'docile body' of the domesticated woman of privilege:

> To preserve personal beauty, woman's glory! the limbs and faculties are cramped with worse than Chinese bands, and the sedentary life which they are condemned to live, whilst boys frolic in the open air, weakens the muscles and relaxes the nerves. As for Rousseau's remarks, which have since been echoed by several writers, that they have naturally, that is since birth, independent of education, a fondness for dolls, dressing, and talking – they are so puerile as not to merit a serious refutation. That a girl, condemned to sit for hours together listening to the idle chat of weak nurses, or to attend to her mother's toilet, will endeavour to join the

> conversation, is, indeed, very natural; and that she will imitate her mother and aunts, and amuse herself by adorning her lifeless doll, as they do in dressing her, poor innocent babe! is undoubtedly a most natural consequence ... genteel women are, literally speaking, slaves to their bodies, and glory in their subjection. ... Women are everywhere in this deplorable state. ... Taught from their infancy that beauty is woman's sceptre, the mind shapes itself to the body and, roaming round its gilt cage, only seeks to adorn its prison. (Wollstonecraft 1988: 55–7)

A more activist generation urged escape from the gilt prison, arguing that the most mundane, 'trivial' aspects of women's bodily existence were in fact significant elements in the social construction of an oppressive feminine norm. In 1914, the first Feminist Mass Meeting in America – whose subject was 'Breaking into the Human Race' – poignantly listed, among the various social and political rights demanded, 'the right to ignore fashion' (Cott 1987: 12). Here already, the material 'micro-practices' of everyday life – which would be extended by later feminists to include not only what one wears, but who cooks and cleans and, more recently, what one eats or does not eat – have been taken out of the realm of the purely personal and brought into the domain of the political. Here, for example, is a trenchant 1971 analysis, presented by way of a set of 'consciousness-raising' exercises for men, of how female subjectivity is normalised and subordinated by the everyday bodily requirements and vulnerabilities of 'femininity':

> Sit down in a straight chair. Cross your legs at the ankles and keep your knees pressed together. Try to do this while you're having a conversation with someone, but pay attention at all times to keeping your knees pressed tightly together.
>
> Run a short distance, keeping your knees together. You'll find you have to take short, high steps if you run this way. Women have been taught it is unfeminine to run like a man with long, free strides. See how far you get running this way for 30 seconds.
>
> Walk down a city street. Pay a lot of attention to your clothing: make sure your pants are zipped, shirt tucked in, buttons done. Look straight ahead. Every time a man walks past you, avert your eyes and make your face expressionless. Most women learn to go through this act each time we leave our houses. It's a way to avoid at least some of the encounters we've all had with strange men who decided we looked available. (Willamette Bridge Liberation News Service 1971)

Until I taught a course in the history of feminism several years ago, I had forgotten that the very first public act of second-wave feminist protest in the United States was the 'No More Miss America' demonstration in August 1968. The critique presented at that demonstration was far from the theoretically crude, essentialising programme that recent caricatures of that era's feminism would suggest. Rather, the position paper handed out at the demonstration outlined a complex, non-reductionist analysis of the intersection of sexism, conformism, competition, ageism, racism, militarism and consumer culture as they are constellated and crystallised in the pageant.[5] The 'No More Miss America' demonstration was the event which earned 'Women's Libbers' the reputation for being 'bra-burners', an epithet many feminists have been trying to shed ever since. In fact, no bras *were* burned at the demonstration, although there was a huge 'Freedom Trash Can' into which were thrown bras, as well as girdles, curlers, false eyelashes, wigs, copies of *The Ladies' Home Journal, Cosmopolitan, Family Circle*, and so on. The media, sensationalising the event, and also no doubt influenced by the paradigm of draft-card burning as the act of political resistance *par excellence*, misreported or invented the burning of the bras. It stuck like crazy glue to the popular imagination; indeed, many of my students today still refer to feminists as 'bra-burners'. But whether or not bras were actually burned, the uneasy public with whom the image stuck surely had it right in recognising the deep political meaning of women's refusal to 'discipline' our breasts – culturally required to be so completely 'for' the other – whether as symbol of maternal love, wet-nurse for the children of the master's house, or erotic fetish.

'Whither the bra in the '90s?' asks Amy Collins, writing for *Lear's* magazine. She answers:

> Women are again playing up their bust lines with a little artifice. To give the breasts the solid, rounded shape that is currently desirable, La Perla is offering a Lycra bra with pre-formed, pressed-cotton cups. To provide a deeper cleavage, a number of lingerie companies are selling side-panel bras that gently nudge the breasts together. Perhaps exercising has made the idea of altering body contours acceptable once more. In any case, if anatomy is destiny, women are discovering new ways to reshape both. (Collins 1991: 80)

Indeed. In 1992, with the dangers of silicone implants on public trial, the media emphasis was on the irresponsibility of Dow, and the personal sufferings of women who became ill from their implants. To my mind, however, the most depressing aspect of the disclosures was the *cultural* spectacle: the large numbers of women who are having implants purely to enlarge or re-shape their breasts, and who consider any health risk worth the resulting boon to their 'self-esteem' and market value. These women are not 'cultural dopes'; usually, they are all too conscious of the system of values and rewards that they are responding to and perpetuating. They know that Bally Matrix Fitness is telling the *truth* about our culture when it tells them that 'You don't just shape your body. You shape your life'. They may even recognise that Bally Matrix is also *creating* that culture. But they insist on their right to be happy on its terms. In the dominant ethos, that right is the bottom line; proposals to ban or even regulate silicone breast implants are thus often viewed as totalitarian interference with self-determination and choice. Many who argue in this way consider themselves feminists, and many feminist scholars today theorise explicitly *as* feminists on their 'behalf'. A recent article in the feminist philosophy journal *Hypatia* for example, defends cosmetic surgery as '*first and foremost* ... about taking one's life into one's own hands' (Davis 1991: 23).

I will return to this contemporary construction later. For now, I would only highlight how very different it is from the dominant feminist discourse on the body in the late sixties and seventies. *That* imagination of the female body was of a *socially* shaped and historically 'colonised' territory, not a site of individual self-determination. Here, feminism inverted and converted the old metaphor of the 'body politic', found in Plato, Aristotle, Cicero, Seneca, Macchiavelli, Hobbes and many others, to a new metaphor: 'the politics of the body'. In the old metaphor of the body politic, the state or society was imagined as a human body, with different organs and parts symbolising different functions, needs, social constituents, forces and so forth – the head or soul for the sovereign, the blood for the will of the people, the nerves for the system of reward and punishments, and so forth. Now, feminism imagined the human *body* as itself a politically inscribed entity, its physiology and morphology shaped and marked by histories and practices of containment and control – from foot-binding and corseting to rape and battering, to compulsory heterosexuality, forced sterilisation, unwanted pregnancy and (in the case of the African-American slave woman) explicit commodification:[6]

> Her head and her heart were separated from her back and her hands and divided from her womb and vagina. Her back and muscle were pressed into field labour where she was forced to work with men and work like men. Her hands were demanded to nurse and nurture the white man and his family as domestic servant whether she was technically enslaved or legally free. Her vagina, used for his sexual pleasure, was the gateway to the womb, which was his place of capital investment – the capital investment being the sex act and the resulting child the accumulated surplus, worth money on the slave market. (Omolade 1983: 354)

One might rightly object that the body's actual bondage in slavery is not to be compared to the metaphorical bondage of privileged nineteenth-century women to the corset, much less to twentieth-century women's 'bondage' to the obsession with slenderness and youth. I think it is crucial, however, to recognise that a staple of the prevailing sexist ideology against which the new feminist model protested was the notion that, in matters of beauty and femininity, it is *women* who are responsible for whatever 'enslavement' they suffer from the whims and bodily tyrannies of 'fashion'. According to that ideology, men's desires have no responsibility to bear, nor does the culture which subordinates women's desires to those of men, sexualises and commodifies women's bodies, and offers them little other opportunity for social or personal power. Rather, it is in our essential feminine nature to be (delightfully if incomprehensibly) drawn to such trivialities, and to be willing to endure whatever physical inconvenience is required. In such matters, whether having our feet broken and shaped into 4-inch 'lotuses', or our waists strait-laced to 14 inches, or our breasts surgically stuffed with plastic, we 'do it to ourselves', are our 'own worst enemies'. Set in cultural relief against this 'thesis', the feminist 'anti-thesis' was the insistence that women are the *done to* not the *doers* here, that *men* and *their* desires (not ours) are the 'enemy', and that our obedience to the dictates of 'fashion' is better conceptualised as bondage than choice. This was a crucial historical moment in the developing articulation of a new understanding of the sexual politics of the body. The limitations of that understanding at this early stage are undeniable. But a new and generative paradigm had been put in place, for later feminist thinkers to develop and critique. It is to this criticism that I now turn in the next section of this chapter.

FOUCAULT'S RE-CONCEPTUALISATION OF THE POLITICS OF THE BODY: NORMALISATION AND RESISTANCE

The initial feminist model of body politics presented various problems for later feminist thought. The 'old' feminist model, for one thing, had tended (although not invariably) to subsume all patriarchal institutions and practices under an oppressor/oppressed model which theorised men as 'possessing' and wielding power over women, who are viewed correspondingly as being utterly power-*less*. Given this model, the woman who has a breast enlargement operation 'to please her man' is as much the victim of his 'power' over her as the slave woman who submits to her owner's desires. Moreover, the oppressor/oppressed model provides no way in which to theorise adequately the complexities of the situations of men, who frequently find themselves implicated in practices and institutions which they (as individuals) did not create, do not control and may feel tyrannised by. Nor does this model acknowledge the degree to which women may 'collude' in sustaining sexism – for example, in our willing (and often eager) participation in cultural practices which objectify and sexualise us.

When I first read Foucault, I remember thinking: 'finally, a male theorist who understands western culture as neither a conversation among talking heads nor a series of military adventures, but as a history of the body!' What fascinated me most about Foucault's work were the historical genealogies themselves. But what I ultimately found most useful to my own work was Foucault's re-conceptualisation of modern 'power'. For Foucault, modern power (as opposed to sovereign power) is non-authoritarian, non-conspiratorial, and indeed non-orchestrated; yet it none the less produces and normalises bodies to serve prevailing relations of dominance and subordination. The key 'moments' of this conception (as Foucault initially theorised it and which I will now attempt to characterise) are found in 'The eye of power' (1977), *Discipline and Punish* (1979), and *The History of Sexuality*, vol. I (1980); later revisions concerning resistance are discussed in 'The subject and power' (1983). Understanding how modern power operates requires, according to Foucault: first, that we cease to imagine 'power' as the *possession* of individuals or groups – as something people 'have' – and instead as a dynamic or network of non-centralised forces. Secondly, we recognise that these forces are *not* random or haphazard, but configure to assume particular historical forms (for example, the mechanisation and later scientisation of 'man'). The dominance of those forms is achieved, however, not from magisterial decree or design 'from above' but through multiple 'processes, of different origin and scattered location', regulating the most intimate and minute elements of the construction of space, time, desire, embodiment (Foucault 1979: 138). Thirdly, (and this element became central to later feminist appropriations of Foucault) prevailing forms of selfhood and subjectivity are maintained not through physical restraint and coercion, but through individual self-surveillance and self-correction to norms. Thus, as Foucault writes,

> there is no need for arms, physical violence, material constraints. Just a gaze. An inspecting gaze, a gaze which each individual under its weight will end by interiorising to the point that he is his own overseer, each individual thus exercising this surveillance over, and against himself. (Foucault 1977: 155)

I would also argue (not all feminists would agree[7]) that this 'impersonal' conception of power does *not* entail that there are no dominant positions, social structures or ideologies emerging from the play of forces; the fact that power is not held by *anyone* does not entail that it is equally held by *all*. It is 'held' by no one; but people and groups *are* positioned differently within it. No one may control the rules of the game. But not all players on the field are equal. (I base my interpretation here less on Foucault's explicitly theoretical statements than on his historical genealogies themselves.)

Such a model seemed to many of us particularly useful to the analysis of male dominance and female subordination, so much of which, in a modern western context, is reproduced 'voluntarily', through self-normalisation to everyday habits of masculinity and femininity.[8] In my own work, Foucault's ideas were extremely helpful both to my analysis of the contemporary disciplines of diet and exercise (1990a) and to my understanding of eating disorders as arising out of and reproducing normative feminine practices of our culture. These are practices which train the female body in docility and obedience to cultural demands while at the same time being *experienced* in terms of 'power' and 'control' (Bordo 1985, 1990a).

Within a Foucauldian framework, power and pleasure do not cancel each other. Thus, the heady experience of feeling powerful, or 'in control', far from being a necessarily accurate reflection of one's actual social position, is always suspect as itself the product of power relations whose shape may be very different. Within such a framework, too, one can acknowledge that women are not always passive 'victims' of sexism, but that we may contribute to the perpetuation of female subordination, for example, by participating in industries and cultural practices which represent women as sexual enticements and rewards for men – without this entailing that we have 'power' (or are equally positioned with men) in sexist culture. While men cease to be constructed as 'the enemy' and their often helpless enmeshment in patriarchal culture can be acknowledged by a Foucauldian model, this does not mitigate the fact that they often may have a higher stake in maintaining institutions within which they have historically occupied dominant positions *vis-à-vis* women. That is why they have often *felt* (and behaved) like 'the enemy' to women struggling to change those institutions. (Such a dual recognition seems essential, in particular, to theorising the situation of men who have been historically subordinated *vis-à-vis* their 'race', class and sexuality.)

Foucault also emphasised, later in his life, that power relations are never seamless, but always spawning new forms of culture and subjectivity, new openings for potential resistance to emerge. Where there is power, he came to see, there is also resistance (1983). I would add to this that prevailing norms themselves have transformative potential. While it is true that we may experience the illusion of 'power' while actually performing as 'docile bodies' (for example, my analysis of the situation of the anorectic), it is also true that our very 'docility' can have consequences that are personally liberating and/or culturally transforming. So, for example, (to construct some illustrations not found in Foucault), the woman who goes on a rigorous weight-training programme in order to achieve a currently stylish look may discover that her new muscles also enable her to assert herself more forcefully at work. Or – a different sort of example – 'feminine' decorativeness may function 'subversively' in professional contexts which are dominated by highly masculinist norms (such as academia). Modern power relations are thus unstable; resistance is perpetual and hegemony precarious.

The 'old' feminist discourse, whose cultural work was to expose the oppressiveness of femininity, could not be expected to give much due to the *pleasures* of shaping and decorating the body or their subversive potential. That was left to a later generation of feminist theorists, who have found both Foucault and deconstructionism to be useful in elaborating such ideas. Deconstructionism has been helpful in pointing to the many-sided nature of meaning; for every interpretation, there is always a reading 'against the grain'. Foucault has been attractive to feminists for his later insistence that cultural resistance is ubiquitous and perpetual. While an initial wave of Foucauldian-influenced feminism had seized on concepts such as 'discipline', 'docility', 'normalisation' and 'bio-power', a second, more 'postmodern'[9] wave has emphasised 'intervention', 'contestation', 'subversion'. The first wave, while retaining the 'old' feminist conception of the 'colonised' female body, sought to complicate that discourse's insufficiently textured, good guys/bad guys conception of social control. Postmodern feminism, on the other hand, criticises *both* the 'old' discourse *and* its reconstruction for over-emphasising such control, for failing to acknowledge adequately the creative and resistant responses that continually challenge and disrupt it.

From this postmodern perspective, both the earlier emphasis on women's bodies as subject to 'social conditioning', and the later move to 'normalisation', under-estimate the unstable nature of subjectivity and the creative agency of individuals – 'the cultural work' (as one theorist puts it) 'by which nomadic, fragmented, active subjects confound dominant discourse'.[10] In this view the dominant discourses which define femininity are continually allowing for the eruption of 'difference', and even the most subordinated subjects are therefore continually confronted with opportunities for resistance, for making meanings that 'oppose or evade the dominant ideology'. There is power and pleasure in this culture, television critic John Fiske insists, 'in being different'. (He then goes on to produce examples of how *Dallas*, *Hart to Hart* and other shows have been read by various sub-cultures to make their own empowering meanings out of the 'semiotic resources' provided by television (Fiske 1987: 11).) In a similar vein, Judith Butler (1990: 137–8) suggests that by presenting a mocking enactment of how gender is artificially constructed and 'performed', drag and other 'parodic practices' (such as cross-dressing and lesbian 'butch/femme' identities) that are proliferated *from within* gender-essentialist culture effectively expose and subvert that culture and its belief in 'the notion of a true gender identity'.

In terms of the very general overview presented in this section, there are thus 'two' Foucaults for feminism, and in some ways they are the mirror-image of one another. The 'first' Foucault, less a product of postmodern culture than a direct descendant of Marx, and sibling to 1960s and 1970s feminism, has attracted feminists with his deep and complex understanding of the 'grip' of systemic power on the body. The appeal of the 'second' Foucault, in contrast, has been his later, postmodern appreciation, for the creative 'powers' of bodies to *resist* that grip. Both perspectives, I would argue, are essential to a fully adequate *theoretical* understanding of power and the body. Yet the question remains as to which emphasis (for we are always and of necessity selective in our attention and emphases) provides the greater insight into the specific historical situations of women today. In the next section of this chapter, focusing on the politics of appearance, I will consider this question.

[. . .]

NOTES

1. Portions of this chapter are based on material from the introduction and conclusion to my book, *Unbearable Weight: Feminism, Western Culture and the Body* (Bordo 1993). Other portions were taken from talks that I delivered at the University of Rochester and Hobart and William Smith Colleges. I offer my thanks to all those who participated in discussions at those presentations.
2. By social 'normalisation' I refer to all those modes of acculturation which work by setting up standards or 'norms' against which individuals continually measure, judge, 'discipline' and 'correct' their behaviour and presentation of self. By social 'resistance' I refer to all behaviours, events and social formations that challenge or disrupt prevailing power relations and the norms that sustain and reproduce them.
3. The postmodern has been described and re-described with many different emphases and points of departure, some critical and some celebratory of the 'postmodern

condition' (see Bordo 1991). Without entering into a lengthy and diverting discussion, for my purposes here I employ the term 'postmodern' in the most general cultural sense, as referring to the contemporary inclination towards the unstable, fluid, fragmented, indeterminate, ironic and heterogeneous, for that which resists definition, closure and fixity. Within this general categorisation, ideas that have developed out of poststructuralist thought – the emphasis on semiotic indeterminacy, the critique of unified conceptions of subjectivity, fascination with the instabilities of systems, and the ability to focus on cultural resistance rather than dominant forms – are decidedly 'postmodern' intellectual developments. But not all poststructuralist thought is 'postmodern'. Foucault, as I read him, has both 'modern' and 'postmodern' moments. In his discussions of the discipline, normalisation and creation of 'docile bodies' he is very much the descendant of Marx; later revisions to his conception of power emphasise the ubiquity of resistance – a characteristic 'postmodern' theme.

4. A final introductory note: The 'stream' of feminist body-politics which is my chief focus in this chapter is the politics of appearance. Even though Foucault himself had little to say about this – or about women – I construct most of my examples and illustrations of Foucault's ideas from this domain, to which I view his ideas as particularly applicable. (For the same reason, I use Foucauldian terminology in describing early feminist perspectives of the body, even though that terminology was unknown to the writers themselves.) This choice of focus should not be taken as implying that I view issues concerning work, sexuality, sexual violence, parenting and reproductive rights as less illustrative of, or important to, a feminist politics of the body. It also explains my omission of any discussion of French feminism, whose contribution to feminist perspectives on the body has been significant, but which has not theorised the politics of beauty and appearance as central to the construction of femininity.
5. The Ten Points of protest listed were: 'The Degrading Mindless-Boob-Girlie Symbol'; 'Racism with Roses'; 'Miss America as Military Death Mascot'; 'The Consumer Con-Game'; 'Competition Rigged and Unrigged'; 'The Woman as Pop Culture Obsolescent Theme'; 'The Unbeatable Madonna-Whore Combination'; 'The Irrelevant Crown on the Throne of Mediocrity'; 'Miss America as Dream Equivalent to——'; and 'Miss America as Big Sister Watching You' (in Morgan 1970: 522–4).
6. Among the 'classics': Susan Brownmiller, *Against Our Will* (1975); Mary Daly, *Gyn/Ecology* (1978); Angela Davis, *Women, Race and Class* (1983); Andrea Dworkin, *Woman-Hating* (1974); Germaine Greer, *The Female Eunuch* (1970); Susan Griffin, *Rape: The Power of Consciousness* (1979) and *Woman and Nature* (1978); Adrienne Rich, 'Compulsory heterosexuality and lesbian existence' (1980). See also the anthologies *Sisterhood is Powerful* (Robin Morgan, ed., 1970) and *Woman in Sexist Society* (Vivian Gornick and Barbara Moran, eds, 1971).
7. See Nancy Fraser (1989) and Nancy Hartsock (1990) for a very different view, which criticises Foucault's conception of power for failing to allow for the sorts of differentiations I describe here.
8. See the section on 'Discipline and the female subject' in Diamond and Quinby (1988), especially Sandra Bartky's piece 'Foucault, femininity, and the modernisation of patriarchal power'. See also Kathryn Pauly Morgan (1991).
9. For my use of 'postmodernism', see note 3.
10. This was said by Janice Radway in an informal presentation of her work, Duke University, spring 1989.

REFERENCES

Bartky, S. L. (1988) 'Foucault, femininity and the modernization of patriarchal power', in I. Diamond and L. Quinby (eds), *Feminism and Foucault: Reflections on Resistance*, Boston: Northeastern University Press.

Bordo, S. (1980) 'Organized sex', *Cross Currents* XXX (3): 194–8.
Bordo, S. (1985) 'Anorexia nervosa: psychopathology as the crystallization of culture', *Philosophical Forum* 17 (2): 73–103.
Bordo, S. (1990) 'Reading the slender body', in M. Jacobus, E. Fox Keller and S. Shuttleworth (eds), *Body/Politics: Women and the Discourses of Science*, New York and London: Routledge.
Bordo, S. (1991) 'Postmodern subjects, postmodern bodies: a review essay', *Feminist Studies* 18 (1): 159–76.
Bordo, S. (1993) *Unbearable Weight: Feminism, Western Culture and the Body*, Berkeley: University of California Press.
Brownmiller, S. (1975) *Against Our Will*, New York: Bantam.
Butler, J. (1990) *Gender Trouble: Feminism and the Subversion of Identity*, London: Routledge.
Collins, A. (1991) 'Abreast of the bra', *Lear's* 4 (4): 76–81.
Cott, N. (1987) *The Grounding of Modern Feminism*, New Haven, CT: Yale University Press.
Daly, M. (1978) *Gyn-Ecology*, Boston: Beacon.
Davis, A. (1983) *Women, Race and Class*, New York: Vintage.
Davis, K. (1991) 'Remaking the she-devil: a critical look at feminist approaches to beauty', *Hypatia* 6 (2): 21–43.
Diamond, I. and Quinby, L. (eds) (1988) *Feminism and Foucault: Reflections on Resistance*, Boston: Northeastern University Press.
Dworkin, A. (1974) *Woman-Hating*, New York: Dutton.
Fiske, J. (1987) *Television Culture*, New York: Methuen.
Foucault, M. (1977) 'The eye of power', in C. Gordon (ed. and trans.), *Power/Knowledge*, New York: Pantheon.
Foucault, M. (1979) *Discipline and Punish*, New York: Vintage.
Foucault, M. (1980) *The History of Sexuality*, vol I: *An Introduction*, New York: Vintage.
Foucault, M. (1983) 'The subject and power', L. Sawyer (trans.), in H. L. Dreyfus and P. Rabinow (eds), *Michel Foucault: Beyond Structuralism and Hermeneutics*, Chicago: University of Chicago Press.
Fraser, N. (1989) 'Foucault on modern power: empirical insights and normative confusions', in N. Fraser, *Unruly Practices: Power, Discourse and Gender in Contemporary Social Theory*, Minneapolis: University of Minnesota.
Gornick, V. and Moran, B. (eds) (1971) *Woman in Sexist Society*, New York: Mentor.
Greer, G. (1970) *The Female Eunuch*, New York: McGraw-Hill.
Griffin, S. (1978) *Woman and Nature: The Roaring Inside Her*, New York: Harper Colophon.
Griffin, S. (1979) *Rape: The Power of Consciousness*, New York: Harper & Row.
Hartsock, N. (1990) 'Foucault on power: a theory for women?', in L. Nicholson (ed.), *Feminism/Postmodernism*, New York and London: Routledge.
Morgan, K. P. (1991) 'Women and the knife: cosmetic surgery and the colonization of women's bodies', *Hypatia*, 6 (3): 25–53.
Morgan, R. (ed.) (1970) *Sisterhood is Powerful: An Anthology of Writings from the Women's Liberation Movement*, New York: Vintage.
Omolade, B. (1983) 'Hearts of darkness', in A. Snitow, C. Stansell and S. Thompson (eds), *Powers of Desire*, New York: Monthly Review Press.
Rich, A. (1980) 'Compulsory heterosexuality and lesbian existence', *Signs* 5 (4): 631–60.
Willamette Bridge Liberation News Service (1971) 'Exercises for men', *The Radical Therapist*, December–January.
Wollstonecraft, M. (1988) 'A vindication of the rights of women', in A. Rossi (ed.), *The Feminist Papers*, Boston: Northeastern University Press.

4.5

WRITE YOUR BODY *AND* THE BODY IN THEORY

Trinh T. Minh-ha

WRITE YOUR BODY

It wrote itself through me. 'Women must write through their bodies.' Must not let themselves be driven away from their bodies. Must thoroughly rethink the body to re-appropriate femininity. Must not however exalt the body, not favor any of its parts formerly forbidden. Must perceive it in its integrity. Must and must-nots, their absolution and power. When armors and defense mechanisms are removed, when new awareness of life is brought into previously deadened areas of the body, women begin to experience writing/the world differently. This is exciting and also very scary. For it takes time to be able to tolerate greater aliveness. Hence the recurrence of musts and must-nots. As soon as a barrier is destroyed, another is immediately erected. Call it reform or expansion. Or else, well-defined liberation revolution. Closure and openness, again, are one ongoing process: we do not *have* bodies, we *are* our bodies, and we are ourselves while being the world. Who can endure constant open-endedness? Who can keep on living completely exposed? We write – think and feel – (with) our entire bodies rather than only (with) our minds or hearts. It is a perversion to consider thought the product of one specialized organ, the brain, and feeling, that of the heart. The past convention was that we desire because we are incomplete, that we are always searching for that other missing half. More recently, we no longer desire-because, we simply desire, and we desire as we are. 'I am a being of desire, therefore a being of words', said Nicole

From: T. T. Minh-ha, *Woman, Native, Other. Writing Postcoloniality and Feminism*, Bloomington: Indiana University Press, 1989.

Brossard, 'a being who looks for her body and looks for the body of the other: for me, this is the whole history of writing.'[1] Gathering the fragments of a divided, repressed body and reaching out to the other does not necessarily imply a lack or a deficiency. In writing themselves, women have attempted to render noisy and audible all that had been silenced in phallocentric discourse. 'Your body must be heard', Hélène Cixous insists, '[Women] must invent the impregnable language that will wreck partitions, classes and rhetorics, regulations and codes.'[2] Touch me and let me touch you, for the private is political. Language wavers with desire. It is 'the language of my entrails', a skin with which I caress and feel the other, a body capable of receiving as well as giving: nurturing and procreating. Let it enter and let it go; writing myself into existence also means emptying myself of all that I can empty out – all that constitutes Old Spontaneous/Premeditated Me – without ceasing from being. 'Every woman is the woman of all women' (Clarice Lispector). Taking in any voice that goes through me, I/i will answer every time someone says: I. One woman within another, eternally. 'Writing as a woman. I am becoming more and more aware of this', notes Anaïs Nin, 'All that happens in the real womb, not in the womb fabricated by man as a substitute ... woman's creation far from being like man's must be exactly like her creation of children, that is it must come out of her own blood, englobed by her womb, nourished by her own milk. It must be a human creation, of flesh, it must be different from man's abstractions.'[3] Man is not content with referring to his creation as to his child, he is also keen on appropriating the life-giving act of childbearing. Images of men 'in labor' and 'giving birth' to poems, essays, and books abound in literature. Such an encroachment on women's domain has been considered natural, for the writer is said to be either genderless or bisexual. He is able to chat with both man's and woman's voices. This is how the womb is fabricated. Women began to be spoken of as if they were wombs on two feet when the fetus was described as a citizen, the womb was declared state property, legislation was passed to control it, and midwifery was kept under continual medical supervision – in other words, when women were denied the right to create. Or not to create. With their bodies. 'All that happens in the real womb': writing as an 'intrinsic' child/birth process takes on different qualities in women's contexts. No man claims to speak from the womb, women do. Their site of fertilization, they often insist, is the womb, not the mind. Their inner gestation is in the womb, not in the mind. The mind is therefore no longer opposed to the heart; it is, rather, perceived as part of the womb, being 'englobed by it'. Men name 'womb' to separate a part of woman from woman (to separate it from the rest that forms her: body and mind), making it possible to lay legal claim to it. By doing so, they create their own contradictions and come round to identifying her with their fabrication: a specialized, infant-producing organ. Women use 'womb' to re-appropriate it and re-unite (or re-differ) themselves, their bodies, their places of production. This may simply mean beating the master at his own game. But it may also mean asserting difference on differences. In the first case, the question is chiefly that of erecting inverted images and defying prohibitions. Annie Leclerc wrote:

> Let me first tell you where I get what I'm saying from, I get it from me, woman, and from my woman's belly. . . . Who would have told me, will I ever be able to tell, from what words shall I weave the bewildering happiness of pregnancy, the very rending, overwhelming happiness of giving birth. . . .
>
> So much the worst for him, I will have to speak of the joys of my sex, no, no, not the joys of my mind, virtue or feminine sensitivity, the joys of my woman's belly, my woman's vagina, my woman's breasts, sumptuous joys of which you have no idea at all.
>
> I will have to speak of them since it is only from them that a new, woman speech will be born.
>
> We will have to divulge what you have so relentlessly put in solitary confinement, for that is what all our other repressions build themselves upon.[4]

Woman's writing becomes 'organic writing', 'nurturing-writing' (*nourricriture*), resisting separation. It becomes a 'connoting material', a 'kneading dough', a 'linguistic flesh'. And it draws its corporeal fluidity from images of water – a water from the source, a deep, subterranean water that trickles in the womb, a meandering river, a flow of life, of words running over or slowly dripping down the pages. This keeping-alive and life-giving water exists simultaneously as the writer's ink, the mother's milk, the woman's blood and menstruation.[5] Logical backlash? An eye for an eye, a tooth for a tooth. Not quite, it seems. A woman's ink of blood for a man's ink of semen (an image found, for example, in Jacques Derrida's hymeneal fable: a sexual union in which the pen writes its in/dis/semination in the always folded/never single space of the hymen). In the second case – that of asserting difference on differences – the question of writing (as a) woman is brought a step further. Liquid/ocean associated with woman/mother is not just a facile play on words inherited from nineteenth-century Romantics (*mer-mère* in French). Motherhood as lived by woman often has little to do with motherhood as experienced by men. The mother cannot be reduced to the mother-hen, the wet-nurse, the year-round cook, the family maid, or the clutching, fear-inspiring matron. Mother of God, of all wo/mankind, she is role-free, non-Name, a force that refuses to be fragmented but suffocates codes (Cixous). In her maternal love, she is neither possessed nor possessive, neither binding nor detached nor neutral. For a life to maintain another life, the touch has to be infinitely delicate: precise, attentive, and swift, so as not to pull, track, rush, crush, or smother.[6] Bruised, half-alive, or dead is often the fate of what comes within the masculine grip. Woman, as Cixous defines her, is a whole – 'whole composed of parts that are wholes' – through which language is born over and over again. (The One is the All and the All is the One; and yet the One remains the One and the All the All. Not two, not One either. This is what Zen has been repeating for centuries.) To the classic conception of bisexuality, the self-effacing, merger-type of bisexuality, Cixous opposes 'The *other bisexuality* ... that is, each one's location in self (*repérage en soi*) of the presence – variously manifest and insistent according to each person, male or female – or both sexes, nonexclusion either of the difference or of one sex, and, from this "self-permission", multiplication of the effects of the inscription of desire, over all parts of my body and the other body.'[7] The notion of 'bisexual, hence neuter' writing together with the fantasy of a 'total' being are concepts that many men have actively promoted to do away with differentiation. Androgyny is another name for such a co-optation. Saying that a great mind is androgynous (and *God knows* how many times we have heard this line – supposedly from Coleridge – and in how many disguises it appears) is equivalent to saying that 'the mind has no sex' (also read 'no gender'). In the salvation theme of androgyny, the male is still seen as the active power of generation and the female as the passive one (a defective male, due to the absence of androgen). Thus Janice Raymond suggests as a substitute the word 'integrity'; she expands it and redefines it as 'an unfolding process of becoming. It contains within itself an insatiable generativeness, that is, a compulsion to reproduce itself in every diverse fashion.'[8] In every diverse fashion ... Laying claim to the specificity of women's sexuality and the rights pertaining to it is a step we have to go through in order to make ourselves heard; in order to beat the master at his own game. But reducing everything to the order of sex does not, obviously, allow us to depart from a discourse directed within the apparatuses of sexuality. Writing does not translate bisexuality. It (does not express language but) fares across it.

THE BODY IN THEORY

It must be different from man's abstractions. Different from man's androgenization. Man's fragmentization. Ego is an identification with the mind. When ego develops, the head takes over and exerts a tyrannical control over the rest of the body. (The world created must be defended against foreign infiltration.) But thought is as much a product of the eye, the finger, or the foot as it is of the brain. If it is a question of fragmenting so as to decentralize instead of dividing so as to conquer, then what is needed is perhaps not a clean erasure but rather a constant displacement of the two-by-two system of division to which analytical thinking is often subjected. In many cases emphasis is necessarily placed upon a reversal of the hierarchy implied in the opposition between mind and body, spiritual and material, thinking and feeling, abstract and concrete, theory and practice. However, to prevent this counter-stance from freezing into a dogma (in which the dominance-submission patterns remain unchanged), the strategy of mere reversal needs to be displaced further, that is to say, neither simply renounced nor accepted as an end in itself. In spite of the distant association, one example that comes to mind are the procedures which in Asia postulate not one, not two, but three centers in the human being: the intellectual (the *path*), the emotional (the *oth*), and the vital (the *kath*). The martial arts concentrate on developing awareness of the latter, which they call the *tantien* or the *hara*. This center, located below the navel (the oth being connected with the heart, and the path with reason), radiates life. It directs vital movement and allows one to relate to the world with instinctual immediacy. But instinct(ual immediacy) here is not opposed to reason, for it lies outside the classical realm of duality assigned to the sensible and the intelligible. So does certain women's womb writing, which neither separates the body from the mind nor sets the latter against the heart (an attitude which would lead us back to the writing-as-birth-delivering-labor concept and to the biologico-metaphorization of women's bodies previously discussed) but allows each part of the body to become infused with consciousness. Again, bring a new awareness of life into previously forgotten, silenced, or deadened areas of the body. Consciousness here is not the result of an accumulation of knowledge and experience but the term of an ongoing unsettling process. The formula 'Know thyself' has become obsolete. We don't want to observe our organism from a safe distance. We do not just write about our body, whether in a demonstrative (objectivist) or a submissive (subjectivist) discourse. Knowledge leads no more to openings than to closures. The idealized quest for knowledge and power makes it often difficult to admit that enlightenment (as exemplified by the West) often brings about endarkenment. More light, less darkness. More darkness, less light. It is a question of degrees, and these are two degrees of one phenomenon. By attempting to *exclude* one (darkness) for the sake of the other (light), the modernist project of building universal knowledge has indulged itself in such self-gratifying oppositions as civilization/ primitivism, progress/backwardness, evolution/stagnation. With the decline of the colonial idea of advancement in rationality and liberty, what becomes more obvious is the necessity to reactivate that very part of the modernist project at its *nascent* stage: the radical calling into question, in every undertaking, of everything that one tends to take for granted – which is a (pre- and post-modernist) stage that should remain constant. No Authority no Order can be safe from criticism. Between knowledge and power, there is room for knowledge-without-power. Or knowledge at rest – 'the end of myths, the erosion of utopia, the rigor of taut patience',[9] as Maurice Blanchot puts it. The terrain remains fresh for it cannot be occupied, not even by its specific creator. The questions that arise continue to provoke answers, but none will dominate as long as the ground-clearing activity is at work. Can knowledge circulate without a position of mastery? Can it be conveyed without the exercise of power? No, because there is no end to understanding power relations which are rooted deep in the social nexus – not merely added to society nor easily locatable so that we can just radically do away with them. Yes, however, because in-between grounds always exist, and cracks and interstices are like gaps of fresh air that keep on being suppressed because they tend to render more visible the failures operating in every system.

Perhaps mastery need not coincide with power. Then we would have to rethink mastery in terms of non-master, and we would have to rewrite women's relation to theory. 'Writing the body' may be immediately heard by a number of male or genderless writers as 'imitating' or 'duplicating the body'. It may be further read as 'female self-aggrandizement' or 'female neurosis'. (It falls on deaf ears, most likely . . .) *There is no such thing as a direct relation to the body*, they assert. For them writing the body means writing *closer* to the body, which is understood as being able to express itself directly without any social mediation. The biological remains here conveniently separate from the socio-historical, and the question 'where does the social stop in the biological?' and vice versa, is not dealt with. They read 'writing the body' as 'the (biological) body writing itself'. They either can't hear the difference or believe there is only *one way* of hearing it: the way they define it. Putting an end to further explorations they often react with anger: *that's not deep enough! What is it supposed to mean?* If we take the case of Roland Barthes, who also passionately writes the (impetus of the) body, the question of depth and meaning no longer exerts its tyranny: he might be fetishist, some of them admit, but We love his fetishism; it's *intelligently* written! and theoretically sound! Through the works of a number of male writers, 'writing the body' may be accepted as a concept because attempts at theorizing it have been carried out, and it has its own place as theoretical object. But when this concept-practice is materialized, the chance of its being understood or even recognized as such never goes without struggle. On the one hand it is a commonplace to say that 'theoretical' usually refers to inaccessible texts that are addressed to a privileged, predominantly male social group. Hence, to many men's ears it is synonymous with 'profound', 'serious', 'substantial', 'scientific', 'consequential', 'thoughtful', or 'thought-engaging'; and to many women's ears, equivalent to 'masculine', 'hermetic', 'elitist', and 'specialized', therefore 'neutral', 'impersonal', 'purely mental', 'unfeeling', 'disengaging', and – last but not least – 'abstract'. On the other hand, it is equally common to observe that theory threatens, for it can upset rooted ideologies by exposing the mechanics of their workings. It shakes established canons and questions every norm validated as 'natural' or 'human'. And it undermines a powerful tradition of 'aesthetics' and 'scholarship' in the liberal arts, in the humanities as well as in the social sciences. To say this is also to say that theory is suspicious, as long as it remains an occupied territory. Indeed, theory no longer is theoretical when it loses sight of its own conditional nature, takes no risk in speculation, and circulates as a form of administrative inquisition. Theory oppresses, when it wills or perpetuates existing power relations, when it presents itself as a means to exert authority – the Voice of Knowledge. In the passage from the heard, seen, smelled, tasted, and touched to the told and the written, language has taken place. Yet in the articulation of language, what is referred to, phenomenally and philosophically, is no more important than what is at work, linguistically, in the referring activity. To declare, for example, that so-and-so is an authority on such-and-such matter (implying thereby that s/he has written with authority on the subject concerned and that this authority is recognized by his/her peers) is to lose sight of the radicalness of writing and theorizing. It is to confuse the materiality of the thing named – or the object of discussion – with the materiality of the name – the modalities of production and reception of meaning – and to give up all attempt at understanding the very social and historical reality of the tools one uses to unmask ideological mystifications – including the mystification of theory. What is at stake is not so much the referential function of language as the authority of language as a model for natural cognition and a transparent medium for criticizing and theorizing. The battle continues, as it should, on several fronts. If it is quite current today to state that language functions according to principles that are not necessarily (like) those of the phenomenal world, it is still unusual to encounter instances where theory involved the voiding, rather than the affirming or even reiterating, of theoretical categories. Instances where poeticalness is not primarily an aesthetic response, nor literariness merely a question of pure verbalism. And instances where the borderline between theoretical and non-theoretical writings is blurred and questioned, so that theory and poetry necessarily mesh, both determined by an awareness of the sign and the destabilization of the meaning and writing subject. To be lost, to encounter impasse, to fall, and to desire both fall and impasse – isn't this what happens to the body in theory? For, in theorizing, can women afford to forget, as Marguerite Duras puts it, that 'men are the ones who started to speak, to speak alone and for everyone else, on behalf of everyone else. . . . They activated the old language, enlisted the aid of the old way of theorizing, in order to relate, to recount, to explain this new situation'?[10] Indeed, women rarely count among those whom Catherine Clément describes as being 'greedy of the slightest theoretical breaking of wind that is formulated, eager not to miss any coach of a passing train in which they hop in flocks behind their chief, whoever he may be, provided that they are not alone and that it is a question of things of the mind.'[11] Difference needs not be suppressed in the name of Theory. And theory as a tool of survival needs to be rethought in relation to gender in discursive practice. Generally speaking, it is not difficult to agree with Duras that 'men don't translate. They begin from a theoretical platform that is already in place, already elaborated. [Whereas] The writing of women is really translated from the unknown, like a new way of communicating rather than an already formed language. But to achieve that, we have to turn away from plagiarism.' More specifically speaking, however, it is difficult to be content with statements she puts forth such as: 'Reverse everything. Make women the point of departure in judging, make darkness the point of departure in judging what men call light, make obscurity the point of departure in judging what men call clarity. . . .'[12] Unless 'point of departure' is constantly re-emphasized so that, again, reversal strategies do not become end points in themselves. Language defying language has to find its own place, in which claiming the right to language and disqualifying this same right work together without leading to the mystical, much-indulged-in angst that pervades many men's works. By its necessary tautness, writing the body in theory sometimes chokes to the breaking point. But the break, like the fall and impasse mentioned earlier, is desired. I do not write simply to destroy, conserve, or transmit. To re-appropriate a few sentences of Blanchot's, I write in the thrall of the impossible (feminine ethnic) real, that share of the detour of inscription which is always a de-scription.[13] From jagged transitions between the analytical and the poetical to the disruptive, always shifting fluidity of a headless and bottomless storytelling, what is exposed in this text is the inscription and description of a non-unitary female subject of color through her engagement, therefore also disengagement, with master discourses. Mastery ensures the transmission of knowledge; the dominant discourse for transmitting is one 'that annihilates sexual difference – where there is no question of it.' 'Her discourse, even when "theoretical" or political, is never simple or linear or "objectivized", universalized; she involves her story in history' (Cixous).[14] Like Monique Wittig's and Sande Zeig's bearers of fables, women 'are constantly moving, they recount, among other things, the metamorphosis of words from one place to another. *They* themselves *change* versions of these metamorphoses, *not in order to further confuse the matter but because they record the changes*. The result of these changes is an avoidance of fixed meanings. . . . They agree upon the words that they do not want to forgo. Then they decide, according to their groups, communities, islands, continents, on the possible tribute to be paid for the words. When that is decided, they pay it (or they do not pay it). Those who do so call this pleasantly "to write one's life with one's blood", this, they say, is the least they can do.'[15] 'Writing the body' is that abstract-concrete, personal-political realm of excess not fully contained by writing's unifying structural forces. Its physicality (vocality, tactility, touch, resonance), or edging and margin, exceeds the rationalized 'clarity' of communicative structures and cannot be fully explained by any analysis. It is a way of making theory in gender, of making of theory a politics of everyday life, thereby re-writing the ethnic female subject as site of differences. It is on such a site and in such a context that resistance to theory yields more than one reading. It may be a mere form of anti-intellectualism – a dis-ease that dwells on the totalizing concept of theory and practice and partakes in the Master's norms of clarity and accessibility while perpetuating the myth of the elite versus the mass, of those who think versus those who do not think. It may also be a distrust of the use of language about language and could be viewed in terms of both resistance and attraction to language itself. For to say that language is caught within a culturally and sexually dominent ideology is not to deny the heterogeneous history of its formation or, in other words, to refuse to 'see race, class *and* gender determinations in the formation of language' (Gayatri C. Spivak).[16] Woman as subject can only redefine while being defined by language. Whatever the position taken ('no position' is also a position, for 'I am not political' is a way of accepting 'my politics is someone else's'), the love–

hate, inside–outside, subject-of–subject-to relation between woman and language is inevitably always at work. That holds true in every case – whether she assumes language is a given, hence the task of the writer is merely to build vocabularies and choose among the existing possibilities; whether she decides to 'steal' demonstrative and discursive discourse from men (Clément), since language cannot free itself from the male-is-norm ideology and its subsuming masculine terms; whether she asserts that language is primarily a tool for transmitting knowledge, therefore there is no attributable difference between masculine and feminine writing and no shift needs to be made (in language and) in metalanguage, whose repressive operations 'see to it that the moment women open their mouths – women more often than men – they are immediately asked in whose name and from what theoretical standpoint they are speaking, who is their master and where they are coming from: they have, in short, to salute ... and show their identity papers',[17] whether she affirms that language is heterogeneous, claims her right to it, and feels no qualms in reproducing existing power relations because she has discoursed on this issue; or whether she insists that the production of woman-texts is not possible without a writing that inscribes 'femininity', just as writing woman cannot address the question of difference and change (it cannot be a political reflection) without reflecting and working on language.

NOTES

1. Quoted by Karen Gould in 'Setting Words Free: Feminist Writing in Quebec, *Signs* 6, no. 4 (Summer 1981), p. 629.
2. Hélène Cixous, 'The Laugh of the Medusa', tr. K. Cohen and P. Cohen, *Signs* 1, no. 4 (Summer 1976), pp. 880, 886.
3. Anaïs Nin, 'Diary', *By A Woman Writt*, pp. 294, 299.
4. Annie Leclerc, *Paroles de femme* (Paris: Grasset, 1974), pp. 11–12 (my translation).
5. For a detailed analysis of these images, see Irma Garcia, *Promenade femmilière*, vol. 2 (Paris: Des Femmes, 1981).
6. I have discussed the relationship between body, mother, cry (voice), hand, and life in Trinh T. Minh-ha, 'L'innécriture: Féminisme et littérature', *French Forum* 8, no. 1 (January 1983), pp. 45–63.
7. Cixous, 'The Laugh of the Medusa', p. 884.
8. Janice Raymond, 'The Illusion of Androgyny', *Ouest* 2, no. 1 (Summer 1975), p. 66.
9. Maurice Blanchot, *The Writing of the Disaster*, tr. A. Smock (Lincoln: Univ. of Nebraska Press, 1986), p. 38.
10. Marguerite Duras, 'Smothered Creativity', *New French Feminism*, ed. E. Marks and I. de Courtivon (Amherst: Univ. of Massachusetts Press, 1980), p. 111.
11. Catherine Clément, *Les Fils de Freud sont fatigués* (Paris: Grasset, 1978), p. 137 (my translation).
12. Duras, in *New French Feminism*, pp. 174–75.
13. Blanchot, *The Writing of the Disaster*, p. 38.
14. Hélène Cixous and Catherine Clément, *The Newly Born Woman*, tr. Betsy Wing (Minneapolis: Univ. of Minnesota Press, 1986), pp. 146, 92.
15. Monique Witting and Sande Zeig, *Lesbian Peoples. Material for a Dictionary* (New York: Avon, 1979), p. 166.
16. Gayatri Chakravorty Spivak, Interview, *Art Network* 16 (Winter 1985), p. 26.
17. Cixous, 'Castration or Decapitation?' *Signs* 7, 1 (Autumn 1981), p. 51.

4.6

PSYCHOANALYSIS AND THE BODY

Elizabeth Grosz

Although psychoanalysis is largely oriented to analysis and interpretation of psychical activities, and although the psyche is generally considered to be allied with mind, and opposed to the body, Freud and a number of other psychoanalysts have devoted considerable attention to the body's role in psychical life. Freud's own background in neurology, and his pre-psychoanalytic writings, especially *Project for a Scientific Psychology* (1895a), make explicit his fascination with finding neurophysiological and biological bases of and correlations with psychical events. Even when his neurological orientation is gradually transformed into a more properly psychological project, starting with the *Studies on Hysteria* (1895b), Freud retains a commitment to a kind of psycho-physical dualism inherited from Cartesian philosophy, in which chemical and neurological processes are neither causes nor effects of psychological processes, but are somehow correlated with them. However, in spite of an avowed commitment to dualism – Freud retained the hope that one day medical science would discover the 'chemistry' of libido but believed it could never replace psychological investigation – there are moments in his writings when he presents an (often implicit) critique and undermining of the assumptions governing dualism. These moments of (self-)critique can be located in his notions of the sexual drives and erotogenic zones, and in his understanding of the ego.

Freud does not assume a givenness or naturalness for the body: on his understanding, the biological body is rapidly overlaid with psychical and social

From: E. Wright (ed.), *Feminism and Psychoanalysis: A Critical Dictionary*, Oxford: Blackwell, 1992.

significance, which displaces what may once (mythically) have been a natural body. Much of Freud's work, particularly on female sexuality, makes biologistic presupppositions, yet he also presents alternative accounts of a socially, historically and culturally sexed body.

Both Freud and Lacan link the genesis of the ego in primary narcissism – the mirror stage – to two distinct but complementary processes. First, the ego is the product of a series of identifications with and introjections of the image of others, most especially the mother. These images are introjected into the incipient ego as part of its ego-ideal (the ego-ideal always being a residue of the subject's identificatory idealizations of the other). Second, the ego is an effect of a re-channelling of libidinal impulses in the subject's own body. The body is thus the point of junction of the social and the individual, the hinge which divides the one from the other. The ego is formed out of a blockage of libido that, until this time, had circulated in an untrammelled, objectless, formless, pleasure-seeking way. Until around six months, the infant does not have a unified, hierarchical relation to its body. It is not yet a subject over and above its various bodily experiences; it does not yet occupy a fixed and bounded space, the barrier provided by its own skin. In Lacan's terms, the child experiences its body as disunified and disorganized, a body in bits and pieces. It is a fragmented body, not yet organized by the distinctions between inside and outside, self and other, or active and passive. Lacan describes the infant at this stage as an 'hommelette', a subject-to-be, a psychical scrambled egg whose processes remain anarchical and chaotically unintegrated.

The child gradually becomes able to distinguish the self from the other, its body from the maternal body, its insides (bounded by the skin) from the outside. At around the same time, the sexual drives begin to emerge and to distinguish themselves in their specificity, first according to the particular sites or erotogenic zones of the body (oral, anal, phallic, scopophilic, etc.) from which they emanate; and then with particular sources, aims and objects. Only now do objects acquire a status independent of the subject who is capable of locating its objects, including its own body, in space and time. The ego is not simply a condition of the separation of subject and object or self and other; it is a product of and not a precondition for the child's relations to its own body, the other and the socio-symbolic order.

The ego is the meeting-point of two different psychical processes: a process of mapping the body and the circulation of libido on to the psyche; and a process of identification with the image of another (or the image of itself as an other, as occurs in the mirror stage). In *The Ego and the Id* Freud claims that the ego must be considered a 'bodily ego', a 'surface projection' of the libidinal body, which he compares with the 'cortical homunculus' described by neurophysiologists (1923: 126).

The ego is like an internal screen on to which the illuminated images of the body's outer surface are projected. It is not a veridical map, a photograph, but a representation of its degrees of erotogenicity, of the varying intensities of libidinal investment in different body parts. The ego is an image of the body's significance or meaning for the subject; it is as much a function of fantasy and desire as of sensation and perception. Freud illustrates this with the example of hypochondria, in which the subject's attitude to the body is not congruent with the body's physiological condition. The subject libidinally over-invests a psychically significant region of the body, withdrawing libido from the erotogenic zones.

Lacan's account of the mirror stage describes the formative effect on the child's ego of the introjection of the externalized image of the child's own body. The ego is regarded as a tracing of the subject's perceived corporeality; it is not an outline or projection of the real body, the body of anatomy and physiology, but of an 'imaginary anatomy', here following the work of the neurophysiologists on the body schema (Schilder 1978). The imaginary anatomy is an internalized image of the meaning that the body has for the subject, for others in its social world and for its culture as a whole. It is a shared and/or individualized fantasy of the body's form and modes of operation. This, Lacan claims, helps to explain the peculiar, non-organic connections formed in hysteria and in such phenomena as the phantom limb. Hysterical paralyses, for example, do not follow the structure and form of organic paralyses, but popular or everyday concepts of how the body functions (see Lacan 1953: 13). In the phenomenon of the phantom limb, a limb that has been surgically removed continues to induce sensations of pain. This pain clearly cannot be located in any real anatomy, for the limb to which it refers is no longer there. The absence of a limb is in this case as psychically invested as its presence. The phantom is a form of nostalgia, a libidinal memorial to the lost limb. Similarly, the hysterical symptom also conforms to an image of anatomy and the body that may often be at odds with physiology. This is strikingly clear in the case of anorexia nervosa, in which the subject's overweight body image is discordant with the reality of the body's real under-nourishment. Indeed, as Schilder (1978) and others argue, the body image extends far beyond the subject's skin, for it includes clothing, the surrounding environment, tools and implements, and even vehicles.

Freud's and Lacan's work on the psychical representations of the body, the cortical homunculus and the body image imply that the body is a pliable 'object', one which in its very biological form is dependent on the acquisition of a psychical image for voluntary behaviour. Human subjects need the mediation of such a psychical image of the body in order to link their wishes and desires to their corporeal capacities, making concerted action and rational thought possible.

If narcissistic identification with the image of the body is one major facet of the acquisition of bodily boundaries and a sense of stable, abiding identity, the other psychical process crucial to the child's growing sense of bodily autonomy derives from the various processes of sexualization that specific zones of the body undergo. Certain parts of the child's body are more eroticized than others. This is partly a function of biological privilege (for example, the sense organs and orifices are particularly well suited to act as erotogenic zones because they are thresholds of the body's interior and its exterior, sites for the perception of the external world and the reception of information); and partly a result of the individual's life history. Freud refers to this as 'somatic compliance': there are zones of the body privileged for the reception and transmission of sexual impulses through accident and contingency. A wound or scar, a history of various organic diseases, or even family myths about body parts all render these bodily zones more susceptible to sexualization, hystericization and fantasy. In a sense, one's psychical life history is written on and worn by the body, just as, in turn, the psyche bears the history of the lived body, its chance encounters, its punctures, transformations and extensions. The child's body becomes marked, privileged, established as a whole, libidinally intensified in an uneven layer across the body's edges and surfaces. Oral, anal and phallic drives are not biologically determined stages of human development (this reduces the drive to a form of instinct), but the results of processes of libidinal intensification which correlate with the acquisition of labile meanings for various body components. The establishment of imaginary sites and locations plays a central role in the constitution of the imaginary anatomy but this libidinal process must not be confused with an identificatory relation: the libidinal site comes to function as an erotogenic rim, a loop seeking an object (or several) to fill or satisfy it. There is no limit to the number of erotic zones in and on the body, nor any way of predicting which ones will dominate an individual's life.

Psychoanalytic theory has enabled feminists and others to reclaim the body from the realms of immanence and biology in order to see it as a psycho-social product, open to transformations in meaning and functioning, capable of being contested and re-signified. Feminists have stressed that the generic category '*the* body' is a masculinist illusion. There are only concrete bodies, bodies in the plural, bodies with a specific sex and colour. This counterbalances psychoanalysis's tendency to phallocentrism, especially the ways it understands the female body. If the female body is castrated, this is not, contrary to Freud, simply a matter of seeing. Rather, we learn how to see and understand according to prevailing systems of meaning and value, not nature. If the body is plastic, malleable and amenable to social re-inscription, this means that the female body is *a priori* capable of being seen and understood outside the notion of castrated privation. This is only one of a number of possible meanings, but the very one men and women have up to now had little possibility of refusing.

Although psychoanalysts have discussed the body and the body image, Freud and Lacan do so largely in passing. Feminists, by contrast, have long disdained analysing and theorizing the body in so far as the body was the site of patriarchy's most entrenched investments and, many feminists believed, had been so contaminated that women needed to aspire to intellectual, conceptual and transcendental positions rather than to reclaim the bodies by which patriarchy had constrained them. However, more recently, spurred at least in part by psychoanalytic theory itself, feminists have sought to re-evaluate the body beyond biologistic, essentialist and universalist presuppositions. While strongly influenced by psychoanalytic theory, feminists such as Irigaray and Gallop develop different understandings of corporeality, although both insist on two autonomous, sexually specific models of the body. Where Gallop tends to see the female body as a site of resistance, a kind of recalcitrance to patriarchal recuperation, but thereby silenced and refused representation, Irigaray sees the female body (like the male body) as sites for the inscription of social significances. Here it is not that the female body is silenced, but rather that it is 'spoken through', produced as such, by a wide variety of forces of social representation.

Where psychoanalysis has always seen the two sexes on a single model, in which the presence or absence of the phallus signified one's psycho-social and sexual position, feminists have insisted on the necessity of conceiving of (at least two) distinct types of imaginary anatomy, two sexually specific types of corporeal experiences, two modes of sexuality, two points of view and sets of interests, only one of which has been explored in its own terms in our history of thought. Both negatively and positively, psychoanalysis has provided a crucial moment in the recognition of women's corporeal submersion in phallocentric models – negatively in so far as it participated in and legitimized models of female corporeality as castrated; positively in so far as its insights provide a challenge to the domination of biology in discourses of the body.

REFERENCES

Anzieu, Didier (1990) *A Skin For Thought: Interviews with Gilbert Tarrab*, London: Karnac Books.

Freud, Sigmund (1895a) *The Project for a Scientific Psychology, SE*, 1, pp. 295–387.

Freud, Sigmund (1895b) *Studies on Hysteria* (with Joseph Breuer), *SE*, 2.

Freud, Sigmund (1905) *Three Essays on the Theory of Sexuality, SE*, 7, pp. 123–246.

Freud, Sigmund (1914a) *On Narcissism: An Introduction, SE*, 14, pp. 67–104.

Freud, Sigmund (1914b) *Instincts and their Vicissitudes, SE*, 14, pp. 109–40.

Freud, Sigmund (1923) *The Ego and the Id, SE*, 19, pp. 12–59.

Gallop, Jane (1988) *Thinking Through the Body*, New York: Columbia University Press.

Irigaray, Luce (1985) *Speculum of the Other Woman*, trans. Gillian Gill, Ithaca: Cornell University Press.

Lacan, Jacques (1953) 'Some reflections on the ego', *International Journal of Psychoanalysis*, 34: 11–17.

Lacan, Jacques (1977) 'The Mirror Stage as Formative of the Function of the I . . .', *Écrits: A Selection*, trans. Alan Sheridan, London: Tavistock, pp. 1–7.

Schilder, Paul (1978) *The Image and Appearance of the Human Body*, Oxford: Basil Blackwell.

Voyat, Gilbert (ed.) (1984) *The World of Henri Wallon*, New York and London: Jason Aronson.

SECTION 5
ALTER/ED BODIES

INTRODUCTION

What this section investigates is the way in which the body, far from being given once and for all, is endlessly subject to alteration by an array of cicumstances – both organic and inorganic – including congenital and aquired illness, technology, terror, and the processes of ageing. All of these and many others shape and inscribe the body, not necessarily from the outside, but in ways that are often beyond the subject's conscious control. While there is clearly an overlap with performing the body (see Section 7), as for example in the practices of body-building and keep fit, or even the simple act of applying make-up, there is a sense here that the alter/ed body is both excessive and lacking. It continues to confound western normativities in being other to an apparently more settled standard, by displaying a morphological difference that cannot easily be undone. On this reading, practices like cosmetic surgery, or the kinds of surgical interventions that performance artists like Orlan or Stelarc make into their own bodies, exemplify elements germane to both Sections 5 and 7. It would, in any case, be quite misleading to attempt closure, and we urge readers to use sections in relationship with one another, rather than seeing them as thematically distinct.

By using the term 'alter/ed' to describe the bodies in this section, we mean to suggest both the notion of alterity, of a corporeal otherness that is always already alien to us, and alteration as that which may befall our own bodies. At an early stage of putting selections together we considered the umbrella of 'unnatural' bodies, but it quickly became clear that the bodies in question are no more or less natural than those that represent the normative standard. Moreover, destabilising the link between women and nature has been an important focus of feminist theorising. Nonetheless it does make some sense to acknowledge a distinction between organic and inorganic bodies, and how those boundaries have become blurred. In the era of postmodernity, the technological ability to construct new corporeal forms which embody machinic elements, notably in the living body of the cyborg, throw up such multiple questions as those concerning gender, race, nature, identity, life and death. Donna Haraway's ground-breaking 'Manifesto for Cyborgs' of 1985 – already much anthologised (see Haraway 1991) and therefore not included here – led the way for feminists to take seriously the impact of radically altering the familiar forms of the body. Anne Balsamo's piece, for example, looks at some of the implications of cybertechnologies, while Rosi Braidotti links her essay on historical monsters to the bodies produced by advanced reproductive technologies.

As Braidotti's contribution makes clear, alter/ed bodies are nothing new. The transhistorical category of monstrosity has become of great interest to feminists during the last few years, both because it uncovers the theoretical links between the feminine and the monstrous (both are the excluded/abjected other), and because it rehearses the breakdown of the binaries between human/animal, between male/female, and ultimately between organism/machine. The ubiquitous influence of Aristotle, who famously characterised women in general as monstrous, deformed, sets up the paradigm of the female body as in need of control and alteration. The notion that the bodies not only of women but of other others could be somehow physically and morally improved is thus of long-standing, and shows itself as much in the issue of genital mutilation – discussed here in the extracts by Pratibha Parmar and Alice Walker, and by Awa Thian – as in high-tech genetic manipulation. In a related register, but moving away from empirical report to discursive theory, Bibi Bakare-Yusuf explores the ontological and epistemological meaning of the violence, mental and physical, directed at black bodies during the period of slavery. The intense rejection of alterity shows too in Susan Wendell's reflection on the implications of changes to her own corporeality. Whilst deploring the reluctance of feminists to come to terms with disability, Wendell nevertheless sees her own way forward in effecting a transcendence of the altered body. In so doing she poses questions as to the limits of efficacy of deconstructing the conventional mind/body split which might give feminist postmodernists cause to ponder.

For Wendell, the opposition between the normal and abnormal body is the issue that matters, and overall, the section and its related readings are intended to challenge a series of familiar binaries. The questions of: what happens to identity and embodiment when the material structure or function of the body is changed? what kind of body can exist in cyberspace? are monsters, hybrids and cyborgs the bodies of the future? are there any *un*altered bodies? – are all ones that is the task of feminist theory to consider.

REFERENCES AND FURTHER READING:

Balsamo, Anne (1995) *Technologies of the Gendered Body. Reading Cyborg Women*, Durham, NC: Duke University Press.

Davis, Kathy (1995) *Reshaping the Female Body: The Dilemma of Cosmetic Surgery*, London: Routledge.
Doane, Mary Ann (1993) 'Technology's Body: Cinematic Vision in Modernity' *differences* 5 (2): 1–23.
Dorkenoo, Efua (1995) *Cutting the Rose*, London: Minority Rights Publications.
Epstein, Julia (1995) *Altered Conditions*, London: Routledge
Garber, Marjorie (1989) 'Spare Parts. The Surgical Construction of Gender', *differences* 1 (3): 137–59.
Greer, Germaine (1992) *The Change: Women, Ageing and the Menopause*, Harmondsworth: Penguin.
Haraway, Donna (1989) 'The Biopolitics of Postmodern Bodies', *differences* 1 (1): 3–43.
Haraway, Donna (1991) 'A Cyborg Manifesto' in *Simians, Cyborgs and Women*, London: Free Association Books.
Haraway, Donna (1992) 'The Promises of Monsters: A Regenerative Politics for Inappropriate/d Others' in Lawrence Grossberg, Cary Nelson and Paula Treichler (eds), *Cultural Studies*, London: Routledge,
Kirby, Vicky (1997) 'Reality Bytes: Virtual Incarnations' in *Telling Flesh: The Substance of the Corporeal*, London: Routledge.
Lupton, Deborah (1996) 'Constructing the Menopausal Body: The Discourses on Hormone Replacement Therapy' *Body and Society* 2 (1): 91–7.
Potts, Tracy and Price, Janet (1995) 'Out of the Blood and Spirit of Our Lives' in L. Morley and V. Walsh (eds), *Feminist Academics: Creative Agents for Change*, London: Taylor and Francis.
Rajan, Rajaswani Sunder (1993) 'The Subject of Sati' in *Real and Imagined Women*, London: Routledge.
Rothschild, Joan (1992) 'Engineering the "Perfect Child": Feminist Responses' in Maja Pellikaan-Engel (ed.), *Against Patriarchal Thinking*, Amsterdam: VU University Press.
Shildrick, Margrit (1996) 'Posthumanism and the Monstrous Body', *Body & Society* 2 (1): 1–15.
Stanworth, Michele (1990) 'Birth Pangs: Conceptive Technologies and the Threat of Motherhood' in Marianne Hirsch and Evelyn Fox Keller (eds), *Conflicts in Feminism*, London: Routledge.
Stone, Allucquere Roseanne (1992) 'Virtual Systems' in J. Crary and S. Kwinter (eds), *Incorporations*, New York: Urzone.

5.1

FORMS OF TECHNOLOGICAL EMBODIMENT: READING THE BODY IN CONTEMPORARY CULTURE

Anne Balsamo

The point of this paper is to annotate a taxonomy of the ways in which the techno-body is constructed in contemporary culture. In contrast to those who would argue that there is a dominant – singular – form of the postmodern techno-body, I argue that when starting with the assumption that bodies are always gendered and marked by race it becomes clear that there are multiple forms of technological embodiment that must be attended to in order to make sense of the status of the body in contemporary culture. [. . .] The polemic in this paper argues that the body can never be constructed as a purely discursive entity. In a related sense, it can never be reduced to a pure materialist object. Better to think of the dual 'natures' of the body in terms of its 'structural integrity' to use Evelyn Fox Keller's (1992) term. This is to assert that the material and the discursive are mutually determining and non-exclusive.

[. . .]

Cyberpunk Techno-Bodies

Here is a story about the postmodern body that is abstracted from contemporary science fiction. As a work of the feminist imaginary, this narrative extracted from Pat Cadigan's (1991) cyberpunk novel, *Synners*, explicitly discusses an often-repressed dimension of the information age: the material identity of the techno-body. True to its genre determinations, *Synners* concerns a loosely identified 'community' of computer users, each of whom is differently, albeit

From: M. Featherstone and R. Burrows (eds), *Cyberspace/Cyberbodies/Cyberpunk*, London: Sage, 1996.

thoroughly, engaged with the technologies of cyberspace – simulation machines, global communication networks, corporate databases and multimedia production systems. What we encounter in the Cadigan novel is the narrativization of four different versions of postmodern embodiment: the laboring body, the marked body, the repressed body and the disappearing body. In this sense, the four central characters symbolize the different embodied relations one can have, in fiction and in practice, to a technological formation. The following figure roughly illustrates how Sam, Gabe, Gina and Visual Mark (the four main characters) represent four corners of an identity matrix constructed in and around cyberspace.

Figure 1

Sam (the body that labours)	*Gina* (the marked body)
Gabe (the repressed body)	*Visual Mark* (the disappearing body)

Where Sam hacks the net through a terminal powered by her own body, Visual Mark actually inhabits the network as he mutates into a disembodied, sentient artificial intelligence (AI). Although both Gina and Gabe travel through cyberspace on their way to someplace else, Gabe is addicted to cyberspace simulations and Gina merely endures them. Each character plays a significant role in the novel's climactic confrontation in cyberspace: a role determined, in part, by their individual relationships to Diversifications (the genre-required evil multinational corporation) and, in part, by their bodily identities.

[...]

In one sense, Cadigan writes fiction implicitly informed by Donna Haraway's cyborg politics: the gendered distinctions among characters hold true to a cyborgian figuration of gender differences whereby the female body is coded as a body-in-connection and the male body as a body-in-isolation. It illuminates the gendered differences in the way that the characters relate to the technological space of information. Sam and Gina, the two female hackers, actively manipulate the dimensions of cybernetic space in order to communicate with other people. Gabe and Visual Mark, on the other hand, are addicted to cyberspace for the release it offers from the perceived limitations of their material bodies. Even as the novel's characters illuminate the gendered distinctions among computer network users, its racial characterizations are less developed. The racial distinctions between characters are revealed through the representation of sexual desire. Gina is the only character to be identified by skin color. She is also the focal object and subject of heterosexual desire, for a moment by Mark, and more frequently by Gabe; and, we know both men's racial identities by their marked difference from Gina's. The unmarked characters are marked by the absence of identifying marks. In different ways then and with different political inflections, the novel reasserts that gender and race are critical elements of post-human identity.

Throughout the book, the characters' material bodies are invoked through descriptions of sexual encounters, bathroom breaks, food consumption, intoxication effects and physical death. The key insight to emerge from the novel is that the denatured techno-body remains a material entity. Although it may be culturally coded and semiotically marked, it is never merely discursive. This is to say that even as *Synners* discursively represents different forms of technological embodiment, it also reasserts the critical importance of the materiality of bodies in any analysis of the information age.

Expanding upon this delimited reading of *Synners* as both cultural landmark and cognitive map yields another version of the matrix above that offers a taxonomy of forms of technological embodiment. The ground upon which this matrix is constructed is the shifting table of body theory: each quadrant is a multidimensional space for filing body snapshots, art installations, performances, readings, enactments and corporeal forms. The qualities that mark each form of embodiment are illustrated by various incarnations of the techno-body. [...] The aim is to illustrate how material bodies are both discursively constructed and culturally disciplined at a particular historical moment.

Figure 2

Postmodern Forms of Technological Embodiment

The LABORING Body Mothers as Wombs Microelectronics	*The DISAPPEARING Body* Bio-engineering Bodies and Databases
The REPRESSED Body Virtual Reality Computer Communication	*The MARKED Body* Multi-cultural Mannequins Cosmetic Surgery

THE MARKED BODY

The marked body signals the fact that bodies are eminently cultural signs, bearing the traces of ritual and mythic identities. In similar ways, both the fashion industry and the cosmetic surgery profession have capitalized on the role of the body in the process of 'identity semiosis' – where identities become signs and signs become commodities. The consequence is the technological production of identities for sale and rent. Material bodies shop the global marketplace for cultural identities that come in different forms, the least permanent as clothes and accessories worn once and discarded with each new fashion season, the most dramatic as the physical transformation of the corporeal body accomplished through surgical methods. Thus the natural body is technologically transformed into a sign of culture.

High fashion – as one technology of urban corporeal identity – is preoccupied with multiculturalism. One of the consequences is that in reading the body displayed on the glossy pages of American fashion magazines, it is evident that the politics of representation are very confused. For example, from its very first US issue in January 1987, *Elle* magazine regularly included photographic layouts that featured black and other non-white models wearing various 'deconstructed' fashions. The May 1988 cover of *Elle* that showcased the faces of two models, one white in the background, made-up in conventional fashion, the other black in the foreground, wearing no discernible make-up, betrays the cultural politics at the center of the worldview of the high fashion industry where 'black bodies' serve as mannequins for designer messages intended for affluent white readers. The narrative constructed around *Elle*'s black bodies and white bodies concerns the fashion industry's appropriation of the trope of primitivism as a seasonal fashion look. In this case, the fashion apparatus deploys signs of the 'primitive' in the service of constructing an anti-fashion high fashion look. While another *Elle* article explained that the American 'love of the exotic' has translated into career success for several new multicultural supermodels,[1] it was rarely mentioned that they were few of the women who could actually afford the clothes featured as items in the new primitivism line that cost upwards of $1000. Thus an interesting paradox takes shape: the black bodies of supermodels are used as billboards for designer messages about the fetishization of black identity as the cultural sign of the ethnic primitive. Just as they are admitted to the elite club of well-paid supermodels, black models are coopted to a cultural myth of racial subordination.

This appropriation of the signs of cultural primitivism for the visual consumption of mostly white readers illuminates the mass-mediated rehearsal of the construction of cultural identity, where what is reviled and despised is projected onto the body of the 'other' such that the identity of the 'One' is established as that which is good and pure and sacred. At the same time, the body of the 'other' is fetishized and eroticized in its object form. In this way, the proper hierarchy of white bodies over black bodies is subtly, and compulsively, reinscribed in each season's look. The recuperative power of corporate culture and its premier technology, mass media advertising, extends far beyond the appropriation of black identity and tropes of primitivism. As the recent trend of deconstructionism as high fashion look attests, even the markers of poverty are able to be rearticulated to a different economic logic.[2] The focal figure and preferred mannequin of these fashion campaigns is the eroticized dark brown female body, but the valorized subject is the white, Western woman, whose white body can be liberated, temporarily, from the debasement of everyday life through her consumption and mimicry of anti-fashion style.

Much like those physicians who use sonograms and laparoscopes to look through the maternal body, cosmetic surgeons also make use of new visualization technologies to exercise a high-tech version of Foucault's scientific bio-power that effects, first, the objectification of the material body and, second, the subjection of that body to the discipline of a normative gaze. In the past five years, several cosmetic surgeons have begun using a new high-tech video imaging program as a patient consultation device. In the process, the medical gaze of the cosmetic surgeon is transformed into the technological perspective of the video camera. Using computer rendering tools, such as erasers, pencils and 'agenic cursors', the cosmetic surgeon manipulates the digitized image of the prospective patient in order to visually illustrate possible surgical transformations. One of the consequences is that the material body is reconfigured as an electronic image that can be technologically manipulated on the screen before it is surgically manipulated in the operating room. In this way, the video consultation enables the codification of surgical 'goals' – goals which effect, in short, the inscription of cultural ideals of Western beauty.[3]

[...]

THE LABORING BODY

Bodies that labor include a full range of working bodies as well as maternal bodies. In the broadest sense these are all reproductive bodies involved in the continuation of the human race in its multiple material incarnations. Such bodies are often invisible in postmodern discourse. But, because they are centrally involved in the reproduction of various technological formations, including now the 'natural' family unit, they must be counted as key postmodern cultural forms.

Perhaps the most obvious form of the laboring body is the maternal body which is increasingly treated as a technological body – both in its science fictional and science factual form as 'container' for the fetus, and in its role as the object of technological manipulation in the service of human reproduction. How, specifically, are the material bodies of pregnant women affected by cultural discourses? In a discussion of the politics of new reproductive technologies, Jennifer Terry (1989) examines how these technologies are deployed in the service of institutionalized practices of surveillance, whereby pregnant women are watched in the name of 'public health' to determine whether they are taking drugs or alcohol while pregnant. In 1989, newspapers across the USA reported on the spectacle of 'cocaine mothers' – the mediated identity of women who deliver babies who show traces of cocaine in their systems at birth. Since 1990, several women, branded thus by the media, have been charged with criminal child neglect for the delivery of a controlled substance to a minor. One of the consequences is that maternal rights of body privacy are set against the rights of a fetus to the state's protection. The personification of the fetus as an entity with 'rights' is made possible, in part, because of the use of visualization technologies such as sonograms and laparascopes. In the application of these technologies, the material integrity of the maternal body is technologically deconstructed, only to be reconstructed as a visual medium to look through to see the developing fetus who is now, according to some media campaigns, 'the most important obstetrics patient'. In short, the use of these technologies of visualization creates new cultural identities and enables new agents of power that, in turn, create new possibilities for the discipline of maternal bodies. In this way, cultural discourses not only establish the meaning of material bodies, but also significantly delimit the range of freedom of some of those bodies.

Attending to laboring bodies also suggests the need to investigate the material conditions of the production of the cheap, high-tech devices purchased in bulk by US consumers of electronic commodities. For example, silicon chips are relatively inexpensive to manufacture and assemble because of the use of cheap labor in south-east Asia. In the course of his investigation of Malaysia's labor contribution to the microelectronics industry, Les Levidow (1991) discovered that women workers were the preferred employees for electronic firms because it is believed that 'they are naturally suited to the routinized work of the electronics assembly line: nimble fingers, acute eyesight, greater patience' (1991: 106). Although these women are compensated monetarily, daughters contribute a significant percentage of their earnings to their families. Other forms of compensation are more intangible, and double-edged. Women factory workers experience a measure of independence from 'village elders', who would have bound them to traditional Islamic practices and values but, at the same time, they risk significant health problems in the form of blindness, respiratory disease and psychological and sexual manipulation. The point is not to argue that these are the only laboring bodies that bear the brunt of technologically assisted disciplinary actions, but rather to assert that women are far more likely to be the targets of such discipline. This is in part due to their historically designated position within labor networks tied to their physiology, that is, as possessing nimble fingers suited for detailed handwork. It is also a consequence of the gendered division of labor whereby women occupy the lowest paying positions because of active discrimination, beliefs about women's inferiority, their socialization to service roles, and the social and cultural pressures to marry, bear children and forego compensated employment in favor of unpaid domestic labor.

THE REPRESSED BODY

Repression is a pain management technique. The technological repression of the material body functions to curtail pain by blocking channels of sensory awareness. In the development of virtual reality applications and hardware, the body is redefined as a machine interface. In the efforts to colonize the electronic frontier – called cyberspace or the information matrix – the material body is divorced from the locus of knowledge. The point of contact with the interior spaces of a virtual environment – the way that the computer-generated scene makes sense – is through an eye-level perspective that shifts as the user turns her head; the changes in the scene projected on the small screens roughly corresponds with the real-time perspectival changes one would expect as one normally turns the head. This highly controlled gaze mimes the movement of a disembodied camera 'eye' – a familiar aspect of a filmic phenomenology where the camera simulates the movement of perspective that rarely includes a self-referential visual inspection of the body as the vehicle of that perspective.

Although some VR users report a noticeable lag time in the change of scene as the head turns, that produces a low-level nauseous feeling, for the most part the material body is visually and technologically repressed. This repression of the body is technologically naturalized in part because we have internalized the technological gaze to such an extent that 'perspective' is a naturalized organizing locus of sense knowledge. As a consequence, 'the body' as a sense apparatus, is nothing more than excess baggage for the cyberspace traveller.

In short, what these VR encounters really provide is an illusion of control over reality, nature and, especially, over the unruly, gender and race-marked, essentially mortal body. There is little coincidence that VR emerged in the 1980s, during a decade when the body was understood to be increasingly vulnerable (literally, as well as discursively) to infection, as well as to gender, race, ethnicity and ability critiques. At the heart of the media promotions of virtual reality is a vision of a body-free universe. In this sense, these new technologies are implicated in the reproduction of at least one very traditional cultural narrative: the possibility of transcendence whereby the physical body and its social meanings can be technologically neutralized. In the speculative discourse of VR, we are promised whatever body we want, which doesn't say anything about the body that I already have and the economy of meanings I already embody. What forms of embodiment would people choose if they could design their virtual bodies without the pain or cost of physical restructuring? If we look to those who are already participating in body reconstruction programs, for instance, cosmetic surgery and bodybuilding, we would find that their reconstructed bodies display very traditional gender and race markers of beauty, strength and sexuality. There is plenty of evidence to suggest that a reconstructed body does not guarantee a reconstructed cultural identity. Nor does 'freedom from a body' imply that people will exercise the 'freedom to be' any other kind of body than the one they already enjoy or desire. This is to argue that, although the body may disappear representationally in the virtual worlds of cyberspace and, indeed, we may go to great lengths to repress it and erase its referential traces, it does not disappear materially in the interface with the VR apparatus, or in its engagement with other high-tech communication systems.

In the Jargon File,[4] the entry on 'Gender and Ethnicity' claims that although 'hackerdom is still predominantly male', hackers are gender and color-blind in their interactions with other hackers due to the fact that they communicate (primarily) through text-based network channels. This assertion rests on the assumption that 'text-based channels' represent a gender-neutral medium of exchange, and that language itself is free from any form of gender, race or ethnic determinations. Both of these assumptions are called into question not only by feminist research on electronic communication and interpretive theory, but also by female network users who participate in the virtual subcultures of cyberspace.[5] Studies of the new modes of electronic communication indicate that the anonymity offered by the computer screen empowers antisocial behaviors such as 'flaming' and borderline illegal behaviors such as trespassing, e-mail snooping and MUD-rape.[6] And yet, for all the anonymity they offer, many computer communications reproduce stereotypically gendered patterns of conversation.[7] Hoai-An Truong, a member of the Bay Area Women in Telecommunications (BAWIT) writes:

> Despite the fact that computer networking systems obscure physical characteristics, many women find that gender follows them into the on-line community, and sets a tone for their public and private interactions there – to such an extent that some women purposefully choose gender neutral identities, or refrain from expressing their opinions.[8]

This is a case where the false denial of the body requires the defensive denial of the body in order to communicate. For some women, it is simply not worth the effort. Most men never notice. The development of and popular engagement with cybernetic networks allows us the opportunity to investigate how myths about identity, nature and the body are rearticulated with new technologies such that traditional narratives about the gendered, race-marked body are socially and technologically reproduced.

THE DISAPPEARING BODY

Of all the forms of technological embodiment, the disappearing body is the one that promises most insistently the final erasure of gender and race as culturally organized systems of differentiation. Bio-engineered body components are designed to duplicate the function of material body parts; bit by bit the 'natural' body is literally reconstructed through the use of technological replacement parts. But even as the material body is systematically replaced piece by piece, system by system, gender identity does not entirely disappear. 'Sexy Robots' and war machines still bear the traces of conventional gender codings. In the case of bodies that more literally disappear into cyberspace – here I'm talking about the technological coding of bodies as part of electronic databases – racial identity functions as a submerged system of logic to organize body information, even as it is coded in bits and bytes.

As part of a special preview of the year 2000 and beyond, the 1988 February issue of *Life* magazine featured an article called 'Visions of Tomorrow' that included a report on the replaceable body parts that were already 'on the market'... elbow and wrist joints, and tendons and ligaments.[9] We are told how succeeding generations of artificial 'devices' will be even more complex than the ones we have today, aided by research in microelectronics and tissue engineering. For example, glass eyes will be replaced with electronic retinas, pacemakers with bionic hearts, and use of the already high-tech insulin dispenser will soon become obsolete in favor of an organically grown biohybrid system that could serve as an artificial pancreas. The availability of manufactured body parts has subtly altered the cultural understanding of what counts as a natural body. Even as these technologies provide the realistic possibility of replacement body parts, they also enable a fantastic dream of immortality and control over life and death.

[. . .]

The relationship between material bodies and the information collected about those bodies is of central concern to people who ask the question, 'Who counts?' This leads to the investigation of both those who determine who counts as instances of what identities, and also those who are treated as numbers or cases in the construction of databases. The politics of databases will be a critical agenda item for the 1990s as an increasing number of businesses, services and state agencies go 'on-line'. Determining who has access to data, and how to get access to data that is supposedly available to the 'public', is a multidimensional project that involves the use of computers, skill at network access and education in locating and negotiating government-regulated databases. Even a chief data coordinator with the US Geological Survey asserts that 'data markets, data access, and data dissemination are complicated, fuzzy, emotional topics right now'. She 'predicts that they likely will be the major issues of the decade'.[10] Questions of public access and of the status of information collected on individuals are just now attracting public attention.

For those who monitor the insurance industry's interest in database development, there are already several warning signs about the material consequences of digitizing bodies. The Human Genome Project (HGP) is a big-science enterprise that promises to deliver an electronically accessible map of all 100,000 genes found on human chromosomes. This electronic map could be used for diagnostic screening; all that is needed is a sample of genetic material collected as part of the application process for life insurance. One insurance company in the US has already petitioned for the right to require a sample of saliva with each new client's application. The ethics of constructing electronic maps of the human genetic code is a critical concern for those involved in the HGP; in fact, in 1994, 5 per cent of the Project's budget was set aside for ethical research. This concern with ethics has not deterred the development of related projects whose ethical implications are more directly contested, and whose methods of gene recording are less digital and more biological. For example, the aim of the Human Genome Diversity project (a project related to the HGP) is to record 'the dwindling genetic diversity of Homo sapiens by taking DNA samples from several hundred distinct human populations and storing them in gene banks' (Gutin, 1994).

Researchers could then examine the DNA for clues to the evolutionary histories of the populations and to their resistance or susceptibility to particular diseases. Even though this seems like an entirely benevolent project on the surface, members of the populations targeted to be archived think otherwise.

[. . .]

The issue at stake is the creation of a 'bank' of information about a certain population whose members have no official right to assess the accuracy of the information collected about them, let alone to monitor the intended or potential use of such information. This raises questions about intellectual property rights, informed consent and rights of privacy. [...] Such widely distributed research programs as the Human Genome Project and the Human Genome Diversity Project are in fact re-tooling notions of privacy and corporeal identity. They demonstrate how the reality of the material body is very much tied to its discursive construction and institutional situation. Digitized representations of corporeal identity impact material bodies. The politics of representation in these cases are doubly complex in that it is difficult not only to determine how the body is being represented, but also who is the agent of representation. Given the 'truth status' of scientific discourse, it is difficult to assert that a representation has been constructed.

CONCLUSION

Postmodern embodiment is not a singularly discursive condition. Failure to consider the multiple ways in which bodies are technologically engaged is to perpetuate a serious misreading of postmodernity as structured by a uniformly dominant cultural logic. Although, if pressed, most critics would probably assert that they don't believe in a 'uniformly' dominant culture logic, such claims implicitly inform both Arthur Kroker's work on 'Body Invaders' and Fredric Jameson's elaboration of the cultural logic of postmodernity. Such a reading obscures the consideration of the diverse range of political forces that determine the reality of material bodies. In offering the matrix of forms of technological embodiment, I argue that the material body cannot be bracketed or 'factored out' of postmodern body theory. This is not an argument for the assertion of a material body that is defined in an essentialist way – as having unchanging, trans-historical gender or race characteristics. Rather, it is to argue that the gender and race identity of the material body structures the way that body is subsequently culturally reproduced and technologically disciplined. What becomes obvious through the study of new reproductive technologies that enable the visualization of the fetus in the womb, there is no blank page of gender identity. That unsigned moment before the birth certificate is marked with an 'F' or an 'M' is an artifact of a mythic era; we are born always already inscribed.

In Pat Cadigan's narrative *Synners*, the female body is symbolically represented as a material body and as a body that labors. The male body, in contrast, is repressed or disappearing. This suggests two points of disagreement with Kroker's theory of 'the disappearing body': that there is no singular form of postmodern embodiment, and that 'disappearing body' is not a post-human body-without-gender. In contrast, I argue that the 'disappearing body' is a gendered response to cultural anxieties about body invasion. Masculinist dreams of body transcendence and, relatedly, masculinist attempts at body repression, signal a desire to return to the 'neutrality' of the body, to be rid of the culturally marked body.

The technological fragmentation of the body functions in a similar way to its medical fragmentation: body parts are objectified and invested with cultural significance. In turn, this fragmentation is articulated to a culturally determined 'system of differences' that not only attributes value to different bodies, but 'processes' these bodies according to traditional, dualistic gendered 'natures'. This system of differentiation determines the status and position of material bodies which results in the reification of dualistic codes of gender identity. So, despite the technological possibilities of body reconstruction, in the discourses of biotechnology the female body is persistently coded as the cultural sign of the 'natural', the 'sexual' and the 'reproductive', so that the 'womb', for example, continues to signify female gender in a way that reinforces an essentialist identity for the female body as the 'maternal body'. In this sense, an apparatus of gender organizes the power relations manifest in the various engagements between bodies and technologies. I offer the phrase 'technologies of the gendered body' as a way of describing such interactions between bodies and technologies.[11] Gender, in this schema, is both a determining cultural condition and a social consequence of technological deployment. My intent is to illuminate the ways that contemporary discourses of technology rely on a logic of binary gender identity as an underlying organizational framework to structure the possibilities of technological engagement, and ultimately to limit the revisionary potential of such technologies.

NOTES

1. For a discussion of the appeal of 'the exotic woman' and the rise of new multicultural supermodels see Glenn O'Brien (1991).
2. Runway grunge, as the deconstructionist fashion of Belgian designer Martin Margiela is sometimes called, is merely the latest movement in rhythmic modulation of the fashion system, designed not so much to express an aesthetic or cultural theory, but rather to keep the fashion apparatus well supplied with consumable and ultimately disposable sign commodities.
3. For an extended discussion of the way in which cosmetic surgeons rely on western ideals of feminine beauty see Balsamo (1992).
4. The Jargon File, version 2.0.10, 01 July 1992. Available on-line from: ftp.uu.net. Also published as *The Hacker's Dictionary*.
5. See especially: Sherry Turkle and Seymour Papert (1990) and Dannielle Bernstein (1991).
6. For a discussion of the ethical/policy dimensions of computer communication see Jeffrey Bairstow (1990), Bob Brown (1990), Pamela Varley (1991), Laurence H. Tribe (1991) and Willard Uncapher (1991).
7. For a discussion of the gendered nature of communication technologies see especially Lana Rakow (1988). For other studies of the gendered nature of computer use see: Sara Kiesler, Lee Sproull and Jacquelynne Eccles (1985) and Sherry Turkle and Seymour Papert (1990).
8. Hoai-An Truong, 'Gender Issues in Online Communication', CFP 93 (Version 4.1). Available on-line from: ftp.eff.org. No date given.
9. The man in the cover photograph wears liquid screen glasses that come with a hand-held computer and earphones. For his viewing pleasure, the computer can transmit video images and other audiovisual material to the eyeglass screens. The cover photo is a striking visual emblem of the future high-tech body which, from *Life*'s point of view, is gendered male. 'The Future and You', a 30-page preview: 2000 and beyond, *Life*, February 1989.
10. The quotation is from Nancy Tosta, chief of the Branch of Geographic Data Coordination of the National Mapping Division, US Geological Survey in Reston, Virginia ('Who's Got the Data?', *Geo Info Systems*, September 1992: 24–7).

Tosta's prediction is supported by other statements about the US government's efforts to build a Geographic Information System (GIS): a database system whereby 'all public information can be referenced by location'. See Lisa Warnecke (1992). Managing data, acquiring new data and guarding data integrity are issues of concern for GIS managers. Because of the cost of acquiring new data and guarding data integrity, GIS managers sometimes charge a fee for providing information. This process of charging 'has thrown [them] into a morass of issues about public records and freedom of information; the value of data, privacy, copyrights, and liability and the roles of public and private sectors in disseminating information' (see Nancy Tosta, 1991).

11. Here I implicitly draw on Teresa deLauretis's transformation of Foucault's notion of the 'technology of sex' into the 'technologies of gender'. She uses this phrase to name the process by which gender is 'both a representation and self-representation produced by various social technologies, such as cinema, as well as institutional discourses, epistemologies, and critical practices' (Teresa deLauretis, 1987: ix).

REFERENCES

Bairstow, Jeffrey (1990) 'Who Reads Your Electronic Mail?' *Electronic Business* 16 (11): 92.

Balsamo, Anne (1992) 'On the Cutting Edge: Cosmetic Surgery and the Technological Production of the Gendered Body' *Camera Obscura* 28: 207–37.

Bernstein, Dannielle (1991) 'Comfort and Experience with Computing: Are They the Same for Women and Men?' *SIGCSE Bulletin* 23 (3): 57–60.

Brown, Bob (1990) 'EMA Urges Users to Adopt Policy on E-mail Privacy' *Network World* 29 Oct.: 7.44.2.

Cadigan, Pat (1991) *Synners*, New York: Bantam.

deLauretis, Teresa (1987) *Technologies of Gender: Essays on Theory, Film, and Fiction*, Bloomington: Indiana University Press.

Feher, Michel (1987) 'Of Bodies and Technologies', pp. 159–65 in Hal Foster (ed.), *Discussions in Contemporary Culture*, DIA Art Foundation. Seattle, WA: Bay Press.

Feher, Michel (1989) 'Introduction', pp. 11–17 in *Zone 3: Fragments for a History of the Human Body*, New York: Urzone.

Gutin, Joann C. (1994) 'End of the Rainbow', *Discover*, November: 71–5.

Kiesler, Sara, Lee Sproull and Jacquelynne Eccles (1985) 'Poolhalls, Chips and War Games: Women in the Culture of Computing' *Psychology of Women Quarterly* 9 (4): 451–62.

Kroker, Arthur and Marilouise Kroker (eds) (1987) *Body Invaders: Panic Sex in America*, New York: St Martins Press.

Levidow, Les (1991) 'Women who Make the Chips' *Science as Culture* 2 (10): 103–24.

O'Brien, Glenn (1991) 'Perfect Strangers: Our Love of the Exotic' *Elle*, Sept.: 274–6.

Rakow, Lana (1988) 'Women and the Telephone: The Gendering of a Communications Technology', pp. 207–8 in Cheris Kramarae (ed.) *Technology and Women's Voices: Keeping in Touch*, Boston: Routledge.

Terry, Jennifer (1989) 'The Body Invaded: Medical Surveillance of Women as Reproducers' *Socialist Review* 39: 13–44.

Tosta, Nancy (1991) 'Public Access: Right or Privilege?' *Geo Info System*, Nov./Dec.: 20–5.

Tosta, Nancy (1992) 'Who's Got the Data?' *Geo Info Systems* Sept.: 24–7.

Tribe, Laurence H. (1991) 'The Constitution in Cyberspace' *The Humanist* 51 (5): 15–21.

Turkle, Sherry and Seymour Papert (1990) 'Epistemological Pluralism: Styles and Voices within the Computer Culture' *Signs* 16 (11): 128–57.

Uncapher, Willard (1991) 'Trouble in Cyberspace' *The Humanist* 51 (5): 5–14.

Varley, Pamela (1991) 'Electronic Democracy' *Technology Review*, Nov./Dec.: 40–3.

Warnecke, Lisa (1992) 'Building the National GI/GIS Partnership' *Geo Info Systems*, April: 16–23.

5.2

SIGNS OF WONDER AND TRACES OF DOUBT: ON TERATOLOGY AND EMBODIED DIFFERENCES

Rosi Braidotti

[. . .] A working definition of the term 'monster' has been available since the late eighteenth century, when Geoffroy de Saint-Hilaire organized monsters in terms of *excess, lack* or *displacement* of his/her organs. There can be too many parts or too few; the right ones can be in the wrong places or duplicated at random on the surface of the body. This is the definition I have adopted in my work.[1]

The Epistemophilic Structures of Teratological Discourses

Discourses about monsters are fundamentally 'epistemophilic', in that they express and explore a deep-seated curiosity about the origins of the deformed or anomalous body. Historically, the question that was asked about monsters was: 'How could such a thing happen? Who has done this?' The quest for the *origin* of monstrous bodies has motivated some of the wildest theories about them.

The epistemophilic dimension makes teratology an ideal testing ground for Freudian critiques of scientific theories in terms of displaced sexual curiosity. For psychoanalytic theory, the desire to know, which is the drive that sustains scientific inquiries, is marked by curiosity about one's own origins, and is consequently stamped with libidinal investments. Psychoanalysis teaches us that desire is that which remains ungraspable at the very heart of our thought, because it is that which propels our thinking in the first place. As such, it will evade us in the very act of constituting us as subjects of knowledge. This is why no science can ever be either 'pure' or 'objective' for psychoanalysis. The

From: N. Lykke and R. Braidotti (eds), *Between Monsters, Goddesses and Cyborgs*, London: Zed Press, 1996.

monstrous or hybrid body is perfect evidence of such theory. In discourses about monsters, the scientific and the fantasmatic dimensions intersect constantly.

There is another, more concrete side to this epistemophilic issue. Historically, monstrous bodies have served as material for experimentation in biomedical practices that eventually led to comparative anatomy and embryology. Disposable bodies are useful to science.

Monsters are linked to the female body in scientific discourse through the question of biological reproduction. Theories of conception of monsters are at times extreme versions of the deep-seated anxiety that surrounds the issue of women's maternal power of procreation in a patriarchal society. To say that compulsory heterosexuality is one of the issues at stake in teratology may seem far-fetched until one reads, in a famous treaty on prodigies, a scathing and rather scurrilous account of the monstrous sexual practices attributed to female hermaphrodites – living in far away places like Africa – who take advantage of their monstrosity by indulging in the filthiest of practices: same-sex sex. The year is 1573; the author is Ambroise Paré. Far-fetched?

Historical examples of the epistemophilic structure of teratology abound. Ambroise Paré concentrates his research on monsters entirely on the question of their reproduction and tells the most extraordinary fictions about their origin; however, these fictions are embedded in some of the most serious canonical texts of Western theology and biology, mostly based on Aristotle.

Paré describes the monstrous birth as a sinister sign ('mauvais augure') that expresses the guilt or sin of the parents. The most common forms of parental transgression concern the norms for acceptable sexual practice, which were regulated by the Catholic Church. Thus, the practice of intercourse on a Sunday or on the eve of any major religious holiday, as well as too frequent intercourse, are quoted reasons for monstrous births.

Sexual excess, especially in the woman, is always a factor. Too much or too little semen are quoted as central causes, as is the mixing of sperm from different sources – for instance, intercourse with animals. Hereditary factors are not ruled out. Intercourse during menstruation is fatal. The influence of stars and planets also matters, as does the consumption of forbidden food, or of the right food at the wrong time. But the monster could also be conceived because of bad atmospheric conditions, or by divine or diabolic interventions.

The devil is extremely resourceful when it comes to satanic penetrations and conceptions. Saint Augustine warns us that Satan – the great simulator! – can take different forms. For instance, as succubus (the one who lies at the bottom) he can take the appearance of a beautiful woman; in this guise, he seduces a healthy young man, thus obtaining his sperm. Then he changes into an incubus (the one who lies on top) and, in this guise – as a man – and in full control of the sperm he has just extracted, he seduces and impregnates a chosen woman. Apart from showing the infinite malice of the evil genius, this would have to count as one of the earliest theories of artificial insemination.

CONTINUITIES AND SPECIFICITIES OF THEME IN TERATOLOGICAL DISCOURSES

The striking historical continuity of some themes regarding monsters is partly due to the two main features I have already pointed out: their non-objective status or 'impurity' and their epistemophilic charge. Clearly, the question 'Where do babies – however monstrous – come from?' is as transhistorical a line of inquiry as one is ever likely to get!

Park and Daston (1981) situate the continuity of teratology in a set corpus of canonical texts: first, the biological works of Aristotle and his classical followers, primarily Albertus Magnus; second, the tradition of divination canonized by Cicero in *De Divinatione*; third, the cosmographical and anthropological components. Glenister (1964) suggests instead a relative stability in the *categories* of historical analysis of monstrous births. The following seem to recur quite regularly: supernatural causes, astrological influences, seminal and menstrual factors, hybridity, mental impressions, philosophical and scientific explanations.

It seems clear that a degree of thematic consistency and order does exist. For the sake of convenience, ever since the encyclopaedic work done in the nineteenth century by Geoffrey Saint-Hilaire, the scientific history of monsters has been divided in three major periods: classical antiquity, the pre-scientific and the scientific eras. To these traditional distinctions, I would like to add a fourth one: the genetic turning point in the post-nuclear era, also known as cybernetic teratology, and the making of new monsters due to the effects of toxicity and environmental pollution.

In the rest of this chapter, I will concentrate on two themes, the continuity of which in the history of teratological discourses strikes me as particularly significant: the racialization of monstrous bodies and the question of the maternal imagination – race and gender as marks of difference. I would like to begin with an introductory remark.

As a signpost, the monster helps organize more than the interaction of heaven and earth. It also governs the production of differences here and now.

The traditional – historically constant – categories of otherness are sexual difference and sexual deviation (especially homosexuality and hermaphroditism); race and ethnicity; the non-human, either on an upward trajectory (the divine, or sacred) or a downward one (the natural environment, the animal, the degenerate, the mutant). A case apart is that of the inorganic other; that is, the machine or technological body-double, the relation of which to the monstrous body is strong. Discussion of this topic would require the kind of special attention that, I regret, I cannot give here.

The peculiarity of the organic monster is that s/he is both Same and Other. The monster is neither a total stranger nor completely familiar; s/he exists in an in-between zone. I would express this as a paradox: the monstrous other is both liminal and structurally central to our perception of normal human subjectivity. The monster helps us understand the paradox of 'difference' as a ubiquitous but perennially negative preoccupation.

This mechanism of 'domestic foreignness', exemplified by the monster, finds its closest analogy in mechanisms such as sexism and racism. The woman, the Jew, the black or the homosexual are certainly 'different' from the configuration of human subjectivity based on masculinity, whiteness, heterosexuality, and Christian values which dominates our scientific thinking. Yet they are central to this thinking, linked to it by negation, and therefore structurally necessary to upholding the dominant view of subjectivity. The real enemy is within: s/he is liminal, but dwells at the heart of the matter. With this in mind, let us look at some historical cases.

THE RACIALIZATION OF 'OTHER' BODIES

One of the dominant teratological discourses in antiquity is that of the monstrous races on the edge of civilization. We find a sort of anthropological geography, the study of territories or special lands where the monstrous races live. Homer had written about cyclops and giant races, of course, but it is Herodotus that started the anthropological trend in the fifth century BC. Though he was rather reserved about neighbouring civilizations such as the Egyptian and the Persian, he went quite wild over more distant lands such as India and Ethiopia, which he thought were populated by cannibals, troglodytes and monstrously deformed people. In the fourth century BC, Ktesias described the Indian tribes of *Sciapodes*, who had one single large foot on which they could hop faster than any bipeds; descriptions of *Cynocephali* (dog-headed people) and *Blemmyae* (headless people) also abound. Through the canonization these monstrous races receive in Pliny's *Natural History*, they will become part and parcel of European medieval folklore. Medieval iconography will, of course, accentuate the monstrosity of monsters and provide moral readings of their morphological deformations.

Whence does this geographical and anthropological racist imaginary originate? Bernal (1987) suggests that the foundations for this topographic determinism of races can be found in Aristotle's *Politics*, in the following passage:

> The races that live in cold regions and those of Europe are full of courage and passion but somewhat lacking in skill and brainpower, for this reason, while remaining generally independent, they lack political cohesion and the ability to rule others. On the other hand, the Asiatic races have both brains and skill but are lacking in courage and willpower; so they have remained both enslaved and subject. The Hellenic race, occupying a mid position geographically, has a measure of both. Hence it has continued to be free, to have the best political institutions and to be capable of ruling others given a single constitution.[2]

The politics of climate and the justice of *in media res* were to have a long and rather successful history in European culture. In a set of continuous historical variations, our culture has tended to represent the furthest away as the most monstrous – that is, the least civilized, the least democratic or least law-abiding; though the actual structures of the scientific discourses conveying this idea underwent historical transformations – from the geographical discourse of the Greeks to the concern for jurisprudence in the eighteenth century, down to evolutionary anthropology in the nineteenth. The idea lived on, stubbornly and lethally.

The colonization of the North American continent, for instance, intensified the trend. Greek theories about climatic and geographical determinism of races lived on in the New World, though they underwent significant revisions. A papal bull by Paul II was needed in 1537 to affirm that Native Americans were fully human and therefore in possession of an immortal soul (de Waal Malefijt 1968), but this did not stop the European settlers from capturing them as 'specimens' and shipping them back to Europe to be placed on public display, a phenomenon which grew throughout the eighteenth century and turned into a major entertainment industry in the nineteenth.

It is worth noting the link between the exhibition of freaks and the orientalist and racist imaginary that underlies it. In the side-shows, spectators wanted to be shocked by the unsightly spectacle of primordial races, in order to be confirmed in their assumptions of racial superiority. Colonial narratives were used to aggrandize the human exhibits (Gould and Pyle 1897), using a pseudo-scientific language borrowed from that of natural-scientific imperialist explorations of unknown continents. Ethiopian, Indian, African and Asian 'monsters' came to be inscribed in these narratives of colonialist teratology.

Theories of geographical determination of monstrosity continued to be produced with stunning regularity. In the eighteenth century, the French 'philosophes', in their concern for jurisprudence, were not immune from the influence of such ideas, though on the whole they opposed slavery. Montesquieu in 1748 and Rousseau in 1764 followed the school of geographical determinism by stating that the northern regions were the ones capable of engendering true virtue and a democratic spirit (Bernal 1987).

Maupertuis (1759), on the basis of his analysis of a monstrosity called 'les nègres-blanc' (black albinos), suggests that black babies are more likely to be born to white parents than white are to blacks; it follows that white is the basic human colour, and blackness is an accidental variation which became hereditary for people living in equatorial zones.

In the nineteenth century, as suggested above, experts pointed to organic disease, intemperance and intermarriage as possible causal factors, but they never abandoned anthropological explanations and ethnographic classification systems. Through the later part of the nineteenth and the early twentieth century, the theory that certain forms of mental deficiency were a biological throwback to earlier races of humans, even to apes, was still widely believed, especially in evolutionary anthropology.

A contemporary version of the continuity of the Greek geoclimatic determination of monstrous races can be found in superstitions and legends surrounding the abominable snowman and, more significantly, in speculations about life in outer space and the colonization of other planets. Extraterrestrials, in popular science-fiction literature and films, perpetuate ancient traditions of representing far-away places as monstrously alien. They also highlight, however, the messianic or divine undertones of the monstrous other, thus reflecting the systematic dichotomy of the *teras* as both god and abjection.

That is the optimistic version of the contemporary situation. A less optimistic one was provided on the front page of the *New York Times* on 21 February 1995. In the Austrian city of Oberwart, a neo-Nazi group attacked a Gypsy settlement of 117 people and left behind a placard saying: 'Gypsies go back to India'. In an important article entitled 'Marvels of the East', Wittkower (1942) analyses the history of racialized teratology centring on India: it originates, as stated earlier, with Herodotus. However wrong the neo-Nazis may be, they are certainly accurate in their fantasmatic geography.

I do not wish to suggest that this is all there is to the racialization of monstrous bodies. Specific historical variations obviously exist – for instance, the vehemence of attacks on Jewish monstrosity throughout the sixteenth century. In his *Histoires prodigieuses et mémorables, extraites de plusieurs fameux autheurs, grecs et latins, sacrés et prophanes*, Boaistuau (1598) devotes a whole chapter to the monstrous race of the Jews. Situated between sections devoted to comets, earthquakes and organic, malformed babies, the chapter on the Jews adopts a different tone. Relying on the classical repertoire of European anti-Semitism (the killing of Christ, the poisoning of water wells, and so on), it describes in minute and rather pictorial details the capital punishment that should be inflicted to 'cette malhereuse vermine' (Book I, ch. X: 35). No other chapter in this text displays such unabashed hatred or such dedication to violent retaliation for alleged sins of monstrosity. Clearly, the monstrousness of the European Jews is of the most negative and demonic kind, with little of the divine sense of wonder that accompanies other prodigies.

Later on, the racialization process intensified and shifted from Jews to African and Asian peoples. For instance, Linnaeus, in his classification system of all living beings, assumes a hierarchical relationship between the races, which was to become central to the European world-view. Thus, in the tenth edition of his *Systema naturae* (1759), Linnaeus postulates a race called *homo monstrosus*, which is one of the branches of *homo sapiens*, living in remote regions of the earth. Black men are classified as being at an equal distance between apes and humans (though satyrs and pygmies are closer to the former than to the latter). This will promote the idea of 'the search for a "missing link", a creature half-ape, half-man' (de Waal Malefijt 1968: 118). This creature was generally believed to roam about in Java and Africa.

The point, however, remains that, in the history of the racialization of the monstrous body, the continuity of certain themes intersects with singular and specific historical instances of teratological discourse. What is both surprising and intriguing is the recycling of the same themes and arguments through time, though they get pinned to different racial groups.

THE THEORY OF THE MATERNAL IMAGINATION

There is no doubt, however, that the 'imagination' hypothesis is the longest lasting theory of monstrous births. It attributes to the mother the capacity to undo the living capital she is carrying in her womb; the power of her imagination is such that she can actually kill or deform her creation. It must be borne in mind here that the power of the imagination has been a major issue since the seventeenth century. At that time, it had a double function: to create order through the principle of making connections or spotting resemblances, and yet also to upset that order (*Encyclopédie* 1765). This double function is to be found in Descartes' treatment of the imagination in his metaphysics. It is also fully deployed in the debates about the maternal imagination.

In his study of freak-shows, Bogdan (1988) reminds us that, as late as the nineteenth century, the explanation for the birth of the famous dwarf General Tom Thumb was the theory of maternal impression. Shortly before Tom's birth the family's puppy had drowned. The mother had been distraught and wept hysterically, causing the baby to be 'marked' and shrink.

Boucé (1985) points out that, in popular teratology, the theory of the maternal imagination continues to be used to explain sexual promiscuity. For instance, as far back as 1573 Paré recounts Hippocrates' story of a princess that was accused of adultery because she gave birth to a black baby. She was excused, however, when she pointed to a large portrait of a Moor that was hanging above the bed where she had consummated her normal, lawful and lily-white intercourse with her husband. Just looking at the picture of the black man had been enough.

In 1642 (Darmon 1977) Aldrovandi pointed out the cases of women who, during Charles V's occupation of Picardy, gave birth to dark-haired, Spanish-looking children, strikingly similar to the foreign soldiers, the sight of whom – they claimed most forcefully – had 'startled' them so. Some of these accounts are not without a sense of humour. The anti-imaginationist Blondel tells of a woman who, on 6 January, gave birth to three babies: two white and one black. Darmon quotes the case of a woman who gave birth to a boy who looked very much like the local bishop. She saved her life, however, by saying that every Sunday she had stood in that church, in pious adoration of the man, and that she 'imprinted' his features on the foetus. Swammerdam quotes the case of a pregnant woman who, startled by the sight of a black man on the street, rushed home to wash herself in warm water. Her child was consequently born white, except for the spaces between his fingers and toes. She had been unable to reach these and they had therefore turned out pitch black on her child. The most recent record I found of this sort of imbrication of teratological and racialized accounts of female reproductive powers dates to the period following the landing of the Allied troops in Normandy. The blonde Norman women claimed that they delivered black babies because they had been 'frightened' by the first black soldiers they had seen (Darmon 1977).

Crucial to this theory is the assumption that the child's entire morphological destiny is played out during conception and the period of gestation. Malebranche (1673) cites a spectacular incidence of this in his report of a pregnant woman who had watched a public hanging and gave birth to a still-born baby, strangled by the umbilical cord. It appeared, further, that even looking at a crucifix might be likely to engender a foetus with broken joints.

The case *for* the maternal imagination through the seventeenth century was upheld by Paré, Descartes and Malebranche. One implication of the importance attributed to the maternal powers of disruption is that procreation was not to be taken for granted but rather constituted a real 'art'. Women became especially responsible for the style and the form of their procreative powers, and many medical treatises were devoted to advising them on how to deal with their delicate situation. A great number of these medical texts concentrate on how to reproduce baby boys, and several are devoted exclusively to the reproduction of male geniuses or 'great men'.

According to common belief, pregnant women were to avoid all excitement and cultivate the serenity of their soul. A special warning was issued against reading, which was seen as the activity most likely to influence and inflame their inflammable imagination. Fraisse (1989), in her study of discourses on women during the French Revolution, focuses on the prohibition surrounding women's reading. This activity seems to be fraught with unspeakable dangers, which, in the case of pregnant women, assume catastrophic dimensions. As late as the nineteenth century, the idea that reading could inflame the female imagination and cause irreparable damage to the woman's frail nervous system remained in fashion. I cannot help being reminded of Freud's patient Dora, whose neurotic symptoms were not unrelated to her reading the 'unhealthy' texts of Mantegazza and other sexologists deemed unsuitable for such a young lady.

The key categories in the theory of the maternal imagination are female desire or wishes ('envies'), the imagination, and the optical structure of human emotions. Glenister (1964) argues that the maternal imagination or impression theory is an optical theory; it is about vision and visual powers. It contains a satanic variable in the tradition of the 'evil eye'. All it takes is for a pregnant woman to think ardently about, dream of, or quite simply long for, certain foodstuffs or for unusual or different people for these impressions to be transferred and printed upon the foetus.

In what Boucé (1985) describes as 'a pervasive epistemological haze', this concept covers phenomena as diverse as the sequels of affective traumas, strong emotions, cravings, wild fantasies and simple memories.

The case *against* the maternal imagination was upheld by Blondel, Buffon, Maupertuis and the *Encyclopédie*. The opposition attacked relentlessly the epistemological haze of the maternal theory. Blondel, of the British Royal Academy, wrote a passionate treatise refuting the theory, based on the assumption of the 'neutrality' of the foetus from the mother. He claimed that the foetus is completely isolated from all sensations or emotions experienced by the mother, thereby showing little knowledge of physiology but great rigour in his argument. Maupertuis followed a similar line.

The most systematic attack against this theory, however, comes from the *Encyclopédie*. Contrary to the view of Blondel, it is argued that the imagination is an important faculty which moves us all, especially pregnant women, quite deeply, but that there is no direct link between the movements of the imagination and physiological processes. There is a general understanding that all passions, emotions or sensations are likely to affect and enervate pregnant women. And there is no denying that these passions have bodily counterparts: the heart beating faster, the muscles contracting, and so on. What comes especially under fire is the faculty that Malebranche had called 'sympathie' (the capacity to feel with/suffer with); that is to say, the causal link between emotions and the capacity to act on other objects.

In fact, the eighteenth-century Encyclopaedists take great care to circumscribe the powers of the imagination because, being unruly, it ends up confusing our ideas and is thus an obstacle to true knowledge. In the same vein, they set out to re-educate the poor gullible women who actually believe in the power of their imaginations. They suggest the following experiment: interview pregnant women before they give birth and make a list of all their desires/'envies', and then compare these to what their newly born baby looks like in an attempt to cure them of their superstition. With customary wit, they do admit, however, that whenever they attempted this experiment, the women got very annoyed and still would not change their minds. So, could women cause monstrous births? No, says the *Encyclopédie*. Were women to have the power actually to create – or deform – life, they might use it to manufacture perfect babies for a change, instead of producing monstrous ones. Moreover, they add gingerly, if the women possessed such powers, they would probably conceive many more baby boys than girls, given that all women are at all times affected by their desire for men.

So what produces monsters? Maupertuis goes to some length in trying to provide an elaborate answer to this eternal question. He proposes a theory of magnetic correspondence between mother and foetus: their respective particles exercise a mutual fatal and foetal attraction which sets in train the process whereby babies are formed. Needless to say, whenever the particles are not strong enough, a monster will be produced by lack; in cases of overattraction of the same particles, a monster by excess is likely to be the result.

By the end of the eighteenth century, however, this question must have begun to seem quite redundant to some people, because the *Encyclopédie* responds dismissively, quoting chance or misadventure as the only possible sources of deformity. To the optical-epistemological question: 'Why do some babies look more like their mothers than their fathers?', they answer (Volume VII, entry 'foetus', p. 2): '[I]l faut bien voir que cela a lieu, sans trop nous instruire du comment ni du pourquoi.'

The theme of monstrous births began to lose scientific momentum. Within less than a century, as teratology gained scientific credibility and led to embryology, *homo monstrosus* became of little scientific relevance to embryological debates, though his place in anthropology was assured for centuries to come.

PRELIMINARY CONCLUSIONS

First, to sum up on the subject of the maternal imagination. In the eighteenth century, the pro-imagination lobby did not fail to respond to the criticism; they emphasized the powerful link between the mother and the foetus, ridiculing any suggestions of the latter's 'neutrality'. They extended this into an attack on the limitations of the rationalist approach and also adopted something of a feminist line in stating that they were taking the side of the poor women, who constantly took the blame for monstrous births. By showing that they were overwhelmed by the imagination, they could be exonerated and even helped out.

In a historical perspective, this theory was indeed a step forward for women, as it recognized their active role in the process of generation. However, scientific teratology was instrumental in creating, or strengthening, a nexus of stifling interdicts, imperatives and even, in effect, pressing advice on women. The disciplining of the maternal body that followed from all this – all 'for her own good', of course – runs parallel to the reorganization of the profession of midwifery, which has been amply documented by feminist scholarship (Oakley 1984; Ehrenreich and English 1979).

The fundamental contradiction that lies at the heart of the quarrel about the maternal imagination concerns the understanding of the woman's body. By the end of the eighteenth century, the mother's body seems to be in a position structurally analogous to the classical monster: it is caught in a deep contradiction that splits it within itself. The female, pregnant body is posited *both* as a protective filter and as a conductor or highly sensitive conveyor of impressions, shocks and emotions. It is both a 'neutral' and a somewhat 'electrical' body. There is an insidious assimilation of the pregnant woman to an unstable, potentially sick subject, vulnerable to uncontrollable emotions. This can be linked to the eighteenth-century discourse about the pathologization of woman (Fraisse 1989).

Second, to conclude the notion of the racialization of the monstrous others. The persistence of the racial and racist overtones in teratological discourses intersects with the continuous emphasis on controlling and disciplining the woman's body. Thus, teratology shows the imbrication of genderized and racialized narratives and the role they play in constructing scientific discourses about the female body. Their interconnection is such that any analysis of female embodied experience simply needs to take into account the simultaneous – if often contradictory – effects of racialized and genderized discourses and practices.

Third, to say some final words on monsters as non-scientific objects of research. Any historical account of teratological discourses has to face up to the limitations and aporias of scientific objectivity. Monsters are not just one object of scientific inquiry. They are many objects, whose configuration, structure and content shift historically. If they can be called an object at all, they are one which is the effect of, while being also constitutive of, certain discursive practices: climatic and geographical anthropologies in antiquity; theological divination through the Renaissance; then anatomy; embryology; until we reach today's cybernetic and environmental chimio-teratology.

Clearly, the epistemophilic or imaginary charge surrounding the monster is partly responsible for this paradox of simultaneous complexity or changeability as well as continuity. The monstrous body, more than an object, is a shifter, a vehicle that constructs a web of interconnected and yet potentially contradictory discourses about his or her embodied self. Gender and race are primary operators in this process.

As a way of concluding, I would like to propose a redefinition: the monster is a process without a stable object. It makes knowledge happen by circulating, sometimes as the most irrational non-object. It is slippery enough to make the Encyclopaedists nervous; yet, in a perfectly nomadic cycle of repetitions, the monstrous other keeps emerging on the discursive scene. As such, it persists in haunting not only our imagination but also our scientific knowledge-claims. Difference will just not go away. And because this embodiment of difference moves, flows, changes; because it propels discourses without ever settling into them; because it evades us in the very process of puzzling us, it will never be known what the next monster is going to look like; nor will it be possible to guess where it will come from. And because we *cannot* know, the monster is always going to get us.

NOTES

1. I have developed this idea at some length in my article: 'Mothers, Monsters and Machines' (Braidotti 1994: 75–95).
2. Aristotle, *Politics*, VII. 7, quoted in Bernal 1987.

REFERENCES

Bernal, M. (1987) *Black Athena. The Afroasiatic Roots of Classical Civilization*, London: Vintage.

Blondel, J. (1727) *The Strength of the Imagination in Pregnant Women Examined*, London: J. Peele.

Boaistuau, P. (1598) *Histoires prodigieuses et mémorables*, Paris: Gabriel Buon.

Bogdan, R. (1988) *Freak Show. Presenting Human Oddities for Amusement and Profit*, Chicago: University of Chicago Press.

Boucé, P.-G. (1985) 'Les jeux interdits de l'imaginaire: onanisme et culpabilisation sexuelle au XVIIIème siècle', in J. Céard, ed., *La Folie et le Corps*, Paris: Presses de l'Ecole Normale Supérieure, pp. 223–43.

Braidotti, R. (1994) *Nomadic Subjects*, New York: Columbia University Press.

Darmon, P. (1977) *Le Mythe de la Procréation à l'age baroque*, Paris: Seuil.

Encyclopédie, ou Dictionnaire Raisonné des Sciences, des Arts et des Métiers (1765) Vol. VII, Neufchatel.

Ehrenreich, B. and D. English (1979) *For Her Own Good. 150 Years of Experts' Advice to Women*, London: Pluto Press.

Fiedler, L. (1979) *Freaks: Myths and Images of the Secret Self*, New York: Simon & Schuster.

Fraisse, G. (1989) *Muse de la raison*, Aix-en-Provence: Alinéa.
Glenister, T. W. (1964) 'Fantasies, Facts and Foetuses: The Interplay of Fancy and Reason in Teratology', *Medical History* 8: 15–30.
Gould, G. and W. Pyle (1897) *Anomalies and Curiosities of Medicine*, Philadelphia: W. B. Saunders.
Linnaeus, C. ([1759] 1964) *Systema naturae*, Weinheim: Cramer.
Malebranche (1673) *De la recherche de la vérité*, reprinted, Vrin, Paris, 1946.
Maupertuis, P. L. de ([1759] 1866) *La Vénus Physique: ou Les lois de la génération*, Paris: Office de la Librairie.
Oakley, A. (1984) *The Captured Womb*, Oxford: Blackwells.
Paré, A. (1971) *Des monstres et des prodiges*, Geneva: Librairie Droz.
Park, K. and L. Daston (1981) 'Unnatural Conceptions: The Study of Monsters in 16th and 17th Century France and England', *Past and Present*, no. 92: 20–54.
Saint Augustine (1972) *The City of God*, Harmondsworth: Penguin.
Saint-Hilaire, G. de (1832) *Histoire Génerale et particulière des Anomalies de l'Organization Chez l'Homme et les Animaux*, Paris: J. B. Ballière.
de Waal Malefijt, A. (1968) 'Homo Monstrosus', *Scientific American*, 219 (4): 113–18.
Wittkower, R. (1942) 'Marvels of the East: A Study in the History of Monsters', *Journal of the Warburg and Courtauld Institutes*, London, 5: 159–97.

5.3

INTERVIEW FROM *WARRIOR MARKS*

Pratibha Parmar and Alice Walker

I wanted to do this interview, at Alice's home in northern California, to establish the beginning of her journey with the film Warrior Marks *and with the issue of female genital mutilation. We also filmed Alice telling her own story of being wounded as a child.*

PP: *Alice, why did you decide to write about female genital mutilation? You said that it was your visual mutilation that led you to think about the subject.*

AW: It helped me to see the ways in which women are rather routinely mutilated in most parts of the world and how people tend to think of the pain done to women as somehow less than pain done to men. It's often overlooked, or they think: Well, women can take it. You can see this in attitudes toward childbirth, which is incredibly painful yet is really minimized. It's a heroic thing that women are doing, bringing a child into the world – it's excruciatingly painful and yet people tend to think of it as something that's just routine: 'It's a woman – she can stand it.'

My own visual mutilation occurred when I was eight, and it led me to a place of great isolation in my family and in my community and to a feeling of being oppressed. It was never really explained to me, nor was there sufficient comfort given to me as a child. And I see this mirrored in the rather callous way that little girls are taken to be mutilated. You take

---

From: A. Walker and P. Parmar, *Warrior Marks: Female Genital Mutilation and the Sexual Blinding of Women*, London: Jonathan Cape, 1993.

a little child off and tell her she is going to visit her grandmother. On the way, you divert her attention from the trip to the grandmother's, and instead you hold her down and cut off her clitoris and other parts of her genitalia and leave her to heal from this as best she can. Everybody else is making merry. She is the only one crying. But somehow you don't care, you don't show sensitivity to this child's pain. I made a very strong connection with that. I was able to feel for the child, while understanding that the adults thought that they were doing the right thing.

PP: *When did you first hear about female genital mutilation?*

AW: At the time I was a student and a Westerner, and I didn't understand it. I couldn't imagine, for instance, what it was that they would remove. I mean, what is there to remove? Since obviously there isn't anything removable, I put it out of my mind for a long time. I was reading Kenyatta, who was the great liberator of Kenya, after the Mau Mau struggle. And he was saying that no Kenyan man would marry a woman who had not been mutilated. They didn't call it 'mutilation', of course, but rather circumcision.

This presented a great conflict, because I was very much an admirer of Kenyatta's and very happy that he had been let out of prison – his case is very similar to Nelson Mandela's. And yet this seemed to me, even then, very strange. After all, there were many women in the struggle for liberation. And if you have women attempting to liberate a country and they can't walk, you have a problem.

PP: *How long did it take you to write your book* Possessing the Secret of Joy?

AW: To write it, I went to Mexico. I needed to be in a Third World country, where I could feel more clearly what it would be like to have a major operation without anaesthetics or antiseptics, because that is what happens to little girls when they are genitally mutilated. It shows such contempt for a child's body and such contempt for the clitoris. The actual writing of the book took a year. But it took me twenty-five years since I first heard about female genital mutilation to know how to approach it. To understand what it means to all of us in the world, that you can have this kind of silencing of the pain of millions of women, over maybe six thousand years. I have talked about this in many countries, and even in Africa people stand up and say, 'I have never heard of this. Are you sure that this happens?' And we are literally talking about something which has been performed on that person's mother, that person's aunt, that person's sister, that person's child, and they don't know a thing about it. It's remarkable.

[. . .]

PP: *When we were in London recently at a public meeting where you were talking about the book and about female genital mutilation, I noticed that one of the common responses among women was terror and shock and then a feeling of being unable to do anything. What would you say to women who want to do something about it but feel paralysed by the fear and the shock of what this is?*

AW: I've noticed that they also faint. Everywhere I have talked about this, women have just dropped. It has been the most amazing thing. It's a real literal shock to the system that this can happen, and you do need time to think about it without feeling pressure. I always tell women when they start reading my book that if the going gets rough, they should just put it aside and not try to press on. Because it is a lot to take in if you are really in touch with your own feelings. If you are not in touch with your feelings, then you can read it as a kind of intellectual exercise. But many women really are not able to do that, and I am happy that they are not, because it means that they are alive.

It is a serious thing that we are talking about being done to the human body. Wherever the human body is, we are one body, and that is why women faint when they hear about female genital mutilation being done to another woman. They know it's also their own body. So, first, just take your time. And because time is pretty much all there is, just take it in, let it go in, and think about it, live with it, sit with it. Then just look around and see where you can be useful. There are organizations like FORWARD International in London that can always use money or help. But mostly be aware and try not to see this just as an isolated brutality that happens to women in other countries but to see how it also happens in this country, in different guises.

[. . .]

PP: *Alice, how do you feel about the fact that genital mutilation is practiced by women on women and young girls? Mothers doing it to their daughters, and grandmothers doing it to grandchildren?*

AW: I feel terrible about it. It is true that in some places men do it – for instance, in Egypt. There was a recent photograph of an Egyptian doing it to a fourteen-year-old, who had her head against her mother's chest, looking absolutely scared and in pain, and the mother very forcefully holding her while the mutilation is being done by this man. But generally, you are right, this is something that women do. Like everything to do with children, men turn it over to women. Men don't on the whole feed them, bathe them, or brush their hair either. But the mother's betrayal of the child is one of the cruelest aspects of it. Children place all their love and trust in their mothers. When you think of the depth of the betrayal of the child's trust, this is an emotional wounding, which will never go away. The sense of betrayal, the sense of not being able to trust anyone, will stay with the child as she grows up. I think that is a reason why in a lot of the cultures that we are talking about, there is so much distrust, so much dissension, and so much silence. There is all this unspoken pain, this unspoken suffering, that nobody is really dealing with, nobody is airing, and it goes somewhere, it always does.

PP: *During British rule in Kenya, the indigenous people fought to hold on to traditional practices, but unfortunately that included some which were harmful, such as female genital mutilation. This is something you have referred to in your interviews with women here. What are your own thoughts on the subject?*

AW: When they themselves are being oppressed, people tend to hold on to the practices that they can enforce. As they can most easily enforce things that control women and children, that is what they have tended to do. It is very sad, because the British, in this particular instance, were right about stopping this sort of evil thing. But because they themselves were so evil, and the harm that they had done was so great, it was very difficult for African men and women to really choose what they would like to retain of their culture, since the British were so busy destroying everything else. Colonizers have managed, in so many instances, to destroy worthwhile traditions – they have even managed to stop people from eating their own native food, such as their rice. They have made available only less nutritional foods. They have made people stop wearing their own clothing. They have substituted British-made and French-made clothing and goods. Yet the practice of female genital mutilation remains unchanged. The colonizers managed to force people to change many other traditions, but mutilation of girls was clearly not their priority – it seems that colonizers don't care as long as it is somebody else's child who is being abused and they get their profits anyway.

PP: *There is a very large percentage of women all over Africa who have been either circumcised or genitally mutilated. What do you feel about the fact that it's a culturally specific form of violence against women, yet there are so many other forms of violence against women in other countries? It seems there is a continuum of violence against women that takes specific forms in different cultures. One of the worries I know I have is that it's very easy for Westerners to use genital mutilation as a way of describing Africa as being backward and savage and barbaric, and feeding into all those sort of racist perceptions of Africa.*

AW: Yes, that is a problem. If you think about the Middle East and Malaysia and Indonesia and countries in Africa where this happens, you could easily think that this is an isolated assault on women. But as we know from our television screen, the assault on women is worldwide; it is not isolated. It varies only by degree.

If you live in a culture like ours, which now wants women who are very thin, very white, very blond, with very big breasts, a lot of women have breast implants, they have bleached skin, they have bleached hair. They try to be the ideal woman for some nonspecific but very powerful man.

So it's really about shaping a woman in the image that men think they want. And every country in the world is busily doing that.

As we know, for centuries the Chinese broke the feet of women and forced their toes back underneath the feet so they would only be three inches long, and you had to hobble along on these little feet. Soon this became an aphrodisiac. The fact that your feet were slowly rotting and emitted a peculiarly awful odor became something that men loved. But essentially it was about preventing a woman from getting away. It was about enslaving someone.

And that is what all this is about. It's why we have bras that make it impossible for us to breathe freely, high-heeled shoes that destroy our feet. Some Burmese women wear fifteen or twenty necklaces, designed solely to make it impossible for them to hold up their necks without the support of these necklaces. So if they are ever 'disobedient', all someone needs to do is take away a couple of the necklaces, and the excruciating pain of trying to keep their head up makes them long for this necklace again. And there are cultures where they put heavy weights on the woman's legs, which destroy the muscles of the leg. Over centuries, women begin to think that these things are beautiful, because the mind needs to rationalize what is happening – when in fact it is about enslavement.

PP: *Alice, there have been so many different actions, campaigns, and outcries against female genital mutilation, yet it still continues. Why do you think this is, and what do you think can be done to stop it?*

AW: It continues partly because in many cultures there is no role for a mature woman after puberty except marriage. And if you have a culture in which men will not marry you unless you have been mutilated and there is no other work you can do and you are in fact considered a prostitute if you are not mutilated, you face a very big problem. Women mutilate their daughters because they really are looking down the road to a time when the daughter will be able to marry and at least have a roof over her head and food.

Yet you see women and men continuing this practice in countries like Holland and Britain and the United States, where who's to say what a child will do, who's to say that this child is going to be marrying anybody? Who's to say that if she is from Somalia or from the Sudan or from The Gambia or from Senegal, she is going to marry somebody from a culture where this is wanted? In fact, what is happening is that men from their own culture are actually choosing to marry women who have not been mutilated. In the Sudan and Somalia, a lot of the men marry European women, or they marry African-American women, or they marry Indian women – whoever has not been mutilated. Because often they don't know fully what has been done to women in their own culture, and when they find out, they are horrified.

There are a lot of reasons why it is difficult to stop the mutilation of women, but I also feel that there are a lot of reasons why today we have a greater opportunity to start putting an end to it. It is all about consciousness, and there are now more educated women in the world, more conscious women. In some cultures, men have confused sexuality with torture, so that they only really enjoy sex with a woman if she is literally screaming in pain. But I think that many men are changing, their consciousness is changing, so there will be men, too, who will fight against it, and have been doing so. Sudanese doctors have been against it; in Burkina Faso, Thomas Sankara made a very strong statement against it before he was assassinated. So there are both women and men who are speaking out against it.

[. . .]

PP: *What are your expectations of going to Africa and talking to women in Africa, both those who are campaigning against genital mutilation and those who have had it done to them?*

AW: I am a great believer in solidarity. Nicaraguans say something very beautiful. They say that solidarity is the tenderness of the people and real revolution is about tenderness. The sharing of this tenderness is beautiful. If you can make one person's life free from a particular kind of pain, that's really enough. It may have ripples, it may go here, it may go there; there's no way of knowing. I think it is the nature of life to continue what is put into action. You may not be around to see where your own action goes – people never are, really, because it never ends.

I'm looking forward simply to being with the women we will be talking to. I want to eat with them, dance with them, see who the priestesses are and who the goddesses are. I want to know what's going on with them and how they feel. If they tell me that they would prefer that I didn't intrude into their affairs, I won't talk to them about that. I want to ask them, do they know what African custom made it possible for me to end up on another continent, in the USA? We have been separated by a custom similar to genital mutilation, a custom of slavery. I have been wondering about my ancestor who came from Africa to America: was she mutilated? What did she feel, in chains and mutilated in this way? If she had been pregnant, she might never have made it alive; she might have died trying to give birth on the ship, with nobody to open her if she had been infibulated and sewn together.

PP: *Can you tell me why you wanted to do a film about female genital mutilation and what hope you have of this film making a difference?*

AW: I always think about people who don't read, who cannot read, but who can relate to things visually. This is because I myself come from a culture where not everybody reads, and a family where not everybody reads, so film is very important. It is very important for people to be able to actually look at what it is you are discussing and to understand it because they see it. I hope that most of the women in the West from mutilating cultures will eventually see this film, and I think they will. I think they will be curious. They may not care much for my perspective or what the film is supposed to be about, but I think that out of curiosity they will want to see, because it is about their own experience. And they will be able to enlarge upon it because they will be able to take issue with it, they can criticize it, they can use it, as an object from which to explore their own experience. And that is what art can do. So I have that hope.

I also think that because it will be portable, it will be able to go into countries where there are a few women who are against genital mutilation and who will be courageous enough to show it to women who don't know really what they feel about it. Many of them have forgotten the pain they felt when it was done to them, because they were so small. And now, of course, they do it to much younger children, to babies. And if it is done to you as a baby, you can only imagine what you retain of this assault. If, in fact, you live to know about it. Because in France, a Gambian woman is on trial for mutilating a baby, who bled to death.

It's also a good thing to have a film that shows that the genital mutilation of women is really just a part of the global mutilation of women, the terrorization of women, one of the numerous things done to keep them in their place, under the foot of the dominant patriarchal culture.

5.4

THE TRIALS OF THE BLACK AFRICAN WOMAN

Awa Thiam

Some First-Hand Evidence

P.K.: I had just turned 12 when I was excised. I still retain a very clear memory of the operation and of the ceremony associated with it. In my village, excision was only performed on two days of the week: Mondays and Thursdays. I don't know if this was based on custom. I was to be excised together with all the other girls of my age. Celebrations were held the previous evening. All the young and old people of the village gathered together and stuffed themselves with food. The tomtoms were beating loud, late into the night. Very early the next morning, as my mother was too easily upset to have anything to do with the proceedings, my two favourite aunts took me to the hut where the excisor was waiting with some other younger women. The excisor was an old woman belonging to the blacksmiths' caste. Here, in Mali, it is usually women of this caste who practise ablation of the clitoris and infibulation.

On the threshold of the hut, my aunts exchanged the customary greetings and left me in the hands of the excisor. At that moment, I felt as if the earth was opening up under my feet. Apprehension? Fear of the unknown? I did not know what excision was, but on several occasions I had seen recently excised girls walking. I can tell you it was not a pretty sight. From the back you would have thought they were little bent old ladies, who were trying to walk with a ruler balanced between their ankles, and taking care not to let it fall. My elders had told me that excision was not a painful operation. It doesn't hurt, they repeatedly assured me. But the memory of the expression on the faces of the

From: A. Thiam, *Speak Out, Black Sisters*, London: Pluto Press, 1986. Trans. by Dorothy S. Blair.

excised girls I had seen aroused my fears. Were not these older women simply trying to put my mind at rest and allay my anxieties?

Once I was inside the hut, the women began to sing my praises, to which I turned a deaf ear, as I was so overcome with terror. My throat was dry and I was perspiring though it was early morning and not yet hot. 'Lie down there', the excisor suddenly said to me, pointing to a mat stretched out on the ground. No sooner had I lain down than I felt my thin frail legs tightly grasped by heavy hands and pulled wide apart. I lifted my head. Two women on each side of me pinned me to the ground. My arms were also immobilized. Suddenly I felt some strange substance being spread over my genital organs. I only learned later that it was sand. It was supposed to facilitate the excision, it seems. The sensation I felt was most unpleasant. A hand had grasped a part of my genital organs. My heart seemed to miss a beat. I would have given anything at that moment to be a thousand miles away; then a shooting pain brought me back to reality from my thoughts of flight. I was already being excised: first of all I underwent the ablation of the labia minora and then of the clitoris. The operation seemed to go on for ever, as it had to be performed 'to perfection'. I was in the throes of endless agony, torn apart both physically and psychologically. It was the rule that girls of my age did not weep in this situation. I broke the rule. I reacted immediately with tears and screams of pain. I felt wet. I was bleeding. The blood flowed in torrents. Then they applied a mixture of butter and medicinal herbs which stopped the bleeding. Never had I felt such excruciating pain!

After this, the women let go their grasp, freeing my mutilated body. In the state I was in I had no inclination to get up. But the voice of the excisor forced me to do so. 'It's all over! You can stand up. You see, it wasn't so painful after all!' Two of the women in the hut helped me to my feet. Then they forced me, not only to walk back to join the other girls who had already been excised, but to dance with them. It was really asking too much of us. Nevertheless, all the girls were doing their best to dance. Encircled by young people and old, who had gathered for the occasion, I began to go through the motion of taking a few dance steps, as I was ordered to by the women in charge. I can't tell you what I felt at that moment. There was a burning sensation between my legs. Bathed in tears, I hopped about, rather than danced. I was what is known as a puny child. I felt exhausted, drained. As the supervising women who surrounded us goaded us on in this interminable, monstrous dance, I suddenly felt everything swimming around me. Then I knew nothing more. I had fainted. When I regained consciousness I was lying in a hut with several people around me.

Afterwards, the most terrible moments were when I had to defaecate. It was a month before I was completely healed, as I continually had to scratch where the genital wound itched. When I was better, everyone mocked me as I hadn't been brave, they said.

5.5

THE ECONOMY OF VIOLENCE: BLACK BODIES AND THE UNSPEAKABLE TERROR

Bibi Bakare-Yusuf

In this chapter I want to explore the relationship between the notion that physical brutality and force transformed the African body from a liberated body to a captive one, and Elaine Scarry's idea that the infliction of physical pain unmakes and deconstructs the body, while simultaneously making and reconstructing the world of the perpetrator. I wish to link these two ideas by proposing that under the slave economy and colonization two kinds of bodies were produced: the body of knowledge and the body of labour. These two bodies are missing from Scarry's discourse of the body in pain. They are also absent from most contemporary discourses about the body. Paying attention to the body of labour in particular, I wish to show how I find some of Scarry's arguments useful in relation to reading the slave narrative, *Mary Prince*. I am inspired by the connection Scarry makes between the infliction of pain, embodiment, voice and subjectivity and the ruination of the subject – and certainly *Mary Prince* bears this out. However, I suggest that the economy of violence which characterized the middle passage and the epoch of slavery had as its primary motive the extraction of capital and wealth through slave labour. Following Hortense Spillers's idea in her essay 'Mama's baby, papa's maybe: an American grammar book' that the middle passage and slavery was a brutal disruption of the African kinship system which denied the captive female a gendered position, I suggest that this degendering also entailed a deconstruction and unmaking of the captive's subjectivity. In their different ways both

From: R. Lentin (ed.), *Gender and Catastrophe*, London: Zed Press, 1997.

Scarry and Spillers address the intersection of bodily damage, objectification and assault on the flesh and the difficulty of expressing that assault. Both address the unspeakability of the terror of torture but Spillers actively focuses on the deconstruction of the African captive.

THE BODY OF THEORY

The body has become the most celebrated site for addressing a wide range of cultural configurations; for articulating contemporary experiences among feminists with divergent interests as well as social and cultural theorists. The privileging of the body is evident in the spate of books published with the word 'body' in their title. Among these are Elaine Scarry's *The Body in Pain* (1985), Elizabeth Grosz's *Volatile Bodies* (1995), Judith Butler's *Bodies that Matter* (1993), Moria Gatens's *Imaginary Bodies* (1996), Bryan Turner's *The Body and Society* (1984). Many of these studies address the numerous ways of using the human body and embodiment as a conceptual tool for rethinking an array of issues: the problematic nature of sex, sexuality and gender; of ageing and aesthetics (Featherstone and Hepworth 1991); of disease and illness (Turner 1984; Bordo 1990), pain and self-alienation (Scarry 1985); of deconstructing the dualism in Western metaphysics. The body has replaced such categories as subjects, social agents, and individuals. It is, as Grosz points out, the 'very "stuff" of subjectivity [and] . . . all significant facets and complexities of subjects can be adequately explained using the subject's corporeality as a framework'. Like many current theorists of the body, Grosz calls for the need to fuse the historical specificity of bodies with the biological concreteness of the body since 'there is no body as such: there are only bodies – male and female, black, brown, white, large or small and gradations in between' (Grosz 1995: 19). Much of this 'corporeal feminism' (Grosz 1995) is grounded in linguistics and psychoanalysis where the body is read as discursive or textual (Butler 1993), even as it claims to be 'concerned with the *lived body*, the body in so far as it is represented and used in specific ways in particular culture' (Grosz 1995: 18). This attention to the body sometimes seems a mere flirtation with the idea of the lived body where the experience of lived bodies is constituted as a metaphor that is 'good to think with'.

But what of the dying body; the weeping, living, hurting body; the body as flesh that 'does not escape concealment under the brush of discourse or the reflexes of iconography' (Spillers 1987: 67)? The body that is a site of physical and psychological trauma; enforced sexual practices and that 'Sadian imagination' (Carter 1979), with its self-announcing presence on newsstands, in popular literature and medicine. The body that has served and continues to serve a heuristic purpose for the European (male) construction of subjectivity on the one hand and on the other the 'source of an irresistible, destructive sexuality' (Spillers 1987: 67). The body at whose breasts white males suckle as they spread the expanse of their fluid around the globe. The body that enables certain categories of white women to retain fine, delicate, sickly hands. The body on which black men can at will unleash their rage and frustration. The body whose physical health will determine whether it will become the body of labour or the body of scientific knowledge. The body that nineteenth-century Europeans found so riveting that it became a prized attraction at fashionable Parisian balls and would later be subjected to the surgical instrument of the physician (Gilman 1985; Schiebinger 1994). What of the body that is always under the seduction of death, white racist violence, diseases, perverse heterosexism, pervasive addictions and unemployment? I am talking about the body that is marked by racial, sexual and class configurations. It is this body, this fleshy materiality that seems to disappear from much of the current proliferation of discourses on the body.

It is not enough to show the body as a discursive entity without addressing how different material practices are interwoven with the discursive to affect and shape the materiality of the body. The French theorist Michel Foucault understood the interdependency between the fleshy materiality of the body and its functioning, representation and regulation in discursive fields when he inaugurated, along with the effort of second-wave feminism in the seventies, the topic of the body in contemporary theory. In his numerous philosophical studies, Foucault described in great detail the historical specificities which produce the body in discourses and in everyday practices structure the way experiences of the body are organized. Accordingly the body is always in a political field where 'power relations have an immediate hold upon it; they invest it, mark it, train it, torture it, for it to carry out tasks, to perform ceremonies, to emit signs' (Foucault 1991: 173). These inscriptions and incorporations of power onto the body mark the ending of one type of body and the beginning of a new kind of body: a useful body. But this body comes into being through new modes of subjection. Such a body, Foucault contends, also produces power that facilitates resistance, rebellion, evasions and disruptions. In other words, where there is power, there is resistance; where there is discourse, there is material; and both power and discourse are interconnected. Discourse then becomes just one of the modes in which political power manifests itself.

If current discourse analysis is to be useful for a black feminist project, we need to make connections with matters of the flesh on the lived body. For it is through the corporeal inscription of the black man as uncivilized savage and the black woman as embodying a hyperbolic sexuality which marked her as sexualized animal in the New World (Gilman 1985; Jordan 1982). As Fanon notes in his book *Black Skin, White Mask*, it is the 'corporeal schema' of the black (man) that structures how black people are perceived and 'the fact of blackness' is established: 'I am over-determined from without. I am the slave not of the "idea" that others have of me but of my own appearance . . . I strive for anonymity, for invisibility' (1992: 224). It is the visible physical difference that marked the African as inherently non-human.

By becoming aware of the slave experience as an embodied phenomenon, we can better understand: (1) the relationship between embodiment and subjectivity; (2) the body as a surface, a surface that can experience and be inflicted with pain, tortured and terrorized, but also a surface that can be pleasured and is pleasuring; and (3) how bodies are linked in distinctive ways through capitalist modes of production. Thus the history of the middle passage and slavery is a history of endless assaults on bodies; of bodies forcibly subjugated, in order to be transformed into productive and reproductive bodies.

THE UNSPEAKABLE TERROR

One way to approach the issue of unspeakable terror is to ask: what is the unspeakable? The answer will of course be: the experience of violence against human flesh wherein the body-surface registers and transmits nothing but pain, a pain that produces nothing but horror, a horror which reads, according to Spillers, like a laboratory prose of festering flesh, of limbs torn from sockets, of breasts branded with hot iron, of severed tendons, bruises, exposed nerves, swollen limbs, of missing teeth as the technology of iron, whips, chains, bullets, knives, and canine patrol went to work (Spillers 1987: 67). It is the struggle, the longing to speak these horrors and the inability, the near impossibility of doing so that confers on slave history the experience of horror and what Gilroy has termed the 'Slave Sublime' (1993).

The conflation of the body and the unspeakable draws us into an awareness of our physical mortality and the erasure of the human voice. In her influential book *The Body in Pain: The Making and Unmaking of the World* (1985), Scarry argues poignantly that the presence of physical pain is difficult to express, and also has the capacity to destroy the sufferer's language because it has no referential content in the external world. Unlike other states of consciousness such as psychological and somatic states of being which have referentiality to the external world – love of *x*, fear of *y*, hatred for, being hungry for, and so on – physical pain has no such referentiality. Its non-referentiality prevents and inhibits the transformation of the felt experience of pain, leaving it to reside in the body, where the sufferer reverts back to a pre-linguistic state of incomprehensible wailing, inaudible whisper, inarticulate screeching, primal whispering which destroys language and all that is associated with language: subjectivity, civilization, culture, meaning and understanding (Sa'ez 1992: 137).

While Scarry's model is plausible it is still based on a conception of what the appropriate function of the body is and the appropriate function of the mind. Thus, for as long as we conceive of pain as an activity of the body, and language as the function of the mind, pain will continue to be resistance to language, and its sedimentation in the body will continue to confirm the notion that the body is always outside of culture and pre-language. I suggest that we view pain as not necessarily resisting language, but rather as resisting everyday speech. Pain has its own morphology and its own logic which govern its expression and representation and which produce its own meaning. The body writhing on the ground in agony communicates to the spectator the presence of pain (even if unshareable), using the body as a resource to do so. What cannot be spoken in language is evoked through other cultural representation such as dance.

Gilroy has argued that, while the experience of the middle passage and the diasporic plight might be resistant to (verbal) language, it is not resistant to representation (1993). The cultures of the black diaspora he notes have been an attempt to bring that scene of 'pure physical experience of negation' (Scarry 1985: 52) into the realms of representation. The most elemental expression of this can be found in the music and dance of the black diaspora which produced new cultural meanings of the African past, present and future. Although pain has no referential content in the external world it is not unrepresentable.

Because physical pain is so nearly inexpressible and 'flatly invisible', its presence is often relegated to the status of non-existence. According to Scarry, that which is expressible is often made visible and thus elicits more attention. The mechanisms of torture and war, she argues, establish the presence of intense pain as absence as the regimes translate the infliction of pain into a language of political power. For Scarry torture is language-destroying; therefore, to elicit information from the tortured while fully aware of his/her powerlessness is to deconstruct his/her world:

> Intense pain is world-destroying. In compelling confession, the torturer compels the prisoner to record and objectify the fact that intense pain is world-destroying. It is for this reason that while the content of the prisoner's answer is only sometimes important to the regime, the form of the answer, the fact of his answering, is always crucial. (1985: 29)

Of course not all forms of torture fall under the interrogational mode in the way Scarry describes. During slavery the use of torture was not to elicit information, but was used instead to inspire terror and confirm to the enslaved the incontestable power of their masters and mistresses (Patterson 1982). It also facilitated the extraction of subsistence labour for the development of the mercantile and industrial economy. Scarry's claims for an internal structure of torture, however, have implications for these arguments.

As theorized by Lacan, we become social subjects through our subjection to the laws of language and our capacity to understand and articulate language. But, as Scarry suggests, the body in pain is not able to participate fully in civic life, because pain destroys the capacity of language; the body is denied the facilities that make subjectivity possible. It should, however, be noted that the formative role of language in subject formation is not the only means of constituting subjectivity. However, the ability to verbally express the presence of pain is unavailable to the person in pain. The (near) impossibility of constituting pain in language initiates a splitting, a splitting between the speaking subject (voice) and corporeal subject (body). This separation between the tortured (powerless) and torturer (powerful) means that the torturers are able to circumvent material representation, and are represented and describable through the making present of their voice while corporeality is displaced onto the person in pain. Thus, the person in pain becomes mere flesh and can only experience her own body as the agent of her own agony (Scarry 1985: 47). This way of perceiving the body permits the one who is being inflicted with pain to shift from the position of sufferer to being the agent of his/her own annihilation; the cause of the pain is represented as outside those inflicting that pain. The possession of voice becomes significant for both torturer and tortured. For the torturer, the awareness of voice confirms his power, his existence, the presence of a world; for the sufferer, the absence of a world, the awareness of his/her corporeality, the limit of his/her extension in the world. 'Consequently, to be intensely embodied is the equivalent of being represented ... is almost always the condition of those without power' (Scarry 1985: 207). This has been precisely the claim of feminists and black theorists, who have pointed out that the association of blacks and females with corporeality excludes and debars them from the public sphere that makes subjectivity possible (Gatens 1996; Spelman 1990; Fanon 1992).

THE ECONOMY OF VIOLENCE; THE VIOLENCE OF ECONOMY

In the narrative of *The History of Mary Prince: A West Indian Slave. Related by Herself* (1987) we have a rare example of writing from the perspective of a black female in British anti-slavery discourse. Mary Prince's account provides insight into the atrocities and barbarity of the slave system, its absurdity and unrelenting commitment to brutalization and mutilation of the flesh. It shows the cool, calculated contempt for the flesh and the capacity of this barbaric system to rip apart and expose hidden tissues to the brazen gaze of the violator, violating human decency. As in most abolitionist material, while concern was to show the atrocities and catastrophic terror of the slave system, piety was far more important to the abolitionist than the desire to retell events as they really were. The detailing of atrocities in *Mary Prince* is somewhat circumscribed, especially as it relates to sexual violence. Nevertheless, Mary Prince's account succeeded in providing details of the atrocities, and the near triumph of the violator/enslaver over the flesh.

A persistent and poignant theme throughout the narrative is the grotesque and harrowing detailing of physical brutality, the physical torment which resulted in Mary's near blindness, and rheumatism, her mistress's noxious brutality, and the sadistic treatment of other slaves:

> Sarah, who was nearly past work ... was subject to several bodily infirmities, and was not quite right in her head, did not wheel the barrow fast enough to please him [the master]. He threw her down on the ground, and after beating her severely, he took her up in his arms and flung her among the prickly-pear bushes, which are all covered over with sharp venomous prickles. By this her naked flesh was so grievously wounded, that her body swelled and festered all over and she died in a few days after. (Prince 1987: 65)

It is this flesh, this live tissue that registers 'these lacerations, woundings, fissures, tears, scars, openings, ruptures, lesions, rendings, punctures of the flesh' (Spillers 1987: 67), that feels the hurt, the intense pain, that is continuously attacked, mutilated, that experiences the flesh as a burden – a venomous prickly pear. It is this flesh, so horribly lacerated, that is marked for enslavement, for raw violence and objectification, that serves others' will-to-power and their becoming beings. It is these ineffaceable markings that conveniently invalidate all claim of ownership to her flesh, because it is reserved for her master. Her flesh is the signification of her worth within a system whose organizing principle is premised on a proprietary conception of bodies; a system which deemed it its birthright to legislate on her very humanity, control her movements, her body. As Spillers notes, within this legal system, 'every feature of social and human differentiation disappears in public discourses regarding the African-American person' (Spillers 1987: 78).

As a slave, she is perceived as having no soul, no human speech, no gendered subjectivity, no culture or language to speak of. Therefore, to devise and enact such elaborate mechanisms of torture, to send waves of terror down the spine of the tortured was the protocol of those who regarded themselves as having those attributes. As Taussig remarked of the colonial conquest of Latin America, it is 'terror as usual'. The use of terror was a national sport during the period of slavery and colonialism; it was the logic underpinning the creation of colonial reality and identity: 'it serves as the mediator *par excellence* of colonial hegemony' (Taussig 1987: 5).

If, as Scarry points out, the infliction of violence on the body is also an assault on language, similarly, the insatiable and perpetual infliction of raw violence on the slaves is consolidated by the erasure of the human voice. All verbal forms of communication were severed. The only form of communication was to literally work upon/put to work the body of the enslaved. For Mary Prince, when her body was put to work in harsh conditions, her flesh began to give way to tears and to seizure, rheumatism, excruciating backaches and to a near-blindness. Her body became the site of ejaculatory orgasms that disfigured her 'while the powers that don't be join for a loving circle jerk' (Sapphire 1994: 14). She is unable to speak these ejaculatory horrors even when she is permitted to do so because of Christian piety and the enormity of her pain. Unable to express the presence of her pain, Mary's masters and mistresses communicated their presence, confirmed the absence of pain.

The point of this grotesque torment was to inspire terror in the minds of the enslaved and ensure the absence of their world. It confirmed the presence of the violator, and, therefore, that the body of the enslaved will always be the property of her master and mistress. The enslaved could never claim ownership of her body. As a slave, her subjection is an act of de-subjectification. As Scarry describes, in the structure of torture, the infliction of bodily pain, the rupturing of tissues, the exposure of interior skin results for the sufferer in the dissolution of a world akin to the process of dying. This damage, so unalterable, leaves its mark – for example, her persistent lower back ache and rheumatism – long after the infliction ceased: 'what is remembered in the body is well remembered' (Scarry 1985: 113). All trace of humanity, civilization is deconstructed in the infliction of bodily pain:

> The arms that had learned to gesture in a particular way are unmade; the hands that held within them not just blood and bone but the movements that made possible the playing of piano are unmade; the fingers and palms that knew in intricate detail the weight and feel of a particular tool are unmade; the feet that had within them 'by heart' (that is, as a matter of deep bodily habit) the knowledge of how to pedal a bicycle are unmade; the head and arms and back and legs that contained within them an elaborate sequence of steps in a certain dance are unmade; all are deconstructed along with the tissue itself, the sentient source and site of all learning. (Scarry 1985: 123)

The horror of slavery was that it was an act of active and systematic deconstruction. To be sure, many bodies were destroyed in the process. However, unlike the Nazi atrocities where the aim was the total extermination of a so-called inferior race, the slave system needed the slaves to work. Although a labour system characterized the Jewish Holocaust, destruction was the ultimate aim. Slavery, on the other hand, was first and foremost a system of labour. The slavers did not intend the slaves to die; they were concerned to ensure the survival of the black body – albeit in a demoralized form. The black female body is a useful body because it is both a labouring, sexual and reproducing body and therefore it was necessary to preserve the health of the enslaved woman. The use of violence was therefore necessary to break them in, to fragment them, to destabilize them and to make them cease to be subjects, to transform them into 'docile bodies' (Foucault 1977) that became bodies that labour.

Clearly, the purpose of torture during slavery was not to destroy, but to deconstruct the world of the body in pain. Torture is an imitation of death 'a sensory equivalent, substituting prolonged mock execution for execution' (Scarry 1985: 27), an externalized violation of the body and psyche. To destroy the body in pain would have been tantamount to economic and ideological suicide. For how could the slave system perpetuate itself if the enslaved population was destroyed? The violent subjection of the slaves was a way of transforming their bodies into an entity that could produce and reproduce the property necessary for accumulating wealth. Thus, the enslaver/victimizer 'needs the victim to create truth, objectifying fantasy in the discourse of the other' (Taussig 1987: 8). Destruction would have hampered a European expansionist programme. Therefore, to have deconstructed all traces of civilization, humanity and freedom encapsulated within the bodies of the enslaved assured both their subjection and the enslaver's subjectivity. Thus, Mary Prince was forced to bear witness to her worldlessness – the disintegration of her world, language, nation, voice, body – while being relegated to corporeality.

Imprisoned thus, Prince is able only to experience herself in the extremities of her body; the body literally and metaphorically caved-in, turned-back-onto-itself. There is no liberation from the body. In one episode she described the enormity of her pain as overwhelming her until she wished for death. This desire becomes pronounced during an event when her mistress noticed that she had broken an already cracked jar, and ordered her to strip naked. She whipped Mary's bare flesh with the cow-skin until her flesh gaped with blood: 'I lay there till morning, careless of what might happen, for life was very weak in me, and I wished more than ever to die' (1987: 59). The presence of such prolonged suffering according to Scarry, is what causes the person in pain to experience her world as no world and her body as the source of pain, deconstructing her world and notions of selfhood in the process:

> It is intense pain that destroys a person's self and world, a destruction experienced spatially as either the contraction of the universe down to the immediate vicinity of the body or as the body swelling to fill the entire universe. (Scarry 1985: 35)

For Scarry then, the infliction of extreme physical violence is an attack, a destruction of subjectivity. This of course implies that the outcome of physical torment is the destruction of a pre-existing subject (Sa'ez 1992). This is clearly not the case for the slaves. Their enslavement meant that all right to humanity and subjectivity was stripped at the moment of capture. According to Patterson, a slave had no claim to her person, no right to citizenship; she is the property of her master or mistress. Patterson quotes Henri Wallon:

> The slave was a dominated thing, an animated instrument, a body with natural movements, but without its own reason, an existence entirely absorbed in another. The proprietor of this thing, the mover of this instrument, the soul and the reason of this body, the source of this life was the master. The master was everything for him: his father and his god, which is to say, his authority and his duty. (Quoted in Patterson 1982: 4)

The slave then ceases to be subject, all claims to personhood having been stripped at the time of capture, and she had no social existence: 'The slave was the ultimate human tool, as imprintable and as disposable as the master wished' (Patterson 1982: 7). Thus the use of torture on the body of the enslaved suggests not a deconstruction of a preexisting subject, but a radical desubjectification of a desubjectified subject. And in so far as the individual derives its subjectivity from the collective body, the absence of the apparatus that makes subjectivity possible means that the collective body also had to undergo a process of desubjectification.

DECONSTRUCTING GENDER, PRODUCING PROPERTY

If Scarry's thesis suggests that the infliction of extreme physical violence deconstructs and unmakes the world of the person in pain, Hortense Spillers's essay 'Mama's baby, papa's maybe: an American grammar book', provides an account of the way in which this deconstruction of the subject also entailed a deconstruction of gendered categories. Spillers contends that the transformation of the African captive into a commodified property deconstructed the African kinship structure, flesh and gendered subjectivity. As she sees it, in 'severing the captive body from its motive will, its active desire ... we lose at least gender difference in the outcome, and the female body and the male body become a territory of cultural and political maneuver, not at all gender-related, gender-specific' (1987: 67). Spillers notes that although the spatial organization of the slaver's hull revealed 'the application of the gender rule' (1987: 67), this was based on the logic of commodity, rather than on a conception of subjectivity constituted via gender difference. Using a Lacanian model, she argues that in so far as the captives were 'literally suspended in the "oceanic"' they cannot assume a gendered position because 'gendering' 'takes place within the confines of the domestic, an essential metaphor that then spreads its tentacles for male and female subject over a wider ground of human and social purposes' (1987: 72). Spillers's account of 'ungendering' is grounded within a legal discourse that marked the enslaved as non-human and consequently denied the rights accruing to the body that is considered to be human. Therefore, contrary to the arguments often mounted in second-wave feminism, which thought that gender deconstruction would lead to an androgynous utopia, Spillers's account suggests this is clearly not the case. In fact gender deconstruction results in a stripping away of any claims to personhood.

According to Spillers, in the slavers' hull enslaved women are prohibited from participating in the reproduction of mothering, which within 'this historical instance carries few of the benefits of a patriarchalized female gender, which, from one point of view, is the only female gender there is' (1987: 73). Therefore, the reproduction of mothering has radically different meanings for free white women and enslaved black women. For white women reproduction enables them to define themselves as human subjects since they are able to birth the next generation of the human subject even though they are excluded from full participation in the public realm of citizenship. For the enslaved woman, constituted as property, her reproductive capacity did not free her, in fact it reinstated her role as property. In this instance reproduction is not a reproduction of mothering but of property, because she transmits her unfreedom to her offspring. Spillers points to the way in which for the white enslavers descent is passed through the father's lineage, who then has ownership and control over his children and wife. But they were not regarded as non-human. For the enslaved descent was recognized through the mother and her children inherited their status from their mother. This means, according to Spillers, that the African male captive cannot participate in the social realm of the Law of the Father. Thus, gendered identity for the enslaved carries a double paternity. As Spillers writes, 'under the conditions of captivity, the offspring of the female does not "belong" to the Mother, nor is s/he "related" to the "owner", [but] "possesses" it, ... often fathered it, and, as often, without whatever benefit of patrimony' (1987: 74).

The destruction of kinship structure Spillers sees as a destruction of maternal rights and consequently a deconstruction of gendered subjectivity. The 'ungendering' process reduced the enslaved subject into a productive capital which denies the right to the claims subjectivity and humanity entails: citizenship, gender, name, language, family, marriage and rights to property: 'In this absence from a subject position, the captured sexualities provide a physical and biological expression of "otherness"' (1987: 67).

RECONSTRUCTING THE FLESH

The fact that my focus on the body so far has fixed the experience of the body as a site of extreme physical violence for the purpose of utility is not to suggest that the body was always experienced in that manner, or that the captive body always relates to him/herself and the captive community solely as property. Rather, I am simply trying to capture the experience of the body from a single perspective in a given moment in its history. The body, as it were, is not what it is and it is not yet what it will become. Even though history has been terribly unkind to the African body, the body was and still is capable of being something quite beautiful, quite sensuous, quite joyous. There is always a memory of the 'flesh', of the flesh that was once liberated. So, by way of summary, I want to return to the distinction Spillers makes between the flesh and the body to suggest an emancipatory reading of the body.

For Spillers the 'flesh' that is transformed to body as property is never totally wiped out in this transformation. Rather, it is hidden from the violation of the body. The flesh makes itself known to the body and in this visibility the captive male and female are carried to the frontiers of survival (Spillers 1987: 67). This transformative return reunites the African captives to their ancestral body; it retrieves, recovers the memory of the body's capacity for resistance, for transformation, for healing. It is this somatic retrieval and recollection that facilitates the creation of what George Lipsitz calls 'counter-memory' (1991). Counter-memory enabled the slaves and their descendants to construct a different kind of history, a different kind of knowledge, a different kind of body that is outside the control of the dominant history and knowledge production. The body's return to the flesh is a central site for the production of that counter-memory. We see this in the expressive cultures of the people of the African diaspora. In her essay 'The Site of Memory', Toni Morrison expresses the way memories for diasporean Africans are stored in artifacts, stories and bodies:

> You know, they straightened out the Mississippi River in places, to make room for houses and livable acreage. Occasionally the river floods these places. 'Floods' is the word they use, but in fact it is not flooding; it is remembering. Remembering where it used to be. All water has a perfect

> memory and is forever trying to get back where it was. Writers are like that: remembering where we were, what valley we ran through, what the banks were like, the light that was there and the route back to our original place. It is emotional memory – where the nerves and the skin remember how it appeared. And a rush of imagination is our 'flooding'. (Morrison 1990)

The terrorized body remembers the stories of the flesh and makes every effort to trace its step back to the feel of the flesh, the fecundity, the freedom and the dance of the flesh.

NOTE

I would like to thank Ekow Essuman, Jan Shinebourne and Bunmi Daramola for their stimulating and spirited dialogues, David Dibosa for bringing Toni Morrison's essay to my attention, and special thanks to Terry Lovell and Christine Battersby for their helpful critique.

REFERENCES

Bordo, S. (1990) 'Reading the slender body' in Mary Jacobus, Evelyn Fox Keller and Sally Shuttleworth (eds), *Body/Politics, Women and the Discourses of Science*, New York: Routledge.

Butler, J. (1993) *Bodies that Matter: On the Discursive Limits of Sex*, New York: Routledge.

Carter, A. (1979) *The Sadeian Woman*, London: Virago.

Fanon, F. (1992) 'The fact of blackness' in James Donald and Ali Rattansi (eds), *'Race', Culture and Difference*, London: Sage.

Featherstone, M. and M. Hepworth (eds) (1991) 'The mask of ageing and the post-modern life course.' in M. Featherstone, M. Hepworth and B. Turner (eds), *The Body: Social Process and Cultural Theory*, London: Sage.

Foucault, Michel (1977) *Discipline and Punish: The Birth of the Prison*, London: Allen Lane.

Foucault, Michel (1991) *The Foucault Reader: An Introduction to Foucault's Thought*, ed. P. Rabinow, London: Penguin.

Gatens, Moira (1996) *Imaginary Bodies: Ethics, Power and Corporeality*, London and New York: Routledge.

Gilman, S. (1985) *Difference and Pathology: Stereotypes of Sexuality, Race and Madness*, New York: Cornell University Press.

Gilroy, P. (1993) *The Black Atlantic: Modernity and Double Consciousness*, London: Verso.

Grosz, E. (1995) *Volatile Bodies: Toward a Corporeal Feminism*, Bloomington and Indianapolis: Indiana University Press.

Jordan, W. D. (1982) 'First impressions: initial English confrontations with Africans' in C. Husband (ed.), *'Race' in Britain*, London: Hutchinson.

Lipsitz, G. (1991) *Time Passages*, Minneapolis: University of Minnesota Press.

Morrison, Toni (1990) 'Site of memory' in Russell Ferguson et al. (eds), *Out There: Marginalization and Contemporary Culture*, Cambridge, MA: MIT Press.

Patterson, O. (1982) *Slavery and Social Death: A Comparative Study*, Cambridge, MA: Harvard University Press.

Prince, Mary (1987) *The History of Mary Prince: A West Indian Slave. Related by Herself*, ed. Moira Ferguson. Michigan: University of Michigan Press,

Sa'ez, N. (1992) 'Torture: a discourse on practice' in F. E. Mascia-Lees and P. Sharpe (eds), *Tattoo, Torture, Mutilation and Adornment: The Denaturalization of the Body in Culture and Text*, Albany, NY: State University of New York Press.

Sapphire (1994) *American Dreams*, London: Serpent's Tail.
Scarry, E. (1985) *The Body in Pain: The Making and Unmaking of the World*, New York: Oxford University Press.
Schiebinger, L. (1994) *Nature's Body: Sexual Politics and the Making of Modern Science*, London: Pandora.
Spelman, E. (1990) *Inessential Woman: Problems of Exclusion in Feminist Thought*, London: Women's Press.
Spillers, H. (1987) 'Mama's baby, papa's maybe: an American grammar book', *Diacritics*, vol. 17, no. 2, Summer.
Taussig, M. (1987) *Shamanism, Colonialism and The Wildmen: A Study in Terror and Healing*, Chicago: University of Chicago Press.
Turner, B. (1984) *The Body and Society: Explorations in Social Theory*, London: Sage.

5.6

FEMINISM, DISABILITY, AND THE TRANSCENDENCE OF THE BODY

Susan Wendell

Western feminist attention to women's bodily differences from men began with arguments that, contrary to long scientific and popular traditions, these differences do not by themselves determine women's social and psychological gender (or the more limited 'sex roles' we used to talk about). These arguments still go on, especially among biologists, anthropologists, and psychologists; understandably, they have little or nothing to say about bodily suffering. But the view that gender is not biologically determined has taken a much more radical turn in feminist poststructuralist and postmodernist criticism, where the symbolic and cultural significance of women's bodily differences from men are examined closely. Here 'the body' is often discussed as a cultural construction, and the body or body parts are taken to be symbolic forms in a culture. In this latter development, experience of the body is at best left out of the discussion, and at worst precluded by the theory; here feminist theory itself is alienated from the body. As Carol Bigwood says, 'A body and nature formed solely by social and political significations, discourses, and inscriptions are cultural products, disemboweled of their full existential content. The post-structuralist body . . . is so fluid it can take on almost limitless embodiments. It has no real terrestrial *weight*' (Bigwood 1991: 59). A body experienced has both limitations and weight.[1]

I was particularly struck by the alienation from bodily experience of some recent forms of feminist theorizing about the body when I read Donna Haraway's exciting and witty essay, 'A Manifesto for Cyborgs' (Haraway

From: S. Wendell, *The Rejected Body*, New York: Routledge, 1996.

1990).[2] The view she presents there, of the body as cultural and technological construct, seems to preclude the sort of experience I have had. When I became ill, I felt taken over and betrayed by a profound bodily vulnerability. I was forced by my body to reconceptualize my relationship to it. This experience was not the result of any change of cultural 'reading' of the body or of technological incursions into the body. I was infected by a virus, with debilitating physical and psychological consequences. Of course, my illness occurred in a social and cultural context, which profoundly affected my experience of it, but a major aspect of my experience was precisely that of being forced to acknowledge and learn to live with *bodily*, not cultural, limitation. In its radical movement away from the view that every facet of women's lives is determined by biology, feminist theory is in danger of idealizing 'the body' and erasing much of the reality of lived bodies.[3] As Susan Bordo says: 'The deconstructionist erasure of the body is not effected, as in the Cartesian version, by a trip to "nowhere", but in a resistance to the recognition that one is always *somewhere*, and limited' (Bordo 1990: 145).

Feminism's continuing efforts on behalf of increasing women's control of our bodies and preventing unnecessary suffering tend to make us think of bodily suffering as a socially curable phenomenon. Moreover, its focus on alienation from the body and women's bodily differences from men has created in feminist theory an unrealistic picture of our relationship to our bodies. On the one hand, there is the implicit belief that, if we can only create social justice and overcome our cultural alienation from the body, our experience of it will be mostly pleasant and rewarding. On the other hand, there is a concept of the body which is limited only by the imagination and ignores bodily experience altogether. In neither case does feminist thought confront the experience of bodily suffering. One important consequence is that feminist theory has not taken account of a very strong reason for wanting to transcend the body. Unless we do take account of it, I suspect that we may not only underestimate the subjective appeal of mind-body dualism but also fail to offer an adequate alternative conception of the relationship of consciousness to the body.

THE SUFFERING AND LIMITED BODY

In *The Absent Body*, philosophical phenomenologist Drew Leder argues that the Western tradition of mind–body dualism and devaluation of the body is encouraged and supported by the phenomenology of bodily experience. He describes how the body tends to be absent to consciousness except in times of suffering, disruption, rapid change (as in puberty or pregnancy), or the acquisition of new skills (Leder 1990: 92). Our experiences of bodily absence, he says, 'seem to support the doctrine of an immaterial mind trapped inside an alien body' (1990: 3). Leder does not like or subscribe to Cartesian dualism and claims that it contributes to the oppression of women, animals, nature, and other 'Others'. However, he argues that we must reclaim the experiential truths that have lent it support even as we 'break its conceptual hegemony' (1990: 3).

Other writers comment upon the phenomenon that the body seems to come to conscious awareness with the onset of illness or disability. Robert Murphy said that 'illness negates this lack of awareness of the body in guiding our thoughts and actions. The body no longer can be taken for granted, implicit and axiomatic, for it has become a problem' (Murphy 1990: 12). May Sarton, in her journal *After the Stroke*, writes, 'Youth, it occurs to me, has to do with not being aware of one's body, whereas old age is often a matter of consciously *overcoming* some misery or other inside the body. One is acutely aware of it' (quoted in Woodward 1991: 19).

At the very least, we must recognize that awareness of the body is often awareness of pain, discomfort, or physical difficulty. Since people with disabilities collectively have a great deal of knowledge about these aspects of bodily experience, they should be major contributors to our cultural understanding of the body. I propose to demonstrate this, in a modest way, by discussing some interesting aspects of pain and some of the effects that bodily suffering has on our desire to identify with our bodies. I hope to open a new feminist discussion of transcendence of the body, one that will eventually take full account of the phenomenology of bodily suffering.

PAIN

Virtually everyone has some experience of physical pain. Drew Leder gives a good phenomenological account of acute or nonchronic pain. He points out that our experience of it is episodic, that it always demands our attention, that it constricts our perception of space to the body and of time to the here-and-now, that the goal of getting rid of it becomes the focus of our intentions and actions, that it often renders us alone psychologically by cutting us off from other people's reality, and that it causes some degree of alienation of the self from the painful body (Leder 1990: 70–9). All this seems true to me. Nevertheless, I believe our understanding of pain can be greatly enriched by experiences of chronic pain. By chronic pain I mean pain that is not endured for some purpose or goal (unlike the pain of intense athletic training, for instance), pain that promises to go on indefinitely (although sometimes intermittently and sometimes unpredictably), pain that demands no action because as far as we know, no action can get rid of it.

From my own and other people's experiences of chronic pain, I have learned that pain is an interpreted experience. By this I mean not only that we interpret the experience of pain to mean this or that (we do, as Leder points out, and I shall discuss the meaning of pain later), but also that the experience of pain itself is sometimes and in part a product of the interpretation of sensations. For example, it is a fascinating paradox that a major aspect of the painfulness of pain, or I might say the suffering caused by pain, is the desire to get rid of it, to escape from it, to make it stop. A cultivated attitude of acceptance toward it, giving in to it, or just watching/observing it as an experience like others, can reduce the suffering it usually causes. People with chronic pain sometimes describe this as making friends with their pain; I suspect they have achieved a degree of acceptance that still eludes me, but I think I know what they mean. (See, for example, Albert Kreinheder's description of his relationship to the severe pain of rheumatoid arthritis in Kreinheder 1991, chapter 6.)

I want to make it clear before I continue that my descriptions of living with chronic pain do not apply to everyone and are certainly not prescriptions for anyone else. Living with pain is a very complex and individual negotiation; successful strategies depend on such factors as how intense the pain is, where it is in the body (for instance, I find pain in my head or my abdomen much more demanding than pain in my back, arms, or legs), how much energy a person has, whether her/his energy and attention are drained into worries about money, family, medical treatment, or other things, what kind of work s/he does, whether her/his physicians and friends encourage and help her/him, how much pleasure s/he has, what s/he feels passionate about, and many other factors. (For a sample of strategies, see Register 1987.) In other words, it is important to remember that pain occurs in a complex physical, psychological, and social context that forms and transforms our experience of it.

For me, pain is no longer the phenomenon described by Leder. I have found that when focused upon and accepted without resistance, it is often transformed into something I would not describe as pain or even discomfort. For example, my disease causes virtually constant aching in the muscles of my arms, upper chest, and upper back. I know this, because any time I turn my attention to those parts of my body, I experience pain; I think of this pain as similar to a radio that is always playing, but whose volume varies a great deal. When the volume is low, or when I am doing something that absorbs my attention very fully, I can ignore it; but when the volume is turned up high, it demands my attention, and I cannot ignore it for long. If I focus my attention fully on the pain, in which case I must stop doing everything else, I am usually able to relax 'into it', which is a state of mind difficult to describe except by saying that I concentrate on remaining aware of the pain and not resisting it. Then the experience of being in pain is transformed into something else – sometimes a mental image, sometimes a train of thought, sometimes an emotion, sometimes a desire to do something, such as lying down or getting warmer, sometimes sleep. Perhaps if I remained focused upon it in this way, I would rarely suffer from pain, but I do not want to devote much conscious attention to this process. Other things interest me more, and this, for me, is the problem of pain.

I must balance the frequency of attention to how my body feels that is required by the constant presence of pain with whatever attention is required by something else I am doing. It surprised me to find that I could learn to do this, and that I got better at it with practice. (Of course, it requires structuring my life so that I can rest and withdraw my attention into my body much more than healthy people my age normally do.) But the most surprising thing about it is that my ability to think, my attitudes, and feelings seem to me less, not more, dependent on the state of my body than they were before I became ill.

Thus, before I had ME, I would never have considered setting to work at a difficult piece of writing if I woke up feeling quite sick, not only because I knew that I should rest in order to recover, but because I thought I could not possibly write well, or even think well, unless my body felt fairly good. Now I do it often, not because I 'have to', but because I know how to do it and I want to. This outcome is the opposite of my expectation that paying much more attention to my bodily experience would make every aspect of myself more dependent on its fluctuating states. In a sense I discovered that experiences of the body can teach consciousness a certain freedom from the sufferings and limitations of the body. I shall return to this subject later, after discussing some strategies of disengagement from the body.

SOME STRATEGIES OF DISENGAGEMENT

Attempting to transcend or disengage oneself from the body by ignoring or discounting its needs and sensations is generally a luxury of the healthy and able-bodied. For people who are ill or disabled, a fairly high degree of attention to the body is necessary for survival, or at least for preventing significant (and sometimes irreversible) deterioration of their physical conditions. Yet illness and disability often render bodily experiences whose meanings we once took for granted difficult to interpret, and even deceptive. Barbara Rosenblum described how a 'crisis of meaning' was created by the radical unpredictability of her body with cancer:

> In our culture it is very common to rely on the body as the ultimate arbiter of truth. . . . By noticing the body's responses to situations, we have an idea of how we 'really feel about things'. For example, if you get knots in your stomach every time a certain person walks into the room, you have an important body clue to investigate. . . . Interpretations of bodily signals are premised on the uninterrupted stability and continuity of the body. . . . When the body, like my body, is no longer consistent over time . . . when something that meant one thing in April may have an entirely different meaning in May, then it is hard to rely on the stability – and therefore the truth – of the body. (Butler and Rosenblum 1991: 136–7)

Chronic pain creates a similar (but more limited) crisis of meaning, since, to a healthy person, pain means that something is wrong that should be acted upon. With chronic pain, I must remind myself over and over again that the pain is meaningless, that there is nothing to fear or resist, that resistance only creates tension, which makes it worse. When I simply notice and accept the pain, my mind is often freed to pay attention to something else. This is not the same as ignoring my body, which would be dangerous, since not resting when I need to rest can cause extreme symptoms or a relapse into illness that would require several days' bed rest. I think of it as a reinterpretation of bodily sensations so as not to be overwhelmed or victimized by them. This process has affected profoundly my whole relationship to my body, since fatigue, nausea, dizziness, lack of appetite, and even depression are all caused by my disease from time to time, and thus all have changed their meanings. It is usually, though not always, inappropriate now to interpret them as indications of my relationship to the external world or of the need to take action. Unfortunately, it is often much easier to recognize that something is inappropriate than to refrain from doing it.

For this reason, I have found it important to cultivate an 'observer's' attitude to many bodily sensations and even depressive moods caused by my illness. With this attitude, I observe what is happening as a phenomenon, attend to it, tolerate the cognitive dissonance that results from, for example, feeling depressed or nauseated when there is nothing obviously depressing or disgusting going on, accommodate to it as best I can, and wait for it to pass. This is very different from the reactions that come most easily to me, which have to do with finding the causes of these feelings and acting on them. I find it hardest to adopt an observer's attitude toward depression, since although in the past I had brief illnesses that caused the other symptoms, I had never experienced severe depression without something to be depressed about. Thus, my first, easiest response to depression was to search my life for something that might be depressing me. Since my world (like virtually everyone's) is full of things that, if focused upon, might cause depression, I increased and prolonged my depressions with this habitual response. Learning to regard severe depression (by this I mean, not the lows of everyday living, but the sorts of feelings that make you wish you were dead) as a physical phenomenon to be endured until it is over and not taken seriously has greatly reduced my suffering from it and may have saved my life. Register describes a similar strategy used by a man who suffers from recurring depressive illness (Register 1987: 280).

In general, being able to say (usually to myself): 'My body is painful (or nauseated, exhausted, etc.), but I'm happy' can be very encouraging and lift my spirits, because it asserts that the way my body feels is not the totality of my experience, that my mind and feelings can wander beyond the painful messages of my body, and that my state of mind is not completely dependent on the state of my body. Even being able to say, 'My brain is badly affected right now, so I'm depressed, but I'm fine and my life is going well', is a way of asserting that the quality of my life is not completely dependent on the state of my body, that projects can still be imagined and accomplished, and that the present is not all there is. In short, I am learning not to identify myself with my body, and this helps me to live a good life with a debilitating chronic illness.

I know that many people will suspect this attitude of being psychologically or spiritually naive. They will insist that the sufferings of the body have psychological and/or spiritual meanings, and that I should be searching for them in order to heal myself (Wilber and Wilber 1988). This is a widespread belief, not only in North America but in many parts of the world, and I have discussed some of its consequences for people with disabilities and/or life-threatening illnesses earlier in this book. I do not reject it entirely. I too believe that, if my stomach tightens every time a particular person enters the room, it is an important sign of how I feel about her/him, and I may feel better physically if I avoid or change my relationship to that person. But, having experienced a crisis of meaning in my body, I can no longer assume that even powerful bodily experiences are psychologically or spiritually meaningful. To do so seems to me to give the body too little importance as a cause in psychological and spiritual life. It reduces the body to a mere reflector of other processes and implicitly rejects the idea that the body may have a complex life of its own, much of which we cannot interpret.

When I look back on the beginning of my illness, I still think of it, as I did then, as an involuntary violation of my body. But now I feel that such violations are sometimes the beginnings of a better life, in that they force the self to expand or be destroyed. Illness has forced me to change in ways that I am grateful for, and so, although I would joyfully accept a cure if it were offered me, I do not regret having become ill. Yet I do not believe that I became ill *because* I needed to learn what illness has taught me, nor that I will get well when I have learned everything I need to know from it. We learn from many things that do not happen to us because we need to learn from them (to regard the death of a loved one, for example, as primarily a lesson for oneself, is hideously narcissistic), and many people who could benefit from learning the same things never have the experiences that would teach them.

When I began to accept and give in to my symptoms, when I stopped searching for medical, psychological, or spiritual cures, when I began to develop the ability to observe my symptoms and reduced my identification with the transient miseries of my body, I was able to reconstruct my life. The state of my body limited the possibilities in new ways, but it also presented new kinds of understanding, new interests, new passions, and projects. In this sense, my experience of illness has been profoundly meaningful, but only because I accepted my body as a cause. If I had insisted on seeing it primarily as reflecting psychological or spiritual problems and devoted my energy to uncovering the 'meanings' of my symptoms, I would still be completely absorbed in being ill. As it is, my body has led me to a changed identity, to a very different sense of myself, even as I have come to identify myself less with what is occurring in my body.

People with disabilities often describe advantages of not identifying the self with the body. For those who are ill, the difficulty of living moment to moment with unpredictable, debilitating symptoms can be alleviated by having a strong sense of self that negotiates its ability to carry out its projects with the sick body (Register 1987, chapter 9). This sense of self and its projects provides continuity in lives that would be chaotic if those who led them were highly identified with their bodies. The anthropologist Robert Murphy, who was quadriplegic and studied the lives of people with paralysis, described another motive for disembodying the self: 'The paralytic becomes accustomed to being lifted, rolled, pushed, pulled, and twisted, and he survives this treatment by putting emotional distance between himself and his body' (Murphy 1990: 100–1).

In addition, people with disabilities often express a strong desire not to be identified with their bodily weaknesses, inabilities, or illnesses. This is why the phrase 'people with disabilities' has come to be preferred over 'disabled people'. When the world sees a whole person as disabled, the person's abilities are overlooked or discounted. It is easy to slip into believing other people's perceptions of oneself, and this can take a great toll on the self-esteem of a person with a disability. Those people with disabilities who still have impressive and reliable physical abilities can counteract people's misperceptions by asserting those abilities. For those of us whose remaining physical abilities are unimpressive or unreliable, not to identify ourselves with our bodies may be the best defense. It is good psychological strategy to base our sense of ourselves, and therefore our self-esteem, on our intellectual and/or emotional experiences, activities, and connections to others.

[. . .]

I do not want to give an exaggerated impression of the degree to which people with disabilities rely upon strategies of disembodiment. For all the advantages that some degree of disembodying the self may have in coping with illness or disability, the process of coming to identify with a sick or disabled body can play an important part in adjusting to it. For many of us who became ill or disabled as adults, reconstructing our lives depended upon forging a new identity. An important aspect of this process is what Register calls 'acceptance: ability to regard the illness [*or, I would add, disability*] as your normal state of being' (Register 1987: 31). This could also be described as learning to identify with a new body, as well as, for most of us, a new social role. For me, this had many advantages: I stopped expecting to recover and postponing my life until I was well, I sought help and invented strategies for living with my sick body, I changed my projects and my working life to accommodate my physical limitations, and, perhaps most important, I began to identify with other people with disabilities and to learn from them. Thus, I do identify with my sick body to a significant degree, but I also believe that my thoughts and feelings are more independent from my experiences of it than they ever were from my experiences of my well body.

TRANSCENDENCE

What has all this to do with transcendence of the body? That, of course, depends on what we have in mind when we speak of transcendence. The forms of independence from the body's sufferings that I have described are partial and mundane. They are strategies of daily living, not grand spiritual victories. Some people might even regard them as forms of alienation from physical experience. I think that would be a mistake. Alienation, as we usually understand it, reduces freedom, because it constricts the possibilities of experience. If we spoke of being alienated from suffering, I think we would mean being unable to face up to and undergo some necessary, perhaps purposeful, pain. To choose to exercise some habits of mind that distance oneself from chronic, often meaningless physical suffering increases freedom, because it expands the possibilities of experience beyond the miseries and limitations of the body.

It is because they increase the freedom of consciousness that I am drawn to calling these strategies forms of transcendence. It is because we are led to adopt them by the body's pain, discomfort, or difficulty, and because they are ways of interpreting and dealing with bodily experience, that I call them transcendence of the body. I do not think that we need to subscribe to some kind of mind-body dualism to recognize that there are degrees to which consciousness and the sense of self may be tied to bodily sensations and limitations, or to see the value of practices, available to some people in some circumstances, that loosen the connection. Nor do I think we need to devalue the body or bodily experience to value the ability to gain some emotional and cognitive distance from them. On the contrary, to devalue the body for this reason would be foolish, since it is bodily changes and conditions that lead us to discover these strategies. The onset of illness, disability, or pain destroys the 'absence' of the body to consciousness, described by Leder and others, and forces us to find conscious responses to new, often acute, awareness of our bodies. Thus, the body itself takes us into and then beyond its sufferings and limitations.

As an alternative to the traditional theological concept of transcendence, Naomi Goldenberg calls for a new concept of transcendence 'with body', which would involve feeling and knowing our connection to other lives, human history, and society (Goldenberg 1990: 211–12). Drew Leder offers an understanding of transcendence that rejects Cartesian dualism and distrust of the body in favour of the realization that the lived body is transpersonal, that we form one body with the world, and that we can experience this one-body relation in compassion, aesthetic absorption, and spiritual communion (Leder 1990, chapter 6). I like both of their conceptions, but I also think they are both talking about transcendence of the *ego*, which they see as an ideal to replace transcendence of the body. The ability at least sometimes to transcend the demands of the ego seems to me central to spiritual life and probably to human happiness. So perhaps it is a more important form of transcendence than transcendence of the body. Nevertheless, I suspect that the idea of transcendence of the body may be too easily rejected in an attempt to throw out mind-body dualism, derogation of the body, and all the sins that have been committed in their names.

By defending some notion of transcendence of the body I do not mean to suggest that strategies of disembodying the self should be adopted by people without disabilities. Instead, I want to demonstrate how important it is to consider the experiences of people with disabilities when theorizing about the relationship of consciousness to the body. One thing is clear: We cannot speak only of reducing our alienation from our bodies, becoming more aware of them, and celebrating their strengths and pleasures; we must also talk about how to live with the suffering body, with that which cannot be noticed without pain, and that which cannot be celebrated without ambivalence. We may find then that there is a place in our discussion of the body for some concept of transcendence.

NOTES

1. In 1984, Adrienne Rich wrote: 'Perhaps we need a moratorium on saying "the body". For it's also possible to abstract "the" body. When I write "the body", I see nothing in particular. To write "my body" plunges me into lived experience, particularity ...' (Rich 1986: 215). Clearly, Rich saw the problems coming, and I like her suggestion, but so much has been written about 'the body' both before and since 1984 that I find I must use the term to discuss this work and to locate my own position in relation to it. I try to use more specific references to bodies when speaking outside the context of the literature on 'the body'.
2. I seem to pick on Donna Haraway's work in this book, despite the fact that I enjoy and learn a great deal from her writings. It is because Haraway is one of the postmodernist feminist theorists who talks most explicitly about the body that her work tends to focus my concerns about the limitations of postmodernist feminist theories for understanding disability.
3. Maxine Sheets-Johnstone also criticizes Haraway and other feminist theorists for talking about the body and 'embodiment' while failing to take the body and bodily experience into account (Sheets-Johnstone 1992: esp. 43–4).

REFERENCES

Bigwood, Carol (1991) 'Renaturalizing the Body (With a Little Help from Merleau-Ponty)' *Hypatia* 6 (3) 54–73.

Bordo, Susan (1990) 'Feminism, Postmodernism, and Gender-Scepticism' in Linda Nicholson (ed.), *Feminism/Postmodernism*, New York: Routledge.

Butler, Sandra and Rosenblum, Barbara (1991) *Cancer in Two Voices*, San Francisco: Spinsters Book Co.

Goldenberg, Naomi R. (1990) *Returning Words to Flesh: Feminism, Psychoanalysis, and the Resurrection of the Body*, Boston: Beacon Press.

Haraway, Donna (1990) 'A Manifesto for Cyborgs' in Linda Nicholson (ed.), *Feminism/Postmodernism*, New York: Routledge.

Kreinheder, Albert (1991) *Body and Soul: The Other Side of Illness*, Toronto: Inner City Books.

Leder, Drew (1990) *The Absent Body*, Chicago: Chicago University Press.

Murphy, Robert F. (1990) *The Body Silent*, New York: W. W. Norton.

Register, Cheri (1987) *Living With Chronic Illness. Days of Patience and Passion*, New York: Bantam.

Rich, Adrienne (1986) 'Notes for a Politics of Location' in *Blood, Bread and Poetry: Selected Prose 1979–1985*, New York: W. W. Norton.

Sheets-Johnstone, Maxine (1992) *Giving the Body Its Due*, Albany: State University of New York Press.

Wilber, Ken and Wilber, Treya (1988) 'Do We Make Ourselves Sick?' *New Age Journal* Sept/Oct: 50–54; 85–90.

Woodward, Kathleen (1991) *Aging and Its Discontents: Freud and Other Fictions*, Bloomington: Indiana University Press.

SECTION 6
BODYSPACEMATTER

INTRODUCTION

This section takes up the contemporary feminist concern with embodiment in the sense of lived experience, in which body and world are mutually constitutive. It investigates the relationship between the materiality of bodies in space, the space of our bodies, and ways in which sexual difference is mapped onto body–space–matter. The central issues it addresses concern not only questions of how sexual hierarchies are organised by the differential allocation and use of space, of how women's bodies resist enclosure in/by space, and of what it is like to inhabit a body, but speak more generally to a concern for our embodied subjectivity.

In mainstream discourse – what is in effect a white, masculinist account – agency is either seen as disembodied, or it is located in undifferentiated and universalised bodies. And putting aside such acceptable contingent circumstances as the state of health, it is assumed that we all have a similar relationship both to the bodies of others and to our own body space, whether that relates to the surface boundaries or more rarely to the interior. Indeed, as Lynda Birke argues (Section 1), the space and matter of the interior has been heavily undertheorised even within feminism, somehow relegated to a realm outside culture, inaccessible to analysis. This is particularly surprising at one level, as it is with reference to our bodily interiors that historically women have been found wanting. The image of floating or unruly wombs and their attendant symptom of hysteria has exercised the male imaginary from classical times right up until the nineteenth century. The concern more generally with the issues of boundaries, which figure differentially both masculine containment and feminine fluidity is taken up in a variety of feminist strategies to rethink embodiment.

That notion of the lack of stasis associated primarily with female bodies is taken up within feminist phenomenology, where what is at stake is not a given materiality detached from questions of selfhood, but the changing nature of lived reality. Most particularly, female bodies change throughout our lives – not just in the common process of 'growing up', but in the specifics of menstruation, menopause, and the radical changes associated with pregnancy. In each, what matters is not just the internal and external change of bodily shape and form, but how that affects our relation to world around us, how we move through space (see Marshall in Section 1). In a very different way, Christine Battersby's piece here also considers the inner and outer space of female embodiment. Like many other theorists she intends to expose the contradictions of the dominant historical models and to give a more adequate account for the embodied experiences of women. Where she differs from initially similar postmodernist accounts is in wanting to hold on to a metaphysics of self/not-self whilst explicitly setting out to take the embodied female as the norm for a model of the self.

In terms of dominant norms, however, sexual difference – like cultural/racial difference – is naturalised to the detriment of women and other others. Among its most enduring affects is a restriction on the spaces that women can legitimately occupy. While some 'natural' spheres, like the home, and private life in general, are seen as appropriate to women, other spheres, like politics or the academy for example, operate more like exclusion zones. Through a feminist critique of time-geography, Gillian Rose raises a series of questions related not only to the the body in space – and to the notion of a supposedly undifferentiated disembodied masculine agency – but also to the gendered nature of different spaces through which bodies move. She analyses how and where space is conceptualised as masculine, and considers how the space we occupy, the ways in which we move through the world, the social roles that we take on, are all differentially experienced and limited for men and women. In short, where men experience relative spatial freedom, women are constrained.

In the same way that sexually different bodies stand in a different relation to space, so too changes or breeches in/of our bodily integrity alter the nature of our relation to the world we occupy and move through. As mentioned above, some transformations and breeches to the body, such as pregnancy, may be the result of choice, but those occurring by force, or as the result of threat, speak to very different changes in relationality. Andrea Benton Rushing explores such issues in her piece 'Surviving Rape'. What are raised here are questions of spatial embodiment, questions of agency, of the masculinised violence of public space, and of the fragile nature of the private. The masculinisation of the public sphere, and the policing and regulation of such space reminds us too of the historical parallels made between body, in its ideal bounded order, and the social order or state. The notion of the body politic is a longstanding one in which the organisation of the good state is likened to the rational control of the body by the central power of the mind. Such parallelism is disrupted, however, by Liz Grosz who offers a different take on the relationship between bodies and spaces through her examination of bodies-cities. She sees not a formal link between the physical body and a particular social order, but an interface which results in temporary groupings and alignments. Bodies do not simply move through spaces, geographies, and cities but constitute and are constituted by them, at both a local and global level.

In contrast to the Western notion of the body politic, the links between bodies and the nation are made in quite different ways in some colonial and postcolonial discourses where it is the feminine form that is at stake. Price and Shildrick, for example, look at the authorised discourses of western missionaries to theorise how the mapping of both the bodies of Indian women, and of the state, worked symbiotically to establish the grounds for the instigation and extension of colonial rule. Rather than focusing on the broad structures of the body politic, the piece maps the regulatory mechanisms that were directed at a local level against the bodies of Indian women, and raises questions about the role of female bodies in what has traditionally been seen as a process dependent upon the conquest of one group of male bodies by another. And just as the colonial powers made use of the female body, so too the female figure has become iconic in the struggle for nationhood, in what Ramaswamy (1998) calls 'the somatics of nationalism'. For example, the hybrid goddess, Bhārata Mātā, took on the form of 'Mother India' in the late nineteenth century, and has re-emerged as an appropriated figure of right-wing Hindu nationalism today. In her extract, Nalini Natarajan takes on the questions of postcolonial body/space in her reading of the novel, *Midnight's Children* (Rushdie 1980). She highlights the differing ways in which women's bodies have served as symbols of resistance to colonialism, as 'a site for testing out modernity', as figures of mythic unity in the face of fragmentation, and for countering the threat of Westernisation. As with all the readings, the body in question instantiates matter and inhabits space in a highly complex way.

REFERENCES AND FURTHER READING

Ainley, Rosa (1998) *New Frontiers of Space, Bodies and Gender*, New York: Routledge.

Anzaldúa, Gloria (1987) *Borderlands/La Frontera*, San Francisco: Aunt Lute.

Biddick, Kathleen (1993) 'Genders, Bodies, Borders: Technologies of the Visible' in *Speculum. A Journal of Medieval Studies* 68 (2): 389–418.

Butalia, Urvashi (1995) 'Muslims and Hindus, Men and Women: Communal Stereotypes and the Partition of India' in Tanika Sarkar and Urvashi Butalia (eds), *Women and the Hindu Right*, New Delhi: Kali for Women.

Gatens, Moira (1991) 'Corporeal Representations in/and the Body Politic' in Rosalyn Diprose and Robyn Ferrell (eds), *Cartographies. Poststructuralism and the Mapping of Bodies and Spaces*, Sydney: Allen and Unwin.

Grosz, Elizabeth (1995) *Space, Time and Perversion. Essays on the Politics of Bodies*, London: Routledge.

Irigaray, Luce (1985) 'Volume-Fluidity' in *Speculum of the Other Woman*, trans. Gillian C. Gill, Ithaca, NY: Cornell University Press.

Probyn, Elspeth (1988) 'The Anorexic Body' in A. Kroker and M. Kroker (eds), *Body Invaders: Sexuality and the Postmodern Condition*, Toronto: Macmillan.
Ramaswamy, Sumathi (1998) 'Body Language: The Somatics of Nationalism in Tamil Nadu' in *Gender and History*, 10 (1): 78–109.
Rushdie, Salman (1980) *Midnight's Children*, New York: Avon.
Young, Iris Marion (1990) 'Breasted Experience' and 'Pregnant Embodiment: Subjectivity and Alienation' in *Throwing Like a Girl and Other Essays in Feminist Philosophy and Social Theory*, Bloomington: Indiana University Press.

6.1

HER BODY/HER BOUNDARIES

Christine Battersby

[...]

THE BODY IN THE MIND

In one of the key texts in cognitive semantics, *The Body in the Mind* (1987), Mark Johnson seeks to base strong metaphysical, epistemological and logical claims on an experience of embodiment that is described as universal. He tells us:

> Our encounter with containment and boundedness is one of the most pervasive features of our bodily experience. We are intimately aware of our bodies as three-dimensional containers into which we put certain things (food, water, air) and out of which other things emerge (food and water wastes, air, blood, etc.). (1987: 21)

Johnson then goes on to spell out five characteristics of the containment relationship from which I infer that, for him, the self is inside the body in much the same way that a body is inside a room or a house. Bodies are containers that protect against and resist external forces, whilst also holding back internal forces from expansion or extrusion. All that is other is on the outside, and the inner self is shielded from the direct gaze of others by skin and other non-transparent features of the body-container (1987: 22).

Mark Johnson develops large claims on the basis of this analysis. Containment is made one of a number of underlying structures of embodiment which shape and constrain the imagination via gestalt-type patterns or 'schemata'

From: C. Battersby, *The Phenomenal Woman*, Cambridge: Polity Press and New York: Routledge.

which operate at a preconceptual level. Johnson adds bodies to an essentially Kantian account of the functioning of the understanding. Difference between humans is registered, but only at a superficial level. 'Meaning' is a product of the way an experience, a theory, a word or a sentence is understood by an individual who is 'embedded in a (linguistic) community, a culture, and a historical context' (1987: 190). However, at the most basic level, this understanding rests on schemata of the imagination which arise out of the (universal) experience of embodiment. Cultural differences act merely as an overlay which affects the way meanings are encoded and transformed; underneath there is human sameness. We all inhabit bodies in similar ways. We all experience the body as a container for the inner self.

These claims are elaborated in another book of cognitive semantics: George Lakoff's *Women, Fire, and Dangerous Things* (1987). Because we live in bodies and move between containing spaces via our bodies, we recognize that 'Everything is either inside a container or out of it' (1987: 272). Our grasp of the law of Boolean logic which determines the relationship of classes – of 'P or not P' – is grounded on our experience of being embodied selves. Lakoff refers to Johnson's 'proof' to establish the links between the law of the excluded middle and the intimate experience of bodily containment. 'On our account, the CONTAINER schema is inherently meaningful to people by virtue of their bodily experience' (1987: 273). Lakoff addresses the problem of whether there might be cultures that fail to fit the cognitivists' model of meaning, but secures the 'universal' status by arguing that even the Australian aboriginal language of Dyirbal – which puts together in a single category human females, water, fire and fighting – obeys the logic of containment, despite the strangeness of its classificatory system (1987: 1, 92ff). Just fitting things into categories relies on the fundamentally human experience of embodiment.

For feminist theorists who have long complained of the neglect of the body by western philosophers, the development of cognitive semantics might seem a promising move. However, as I read Johnson's and Lakoff's accounts of embodiment, I register a shock of strangeness: of wondering what it would be like to inhabit a body *like that*. And that is because I do not experience my body as a three-dimensional container into which I 'put various things' – such as 'food, water, air' – and out of which 'other things emerge (food and water wastes, air, blood, etc.)'.

The shock intensifies as I note that the description purports to be of a preconceptual experience that is so immediate and obvious that it would command intuitive assent. I live/have always lived in the west, and am a full – and I hope sophisticated – user of the English language. So why do I feel so alienated as I read these descriptions of what it is like to inhabit a body? How is it that the cultural imperialism of the cognitivists' model can manage to position aboriginal Australians within the 'universal', whilst managing to make me feel singular: odd; a freak; outside the norm?

My reaction raises a dilemma that constantly confronts feminist philosophers.

By the double use of the definite article in the title of his book – *'The' Body in 'the' Mind* – Mark Johnson gestures at a level of the universal which must be registered by all linguists and philosophers interested in the cognitive aspects of language and metaphor. However, feminist philosophers know that in terms of actual practice philosophers throughout the history of the discipline have taken (and continue to take) male life-patterns, personalities and life-experiences as ideals and/or norms. Thus, writing and reading self-consciously as both a philosopher and a woman, a feminist philosopher always confronts a methodological dilemma. She cannot easily know whether her failure to recognize herself as fitting the philosophical paradigms is due to simple idiosyncrasy; to issues of sexual difference; to historical and cultural factors; or to the fact that the theories were false even for males at the time they were propounded.

In order to think which of these explanations is right in this instance, I will explore five competing hypotheses for my 'failure' to recognize myself in the cognitivists' description of what it is like to inhabit a body. In so doing, I will gradually displace the focus of my analysis away from an experientially based semantics and confront current debates within feminist theory regarding the nature of boundaries. I will end not by demanding their deconstruction in the manner of many postmodern and poststructuralist feminist theorists, but by using the perspective of sexual difference in order to reconstitute the inside/outside, self/other, body/mind divides. In so doing, I will turn to the sciences for new models for thinking identity in ways that will, in effect, undermine the semantic and epistemological approaches to the questions of philosophy that have dominated Anglo-American philosophy this century. I will also be countering those who would seek to model woman's 'otherness' in terms of a Lacanian or Derridean understanding of language. [...] However, here it can be noted that looking to the sciences allows us to theorize a 'real' that is not 'unspeakable' or 'ungraspable', but that nevertheless falls outside the framework of the Lacanian symbolic that looks to the (masculinized) self for its model of identity.

These are the five hypotheses that I will examine for my 'failure' to think of my body as a container:

1. That I am idiosyncratic or peculiar: that there is something the matter with my image schematism, perhaps in the manner of Oliver Sacks's freakish case studies in *The Man Who Mistook His Wife for a Hat* (1985). Sacks's subjects are neurologically peculiar in the way they match bodily and mental imaging to concepts, and thus in terms of the way that they use language.
2. That I do think in the bodily containment schemata that Johnson outlines; but that something (too much philosophical introspection? some crisis or trauma?) has made me repress this awareness.
3. That I, as a woman, have a different relationship with my body than does a man, and that the containment model for bodily boundaries and selves might be more typical of male experience.
4. That I, as a westerner living in the last decades of the twentieth century, no

longer think in terms of the outmoded science that this container model presupposes; that something (exposure to media, science fiction, fiction, philosophy) prevented me ever developing (at least in a way that I can recall) the containment model.

5. That there have in the west always been alternative models for thinking the relationship between self and body and self and not-self; and that for contingent reasons (reading? social environment?) exposure to these alternative models prevented the container model from ever really taking hold in my case.

HYPOTHESIS ONE: IDIOSYNCRASY

Taking these hypotheses in order, I wish first to reject the claim that my response might simply be idiosyncratic. I do not (I think!) manifest the bizarre disturbances of correlation of spatial co-ordination and behaviour and word-usage which characterize Oliver Sacks's neurological case studies. I am not like Sacks's 'Disembodied Lady' who has lost proprioception and with it the ability to feel her body. It is not that I – like Sacks's Christina – am unable to remember or even imagine what embodiment is like (1985: 49–50). As support for my claim that I am not merely a freak, I would appeal to Emily Martin's analysis of the way women talk about themselves as bodies in *The Woman in the Body* (1987).

This book is particularly relevant in this context, since Martin is an anthropologist who applies to western female subjects some of the techniques of metaphor analysis which Lakoff and Johnson adopted in their earlier, co-authored, *Metaphors We Live By* (1980). Emily Martin uses the imagery employed by women to describe their reproductive processes (menstruation, pregnancy, childbirth, menopause) to determine the woman-as-body-image of the subjects she interviews. Arguing that different models are prevalent in the language of middle-class and working-class women, her work demonstrates that women of all classes quite frequently talk about their bodies in ways that suggest extreme fragmentation, with the self located 'outside' the body. Working-class women tend to resist altogether the spatialized model of the body as 'container' of eggs, blood, womb, and so on. Although many of the middle-class women do use the language of bodily containment and inner functioning to describe their reproductive processes, a number of them also go on to note that 'this internal model was not relevant to them' (1987: 106). Martin thus interprets the containment imagery as the educated subjects' (unsuccessful) attempts to make body-image coincide with the 'medical' or 'scientific' renditions of female reproductive processes. Her research thus implicitly undermines any simplistic understanding of bodily containment as a universal schema, as well as any attempt to restrict the containment schema to a homogenized 'western' self.

HYPOTHESIS TWO: REPRESSION

The second hypothesis for explaining my cognitive failure to image my body as a container posits that I do (unconsciously/subconsciously) think in this way;

but the sense of my body as a container is subject to repression. Now this might be true. But it would be hard to know how to test this. I have very clear memories of my childhood relationship with my body: of the way the 'I' was 'zoned' into body regions that were at war with other body regions; but I have no memory at all of thinking of my body of some form of containing, safe space that encloses an 'I'. It is true that I use such linguistic expressions as 'pain in the foot'; but are those who infer from this some reference to spatial or territorial containment right so to do? Not all prepositions referring to bodily happenings are ones of containment. We might now think of the preposition 'in' primarily in terms of the mapping of a location, but cartography was only developed as a science with the invention of compasses, globes and other 'objective' means of recording place towards the end of the fifteenth century. In Latin and old English the term 'in' did not necessarily imply spatial containment or territorial inclusion. And nor does it now, as the *Oxford English Dictionary* shows via some quite bewildering examples of diversity of usage: 'in a blaze', 'in tears', 'in all weather', 'in confidence', 'in crayons', 'in the Lord'.

Anorexics carry on referring to things happening 'in' their bodies, even though a number of researchers have shown that typically they 'grow up confused in their concepts about the body and its functions and deficient in their sense of identity, autonomy, and control'. It seems that in many ways anorexics feel and behave as if 'neither their body nor their actions are self-directed', and as if their bodies are 'not even their own' (Bruch 1978: 39). Anorexics typically describe their 'stomach' as rejecting the food – a process based on horror at food or disgust. Helmut Thomä quotes from one treatment session with a female anorexic:

> Bottle – child – disgust, if I think of it – injections – the idea that there is something flowing into me, into my mouth or into the vagina, is maddening – integer, integra, integrum occurs to me – untouchable – he does not have to bear a child – a man is what he is – he need not receive and he need not give. (Thomä, 1977; quoted Bell 1985:16)

This anorexic woman uses disgust to try to produce body boundaries; but the way that she talks about herself makes it simultaneously clear that she does not view herself as a spatial container closed off from the outside.

In *Powers of Horror* (1980) Julia Kristeva ascribes to the mechanism of disgust a constitutive role in the formation of identity during the process of weaning:

> *nausea* makes me balk at that milk cream, separates me from the mother and father who proffer it. 'I' want none of that element, sign of their desire; 'I' do not want to listen, 'I' do not assimilate it, 'I' expel it. But since food is not an 'other' for 'me', who am only in their desire, I expel *myself*, I spit *myself* out, I abject *myself* within the same motion through which 'I' claim to establish *myself*. (1980: 3)

According to Kristeva, the boundary of the body and the distinction between self and not-self is established through the processes of repulsion which occur at a preconceptual stage and before the infant has clearly demarcated the boundaries between self and 'other', self and mother. Inner space is thus not known intuitively or immediately, but is secured only via an expulsion of things that cannot be embraced within its borders.

In the Kristevan model identity is secured at the level of the imaginary; before the child is inducted into language (and hence in Lacanian terms before entrance into the realm of the father). Disgust operates at the level of the pre-oedipal *féminin*. [...] For Lacan there is no necessary connection between femininity and women. And the same is true for Kristeva also. The words of the anorexic filled with disgust at drinking from a bottle might lead us to question this, however. The female anorexic in the above example – and ninety per cent of all anorexics are female – knows that the normalized female body is permeable; penetrable; with the potential of becoming more-than-one body through a gradual process of growth and the labour of separation. The boundaries of the female anorexic's body – in so far as they are establishable at all – are insecure and thus require careful policing. And this observation thus leads on to a consideration of the third hypothesis for my failure to register my body as a container with a self safe 'within' and the dangerous other on the outside: the claim that this is typical of women.

HYPOTHESIS THREE: FEMALE

There is a good deal of work by psychologists, artists and theorists of opposing political persuasions that shows that females in our society employ space differently – even whilst young – and that this can be detected by observing their games, their bodily movements and the projects that they devise for themselves. The problem, however, is in deciding what this difference means, particularly if what we are asking is whether this shows that women typically lack an awareness of their bodies as containing spaces. Thus, on the basis of research conducted on boys' and girls' games in the 1940s and 1950s, Erik Erikson concluded that girls' bodies destine them towards a preoccupation with inner space that fits them for child-bearing. Supposing that all females must experience their bodies as empty containers until filled by pregnancy, Erikson asserts that 'in female experience an "inner space" is at the centre of despair even as it is the very centre of potential fulfilment. Emptiness is the female form of perdition ...' (1964: 121).

This 'inner space' or 'void' is presented as a 'clinical observation' of what determines women's personalities, potentialities and development. As such it becomes a biological 'fact' that each menstruation is experienced – presumably subconsciously – as a form of 'mourning' over lost potentialities, and that menopause effects a permanent 'scar' on women's psyches (1964: 121). With such false (and implausible) universals in the public domain, it is not surprising that many feminist theorists should have gone to the opposite extreme. Like Marianne Wex (1979), they frequently write as if female space-manipulation is entirely a function of socialisation, and can thus be modified, in ways that allow public – and hence also private – body-space to be reclaimed.[2]

Via anorexia, slimming, cosmetic surgery, the fashion industry and simply the way they stand, move or play, women discipline their body boundaries intensively (Bartky 1990: 63–82). In her classic *Throwing Like a Girl* (1990), Iris Marion Young has ascribed the constricted posture adopted by middle-class, white women in western societies when stretching, reaching, catching, throwing and moving generally to the construction of a spatial field surrounding the female body which is experienced as an enclosure, instead of as a field in which her intentionality can be made manifest. But why this occurs and what it shows about the phenomenological body-experience of women is a matter of dispute. Not only is there the obvious difficulty that women exist in a patriarchal reality and hence use the same language as men to describe the body/mind relationship, but there is also a surprising lack of work on masculine embodiment.

Some recent theorists, such as Elizabeth Grosz (1994: 198ff) and Paul Smith (1988), suggest that the 'repressed' of masculine consciousness might be the sense of the flowing out and away of ejaculated bodily substances. On this model, the boundaries of normal male selves are secured against flowing out, in ways that would make sense of Mark Johnson's account of (male) selves as 'three-dimensional containers' for food, water, air and the like: 'out of which other things emerge (food and water wastes, air, blood, etc.)'. Since the same logic would not apply to the construction of female identity, we could indeed have a hypothesis here that would explain why I as a woman would fail to recognize this as a description of what it is like to inhabit a body. My identity is secure precisely because I do not envisage my body-space as a container in which the self is inside: protected from the other by boundaries which protect against and resist external forces, whilst also holding back internal forces from expansion. I construct a containing space around me, precisely because my body itself is not constructed *as* the container. Adopting the container body-ideal and inhabiting a female body would be likely to pathologize me, as the anorexic's commentary on herself as a container lacking integrity would tend to show. The anorexic desires the completion that pertains to the ideal male body-image.

Fascinating though such speculations are, it makes little sense to treat these claims about the phenomenological relevance of male and female body-spaces as testable psychological hypotheses. It is more useful to consider these psychoanalytic claims in terms of the metaphysical boundaries of the self and the not-self as constructed in the history of western philosophy. This is the move that Luce Irigaray makes – most notably in 'Volume without Contours',[3] 'The "Mechanics" of Fluids' (in Irigaray, 1977) and 'Is the Subject of Science Sexed?' (1985). According to her, identities based on spatial containment, substances and atoms belong to the *masculin* imaginary, and what is missing from our culture is an alternative tradition of thinking identity that is based on fluidity or flow. It is important to note that Irigaray is not making an experiential claim: she is not asserting that women's 'true' identity would be expressed in metaphors and images of flow. What she is claiming, by contrast, is that identity as understood in the history of western philosophy since Plato has been constructed on a model that privileges optics, straight lines, self-contained unity and solids. According to Irigaray, the western tradition has left unsymbolized a self that exists as self not by repulsion/exclusion of the not-self, but via interpenetration of self with otherness.

Irigaray's analysis of the history of philosophy can be read as a discourse on boundaries. She explores the way that woman/the mother serves as the protective screen, barrier and the (unobtainable) obscure object of desire that remains always just out of reach. Woman is both the boundary/the Other against which identity is constructed, and that which confuses all boundaries. In the tradition of philosophy that reaches from Plato and Aristotle to Freud and Lacan, woman falls both inside and outside the boundaries of the human, the genus, the self itself. Although I will later argue against Irigaray's tendency to close down the history of philosophy to a trajectory in which sameness is always privileged over difference, viewed as a commentary on woman as boundary *Speculum of the Other Woman* (1974) makes a powerful case.

For Plato woman was – despite the apparent moves towards egality in *Republic* Book V – the state of existing transitional between animal and man. Thus, he tell us in *Timaeus* that a male who failed to live the good life would be reborn as a woman; a second failure would mean that the next rebirth would be as an animal (*c*.350 BC: 42b–c, 90e–91a). Furthermore, for Aristotle woman was literally a monster: a failed and botched male who is only born female due to an excess of moisture and of coldness during the process of conception. A female lacks essence: that which makes an entity distinctively itself and not something else. She is both of the species and not of the species: she is neither the goal of the processes of reproduction, nor is she able to pass on to her offspring characteristics which represent the species.

In Aristotle the female is instead allied with matter – an undifferentiated mass of unshaped material – which can only be formed into an entity by the *logos* or generative power of the male. Change is hylomorphic: an active form (male) imposed on homogeneous matter (female). As such, the female cannot even be said to have an identity of her own. She represents the indefinite: neither one nor two. The female sex is, literally, 'this sex which is not one': the title of Irigaray's book of 1977. Woman is indefinite; she exists at the margins – an enigma – but in ways that problematize the human subject placed at the centre of the traditions of philosophy and psychoanalysis.

Irigaray uses the techniques of philosophical terrorism to mount raids on past philosophers and psychoanalysts. And via these skirmishes an intriguing conjecture emerges: that the privilege given to form, solidity, optics and fixity in the history of the west has, in effect, delayed us from developing alternative models of identity which would treat flow or the indefinite in its own terms, and not simply as a stage en route to a new developmental fixity. In Irigaray's texts the

Lacanian account of the 'fixing' of identity in the mirror of the (M)other's eyes becomes symptomatic of the west's refusal to think a self that is permeated by otherness. Psychoanalysis is presented as a repetition of the philosophical moves of Kant and Hegel, in which self is only established via opposition to, and spatial symmetry with, a not-self. Via insistent questioning and mimicry, Irigaray suggests that this model of identity is the oedipal one of the world of the boy: in which self-identity is established against an oppositional other.

What remains unsymbolized in the whole process according to Irigaray is mother/daughter relations: the formation of a self which can be permeated by otherness, and in which the boundary between the inside and the outside, between self and not-self, has to operate not antagonistically – according to a logic of containment – but in terms of patterns of flow. Irigaray is not claiming, of course, that western philosophy and psychoanalysis have not theorized female identity. Her claim is simply that it has been understood according to what she will term the '*hom(m)osexual* economy of the same' (1974: 102–3). Irigaray puns on the French term '*homme*' – the (supposedly) gender-neutral 'man' – as she points to the fact that woman has been conceived as 'like' the male, only 'different': both lacking and excessive. To put it in my own language: the male has acted as both norm and ideal for what it is to count as an entity, a self or a person.

HYPOTHESIS FOUR: FIN DE MILLÉNIUM

I will return to the problem of how identity might be constructed according to a different logic, but first I want to comment on a problem with Irigaray's position: one that bears on the fourth hypothesis that might be used to explain why I 'fail' to think of my body as a three-dimensional container. This was the hypothesis that my deviance from the container model was not to be ascribed to my sex, but to the fact that I am a westerner living in the last decades of the twentieth century. For although it is surely true that in the history of western science there has indeed been a privilege given to solidity, space as a container and the mechanics of solids, there have been alternatives developed since around the time of the 1914–18 war. Indeed, during the last twenty-five years the mathematics of fluidity and the indefinite have become central to the scientific tradition. If Irigaray wants to characterize an emphasis on the indefinite as 'feminine' science, then we now live – and also kill each other – with the techniques of femininity.

New scientific, mathematical and topographic models now stand alongside hylomorphism, with its attempt to explain change by reference to an active form imposed upon an inert, homogeneous and unshaped matter. Classical science is reductionistic. Although not strictly speaking limited to static models of form, it is limited to conceptualizing and measuring movement or change in linear terms (Kwinter 1992). Modern topological theory became possible on the basis of work produced by the mathematician Henri Poincaré during the years 1889–1909. On the new model, matter is not in any sense homogeneous, but contains an infinity of singularities which may be understood as properties that emerge under certain, but very specific, conditions. What topology does is provide a way of mathematically explaining the emergence of those singularities. Thus, 'ice' and 'water' as well as 'magnetism' and 'diffusion' are forms on the new model, and these forms all depend on singularities for coming into existence. As Sanford Kwinter puts it: 'A singularity in a complex flow of materials is what makes a rainbow appear in a mist, magnetism arise in a slab of iron, or either ice crystals or convection currents emerge in a pan of water' (1992: 37).

Although the classic differential calculus could successfully plot the movements of a body within a system, it did so by regarding the system itself as closed and as incapable of change. By contrast, in the new theory it is possible to measure not only translational changes within the system, but also transformations that the system itself undergoes. Because of this shift of reference, modern topological theory is able to offer a dynamical theory of birth, manufacture and change, in which form is not something that is imposed on matter, but which instead irrupts in matter or is a state of matter. On the new model there is no fundamental difference between states and forms: forms simply are structurally stable moments within the evolution of a system (or space). Indeed, for a new form to emerge on this model the entire space or system is itself subject to transformation or distortion from exterior spaces or systems. We are dealing here with open dissipative systems and with leaks of energy into and out of the system. Spinoza's account of substance [...] comes forcibly to mind. We can also think of Bergson, and my advocacy of his account of fluid 'essences' or 'forms' emerging from within the shifting patterns of flow.

To make the links with Irigaray more explicit, Irigaray claims that the west has been slow to develop a science that can measure and model patterns of the indefinite and of fluidity. However, even if that science was slow to come into existence, it does now exist. The new topologies see form as no more than an apparent and temporary stability in the patterns of potentiality or flow. To quote Kwinter:

> A potential is a simple concept: anything sitting on one's desk or bookshelf bears a potential (to fall to the floor) within a system (vector field) determined by gravity. The floor on the other hand is an attractor because it represents one of several 'minima' of the potential in the system. Any state of the system at which things are momentarily stable (the book on the floor) represents a form. States and forms then are exactly the same thing. If the flow of the book on the shelf has been apparently arrested it is because it has been captured by a point attractor at one place in the system. The book cannot move until this attractor vanishes with its corresponding basin, and another appears to absorb the newly released flows. (1992: 37)

Forms – apparent stabilities – are brought about only because dissipative systems tend to remain in equilibrium or at a state of rest up to a certain threshold of destabilization. However, because energy pervasively leaks to and from the system and is also transported between macro- and micro-levels within the system, such destabilization could occur at any moment and be triggered by an apparently minor deformation in a contiguous system. Catastrophe theory is one of the models for mapping this. Slow and gradual change within a system can suddenly flip over to produce different patterns of behaviour, of turbulence or of random patterning. To borrow a celebrated example from chaos theory (another of the new topologies), something as slight as the flapping of a butterfly's wings in the Pacific might act as the trigger that 'causes' a hurricane at the other side of the world.

For my purposes, what matters is not the details of the theories, but the metaphysics that underlies many of these new topological models. For on at least some of these models – and here it is necessary to register two different traditions in the complex multidisciplinary field that comprises 'chaos theory'[4] – forms are not fixed things, but temporary arrestations in continuous metastable flows, potentialities or evolutionary events. If we think about boundaries, then, from the point of view of the new sciences, the boundaries of our bodies need not be thought as the edges of 'three-dimensional containers into which we put certain things . . . and out of which other things emerge'. The boundary of my body can also be thought as an event-horizon, in which one form (myself) meets its potentiality for transforming itself into another form or forms (the not-self). Such a body-boundary entails neither containment of internal forces nor repulsion of/protection against external forces. Those who are aware of themselves as centred 'inside' an insulated container – free from contamination by the threatening other which is located on the 'outside' – are captured by an illusion generated by the mechanisms of ego-protection, as well as by spatial models inherited from a classical science which is now no longer compelling. To imaginatively construct the self as inhabiting a 3D 'container' is to treat the self as a system that is closed: a form of narcissism. It means blocking out other systems (including other selves) from which and to which energy might leak. It means to refuse to model the self as a dissipative system.

In terms of the way the new topological paradigms impinge on feminist theory, Donna Haraway can be read as seeking to align 'woman' with the new sciences. In 'A Manifesto for Cyborgs', Haraway takes the image of the cyborg – a machine/human hybrid – and evokes an imaginative schematism in which there is no nostalgia for the (illusory) oneness of the autonomous (male) self. What Haraway is doing here can be easily misunderstood and assimilated to just one more form of North American postmodernism in which the impossibility of defining what women have in common is supposed to entail the necessity of abandoning gender as a framework for political organization. But what Haraway is rejecting is not feminism as such, but merely the dialectics of self versus other as figured by those feminists who both recognize the oneness of the self as illusion, and identify with the 'other' rather than the self. This is

Haraway characterizing the type of feminism that she will reject: caught between excess and lack, impaled on a model in which the male provides the standards for self-unity:

> The self is the One who is not dominated. ... To be One is to be autonomous, to be powerful, to be God; but to be One is to be an illusion, and so to be involved in a dialectic of apocalypse with the other. Yet to be other is to be multiple, without clear boundary, frayed, insubstantial. One is too few, but two are too many. (1984: 177)

Haraway is, in effect, rejecting the predominant forms of deconstructive, cultural and psychoanalytic feminist theory. She envisages another form of feminism which charts the self – and disruptions to the 'system' of patriarchal control – by cybernetic schemata made possible by the new topologies.

> Our bodies, ourselves; bodies are maps of power and identity. Cyborgs are no exceptions. A cyborg body is not innocent; it was not born in a garden; it does not seek unitary identity and so generate antagonistic dualisms without end (or until the world ends); it takes irony for granted. One is too few, and two is only one possibility. ... The machine is us, our processes, an aspect of our embodiment. We can be responsible for machines; *they* do not dominate or threaten us. We are responsible for boundaries; we are they. (1984: 180)

Thus, read carefully and in terms of the new scientific paradigms of form and identity, it can be seen that Haraway is embarked on much the same project as Irigaray: of asking that female identity be conceptualized in terms of a different understanding of boundaries. This emerges perhaps best in her 'The Biopolitics of Postmodern Bodies' (1988), in which she analyses discourse relating to the immune system. Haraway uses a biological vocabulary; but interpreting her points back into the language of the new physics, we can understand her bodies as dissipative systems. It is not that all identity disappears on this model; but rather that identity has to be understood not in terms of an inner mind or self controlling a body, but as emerging out of patterns of potentialities and flow.

Coming back to the hypotheses as to why I might not have picked up the 'container' view of my body (with the self safe inside it), I would like to suggest that hypotheses three and four are not mutually exclusive, and perhaps even reinforce one another. Hypothesis three ascribes my 'failure' to schematize my body as a container to sexual difference. I have offered limited support for this thesis by indicating how women in western culture are precariously placed by reference to boundaries. This is both because the primary model of the self is based on that of an individual who does not have to think of himself each month as potentially evolving into two individuals; and also because women are accustomed to seeing 'humanity' and 'persons' described in ways that both include and exclude women. Hypothesis four ascribed my 'failure' to historical factors. The fact that new models for self/other, mind/body relationships are now prevalent in science – and hence, whether we are aware of it or not, in fiction, computer-games, advertising and in the media generally – can provide a resource with which women are able to resist the schemata that would render bodies containers and selves autonomous. Although the new models provided by the sciences by no means constitute a new 'common sense', they do involve alternative models for the phenomenology of the self to those provided by 3D spatiality. However, it is also notable that the richest accounts of phenomenological experience (Sartre, Merleau-Ponty, Beauvoir) also stressed potentiality, force, flow, over stasis and containment. There always has been more than one way of thinking the mind/body, self/other relationship in the history of the west.

HYPOTHESIS FIVE: ALTERNATIVE HISTORIES

This brings me to the fifth and final hypothesis that I wish to consider: that my 'failure' to think of myself in terms of a container might be ascribed to currents in the history of western thought that preceded the new sciences. Again, I want to say that I see this hypothesis not as a contrast to the two preceding explanations, but as an additional factor. Kwinter describes catastrophe theory as a 'fundamentally Heraclitean "science"' (1992: 39). We can also find in Nietzsche a model for individuation that conceptualizes the ego as an (unstable) balance between different, overflowing sources of energy. Thus, Nietzsche claims throughout his mature writings that 'life simply *is* will to power', and that, as such, a 'self' is a collective that 'will strive to grow, spread, seize, become predominant – not from any morality or immorality but because it is *living*' (1886: § 259, 203). The self is 'living' whilst it is expansive, but it expands only until it finds some energy field that it cannot appropriate.

For Nietzsche, health or individuality is not a given, but something that is only maintained via a war of energies. Whatever does not kill a living entity makes it stronger, and resistance is what forms the boundaries of the ego. Thus, the Nietzschean body does not involve a containment which entails protection against external forces or which holds back internal forces from expansion. On Nietzsche's model the boundaries of the self are potentially fuzzy, and hardened only by the processes of appropriation and expansion. The dangers come from within as well as without: decadence and disease simply are the tendency for the parts to exert energies in an anarchic way and act independently of the whole. Death is merely an intensification of this process, and entails dissolution into the micro-organisms that constitute any apparent unity.

Although this antagonistic account of self/other differentiation is not identical to modelling individuality in terms of leakages of energy between levels of system or contiguous systems, it certainly provides a model for thinking the self that involves an acceptance of flux. Furthermore, although antagonism and conflict mark the Nietzschean world into an infinity of sub-systems, in discussing the overman Nietzsche also suggests a less conflictual form of energetic relationships: via the economy of the gift. Thus, the potential overman, Zarathustra, finds a productive 'chaos' within the 'selves' of his followers, and seeks to exploit it: 'one must still have chaos in oneself to be able to give birth to a dancing star' (1883/5: Pt 1, 129). Indeed, Zarathustra's own 'virtues' involve an 'overflowing' of the energies that constitute 'self' and 'other' via the 'will-to-power'.

The move 'back' to a Heraclitean science has not simply occurred in the new physics, but was already present in the writings of Friedrich Nietzsche at the end of the nineteenth century. Henri Bergson can also be read as developing a Heraclitean model in which 'becoming' is privileged over 'being'. Furthermore, both of these writers were important influences on modernist and avant-garde art. Thus, Bergson's emphasis on 'flux' and 'flow' influenced the Cubists, the *Fauves*, and also the Futurists. (See Antliff 1993.) And in this respect it is significant that Sanford Kwinter developed his account of the new topological models of form in the context of an analysis of a futurist triptych by Umberto Boccioni which he also links with Bergson's description of movement as more basic than matter. (See Kwinter 1992: 36; Bergson 1896: 201ff.)

I am not then disagreeing with Irigaray's suggestion that patterns of meaning and inference have in the west been shaped by containment metaphors and relationships. However, I am denying that this sense of the self as housed by a container is common to all socialized western subjects. Indeed, the homogenizing of the west in this fashion is a dangerous political move. The potentiality of thinking beyond the predominant models of the self/other, mind/body schemata has been severely limited by the tendency in poststructuralist and postmodernist feminist theory to look back at that history and find only an 'economy of the same'. It is on this point that I depart most radically from Irigaray, who, despite criticizing Lacan for his synchronic treatment of language and the imaginary (and despite being profoundly influenced by Nietzsche), views the history of western philosophy since Plato as a unitary symbolic system.

Irigaray finds only one oedipalized – and masculinized – model of the self in the history of the west. This model becomes more sophisticated as it evolves, but remains caught within a logic that privileges form, solidity and optics over the indefinite, fluidity and touch. For Irigaray, the history of western philosophy remains the expression of a seamless masculine imaginary. For her, therefore, change can only come from attempting to symbolize that which exists but which has remained unsymbolized in historical times: mother/daughter relations. It is this relationship that forms the 'other of the Other': a position of speaking/viewing from beyond the masculine symbolic that orthodox Lacanians would fiercely deny. Although I agree with Irigaray as to the prevalence of the masculine model for the self, I do not see the history of the west as homogeneous. There have been singularities within it. Openings come from the writings of some familiar philosophers, such as Nietzsche, Diderot, Kierkegaard, Bergson or Foucault. But we need also to look in some unfamiliar places: in texts by past women writers who register that they must count as abnormal, peculiar or singular in terms of the dominant models of the self – and then go on to make imaginative or theoretical adjustments. (See Battersby 1994, 1996; B. Martin, 1991.)

If the emphasis is on generalities (about philosophy, science or 'the west'), singularities (and hence women) can be overlooked: they are merely exceptions to the rule. By contrast, the new topologies work with singularities. Indeed, singularities embody radical transformative potential. By adopting the paradigm of patriarchy as a dissipative system, feminists can register the centrality of masculine models of identity to all existing symbolic codes, without becoming trapped in an impossible dualism: of having to choose between the (over-pessimistic) strategy of deconstruction or the (over-optimistic) strategy of searching for a place to speak from that is beyond the symbolic. Dissipative systems are not closed: there are other systems both without and within. Without a model of the system of patriarchy as itself a dissipative system, speaking from the position of 'Otherness' – even from the position of 'the other of the Other' – has limited political point. Applying insights from the new topologies allows us to see how patriarchy might itself be inherently unstable as a system, and hence how the slight flapping of the wings of a feminist butterfly might – metaphorically – provide the trigger that would enable it to flip over into a state of radical change.

REVOLUTIONARY SPACES

At this point I need also to emphasize again the difference between my own position and more standard treatments of the 'feminine' (or *feminin*). Having registered that it is the male self that is privileged as unitary and contained in the history of the west, postmodernists, deconstructionists and poststructuralists have frequently proceeded to argue for an abandonment, transgression or deconstruction of boundaries. I hope it is clear from the argument that I have mounted here that I would sharply disagree. Without talk of identity (and hence also of boundaries), I do not see that there can be a basis for responsibility or action, including political action. What I have been wanting to stress throughout this chapter is that not all talk of identity involves thinking of the self as unitary or contained; nor need boundaries be conceived in ways that make the identity closed, autonomous or impermeable. We need to think individuality differently, allowing the potentiality for otherness to exist within it, as well as alongside it. We need to theorize agency in terms of patterns of potentiality and flow. Our body-boundaries do not *contain* the self; they *are* the embodied self. And the new sciences give us topological models for imaging the self in these terms.

Some recent uptake of these new topologies has, however, been fundamentally misleading. Thus in a recent book, *Flexible Bodies* (1994), Emily Martin has taken further the model of metaphor analysis that characterized her earlier work, *The Woman in the Body*. Once again she explores the use of metaphors by contemporary American subjects in describing their bodies. This time she explores the language used by both sexes; and this time also her analysis is less satisfactory. Thus, Martin looks to the discourse of disease and health – particularly AIDS and the immune system generally – and argues that there has been a breakdown in the metaphors which are currently employed to understand the relationship between self and not-self. Martin suggests that the metaphors used by present-day Americans show that they are 'coming to think of their bodies as complex systems embedded in other complex systems' (1994: 115). However, the key to the new metaphorics for being an embodied self is represented as 'flexibility'.

Puzzlingly, Martin uses an analysis of the notion of a 'complex system' found within chaos theory to help make her case, whilst then moving speedily from the metaphorics of the new physics to the very different metaphorics of the body as a 'loosely coupled system' taken from manufacturing or service organizations. It is within discussion of the latter framework that the human body is represented as: 'a flexible organization [that] can respond quickly to changes in its environment and can initiate changes in innovative ways' (1994: 144). The argumentative slide shows that Martin has failed to register the power of the new physics to disrupt that which gets characterized as an identity or a 'form'. Thus, she seeks to combine an ontology of flux with an ontology of flexibility, as she charts 'the delicate outlines of an emerging new common sense, entailing changes in notions of identity, groups, wholes or parts' (1994: 13). But despite the paragraph (1994: 144) which juxtaposes these two conflicting conceptual networks, there could not be anything more at odds with an ontology based on flux than one that is based on flexibility.

Flexibility involves an adaptation of 'the same'; by contrast, an identity based on flux entails a far more profound morphological irruption. Thus, on the one hand, the new metaphorics that Martin detects in contemporary American culture generally privileges a self that has a pre-existent identity (and then flexibly 'responds' or 'reacts' to changes acting within and without). On the other hand, the Heraclitean model of identity suggested by the order-out-of-disorder stand of chaos theory, by Nietzsche, Irigaray and also by Bergson, entails not flexibility, but generation (and degeneration and regeneration) based on singularities.[5] Here, as Martin herself puts it, the complex system is generated, is subject to decomposition and recomposition 'in a continuous flow of its components' in a kind of Heraclitean 'dying and becoming' (1994: 118). Thus, in the case of fluidity (but not in the case of flexibility), new identities are born out of difference; self emerges from not-self; and identity emanates from heterogeneity via patterns of relationality.

What Martin's research shows is not a new 'consensus' which represents a new 'common sense', but traces of conflicting ontologies currently operational within everyday discourse. Furthermore, the notion that this is 'new' is also too simple. The strand of contemporary thinking about the relation between self and not-self that privileges flux has been prefigured by various (fairly isolated) voices in western modernity and pre-modernity. Thus, my own turn to the sciences is not intended to provide some new model of 'consensual' reality that reflects the needs of this 'postmodern' age. Nor am I claiming that all women *do*, a matter of empirical fact, fail to think of themselves in terms of the containment schemata. It is rather that feminists *need* – for political ends – to exploit the difficulties of containing female identity within the schemata provided by classical science and metaphysics, and use the resources provided by contemporary science and the history of philosophy to think selves, bodies and boundaries in more revolutionary terms.

The rethinking of the female self that I am demanding does not entail a cognitivist claim: it is not necessary to essentialise a specifically female experience. Instead, looking towards the new sciences for a different understanding of the mind/body and form/matter relationships allows us to theorize a 'real' beyond the universals of an imagination or a language which takes the male body and mind as ideal and/or norm. Positing this 'real' in terms of a fluid force-field and a network of power relations gives us a way to move on beyond the deconstruction of boundaries, and towards the reconstruction of the subject/object and self/other divides. In this respect I agree with Donna Haraway in her alliance with the cyborg. It is time to turn our backs on those forms of feminist theory that castigate all science and all western philosophy as rationalist, masculinist and weapons of the enemy. It is time to investigate the imaginative schemata that old philosophies and new sciences offer us for revisioning the female self.

[. . .]

NOTES

1. A version of this chapter was published in A. E. Benjamin (1993), pp. 30–9. It was originally given at the first European Humanities Research Centre Interdisciplinary Seminars on Boundaries, University of Warwick, December 1992.
2. This book contains over two thousand photos of male and female body-posture and gesture. Its historical and causal analysis is badly flawed. However, the majority of the images are from the 1970s, and clearly show a differential use of space by male and female subjects.
3. First translated as 'Volume-Fluidity' in Irigaray's *Speculum of the Other Woman* (1974). A more accurate translation is included in Irigaray (1991), pp. 53–67.
4. One tradition within chaos theory charts the emergence of order out of chaos via dissipative structures; the other examines the hidden order within apparently disordered systems. It is the first of these two traditions, as exemplified by Prigogine and Stengers (1984), that is relevant to the argument of this chapter. See Hayles (1990), pp. 9–17.
5. See 4. above. Irigaray (1985) also references Prigogine and Stengers (1984).

REFERENCES

Antliff, Mark (1993) *Inventing Bergson*, Princeton University Press.

Bartky, Sandra Lee (1990) *Femininity and Domination: Studies in the Phenomenology of Oppression*, Routledge.

Battersby, Christine (1994) 'Unblocking the Oedipal: Karoline von Günderode and the Female Sublime' in Sally Ledger, Josephine McDonagh and Jane Spencer (eds), *Political Gender*, Harvester Wheatsheaf, 129–43.

Battersby, Christine (1996) 'Her Blood and His Mirror: Mary Coleridge, Luce Irigaray and the Female Self' in Richard Eldridge (ed.), *Beyond Representation: Philosophy and Poetic Imagination*, Cambridge University Press, 249–72.

Bell, Rudolph M. (1985) *Holy Anorexia*, University of Chicago Press.

Benjamin, A. E. (ed.) (1993) *The Body: Journal of Philosophy and the Visual Arts*, no. 4. Academy Editions; Ernst and Sohn.

Bergson, Henri (1896) *Matter and Memory*, trans. Nancy Margaret Paul and W. Scott Palmer. Based on French 1908 edn. Zone Books, 1994.

Bruch, Hilde (1978) *The Golden Cage*, Open Books.

Erikson, Erik H. (1964) 'Inner and Outer Space: Reflections on Womanhood' in Patrick C. Lee and R. S. Stewart (eds), *Sex Differences*, Urizen, 1976, 104–32.

Grosz, Elizabeth (1994) *Volatile Bodies: Toward a Corporeal Feminism*, Indiana University Press.

Haraway, Donna (1984) 'A Cyborg Manifesto' in *Simians, Cyborgs and Women: The Reinvention of Nature*, Free Association Books, 1991, 149–83.

Haraway, Donna (1988) 'The Biopolitics of Postmodern Bodies' in *Simians, Cyborgs and Women: The Reinvention of Nature*, Free Association Books, 1991, 203–30.

Hayles, N. Katherine (1990) *Chaos Bound: Orderly Disorder in Contemporary Literature and Science*, Cornell University Press.

Irigaray, Luce (1974) *Speculum of the Other Woman*, trans. Gillian C. Gill, Cornell University Press, 1985.

Irigaray, Luce (1977) *This Sex which is Not One*, trans. Catherine Porter and Carolyn Burke, Cornell University Press, 1985.

Irigaray, Luce (1985) 'Is the Subject of Science Sexed?' in Nancy Tuana (ed.), *Feminism and Science*, Indiana University Press, 1989, 58–68.

Irigaray, Luce (1991) *The Irigaray Reader*, ed. Margaret Whitford, Blackwell.

Johnson, Mark (1987) *The Body in the Mind*, University of Chicago Press.

Kristeva, Julia (1980) *Powers of Horror: An Essay on Abjection*, trans. Leon S. Roudiez., Columbia University Press, 1982.

Kwinter, Sanford (1992) '"*Quelli che Partono*" as a General Theory of Models' in A. E. Benjamin (ed.), *Architecture, Space, Painting: Journal of Philosophy and the Visual Arts*, no. 3. Academy Editions; St Martin's Press, 1992, 36–44.

Lakoff, George (1987) *Women, Fire, and Dangerous Things*, University of Chicago Press.

Lakoff, George and Mark Johnson (1980) *Metaphors We Live By*, University of Chicago Press.

Martin, Biddy (1991) *Woman and Modernity: The (Life)Styles of Lou Andreas-Salomé*, Cornell University Press.

Martin, Emily (1987) *The Woman in the Body*, Open University Press, 1989.

Martin, Emily (1994) *Flexible Bodies*, Beacon Press.

Nietzsche, Friedrich (1883/5) *Thus Spoke Zarathustra*. In *The Portable Nietzsche*, ed. Walter Kaufmann, Viking, 1968.

Nietzsche, Friedrich (1886) *Beyond Good and Evil*, trans. Walter Kaufmann, Vintage, 1966.

Plato (*c*.350 BC) *Timaeus*, trans. Desmond Lee, Penguin, 1977.

Prigogine, Ilya and Isabelle Stengers (1984) *Order out of Chaos*, Flamingo, 1985.

Sacks, Oliver (1985) *The Man Who Mistook His Wife for a Hat*, Duckworth, 1986.

Smith, Paul (1988) '*Vas*' in Robyn R. Warhol and D. P. Herndl (eds), *Feminisms*, Rutgers University Press, 1991, 1011–29.

Thomä, Helmut (1977) 'On the Psychotherapy of Patients with Anorexia Nervosa' *Bulletin of the Menninger Clinic* 41: 437–52.

Wex, Marianne (1979) *Let's Take Back Our Space*, trans. Johanna Albert and Susan Schultz, Frauenliteraturverlag Hermine Fees.

Young, Iris Marion (1990) *Throwing Like a Girl*, Indiana University Press.

6.2

WOMEN AND EVERYDAY SPACES

Gillian Rose

Time-Geography and Hegemonic Masculinity

Bodies seem central to time-geography [T. Hägerstrand's notion of the temporo-spatial structuring of social life], because the routine actions of individual human agents in time and space, producing and reproducing social structures, are represented by the paths that their bodies follow. But it is these paths that define Hägerstrand's oddly minimalist account of the body. In reference to the body, he notes only that an individual cannot be in two places at once, and that certain constraints are imposed by the need to eat and sleep, comments so obvious as to be unobjectionable.[1] Movements of bodies which cannot be explained with reference to these inherent limits to a body's possibilities of locating itself in time-space are assumed to stem from social, cultural or economic causes. This, of course, is one reason for feminist geographers' use of time-geography to reveal the restrictions that women face on their mobility: it allows masculinism to reveal itself as an unnatural constraint on women's lives. However, when our attention is directed towards social constraints in this way, the body itself is rendered unproblematic. Indeed, it virtually disappears altogether, for the body in Hägerstrand's account becomes its path – it is reduced to its movement. As Hägerstrand says, 'people are not paths, *but they cannot avoid drawing them in space-time*'.[2] In this context, I can only echo Riley's comment that 'the queer neutrality of the phrase "the body" in its strenuous colourlessness suggests that something is up'.[3] Time-geography tries to ignore the body; the next subsection rescues it from its invisibility.

From: G. Rose, *Feminism and Geography: The Limits of Geographical Knowledge*, Cambridge: Polity Press and Minneapolis: University of Minnesota Press, 1993.

Western Bodies: Possessed and Repressed

I can begin by noting that the body/path of time-geography is undifferentiated: all bodies are the same because no body is specified; and these bodies are any bodies, or so they claim. But their very lack of defining characteristics begins to specify them. They are literally colourless, for example; the trace that they leave does not tell whether the body is white or black. Skin does matter to these bodies though, since a corporeal boundary is assumed by time-geographers in their claim that external (to the body) social relations are internalized by human agents in the course of their life-path.[4] This sense of a bounded body has implications for its biology. This biology is a peculiarly selective one, since bodily processes which transgress the boundary between inside and outside the body – childbirth, say, or menstruation – are ignored as characteristics of the body when it is reduced to its path. To emphasize other ways of imagining the body, Iris Marion Young has described childbirth for women who have chosen to become pregnant and can give birth safely precisely in terms of bodily boundary confusions:

> ... the birthing process entails the most extreme suspension of the bodily distinction between inner and outer. As the months and weeks progress, increasingly I feel my insides, strained and pressed, and increasingly feel the movement of a body inside me. Through pain and blood and water this inside thing emerges between my legs, for a short while both inside and outside me.[5]

In contrast, the agency of time-geography is clearly delimited and bounded – its paths mesh but never merge, always individual. There is no bodily passion or desire.

But whose, then, is the minimalist, colourless, bounded body/path that represents human agency? Feminist historians offer an answer in the context of their arguments that two of the most important ways of encoding bodies now are through their gender and sexuality. Some of the historical shifts towards this interpretation of the body have been traced, and Riley claims that from the seventeenth century onwards it has become more and more difficult to speak of 'the' body, because since then bodies have become read more and more by masculinist science through a bipolar understanding of gender and sexual orientation.[6] Poovey has drawn on the arguments of Laqueur to suggest that the nineteenth century in particular witnessed an enormous amount of ideological work which strengthened the masculine/feminine dualism,[7] both establishing gender difference and assuming heterosexuality. Medical discourse in general, and gynaecology in particular, argued that women's spontaneous ovulation meant that they were dominated by their reproductive system. This particular reading of their bodies meant that women were represented as natural creatures, beyond culture and society, compelled to remain in the private domestic sphere by their natural maternal instinct. Victorian racism also legitimated its assertions about black sexuality and white superiority by citing biological difference.[8] Far from being natural, then, bodies are 'maps of power and identity';[9] or, rather, maps of the relation between power and identity.

However, the construction of imaginary bodies involves what Wolff describes as both the repression and possession of 'the' body.[10] While white bourgeois men classify others through oppressive interpretations of Others' embodiment (possession), they assume that they themselves are only contained by their body, not controlled by it (repression). As Simone de Beauvoir sardonically notes at the beginning of *The Second Sex*:

> ... there is an absolute human type, the masculine. Woman has ovaries, a uterus: these peculiarities imprison her in her subjectivity, circumscribe her within the limits of her own nature. It is often said that she thinks with her glands. Man superbly ignores the fact that his anatomy also includes glands, such as the testicles, and that they secrete hormones. He thinks of his body as a direct and normal connection with the world, which he believes he apprehends objectively, whereas he regards the body of woman as a hindrance, a prison, weighed down by everything peculiar to it.[11]

A history of the white masculine heterosexual bourgeois body in Euro-America can therefore be told in terms of a series of denials of its corporeality. Elias has traced that body's loss of vulgar and feminine orifices and excretions from the seventeenth century onwards, for example: the civilized body was one with limited and carefully controlled passages between its inside and outside.[12] This corporeality was merely a container for a consciousness capable of classifying others, for the Enlightened masculine mind was argued to be clearly separate from and untainted by its body. This Enlightenment dualism between mind and body, which saw the mind as rational and the body as the place of emotion, passion and confusion, has been discussed by Bordo. She examines the importance of a 'complete intellectual transcendence of the bodily' in the work of Descartes.[13] The concern with clear boundaries is an expression of separation not only from Man's own body, then, but also from others represented as less able to overcome theirs. And, as Bordo has also argued, the denial of the body is still central to Western masculinity.[14]

The notation of the body in time-geography as a path which does not merge depends on this particular masculine repression of the bodily. This bounded body and its role as a neutral container of rationality both contribute to the idea that we are socialized by internalizing lessons which the 'outside' world teaches us when we act in it. This is the model of socialization used in time-geographic accounts, as we have seen. The unbroken border between inside and outside which this assumes is not only masculinist, however; it is also racist. It represents itself as colourless skin, but in racist discourse, 'the body could not be separated from its colour'.[15] Colour is a key signifier of difference, but only those seen as different from the master subject are designated 'coloured'. Whiteness retains its hegemonic position by denying its own colour and so becoming transparent to the critical gaze.[16] Critiques of whiteness stress the importance of absence to the representation of the 'white' body, an absence of colour. Yet in time-geography there are apparently colourless bodies. This white masculinist self-representation as a number of denials of its own embodiment accounts for the minimalism of time-geography's account of embodied agency: it is an effort to be limited in as few ways as possible by corporeality.

It is now possible to understand more fully why Ryan has to add the emotions and passions of the body on at the end of her account, instead of being able to integrate them within it. Her stories of women's anger and frustration, of the domestic grief of wives and mothers, speak of relations with others through love or maternity or desire. So do feminist geographers' accounts of mothers and their time-space zoning, with their implicit stories of childbirth and love: 'I'm very involved with my kids – they come first before anything else'.[17] It is these kinds of emotional and physical fusion between people which time-geography cannot admit in its reduction of human agency to a path and its consequent masculinist, bourgeois and racist repressions of the body. Ryan has to add these in almost as a postscript because the agency of time-geography embodies masculinity to the exclusion of passion; as Judith Butler suggests, 'the denial of the body ... reveals itself as nothing other than the embodiment of denial'.[18]

Public Life, Public Space, Public Theory

Time-geography embodies an agency that purports to be human but, as we have seen, this agency inhabits a masculine (no)body. This subsection considers whether the space that these agents travel through is also masculine.

Like the embodiment of its agency, there is little discussion in the time-geography literature about space itself: it is taken for granted as the medium of social life. However, a rare attempt to articulate its sense of space is revealing. It emphasizes space as infinitude and unboundedness, transparency; it is simply everywhere, and what is stressed above all is the liberty possible in this space: 'it is freedom to run, to leap, to stretch and reach out without bounds – and without constraint'.[19] And even though time-geography focuses on constraints, its language is untouched by the experiences of being constrained, by the feelings that come with the knowledge that spaces are not necessarily without constraint. Sexual attacks warn women every day that their bodies are not meant to be in certain spaces, and racist and homophobic violence delimits the spaces of black, lesbian and gay communities. Thinking about bodies and emotions against their repression by time-geography, then, does not only invoke the pleasures and desires, lovers and children, of the previous section. It can also invoke violence and horror, brutality and fear. In its erasure of these experiences, time-geography speaks the feeling of spatial freedom which only white heterosexual men usually enjoy.

Many feminists have looked at women's unease in and fear of public spaces, and many argue that 'women's sense of security in public spaces is profoundly shaped by our inability to secure an undisputed right to occupy that space'.[20] Feminist geographers Gill Valentine and Rachel Pain have examined the effects of women's fear of attack on their mobility.[21] June Jordan argues that there is:

> ... a universal experience for women, which is that physical mobility is circumscribed by our gender and by the enemies of our gender. This is one of the ways they seek to make us know their hatred and respect it. This holds throughout the world for women and literally we are not to move about in the world freely. If we do then we have to understand that we may have to pay for it with our bodies. That is the threat. They don't ask you what you are doing in the street, they rape you and mutilate you bodily to let you remember your place. You have no rightful place in public.[22]

Following these arguments, the group of feminist designers called Matrix note that 'many men still perceive women's sexuality as partly defined by their location'.[23] Valentine has noted the connection between the public and the private which underlies this masculinist perception of women's place.[24] She argues that women are seen as properly belonging to the domestic sphere, and she notes how vulnerable to men's violence this makes women, both inside and outside the home: inside, it is no-one else's concern; outside, she deserved it.

The most sustained elaboration of the masculinity of public space is found in feminist critiques of arguments about the ability to undertake political action in the public sphere. Although, in political theory since Plato, 'the existence of a distinct sphere of private, family life, separated off from the realm of public life, leads to the exaggeration of women's biological differences from men, to the perception of women as primarily suited to fulfill special "female" functions within the home, and consequently to the justification of the monopoly by men of the whole outside world',[25] feminists have detected historical inflections in the justifications for the exclusion of women from politics. The key period in these discussions is the seventeenth and eighteenth centuries, and the development of ideologies of nationalism and individualism which allowed only certain people the ability to be active individuals in the national polity. The realm of the public and the political was constructed as one of rationality, individuality, self-control and hence masculinity, since only men could be fully rational individuals, free from passionate attachments. Citizenship, the ability to participate in politics and public life, was limited to (property-owning) men, and feminist geographer Sallie Marston has explored the political exclusions of such discourses in the context of citizenship in the USA.[26] The body politic was masculine, but this individualism did not preclude certain forms of collective action: indeed, the public is the sphere of collectivities, and Pateman has explored in detail the form that this collectivity takes in order not to lose the individualism of its components – the contract. Pateman focuses on the seventeenth-century contract theorists, and renames their social contract a 'fraternal social contract'.[27] The term 'fraternal' comes from her reading of Freud's story of the overthrow of the primal father by his sons, and she uses it to distinguish this 'new, *specifically modern*' form of patriarchy from the earlier political theory based on the powerful father. Both the citizen and the contract are explicitly opposed in classical political philosophy to the particularistic bonds of the feminine family and private life; and both become meaningful through the exclusion of the domestic as the world of unreason:

> The separation of 'paternal' from political rule, or the family from the public sphere, is also the separation of women from men through the subjection of women to men ... the fraternal social contract creates a new, modern patriarchal social order that is presented as divided into two spheres: civil society, or the universal sphere of freedom, equality, individualism, reason, contract and impartial law – the realm of men or 'individuals'; and the private world of particularity, natural subjection, ties of blood, emotion, love and sexual passion – the world of women, in which men also rule.[28]

Riley notes too that femininity became 'intimate, particular, familial, pre-rational, extra-civic, soaked in its sexual being' from the late seventeenth century onwards.[29] Other writers such as Okin and Elshtain focus on the eighteenth century, especially Rousseau's emphasis on women's innate domestic nature and its importance to affective rather than to public political life.[30] By 1785, when Jacques-Louis David painted the 'Oath of the Horatii', masculine and feminine bodies were starkly differentiated in relation to the public. On the left and in the centre of the canvas the men stand erect and rigid, caught in the act of swearing loyalty to the greater good of the state; their spoken oath binds them to action and battle. Citizenship, the formal right of entry into political discussion, is represented by the ideals of autonomy and selfhood which constitute masculinity and masculine bodies. In contrast to this bounded masculine body, on the right of the canvas swoon a group of women; silent, passive, grieving and intertwined, their softness, emotionality and marginality to the action embodies their exclusion from the masculine, public and political sphere.[31] Through the masculinization of the body politic, public space was also represented as a masculine arena.

This construction of public space as masculine does not go uncontested, nor is it without contradictions. Some women – from the Communard *citoyennes* to Take Back the Night marches – have struggled to reconstruct public space by demanding equal rights in it for, as Jones remarks, a 'new claim on public space' also implies 'a new social form'.[32] Feminist geographers using time-geography offer a similar challenge to its space. Just as they implicitly challenge its disembodied human agent, so have they hinted that there are more spaces than meets its eye. Dyck argues that the construction of motherhood through the everyday negotiation of meanings in particular spaces means that women's conceptions of space may alter; 'women generate definitions and understandings of appropriate modes of mothering *and the spaces within which this takes place* through the recurrent practices of mothering work beyond the immediate confines of the home'.[33] By watching their children playing in the street, women get to talk with other mothers – their neighbours – and networks develop which establish a safe place for their children beyond the confines of the home, as well as renegotiating the meaning of motherhood. Dyck is suggesting that the social constitution of different identities may also imply different kinds of space. This implies that everyday space is not only not self-evidently innocent, but also bound into various and diverse social and psychic dynamics of subjectivity and power. The possible contradictions created by this complexity have been the subject of several recent studies by feminist historians of the public/private distinction.[34]

Outram suggests that the masculine claim to public space is potentially fragile not only because of these contestations of its meaning, but also because what it excludes can erupt into it; moreover, its masculinization through a certain policing of bodies means that every new body requires disciplining in order to guarantee its reconstitution.[35] This policing can be violent. In her discussion of the costs of the fraternal contract, Pateman stresses that the military exemplifies fraternities in action, and espies 'the figure of the armed man in the shadows behind the civil individual'.[36] Thus she makes a link between the contract and the violence with which public spaces are kept as white masculine, heterosexual spaces. This bounded individualism, with its violence, remains as a condition of hegemonic masculinity and citizenship today, as Watney makes clear in his discussion of men with AIDS in the UK.[37] He argues that they are not seen as worthy of the same rights as full citizens because their sexuality transgresses the boundaries of acceptable, masculine behaviour; gay men too are victims of violence in the public streets of this masculine individual, of course.

To conclude this section: in Haraway's description of the individualism of masculine subjectivity, its particular geography is also revealed. She speaks of the '"West's" escalating dominations of abstract individuation, an ultimate self untied at last from all dependency, a man in space'.[38] Haraway's 'space' is the outer space in which the male astronaut survives alone supported only by his technology, entirely distant from other people and from Mother Earth, and this is precisely the space which Gould suggests is used in time-geography.[39] The space of time-geography, then, seems to be the dominant space of patriarchy: public, masculine space, which fully acknowledges only (repressed) white heterosexual bourgeois masculine bodies. The ability to act in the public sphere, as opposed to breed in the private, is a privilege violently reserved for men, and the human agency produced in time-geography also speaks only of this masculine sociality and its public spaces. Buttimer's description of time-geography as a *danse macabre* seems entirely appropriate.[40]

THE CONSTITUTION OF A MASCULINITY THROUGH TIME-GEOGRAPHY

This chapter has argued so far that both the human agency and the space through which it moves in time-geography are masculine; they are constructed in the image of the master subject. For all its claims to self-evident reality, to be 'anchored in certain basic facts of life',[41] and to represent those facts objectively, time-geography assumes unproblematized but in fact highly specific theorizations of society and space, and of the bodies which constitute human agency, and this specificity excludes other socialities, spaces and bodies from knowledge. Its notions of agency and space are taken for granted as universal, and other understandings are thereby refused. This is masculinism's false exhaustiveness of the Same. Although it is difficult to think of anywhere beyond the mapping capabilities of this kind of geographer, it is not impossible, as I have argued. There are other spaces, and other kinds of subjectivity. The erasure of such different subjectivities and socialities means that time-geography makes a claim to power through its knowledge. I now want to address this power/knowledge relation more directly. In particular, I want to suggest that this masculinism is not only an effect of a specific masculinity; it also constitutes that masculinity.

Geographers believe that space can always be known and mapped; space is understood as absolutely knowable. That is what its transparency, its innocence, signifies: it is infinitely knowable; there are no hidden corners into which time-geography cannot penetrate. This is a necessary consequence of its search for totality; for if the societies structured through space are understood as wholly visible, as they are in time-geography, then space must be wholly representable. It is real, natural and unproblematic: time-geography's space clearly presents no problem to its theorizers. The visual has always been central to masculinist claims to know. Seeing was certainly important to the emergence of the social sciences towards the end of the nineteenth century. Philanthropists, journalists, early social scientists and voyeurs of every sort then went into the city to gaze at its horrors and systematize its dreadful spaces.[42] They wanted to see completely and so to produce and control knowledge of urban social life. Contemporary cities are subjected to the same heroic feats of interpretation: Los Angeles is probably the best example of a city interpreted by great men from their lofty vantage points.[43]

Implicit in this claim to see all and know all are the subjectivity and compulsions of the bounded body that I have already described as the object of time-geography. I want to suggest that this imaginary body is also the author of time-geography. Remember that the construction of imaginary bodies involves both the repression and possession of the body. Fundamental to its construction and possession of other imaginary bodies is the masculinist denial of the male body; others are trapped in their brute materiality by the rational minds of white men. This erasure of his own specificity allows the master subject to assume that he can see everything. In our own time, Haraway has talked about the contemporary escalation of this 'unmarked category' through the proliferation of visual technologies: 'vision in this technological feast becomes unregulated gluttony; all perspective gives way to infinitely mobile vision, which no longer seems just mythically about the god-trick of seeing everything from nowhere, but to have put the myth into ordinary practice'.[44] As Ryan implies at the end of her representation of society as a piece of well-oiled machinery, no cog or gear snagging or grating, everything accounted for, time-geographers seem to be pulling the god-trick hard and fast. Their masculine consciousness peers into the world, denying its own positionality, mapping its spaces in the same manner in which Western white male bodies explored, recorded, surveyed and appropriated spaces from the sixteenth century onwards: from a disembodied location free from sexual attack or racist violence. Space for them is everywhere; nowhere is too threatening or too different for them to go. Time-geographers become the invisible observers of social life, tracing its patterns and making sense of it all, its reproduction, resistance and contradiction.

The contemporary character of this particular masculinity can be caught by paraphrasing Haraway: objects come to us simultaneously as indubitable recordings of what is simply there and as heroic feats of social-scientific production.[45] The heroism of being able to know what really exists both depends on a certain masculinity and constitutes it. Time-geographers' particular masculinity is established through their assumption that all space is white, bourgeois, heterosexual masculine public space. They deny other possibilities, including an Other; the domestic is not addressed as the Other of public space – it is ignored. The costs of its claim to truth through privileged position and universalised categories have been summarized by Deutsche:

> In the act of denying the discursive character of those objects, such depictions also disavow the condition of subjectivity as a partial and situated *position*, positing instead an autonomous subject who observes social conflicts from a privileged and unconflicted place. As this total vantage point can be converted from fantasy into reality only by denying the relational character of subjectivity and by relegating other viewpoints – different subjectivities – to invisible, subordinate, or competing positions, foundational totalizations are systems that seek to immunize themselves against uncertainty and difference.[46]

I will call this denial of the Other (as well as the claims of others) in order to establish a claim to know what is really there 'social-scientific masculinity'. Transparent space, as an expression of social-scientific masculinity's desire for total vision and knowledge, denies the possibility of different spaces being known by other subjects.

However, feminist geographers working with time-geography do refer to different spaces and other worlds; they focus on women's everyday world and the centrality of women's embodiment. This chapter has elaborated their implicit references to a feminized realm of mothering and bodies, of blood spilt for love and in violence, of passion, desire and hate, in order to reveal the specificity of time-geography. The aim of that elaboration of time-geography's repressed Other was to mark the unproblematized universal spaces and bodies of time-geography as masculine (and white and straight and middle-class), and to that extent my strategy succeeded, I think. But that strategy also has its risks.

NOTES

1. D. Parkes and N. Thrift, *Times, Spaces, and Places: a Chronogeographic Perspective* (John Wiley, Chichester, 1980), pp. 247–8. This erases the body by relegating it to the natural as opposed to the social or cultural; it thus depends on the opposition between Nature and Culture.
2. T. Hägerstrand, 'Diorama, path and project', *Tijdschrift voor Economische en Sociale Geografie*, 73 (1982), pp. 323–39, p. 324.
3. D. Riley, '*Am I That Name?*', *Feminism and the Category of 'Women' in History* (Macmillan, London, 1988) p. 104.
4. Thrift, 'On the determination of social action in space and time', p. 43; Pred, 'Social reproduction and the time-geography of everyday life', pp. 166–70.
5. I. M. Young, *Throwing Like a Girl and Other Essays in Feminist Philosophy and Social Theory* (University of Indiana Press, Bloomington, 1990), p. 163.
6. Riley, '*Am I That Name?*'.
7. M. Poovey, *Uneven Developments: the Ideological Work of Gender in Mid-Victorian England* (Virago, London, 1989), pp. 6–7. See also O. Moscucci, *The Science of Woman: Gynaecology and Gender in England 1800–1929* (Cambridge University Press, Cambridge, 1990).
8. S. L. Gilman, 'White bodies, black bodies: toward an iconography of female sexuality in late nineteenth-century art, medicine and literature', *Critical Inquiry*, 12 (1985), pp. 204–41.
9. D. Haraway, *Simians, Cyborgs, and Women: the Reinvention of Nature* (Free Association Books, London, 1991), p. 180.
10. J. Wolff, *Feminine Sentences: Essays on Women and Culture* (Polity Press, Cambridge, 1990), p. 121.
11. S. de Beauvoir, *The Second Sex*, tr. H. M. Parshley (Picador, London, 1988), p. 15.
12. N. Elias, *The Civilising Process. Volume 1: The History of Manners* (Blackwell, Oxford, 1978).
13. S. Bordo, 'The Cartesian masculinization of thought', *Signs*, 11 (1986), pp. 439–56, p. 450.
14. S. Bordo, '*Anorexia nervosa:* psychopathology as the crystallisation of culture', in *Feminism and Foucault: Reflections on Resistance*, eds I. Diamond and L. Quinby (Northeastern University Press, Boston, 1988), pp. 87–117.
15. B. Omolade, 'Hearts of darkness', in *Desire: the Politics of Sexuality*, eds A. Snitow, C. Stansell and S. Thompson (Virago, London, 1984), pp. 361–77, p. 365.
16. R. Dyer, 'White', *Screen*, 29 (1988), pp. 44–64.
17. 'Anna', a mother quoted in I. Dyck, 'Space, time and renegotiating motherhood', *Environment and Planning D: Society and Space*, 8 (1990), p. 471.
18. J. Butler, 'Variations on sex and gender: Beauvoir, Wittig and Foucault', in *Feminism as Critique: on the Politics of Gender*, eds S. Benhabib and D. Cornell (University of Minnesota Press, Minneapolis, 1988), pp. 128–42, p. 133.
19. P. Gould, 'Space and rum: an English note on espacien and rumian meaning', *Geografiska Annaler*; 63B (1981), pp. 1–3, p. 2. Gould notes that this sense of freedom is much stronger in the English word 'space' than in the Swedish 'rum', which is the term used by Hägerstrand. See also D. Gregory, 'Presences and absences: time–space relations and social theory', in *Social Theory and Modern Societies: Anthony Giddens and his Critics*, eds D. Held and J. B. Thompson

(Cambridge University Press, Cambridge, 1989), pp. 185–214, p. 194.
20. J. Hamner and S. Saunders, *Well-founded Fear: a Community Study of Violence to Women* (Hutchinson, London, 1984), p. 39.
21. R. Pain, 'Space, sexual violence and social control: integrating geographical and feminist analyses of women's fear of crime', *Progress in Human Geography*, 15 (1991), pp. 415–31; G. Valentine, 'The geography of women's fear', *Area*, 21 (1989), pp. 385–90; G. Valentine, 'Women's fear and the design of public space', *Built Environment*, 16 (1990), pp. 279–87.
22. J. Jordan in P. Parmar, 'Black feminism: the politics of articulation', in *Identity: Community, Culture, Difference*, ed. J. Rutherford (Lawrence and Wishart, London, 1990), pp. 101–26, p. 113.
23. Matrix, *Making Space: Women and the Man-made Environment* (Pluto Press, London, 1984), p. 49.
24. G. Valentine, 'Images of danger: women's sources of information about the spatial distribution of male violence', *Area*, 24 (1992), pp. 22–9.
25. S. M. Okin, *Women in Western Political Thought* (Princeton University Press, Princeton, 1979), p. 275; and see J. B. Elshtain, *Public Man, Private Woman: Women in Social and Political Thought* (Martin Robertson, Oxford, 1981); A. Phillips, *Engendering Democracy* (Polity Press, Cambridge, 1991); K. B. Jones, 'Citizenship in a woman-friendly polity', *Signs*, 15 (1990), pp. 781–812.
26. S. Marston, 'Who are "the people"?: gender, citizenship and the making of the American people', *Environment and Planning D: Society and Space*, 8 (1990), pp. 449–58. For a rare acknowledgement by geographers of the social specificity of public space, see E. Muir and R. F. E. Weissman, 'Social and symbolic places in Renaissance Venice and Florence', in *The Power of Place: Bringing Together Geographical and Sociological Imaginations*, eds J. A. Agnew and J. S. Duncan (Unwin Hyman, London, 1989), pp. 81–103.
27. C. Pateman, *The Disorder of Women* (Polity Press, Cambridge, 1989), p. 35.
28. Pateman, *The Disorder of Women*, p. 43.
29. Riley, '*Am I That Name?*', p. 41.
30. Okin, *Women in Western Political Thought*; Elshtain, *Public Man, Private Woman.*
31. J. Landes, *Women and the Public Sphere in the Age of the French Revolution* (Cornell University Press, Ithaca, NY, 1988), pp. 152–8. See also M. Gatens, 'Corporeal representation in/and the body politic', in *Cartographies: Poststructuralism and the Mapping of Bodies and Spaces*, eds R. Diprose and R. Ferrell (Allen & Unwin, Sydney, 1991), pp. 79–87; D. Outram, *The Body in the French Revolution: Sex, Class and Political Culture* (Yale University Press, New Haven, 1989).
32. Jones, 'Citizenship in a woman-friendly polity', p. 803.
33. I. Dyck, 'Integrating home and wage workplace' *Canadian Geographer*, 33 (1989), p. 330, my emphasis; see also Dyck, 'Space, time and renegotiating motherhood', p. 466. Dyck's arguments track a trajectory in feminist geographers' discussions of everyday space similar to that remarked on by Moore in anthropology: a shift from Ardener's early formulation of space as a result of patriarchal structure to a recent insistence that space is better seen as an enacted and negotiated text; see H. L. Moore, *Space, Text, Gender: an Anthropological Study of the Marakwet of Kenya* (Cambridge University Press, Cambridge, 1986).
34. See, for example, L. Davidoff and C. Hall, *Family Fortunes: Men and Women of the English Middle Class 1780–1850* (Hutchinson, London, 1987); J. Meyerowitz, 'Sexual geography and gender economy: the furnished room districts of Chicago, 1890–1930', *Gender and History*, 2 (1990), pp. 274–96; L. Nead, *Myths of Sexuality: Representations of Women in Victorian Britain* (Blackwell, Oxford, 1988); Poovey, *Uneven Developments*; M. P. Ryan, *Women in Public: Between Banners and Ballots, 1825–1880* (Johns Hopkins University Press, Baltimore, 1990).
35. D. Outram, *The Body in the French Revolution* (Yale University Press, New Haven, 1989) p. 164.

36. Pateman, *The Disorder of Women*, p. 51. See also G. Lloyd, 'Selfhood, war and masculinity', in *Feminist Challenges: Social and Political Theory*, eds C. Pateman and E. Gross (George Allen & Unwin, London, 1986), pp. 63–77.
37. S. Watney, 'Practices of freedom: citizenship and the politics of identity in the age of AIDS', in *Identity: Community, Culture, Difference*, ed. J. Rutherford (Lawrence and Wishart, London, 1990), pp. 157–88, pp. 68–73.
38. D. Haraway, *Simians, Cyborgs, and Women* (Free Association Books, London, 1991), p. 151.
39. P. Gould, 'Space and rum: an English note on espacien and rumian meaning', *Geografiska Annaler*; 63B (1981), pp. 1–3.
40. Quoted in Gregory, 'Suspended animation', p. 335. Lefebvre also stresses the masculinity and the deadness of what he calls 'abstract space'; see H. Lefebvre, *The Production of Space*, tr. D. Nicholson-Smith (Blackwell, Oxford, 1991).
41. Parkes and Thrift, *Times, Spaces, and Places*, p. 243.
42. P. J. Keating, *Into Unknown England, 1866–1913: Selections from the Social Explorers* (Manchester University Press, Manchester, 1976); F. Driver, 'Moral geographies: social science and the urban environment in mid-nineteenth century England', *Transactions of the Institute of British Geographers*, 13 (1988), pp. 275–87; G. Pollock, *Vision and Difference: Femininity, Feminism and Histories of Art* (Routledge, London, 1988), pp. 50–90; J. Tagg, *The Burden of Representation: Essays on Photographies and Histories* (Macmillan, London, 1988), esp. pp. 117–52.
43. M. Davies, *City of Quartz* (Verso, London, 1990); F. Jameson, 'Postmodernism, or the cultural logic of late capitalism', *New Left Review*, 146 (1984), pp. 53–92; E. Soja, *Postmodern Geographies: The Reassertion of Space in Social Theory* (Verso, London, 1989).
44. Haraway, *Simians, Cyborgs and Women*, p. 189. Haraway is in part referring to a National Geographic Society publication.
45. Haraway, *Simians, Cyborgs and Women*, p. 189.
46. R. Deutsche, 'Boys town', *Environment and Planning D: Society and Space*, 9 (1991), pp. 5–30, p. 7. For a brief comment on the visual control of the city in certain geographical models, see L. C. Johnson, 'Gendering domestic space: a feminist perspective on housing', *New Zealand Journal of Geography*, 90 (1990), pp. 20–4, pp. 21–2.

6.3

SURVIVING RAPE: A MORNING/MOURNING RITUAL

Andrea Benton Rushing

For Frank who asked how I endured it, and Audre who asked me to write about it, and those whose love leads me

> I am writing because rape is . . . I am writing to understand. I am writing so I won't be afraid. I am writing so I won't start crying again. . . . I am writing to allow myself to feel the anger. I am writing to keep from running toward it or away from it or into anybody's arms. I am writing to find solutions and pass them on. I am writing to find a language and pass it on.
>
> I am writing, writing, writing, for my life. (Pearl Cleage, *Mad at Miles*, p. 5)

'DON'T MOVE!' yanks my eyes open. Night light's off. Can't see. Not in-the-swimming-pool-without-my-glasses can't see. REALLY can't see. Need to get up and find out what's wrong, but something's pressing me down. 'DON'T MAKE A SOUND OR I'LL HAVE TO HURT YOU!' The barked command's garbled. Have to come out of this nightmare. A man's on top of me. One of his hands pinned both of mine above my head. His breath's warm on my face, but his jacket's cold when he pulls my robe and nightgown up, my underpants down. Short, flabby penis. He is telling me how much I want him, how much he satisfies me. Maybe I can go back to sleep or pass out until it's over. . . . Suppose he can't get inside my vulva or ejaculate. If I make him angry, he'll cut, shoot, strangle me. Did he kill the kitten? Where are Osula and Ann?

From: S. M. James and A. Busia (eds), *Theorizing Black Feminisms: the Visionary Pragmatism of Black Women*, London: Routledge, 1993.

It's been 21,900 hours, 912 days, 130 Saturday nights, 30 months, 3 years since October 16, 1988 when I was stunned awake, straddled by a man I did not know. First I think I'm nightmaring. ... Then try to sink back under sleep's blanket and weave whatever is going on into a dream the way I do when I have to urinate or am hungry and don't want to leave the bed's womb. Then I try to pass out. Can't because I don't know what he'll do, because I am adrenalined, 360 degrees opposite of relaxed. Besides, he wants me to be conscious of his omnipotence and my humiliating powerlessness. It's almost blackstrap-molasses dark. And horribly quiet.

Saying the story, I usually claim, 'All I asked God for was my life. God gave me that and so much more.' But there were no words in my mind when I was being raped. I chose life and did what I thought would keep me in the land of the living. My body responds to his in a vicious parody of intimacy while my brain whirls to read his mind so I can save my life by satisfying him. The sustaining, salvific prayers came from friends and ancestors on the other side of the membrane that separates the living from the dead. While I moan and whimper, some choir is singing, 'I don't know where, but I know that you do. I can't see how, but I know you'll get through. God, please touch somebody right now, right now.'

Will he slash my face when he's through? All he asks for is money. Not a demand. A casual request as if I'm 'his woman', and he's going to run an errand. Same man who has walked through the walls of my apartment to rape me, waits, almost patiently, while I fumble through two purses for my wallet *and* he lets me take the bills out. Doesn't count the seventy dollars I hand over or mention my credit cards. He leaves the room, speaking over his shoulder, 'Stay there, I'll be back', as if he were my honey getting up to fix us drinks or snacks or turn our favorite music on, as if every cell in my body wasn't trying to eject him from my life.

If you'd told me way back then that I'd still be recovering from rape now, I wouldn't have laughed in your face, but I wouldn't have believed you either. I'd faced traumas before – tenure review, major surgery, heartshattering divorce – stumbled through some and transcended others, so I expect rape to slip from me like a boiled beet's rough skin.

[. . .]

Since I'm feeling fine (not a single bruise, broken fingernail, or out-of-place earring), it's a hassle to be in the dry cold of Clayton Hospital's air-conditioned examining rooms waiting the hours it takes doctors to finish treating Saturday night knife slashes and bullet gashes before one can get to me. The crisis counsellor says I seem like a woman accustomed to being in control and, since rape rendered me powerless, I may have a more difficult recovery than a more passive woman.

The next time the counsellor checks up on me, I ask what stages I can expect to go through and what I should do to recover from rape quickly and completely. Suck my teeth and groan to hear, 'Each victim has to find her own way. It's hard to predict. Women mourn and mend differently.' WAIT! STOP! HALT! Every six minutes some girl or woman in the USA is raped. Some recovered victim must have chronicled her journey, published her 12-step program, copywritten a recipe I can improve on. Though I scour, I never find a thing. Three years later all I can offer the next rape victim is two poems by twice-raped (!!) June Jordan. Audre Lorde's *Cancer Journals* have saved minds and lives, but there's no equivalent to guide a rape survivor. And there's no Bessie Smith, Ida Cox, Dinah Washington, Nina Simone, Koko Taylor, Sweet Honey in the Rock sound that testifies about and transcends rape's agonies.

> Way down younger by myself,
> and I couldn't hear nobody pray.

In movie and television versions of rape, *the* problems are that people think you seduced the man, police are sexistly hostile, hospital staff is icily callous, but my ordeal wasn't going that way at all. In my apartment, the Georgia police officers who look and sound like red-necks treat me with a courtesy nothing in my childhood summers in segregated Jacksonville, Florida or Dothan, Alabama prepared me for. I'm questioned gently. Did I recognize the rapist? A boyfriend? Someone who'd stalked me? Was he a college student my daughter and her friend knew? Did we have oral sex? Anal sex? Did he bite me? They accept my word that I've never seen the man before and don't even ask if I tried to fight him off. At the hospital, the in-take clerk, crisis counsellor, lab technician, nurse, doctor, billing clerk are all considerably consoling. At the time I didn't notice, but a week later their behavior upsets me. There is, I tell sympathizers, no plan to end rape. People are just refining their treatment of the inevitable.

When the crisis counsellor leaves, I scan the hospital's 'Recovering from Rape' brochure:

> REMEMBER ... You did nothing to provoke the attack. You are not at fault. You are the victim of a violent crime. Men do not rape because of sexual desire; they rape to humiliate, control, and degrade. Rape is a violent assault on the body and leaves emotional scars which can take weeks, months, or years to resolve. Each person is unique and must work through this emotional trauma in their own individual way. There is no 'right' or 'wrong' way to deal with this stressful time.

The counsellor describes the hospital's exam and treatment: If there's a chance that I might be pregnant, I'll get the morning-after pill; had the rapist bitten me, I'd get a tetanus shot – BITTEN ME; blood will be drawn to see if I've gotten a sexually transmitted disease, but, while waiting for the results, I'm to take antibiotics just in case. . . . The doctor – a nurse by his side – turns off the light so his ultraviolet lamp can look for traces of the rapist's semen between my legs. The second strange man in this long evening standing over my body, focused on my vulva. I've neither seen nor heard about this phase of rape's aftermath. As my nails claw the crisis counsellor's palm, I'm glad she alerted me and glad I said yes to her hand-holding offer. Hairs are pulled, one by one, from my head and vulva for DNA testing. After the lab technician takes my blood and gives me a band-aid with Daffy Duck decorations, I collapse into snuffling tears, 'I was so scared and there was no one to help me.'

The sun's high by the time I finally leave the hospital. So exhausted I *know* I'll go right to sleep as soon as I get home, which isn't what happens. Kay, who has met me in the emergency room with a serene smile and steadying affirmation, alternates between soothing Osula and Ann in the waiting room and chatting with me as I'm shuttled from examining room to examining room, get a douche, Xanax for my expected anxiety, and antibiotic in case I have a sexually transmitted disease (I'm recoiling from gonorrhea and syphilis. It's weeks before herpes occurs to me. Months before AIDS seems possible). Kay assigns the girls to rid the apartment of all signs of police and rapist intrusion, but broad black scuff marks from the rapist's sneakers won't come off the white window ledge. (And, later, I refuse to have them painted over because I want to *remember* why I am suffering so.) Kay suggests a Caribbean vacation. Overhearing, Osula thinks I should spend time with family in Boston, Minneapolis, or NYC. Having just moved to Atlanta, I can't plan, much less pack for, another trip. Besides, I don't intend to let the rapist – a man whose face and name I don't know, whose unschooled and country Southern accent is all I'm sure I can recognize – make me skitter scared. As hard-headed as always, I pooh-pooh advice about getting a second-floor apartment, dog or gun; and I resist all suggestions to go back to safe white Amherst. Not only don't I leave apartment 2401, I even sleep (sedated, the pallet moved from its rape location, and a woman friend spending the night in the living room) in the bedroom I was raped in. Four days later, I rent furniture for my garden apartment.

Throughout the sunny Sunday hours after the hospital, I notify family and friends that I've been raped, assuring them that my body is fine and that I'm exhilarated to be alive. I ask proven intercessors to help me glorify a wonderful God and, since 'the prayers of the saints availeth much', to storm heaven on my behalf. That Sunday, and ever after, my candid talk about being raped surprises because people are accustomed to rape victims' shame and reticence. 'But', I say over and over, 'I am the victim, not the criminal.'

[. . .]

Since I'd felt so good and been so clear-headed and capable in the immediate aftermath of being raped – no signs of the physical exhaustion, disorientation, anxiety, or amnesia that, later, become my almost constant companions – I was as unprepared as everyone else when shock's soft shawl slipped from my shoulders. I thought what went wrong with rape victims was they denied being raped, but I hadn't. I'd called the police, pressed charges, been hospital-examined, begun twice-a-week rape crisis counselling, even seen the pastoral counsellor at Morehouse Medical School to talk about the politics of being raped by a 'brother' and helped by Euro-Americans. In spite of my intentions and efforts, I didn't become my old self again in the weeks and then months I'd set aside for rape recovery. As the hours, days, weeks, months, years marched away from the October rape day, I came to see that I would never be my old self again. 'The only thing about me that's the same', I tell those who compliment how strong my voice sounds and how well I look, 'is my fingerprints'. 'You get better', Evelyn-the-counsellor tells me, 'and, because you feel *more*, feel worse.'

> From a December '88 journal entry ...
> 8:45 a.m ... So it was dreadful to feel my body getting tenser and tenser as I became more and more awake. Tense about what time it was. Tense about when the alarm clock would go off. Tense about what I'd eat. Tense about calling the phone company. Tense about calling Dr D. for a Xanax refill. Tense about the car's possible problems. So I decided to give in and take an anti-anxiety pill, but then I couldn't find my pocketbook. Couldn't even find my glasses so I could look for the pocketbook. And all the while the kitten kept sneezing. I moaned for her to stop. And the sound of me mewling reminded me of the whimpering noises I made while I was being raped.

By January, still in twice-weekly rape crisis counselling, I finally dare to open a few windows in the apartment for the first time since I was raped, but being one of 3,000 people evacuated from an extremely bourgeois (the pastor drives a Jaguar) church near the beginning of a Sunday service has frightened me unbearably. As I watch police dogs bomb-sniff, it feels like no place is safe. I can be hurt by someone I don't know at home in bed and in church which is, as I've said for years, where I'd rather be than any other place. At my request, E. calls a psychiatrist for me. The AfAm doctor diagnoses post-traumatic stress disorder, which I thought you had to go to Vietnam to get. He prescribes Xanax for my anxiety and is surprised that I'd expected to be fully recovered from rape by now. It will, he dismays me by saying, require a year for short-term recovery and five years – FIVE! – for as close to full recovery as one gets.

By March, despite the successes I call myself having – riding MARTA public transportation and mounting a photography exhibit at Spelman – Dr P. stuns me by diagnosing clinical depression. (Only hearing him read textbook definition and having him say that my progress so far reflects my disciplined will power, persuades me.) Rape lugs me around. No sign of light at the end of the tunnel.

Rape makes me doze off at night propped up on pillows so I'll be prepared not so much 'if' as 'when' the rapist comes back. Lights on all over the apartment. I, nervous as my kitten, jump when she gets in or out of bed with me, and wake up almost every night at the 3:00 a.m. rape time. Raped in almost total darkness, I don't sleep soundly until dawn. And I wake up, as though demon-ridden, with crusty saliva lines around my mouth.

Before I was raped, I'd prided myself on waking up an alert that didn't need the caffeine props of coffee and tea. Afterwards, it's a daily struggle to come into consciousness and realize AGAIN that I didn't nightmare being raped. Now, as it did when I was actually being raped, my mind scrabbles for a safe place and, finding none, tries to shut off, but the strategy is no more effective than it was that gruesome night. No idea how I'll get out of bed, much less take the ten steps from the bedroom I was raped in to the bathroom I've become afraid to shower in. Grope for eyeglasses. Turn off the bedside lamp rape has made a night-time necessity. Step over the telephone wire the newly installed burglar alarm is hooked up to. Peer blinking as I did the rape night, into the living room. Tense, heart racing, afraid.

In the western Massachusetts university town I'm on sabbatical from, my days start (weather permitting) with a mile-long up and down hill walk while the Connecticut Valley's air is dewy and still. Atlanta is much warmer than Amherst, so I'd looked forward to being outside much more and alternating bicycle rides with long walks. But family, friends and police think the rapist stalked me, waited until I was alone and defenseless before he came through a living room window, so I'm much too frightened to leave the house and risk being raped again. Besides, though it takes me years to know this, my body's no longer mine. The rapist controls it. Moving feels like Herculean work and distracts my mind from being rape-alert.

'Did you stop and pray this morning?' a song from childhood Florida summer wonders, 'as you started on your way? Did you ask God to guide you, walk beside you all the way? Did you stop and pray this morning? Did you kneel just one moment and say, "Give me comfort for my soul on this old, rugged road?" Did you just remember to pray?' My mind's too centrifugal for prayer. Can neither sing the songs of Zion myself nor decide whether James Cleveland's gospel growl, Aretha Franklin's sanctified melisma, or Marion William's octave-defying witness will encourage my heart, regulate my mind, relax my forehead's accordion pleats. Exhausted from my marrow on out, I'd die if I had to will air into my lungs or work to make my blood flow. Difficult to believe I can choose an ensemble, do laundry, grocery shop, or collect the mail. Impossible to see myself doing anything I came to Atlanta to do. Foolhardy to claim one day at a time. Half-hour by half-hour is all I dare.

But I *can* make a pot of tea and focusing on that task magnetizes my mind's scattered steel filings on a north–south axis. Drop Lemon Verbena or Mellow Mint tea bags into a small pot, sweeten with an exotic honey or plain white sugar, place the pot, a tall clear glass, yogurt, a spoon and a cloth napkin on a thick wooden tray and, emboldened by my success, turn off the lights that have protected me all night and open the blinds that face the thicket behind the apartment and the grass police found the rapist's footprint path on. Then nestle back under the covers. Inhaling the tea's steam relaxes me. First ('So glad I got my religion in time!'), I read *Forward Day by Day*'s meditations; half-dozen prayers from the Book of Common Prayer, and the psalm, Old Testament reading, gospel and epistle for the day. Next, notes about my health: medicine, pains and aches, mood swings, menu, exercise. Then, as an outward and visible sign that, though horribly helpless while I was raped, I can control some things, as evidence that although I have no idea when-where-why the rapist chose me, I can figure some things out, I work a *New York Times* crossword puzzle. Finally, still desperate for motives and solutions, I read a chapter or two in a mystery.

Pre-rape, I escaped into British mysteries, and my favorites all featured men who – no matter how they differed from each other – lived in worlds as far from mine as Tolkien's Middle Earth, C. S. Lewis's Narnia, or Alice's Wonderland. Rolling manicured lawns, visiting cards; butlers, valets, chauffeurs, cooks and housekeepers; port, fine sherry, and dressing for dinner; Eton–Oxford–Cambridge; tea, scones and fairy cakes; witty conversations; country villages and estates; bumblingly well-intentioned vicars and earnest innkeepers; cashmere, titles, charming eccentrics. ... Women are never raped, police don't carry guns, crimes occur off-stage, and there is a logical explanation for every crime.

Raped, I'm a character in a cruder mystery. (File #88747812.) Detective C. L. Butler is in charge of the case. Now stories about British sleuths make my eyes slide off the page. 'No, no, no', the gospel song insists, 'They couldn't do. They didn't have the power that you needed to bring you through ...'

Then my best friend sends me books about two new-style US women detectives, Sue Grafton's California-based Kinsey Milhone and Sara Paretsky's V. I. Warshawski. Week after week I gulp, as if drowningly desperate for air, their plain-spoken stories about acid-tongued, fast-thinking, single and self-employed women who not only dare to live alone, but scoff, sneer, seethe when men try to put them in their 'weaker sex' place. Parched and starved, I read and reread Sue Grafton's alphabet adventures, but Sara Paretsky's books become my favourites because her Chicago-based private investigator is even more bodacious and sassy than my pre-rape self. And, in stark contrast to television and movie renditions of women as powerless victims of men, she both withstands and metes out physical violence in every single book. Murder mysteries restore order to worlds thrown out of balance. ... The police may never find the 'brother' who raped me, and I may never get to read him the seven-foot scroll detailing *all* I've suffered since he slithered into my life.

The morning ritual comforts me. If I don't do another purposeful thing all day, I *have* accomplished something. An hour and a half after my first sip of tea, I am focused enough to turn on radio jazz and churn out the six typewritten journal pages I require of myself daily. On days when I can't do the ritual or fall asleep as soon as I've done it, I know I am, once again, nailed to the dank floor of the abyss.

By the time my lease runs out at the end of June, Atlanta's warm enough for sweet and eat-them-in-the-bathtub-juicy peaches. I'm re-mastering grocery shopping and the basics of cooking. Fear has subsided enough for me to sleep with fewer lights on and ride MARTA in relative calm. Still automatically check the physique, skin color and hair-do of all the AfAm men I see to be sure they aren't the rapist returned. Carry the phone the burglar alarm's hooked up to from room to room, spend hours watching CBS soap operas and reruns of *Murder, She Wrote* and (twice a day) *Miami Vice*. No longer expect my '78 Toyota to collapse around me the way my life has, leaving me clutching the steering wheel on Atlanta's maze of highways. But . . . my sabbatical has shape-shifted into sick leave, and I've filed Social Security and TIAA/CREF claims for total disability since I'm too frightened in groups of people, too easily exhausted and too amnesiac to teach.

Summer '89, back in the college town I fled to Atlanta to escape, I can't recall what's in the drawers, closets and cabinets of a house I've lived in for a decade and am astonished at how little I remember about campus buildings, college routines, colleagues' faces, names, disciplines. When the old farmhouse makes its floor-settling night noises, I panic alert. It feels like I've sunk back to the beginning of rape recovery when, the first Saturday night I spend alone a thousand miles away from the rape site, I put knives, scissors, letter openers, potato peelers and pantyhose away with the same compulsion and dismay that drove me in Atlanta. I am back in the high-walled chasm, buried alive beneath a man in a stocking cap mask, bleeding from internal wounds people can't see when they insist on how good I look and sound.

When fast-track people in Amherst ask about my time away, they expect to hear about my photography, book proposals, chapters written, contracts signed. During my nine-month Atlanta stay, I haven't written one line or edited a single page of 'A Language of Their Own', 'These Wild and Holy Women', or *Birthmarks & Keloids*. Reeling toward healing, I have neither organized the negatives, photographs and slides of my Nigeria research nor mastered the word processor. My Atlanta 'accomplishments' are not screaming when a man next to me in church stood close enough to hold one side of the hymnal we were singing out of and sitting between the two AfAm men AAA sent to tow my car to a Toyota dealership. My writing consists of a few personal letters, two letters of recommendation composed at a hobbled snail's pace, disability claims and my journal. Each time I'm asked how Atlanta was, the scab's scraped from my scar.

ATLANTA IS WHERE RAPE TORTURED ME
WHERE I ALMOST DIED FROM AN INVISIBLE WOUND.

Too sick to teach Fall semester, I plan to spend it in St Croix, healing in the sea and sun, but, for the first time in eighty-two years, St Croix is undone by a hurricane. My *second* away-from-Amherst plan in two years pulverized, sealing me in the tomb Amherst feels like – with no resurrection in sight. I'm Brer Rabbit. And Amherst is my tarbaby.

Winter-bare trees. Short days. Low, leaden skies. Cold. Boots, mittens, scarves, thermal underwear, snow shovels and tires, anti-freeze, rock salt. I cling, as though they could save me – to the covers when the alarm goes off. Regress back to daily naps and coloring books. Exhausted no matter how much I sleep. Even the tiniest task monumental. A kind of suicidal I talk myself out of over and over only by realizing that my Sun-and-moon-and-stars would be undone. For the first time since my ugly-duckling high school days, I cry in outbursts that last an hour and try to see the banked emotions through the prisms Dame Julian of Norwich and Rebecca Jackson (of Philadelphia) did.

My Amherst psychiatrist decides Tofranil isn't affecting my depression, prescribes Prozac and makes seemingly Simon Legree rules: Do not sleep in the clothes you wore all day and wear them again the next day; leave the house, even if it's only to get the mail, every day; go, since you have friends, family and church there, to Boston as often as your scant store of energy will allow; write. When asked, my Atlanta psychiatrist said that though raped at home I'd rather be there than anywhere else because home was where I had most control and that, as a person who 'lived in my mind', getting my memory back was complicated by how many things I know. 'Most people only have one language to get back. You have English, French, Spanish, and smatterings of Yoruba.' My Amherst psychiatrist depicts clinical depression as a cunning illness: If you've prided yourself on the regularity of your schedule, it keeps you from going to sleep at night and getting up in the morning; if you're a sensual person, desire is expunged; if you're an intellectual, depression breaks your mind.

My morning ritual is my life-jacket. Downstairs to bigger teapot, a wider range of teas and the inventiveness to combine Mellow Mint and Tropical Escape. Sliced raw ginger and – the ingredients vary daily – pounded cloves. As the weather chills, yogurt's replaced by pears, and tangerines, warmed apple cider with ginger and vinegar or warm milk with a combination of Ovaltine and Postum alternate with tea. Winter's grip tightens, and fruit gives way to seven grain, challah and anadama bread. Using the same thick circular wooden tray I had in Atlanta, I carry a cloth napkin in a wooden ring and a tall clear glass for my elixir upstairs for myself the way I'd do for an invalid friend or a luscious man. Then burrow back in bed for the morning 'work' I did in Atlanta, with additions.

Leaving Georgia has meant giving up the arc of gracious women I relied on, and I miss them so. Over and over I find myself longing for the web of loving women who surrounded Audre Lorde's mastectomy recovery. Once again, as they were in my friendless adolescence, books become my best friends, and their sister-care feeds me. Before I begin each one, I just *know* reading it will gnarl me with envy and make me even more ashamed of allowing rape to derange me so completely. But I am always wrong. Quilts in *Stitching Memories*. Poetry by Lucille Clifton, and Rita Dove, *Lionheart Gal*'s collection of feisty and backative Jamaican women's testimonies ... Toni Morrison's heart-stopping *Beloved*. Alice Walker's *Living by the Word*. Toni Cade Bambara's *Salt Eaters*. ... These women's stories aren't mine so I know there is still space for my frayed pieces of the patchwork quilt, and their sturdy and magical creations brave me to try. Some raped woman needs my witness. Not emotions recollected in elegant tranquility when I am *finally* out of the tunnel it takes all my faith to believe even exists. She needs to taste my terror, hear my gasps for life, watch me inch through brambles of despair, reaching for life and sanity with bleeding stigmata all over me.

> The story of the people and the spirits, the story of earth, is the story of what moves, what moves on, what patterns, what dances, what sings, what balances, so life can be felt and known. The story of life is the story of moving. Of moving on.
>
> Your place in the great circling spiral is to help in that story, in that work. To pass on to those who can understand what you have learned, what you know.
>
> It is for this reason you have endured ...
>
> ... Pass it on ... That is the story of life ... Grow, move, give, move. (Paula Gunn Allen, *The Woman Who Owned the Shadows*, p. 210)

6.4

BODIES–CITIES

Elizabeth Grosz

For a number of years I have been involved in research on how to reconceive the body as socio-cultural artifact. I have been interested in trying to refine and transform traditional notions of corporeality so that the oppositions by which the body has usually been understood (mind and body, inside and outside, experience and social context, subject and object, self and other – and underlying them, the opposition between male and female) can be problematized. Corporeality can be seen as the material condition of subjectivity, and the subordinated term in the opposition, can move to its rightful place in the very heart of the dominant term, mind. Among other things, my recent work has involved a kind of turning inside out and outside in of the body. I have been exploring how the subject's exterior is psychically constructed; and conversely, how the processes of social inscription of the body's surface construct a psychical interior: i.e., looking at the outside of the body from the point of view of the inside, and looking at the inside of the body from the point of view of the outside, to reexamine the distinction between biology and culture and explore the way in which culture constructs the biological order in its own image. Thus, what needs to be shown is how the body is psychically, socially, sexually, and representationally produced.

One area that I have neglected for too long is the constitutive and mutually defining relations between bodies and cities. The city is one of the crucial factors in the social production of (sexed) corporeality: the built environment

From: E. Grosz, *Space, Time and Perversion. Essays on the Politics of Bodies*, London: Routledge, 1995.

provides the context and coordinates for contemporary forms of body. The city provides the order and organization that automatically links otherwise unrelated bodies: it is the condition and milieu in which corporeality is socially, sexually, and discursively produced. But if the city is a significant context and frame for the body, the relations between bodies and cities are more complex than may have been realized. My aim here will be to sketch out the constitutive and mutually defining relations between corporeality and the metropolis.

It may be useful to define two key terms: body and city. By 'body' I understand a concrete, material, animate organization of flesh, organs, nerves, and skeletal structure, which are given a unity, cohesiveness, and form through the psychical and social inscription of the body's surface. The body is, so to speak, organically, biologically 'incomplete'; it is indeterminate, amorphous, a series of uncoordinated potentialities that require social triggering, ordering, and long-term 'administration'. The body becomes a human body, a body that coincides with the 'shape' and space of a psyche, a body that defines the limits of experience and subjectivity only through the intervention of the (m)other and, ultimately, the Other (the language- and rule-governed social order). Among the key structuring principles of this produced body is its inscription and coding by (familially ordered) sexual desires (i.e., the desire of/for the other), which produce (and ultimately repress) the infant's bodily zones, orifices, and organs as libidinal sources; its inscription by a set of socially coded meanings and significances (both for the subject and for others), making the body a meaningful, 'readable', depth entity; its production and development through various regimes of discipline and training, including the coordination and integration of its bodily functions so that not only can it undertake general social tasks, but also become part of a social network, linked to other bodies and objects.

By 'city', I understand a complex and interactive network that links together, often in an unintegrated and ad hoc way, a number of disparate social activities, processes, relations, with a number of architectural, geographical, civic, and public relations. The city brings together economic flows, and power networks, forms of management and political organization, interpersonal, familial, and extra-familial social relations, and the aesthetic/economic organization of space and place to create a semi-permanent but everchanging built environment or milieu.

I will look at two pervasive models of the interrelation of bodies and cities and, in outlining their problems, I hope to be able to suggest alternatives.

In the first model, the body and the city have a de facto or external relation. The city is a reflection, projection, or product of bodies. Bodies are conceived in naturalistic terms, pre-dating the city, the cause and motivation for its design and construction. More recently, we have heard an inversion of this presumed relation: cities have become (or may have always been) alienating environments that do not allow the body a 'natural', 'healthy', or 'conducive' context. Underlying this view in all its variations is a form of humanism: the human subject is conceived as a sovereign and self-given agent who, individually or collectivity, is responsible for all social and historical production. Humans make cities. Cities are reflections, projections, or expressions of human endeavour. On such views, bodies are usually subordinated to and seen as merely a 'tool' of subjectivity, self-given consciousness. The city is a product not simply of the muscles and energy of the body, but of the conceptual and reflective possibilities of consciousness itself.

This view has, in my opinion at least, two serious problems: first, it subordinates the body to the mind while retaining their structure as binary opposites. Second, such a view only posits, at best, a one-way relation between the body or the subject and the city, linking them through a causal relation in which body or subjectivity is conceived as the cause, and the city, the effect. In more sophisticated versions of this view, the city may have a negative feedback relation with the bodies that produce it, thereby alienating them. Implicit in this position is the active causal power of the subject in the design and construction of cities.

A second, also popular, view suggests a parallelism or isomorphism between the body and the city, or the body and the state. The two are understood as analogues, congruent counterparts, in which the features, organization and characteristics of one are also reflected in the other. This notion of the parallelism between the body and the social order (usually the state, but clearly there is a conceptual and historical linkage between the state [the domain of politics] and the city [the polis]) finds it clearest formulations in the seventeenth century when the liberal political philosophers justified their various allegiances (the divine right of kings, for Hobbes; parliamentary representation, for Locke; direct representation for Rousseau, etc.) through its use. The state parallels the body; artifice mirrors nature. The correspondence between the body and the body politic is more or less exact and codified: the King usually represents the Head of State; the populace is usually represented as the body. The law has been compared to the body's nerves; the military to its arms, commerce to its legs or stomach, and so on. The exact correspondences vary from text to text. However, if there is a morphological correspondence between the artificial commonwealth (the Leviathan) and the human body in this pervasive metaphor of the body politic, the body is rarely attributed a sex. What, one might ask, takes on the metaphoric function of the genitals in the body politic? What kind of genitals are they? Does the body politic have a sex?

Again, I have serious reservations with such a model. The first regards the implicitly masculine coding of the body politic, which, while claiming it models itself on the human body, uses the male to represent the human, in other words, its deep and unrecognized investment in phallocentrism.

A second problem is that this conception of the body politic relies on a fundamental opposition between nature and culture, in which nature dictates the ideal forms of culture. Culture is a supersession and perfection of nature.

The body politic is an artificial construct that replaces the primacy of the natural body. Culture is moulded according to the dictates of nature, but transforms nature's limits. In this sense, nature is a passivity on which culture works as male (cultural) productivity supersedes and overtakes female (natural) reproduction.

A third problem concerns the political function of this analogy: it serves to provide a justification for various forms of 'ideal' government and social organization through a process of 'naturalization'. The human body is a natural form of organization that functions not only for the good of each organ but primarily for the good of the whole. It is given in the functional 'perfection' of nature. As a political and hence a social relation, the body politic, whatever form it may take, justifies and naturalizes itself with reference to some form of hierarchical organization modelled on the (presumed and projected) structure of the body.

In such models, which underlie certain conceptions of civic and public architecture, and even more, town planning, there is a slippage from conceptions of the state, which, as a legal entity, raises political questions of sovereignty, to conceptions of the city, a cultural entity whose crucial political questions revolve around commerce. As such, their interests and agendas are separate and at times in conflict: what is good for the nation or state is not necessarily good for the city; conversely, the city may prosper while the state is at war. The state functions to grid and organize, to hierarchize and coordinate the activities of and for the city and its state-produced correlate, the country (side). These are the site(s) for chaotic, deregulated, and unregulatable flows. (The movement of illicit drugs is simply one trail through underground networks of exchange that infiltrate and permeate the city's functioning. The movement of commodities, and of information, are other trails). The city (or town) is formed as a point of transit while the state aims to function as a solidity, a mode of stasis or systematicity:

> The town is the correlate of the road. The town exists only as a function of circulation and of circuits; it is a singular point on the circuits which create it. It is defined by entries and exits; something must enter it and exit from it. It imposes a frequency. It effects a polarisation of matter; inert, living, or human – it is a phenomenon of transconsistency, a network because it is fundamentally in contact with other towns. ...
>
> The state proceeds otherwise; it is a phenomenon of ultraconsistency. It makes points resonate together, points ... of very diverse order – geographic, ethnic, linguistic, moral, economic, technological particulars. The state makes the town resonate with the countryside.
>
> ... the central power of the state is hierarchical and constitutes a civil sector; the center is not in the middle but on top because (it is) the only way it can recombine what it isolates through subordination. (Deleuze and Guattari 1986: 195–7)

The statist representation of the body politic presumes an organized cohesive, integrated body, regulated by reason, as its ideal model. Such a model seems to problematize this cohesive understanding of the ordered body, and to produce instead a deranged body-image, a body frantic to be linked to and part of the network of flows, a body depleted, abandoned, and derelict insofar as it is cast outside these nets (Lingis 1994). The state can let no body outside of its regulations: its demand for identification and documentation relentlessly records and categorizes, though it has no hope of alleviating such dereliction. If the relations between the body and the city are the object of critical focus, the body itself must shake free of this statist investment.

If the relation between bodies and cities is neither causal (the first view) nor representational (the second view), then what kind of relation exists between them? These two models are inappropriate insofar as they give precedence to one term or the other in the body/city pair. A more appropriate model combines elements from each. Like the causal view, the body must be considered active in the production and transformation of the city. But bodies and cities are not causally linked. Every cause must be logically distinct from its effect. The body, however, is not distinct from the city for they are mutually defining. Like the representational model, there may be an isomorphism between the body and the city. But it is not a mirroring of nature in artifice; rather, there is a two-way linkage that could be defined as an *interface*. What I am suggesting is a model of the relations between bodies and cities that sees them, not as megalithic total entities, but as assemblages or collections of parts, capable of crossing the thresholds between substances to form linkages, machines, provisional and often temporary sub- or micro-groupings. This model is practical, based on the productivity of bodies and cities in defining and establishing each other. It is not a holistic view, one that would stress the unity and integration of city and body, their 'ecological balance'. Rather, their interrelations involve a fundamentally disunified series of systems, a series of disparate flows, energies, events, or entities, bringing together or drawing apart their more or less temporary alignments.

The city in its particular geographical, architectural, and municipal arrangements is one particular ingredient in the social constitution of the body. It is by no means the most significant (the structure and particularity of, say, the family is more directly and visibly influential); nonetheless, the form, structure, and norms of the city seep into and affect all the other elements that go into the constitution of corporeality. It affects the way the subject sees others (an effect of, for example, domestic architecture as much as smaller family size), the subject's understanding of and alignment with space, different forms of lived spatiality (the verticality of the city, as opposed to the horizontality of the landscape – at least our own) must have effects on the ways we live space and thus on our corporeal alignments, comportment, and orientations. It also affects the subject's forms of corporeal exertion – and the kind of terrain it must negotiate day-by-day, the effect this has on its muscular structure, its nutritional context, providing the most elementary forms of material support and sustenance for the body. Moreover, the city is also by now the site for the body's cultural saturation, its takeover and transformation by images, representational systems, the mass media, and the arts – the place where the body is representationally reexplored, transformed, contested, reinscribed. In turn, the body (as cultural product) transforms, reinscribes the urban landscape according to its changing (demographic) needs, extending the limits of the city ever towards the countryside that borders it. As a hinge between the population and the individual, the body, its distribution, habits, alignments, pleasures, norms, and ideals are the ostensive object of governmental regulation, and the city is both a mode for the regulation and administration of subjects but also an urban space in turn reinscribed by the particularities of its occupation and use.

Now, to draw out some general implications from this schematic survey:

First: there is no natural or ideal environment for the body, no 'perfect' city, judged in terms of the body's health and well-being. If bodies are not culturally pregiven, built environments cannot alienate the very bodies they produce. However, what may prove unconducive is the rapid transformation of an environment, such that a body inscribed by one cultural milieu finds itself in another involuntarily. This is not to say that there are not unconducive city environments, but rather there is nothing intrinsic about the city which makes it alienating or unnatural. The question is not simply how to distinguish conducive from unconducive environments, but to examine how different cities, different socio-cultural environments actively produce the bodies of their inhabitants.

Second, there are a number of general effects induced by cityscapes, which can be concretely specified in particular cases. In particular, we can say that the city helps to:

1. Orient sensory and perceptual information, insofar as it helps produce specific conceptions of spatiality.
2. Orient and organize familial, sexual, and social relations insofar as the city, as much as the state, divides cultural life into public and private domains, geographically dividing and defining the particular social positions individuals and groups occupy. Lateral or contingent connections between individuals and social groups are established, constituting domestic and generational distinctions. These are the roles and means by which bodies are individuated to become subjects.
3. The city structure and layout also provides and organize the circulation of information and structures social and regional access to goods and services.
4. The city's form and structure provides the context in which social rules and expectations are internalized or habituated in order to ensure social conformity or, failing this, position social marginality at a safe distance (ghettoization). This means that the city must be seen as the most immediate locus for the production and circulation of power.

And third, if the city is, as I have suggested, an active force in constituting bodies, and always leaves its traces on the subject's corporeality, corresponding to the dramatic transformation of the city as a result of the information revolution will have direct effects on the inscription of bodies. In his paper on 'The Overexposed City' (1986), Paul Virilio makes clear the tendency in cities today towards hyperreality: the replacement of geographical space with the screen interface, the transformation of distance and depth into pure surface, the reduction of space to time, of the face-to-face encounter to the terminal screen:

> On the terminal's screen, a span of time becomes both the surface and the support of inscription; time literally ... surfaces. Due to the cathode-ray tube's imperceptible substance, the dimensions of space become inseparable from their speed of transmission. Unity of place without the unity of time makes the city disappear into the heterogeneity of advanced technology's temporal regime. (Virilio 1986: 19)

The implosion of space into time, the transmutation of distance into speed, the instantaneousness of communication, the collapsing of the workspace into the home computer system, will clearly have major effects on the bodies of the city's inhabitants. The subject's body will no longer be disjointedly connected to random others and objects through the city's spatiotemporal layout; it will interface with the computer, forming part of an information machine in which the body's limbs and organs will become interchangeable parts. Whether this results in the 'crossbreeding' of the body and machine – whether the machine will take on the characteristics attributed to the human body ('artificial intelligence', automatons) – or whether the human body will take on the characteristics of the machine (the cyborg, bionics, computer prosthesis) remains unclear. Yet it is certain that this will fundamentally transform the ways in which we conceive both cities and bodies, and their interrelations. What remains uncertain is how.

REFERENCES

Deleuze, Gilles and Felix Guattari (1986) 'City State' *Zone* 1/2: 194–99.
Lingis, Alphonso (1994) *The Community of Those Who Have Nothing in Common*, Bloomington: Indiana University Press.
Virilio, Paul (1986) 'The Overexposed City' *Zone* 1/2: 14–39.

6.5

MAPPING THE COLONIAL BODY: SEXUAL ECONOMIES AND THE STATE IN COLONIAL INDIA

Janet Price and Margrit Shildrick

In the following paper we intend to explore a specific but highly influential area of discourse in the mapping of colonial India. Drawing on extensive archival material emanating mainly from various nineteenth-century and early twentieth-century protestant missionary societies, and particularly those associated with female missionaries, we argue that during this period, the sexuated body of Indian women was a necessary ground for the imposition of colonial state power. And although our detailed analysis does not extend through to Independence, in our conclusion we draw on material unaddressed here to suggest ways in which the sexual economies of the female body were (re)-presented in the power/knowledge nexus of twentieth-century colonialism. Given this emphasis, the issue of racism is not directly addressed here, but is nonetheless implicated in many of the original texts where the discursive constructions of race, sex, and sometimes class are intricately interlinked.

The analysis we offer is not predominantly materialist nor overtly 'historical' but is centred on the production and manipulation of certain discursive meanings which came to centre on India. But to take representation as our focus is to deny neither the reality of certain practices nor to downgrade the substantive political and economic forces at work in Indian society at the time. As Said has said of Orientalism, 'It is not an airy European fantasy about the Orient, but a creative body of theory and practice in which, for many generations, there has been a considerable material investment' (Said 1978: 6).

From: T. Foley, L. Pilkington, S. Ryder and E. Tilley (eds), *Gender and Colonialism*, Galway: Galway University Press, 1995.

The point, nonetheless, is first, that any understanding we may have of historical change springs from knowledge that is always already mediated, and second, that 'reality' as such is inaccessible. And this must be particularly evident in our own position of enunciation as white western feminists attempting to reach a new understanding of the Indian past and present. The very most that we can claim is to reread the maps of the colonial body.

Now there are at least four crucial and interconnected ideas going on here, namely:

1. that maps are interpretive and reinterpretable;
2. that the exercise of mapping is the exercise of control and regulation;
3. that the colonial *body politic* (in a Foucauldian sense) is concerned with the set of material elements and techniques through which power/knowledge relations are mapped onto the individual body of Indian women (Foucault 1975: 28);
4. that the morphology of the body and the morphology of the state meet in an exercise of symbiotic mapping.

In concentrating on the female body as the paradigmatic site of colonial power/knowledge we do not want to suggest that there were not other parallel discourses in play, but rather that Foucault's analysis of the focus of regulatory power in the emergence of the modern *western* society can be equally useful in our understanding of how those societies mapped their domination of others. And in our material about colonial India, what is preeminent is a discursive concentration on female sexuality. If we compare successive maps of the subcontinent over time, noting the increasing and encroaching areas of empire pink, we can imagine too the underlying, irregular morphology of accumulated knowledge about, and intervention in, the lives of Indian women as sexuated beings.

Before looking at specific discourses, the question that needs at least to be flagged if not adequately answered, is why it should be the gendered female rather than the male body which provides the prime, although by no means exclusive, site of regulatory practices.[1] In his own expositions, unless he is talking specifically about the hystericisation of the body, Foucault is most often gender-neutral in his approach, as though 'truths' about the male and female body were really constructed in an undifferentiated way (Foucault 1990: 48). He seems to take for granted that we are all embodied in the same way, and are all equally both subject and subjected in that discourse. In contrast we would say that women are always already in a different relation to their bodies as sexed, not in the sense of that being the source of rigid givens, but in the sense that there is overlap but never identity between the lived experience of women and men. Further, the constructions of meaning through which we 'know' the body consistently privilege the male for his supposed capacity to transcend his embodiment, to become the subject in, rather than of, discourse. Men then are both in and out of their bodies, while women simply are their bodies, 'to be subjected, used, transformed and improved' (Foucault 1975: 136). The Foucauldian category of hystericisation is for them not the special case, but the very condition of being female.

In consequence, then, if the imposition of a regulatory power in any way presupposes both the exercise of a superior and rational will through which meaning can be inscribed on a body (the analysable/intelligible body) and, concurrently, techno-political manoeuvres through which a body which can be controlled and manipulated (the useful/manipulable body), then it is no surprise that women should be identified as the point of least resistance (Foucault 1975: 136). Yet paradoxically, they may simultaneously occupy a point of resistance to this move as, hidden behind metaphorical veils, they remain the unknown and unknowable Other. Nonetheless, the process of colonisation takes many different forms, and it is therefore necessary to look for further reasons why the female body is of such importance in the discourses with which we are concerned.

It cannot be supposed, of course, that, although the construction of Empire generated its own specific but often indiscriminate representations of women in both Hindu and Muslim society, it did so on the basis of a blank field. First, though it is not our focus here, the Indian people could look to the authority of well-established indigenous cultural codes, while the British in India had by the mid-1800s their own counter-history.[2] By the time the imperialist process rapidly accelerated, the colonialists had already long occupied limited territory of the subcontinent in the pursuit of advantageous trading conditions. The dominant force had been the East India Company which only reluctantly ceded its own power and influence to the British state, and whose own hegemony seems to have been accomplished by a combination of military and economic intervention in which the discourses on Indian women were not greatly evident. What seems to be different about the subsequent state takeover is the way in which a repressive force gave way to a yet more effective and widespread dominance explicitly based on moral rights and enacted through the bodies of women. What is perhaps surprising to us now is the scale of influence wielded by the missionary discourses of the nineteenth and early twentieth century which were extensively used, in parliamentary debates for example, to justify the colonial process. At the same time, at least during the early part of the nineteenth century, the moral probity of the East India Company was being undermined by the British civil service in the interests of taking over the field of influence occupied by the Company. And here too the bodies of women were the focus of competing discourses. In both cases – and it must be stressed that the material we refer to is specific and local – the issue is one not of repression, but of rescue and reform. There was a proliferation of discourses in which women were cast as victims – sati, child marriage, temple dancing girls – individually in need of rescue, but *en masse* forming the ground upon which competing colonial discourses of reform could be conducted.

[. . .]

The female body plays no active part in the political skirmishing but is nonetheless the site of justificatory discourses. And this process is most evident in the vast amounts of material which emanated from the various missionary societies for home consumption.

Without necessarily being committed to any view of the accuracy of the representations of Indian women propagated throughout the colonial period, it is clear that the major focus is on their highly subordinate position within both Hindu and Moslem society. The Ordinances of Manu, for example, were very widely cited as the textual basis on which Hindu women were treated not just as inferior, but as morally and physically degraded creatures in a society organised entirely for the benefit of men (Weitbrecht c.1878: 7). From the earliest travellers, settlers and above all missionaries came horrific stories. *The Daughters of India: Their Social Condition, Religion, Literature, Obligations and Prospects* (1860) was written by the Reverend Edward Jewitt-Robinson. He writes,

> the peculiar disgrace of India is female infanticide ... harlotry is a sacred institution ... thousands of mere children who never left the parental roof are treated as widows in India. Is it a wonder that self-immolation has been preferred by some to the sorrowful life of an Indian widow. (Jewitt-Robinson 1860: 97)

What unifies these multifarious texts is not simply the reiteration of similar stories but the actual recirculation in many cases of the very same source story. Many texts used, for example, material that could be traced back to the redoubtable Rev James Peggs who, in the 1840s, had produced two sensationalist books, themselves drawing on much second hand material, called *The Infanticides Cry to Britain* (1844) and *A Cry from the Ganges* (1843). He wrote, for example, on female infanticide amongst the Jahreja Rajputs, 'To render this death, if possible, more horrible, the mother is commonly the executioner of her own offspring' (Peggs 1844: 33). Such books outline in lurid and distressing detail the endless abuses practised on women's bodies which undifferentiated Indian religions and society seemed to sanction. Later accounts are often less geographically specific then Peggs, so that it becomes difficult to avoid the implication that the practices are the universal experience of all Indian women.

Our project is not to question the specific 'truths' of these accounts but always to problematise the production of knowledge; and it is in this that the political significance lies. Our poststructuralist framework does not commit us, however, to the view that there is nothing to choose between one account and another. As a strategic move it is legitimate to prefer any one of the counter-discourses of resistance, but given the material in hand, our initial analysis must focus on the deconstruction of *authorised* discourse. Our intention is to point out how specific texts were incorporated into a cultural code about India which served to promote and justify the expanding map of colonial power.

Even where the issue was expressed more temperately in terms of the purdah conditions under which many women, both Moslem and Hindu, lived, the import of the prevailing discourses was that women *needed to be rescued*.[3] In other words the putative project of liberating the bodies of these women orchestrated a move in which the British state successively appropriated the remaining political freedoms of the Indian people as a whole. And the missionary societies, whose avowed aim was to reclaim the soul by the rescue of the body, were prime agents in establishing the necessary discursive field.

Although some missionary work had begun alongside the very earliest trading ventures, it was formally restricted to settler areas until the 1813 Charter Act (Drummond and Bulloch 1975: 170). Within three decades most societies began, for the first time, to send out women, not simply in the role of accompanying wives, but as missionaries in their own right.[4] This seems to reflect a frequently stated nostrum of the time that it was with Indian women that the opportunity of success lay both in so far as they were responsible for future generations, and in that they presented a major site of resistance to male conversion (Weitbrecht c.1878: 12). Removing the veil, both real and metaphorical, thus became the prime concern of the missionaries. The societies recognised that despite their own preferences for the male approach, only female missionaries would be in a position to enter the closed world of the zenana, where Indian women were

> immured like caged birds, beating their tired wings against the prison walls vainly, yet eagerly longing to learn something of what is beyond, and to hear further of the faint whisper which has been borne in to them of a brighter life somewhere, they know not where. (Warner-Ellis 1883: xii)

And the approach that these Christian women adopted yet again focused explicitly on the female body as the central point of the power/knowledge regime that served both god and the state. The insistent discourse of female degradation behind the veil of purdah – 'our suffering sisters' – lent weight to state moves to extend colonial power through the regulation of the sexed body ('Homes of the East' 1911: 1). The existence of the customs of infanticide and sati, for example, in areas formally outside the control of the British government service was given as one of the major reasons to take over those independent and quasi-independent states.

Throughout the nineteenth century a series of laws was promulgated banning the worse excesses against women and at the same time greatly extending British power. Whether or not such laws as the 1829 Abolition of Suttee Act or the 1856 Widow Remarriage Act had any good effect for women is of course highly debatable. Nonetheless, the discursive positioning of Indian women as victims was a necessary 'truth' in the mapping of the sub-continent.

Now the whole idea of tearing aside the veil (and the word purdah literally means 'curtain') implied of course not that some 'truth' was to be constructed, but that the Truth would be revealed – both about and for the supposedly silent victims whose lives it masked. But for the women missionaries, who for a time were the prime actors, the project of unveiling was the intersection of multiple meanings. On the one hand they remained obsessed with charting the aberrant socio-sexual customs, of emphasising the 'otherness' of the Indian women in whose lives they intervened. 'The daughters of India are unwelcomed at their birth, untaught in childhood, enslaved when married, accursed as widows and unlamented at their death' (Weitbrecht 1880: 3). At the same time they offered rescue and relief, extending the hand of sisterhood to Indian women whom they saw as their special responsibility. In the words of Mrs Weitbrecht again,

> We hardly dare contemplate even our own share of the effort needed, to compass the duties we as Englishwomen owe to our sisters, lying, as alas! the great majority of them still are, in the helplessness of heathenism in that far-off land. (Weitbrecht 1880: 2)

This simultaneous approach to both difference and sameness is the characteristic move of western discourse and flags that Indian *women* at least were already positioned within a normative economy of the same. 'Comparing the Hindu female with the women of England, it must be acknowledged that her darker complexion is only a veil of shadow, more or less dense over equal loveliness' (Jewitt-Robinson 1860: 3).

The specificity of their differences in the sense of multiple diversities – both with regard to the westerners and among themselves – was unacknowledged, and the vast majority of contemporary accounts signally fail to distinguish not simply between geographical location, ethnic grouping, and caste, but even between Hindu and Moslem. What served to justify intervention and to guarantee the self-authorised, self-present voices of the missionaries and their colonial allies was the construction of a unified and different, silenced other who paradoxically speaks/calls out for help and rescue. In other words the discourse on Indian women, and more generally of the Orient, arose as Said puts it, 'according to a detailed logic governed not simply by empirical reality but by a battery of desires, repressions, instruments and projections' (Said 1978: 8).

How then did the interaction between British and Indian women proceed? Who were the 'Eastern sisters' on whose behalf the missionary tracts appealed?[5]

> I was asked to plead with the girls of England to sacrifice themselves for the sake of the girls of ALER. Surely there are two friends who would count it joy to work together, one medical and the other educational, for the neglected women and children in the hundreds of villages around ALER. (*Women's Work on the Mission Field 1858–1908*, 1908: 771–72)

Whilst the mediation of discourse may make it impossible to discern any clear image, we can at least suggest some points of contact. Despite missionary work with Indian prostitutes, for example, the material on which we draw – produced largely for home consumption – portrays early missionary concerns as lying not with 'those of the lower orders [who] have as much liberty as is good for them', but with 'the *ladies* of Bengal [who], as everyone knows, are

Zenana prisoners' ('The Women of Bengal' 1881: 114), and whose relative status most nearly matched those of the European ladies. The institution of purdah which so clearly focused the concern of missionaries was in fact by no means universal, but was geographically varied, religiously diverse, and practised, for economic reasons, mainly by the higher classes. Similarly the abuses which the British catalogued were often specific rather than general experiences, but many were recirculated across the terrain of a universalised upper caste. Child marriage and the treatment of widows, for example, were not the signs of a degraded people as the missionaries thought, but were accepted and, in cultural terms, sometimes highly respected practices.

And yet even in addressing this narrower band of women who, though of higher status, had the least visible independence, it is unclear just whose interests were being served. What was at stake was not so much the physical well-being or even the moral uplift of the Indian women concerned, but the needs of the colonial process. The manipulation of cultural codes, the remapping of British concerns onto the Indian body, could in any case hardly result in any authentic freedoms for women. Rather women remained throughout simply the ground on which the colonial discourses were imposed.[6] They were the currency of the discursive exchanges but never the subjects. And whether they were involved in purdah practices or not, women's space was, as always, confined and manipulated by others. Indeed the delineation of boundaries was a necessary feature of colonial morphology.

During late nineteenth and the early decades of the twentieth century, missionaries, ostensibly in pursuit of souls, developed and refined new techniques of discipline and surveillance for the female body. In the same way that the colonial state regulated and enclosed the landscape of India, through such ventures as the railway system – which 'established in India under the auspices of the government of the country has two chief objects – the moral advancement of the population, and the development of commercial resources' (Clarke 1857: 84) – mission stations were likewise surveying *their* territory, laying out their surroundings in such a way that constant constraints and obligations were imposed on the women and children who lived on or visited them in increasing numbers. Within the mission compound was a whole series of institutions – the church, the hospital, the school, the rescue home for women converts, the home for orphans, the industrial classes and workrooms, the recreation ground – through whose regulatory mechanisms, the female body could be subtly manipulated and organised. Within this space, women could be kept permanently in view, a kind of 'panopticon' in which their every move and whereabouts could potentially be monitored, the subjects of 'an uninterrupted, constant coercion' (Foucault 1975: 201, 136).

The disciplinary mechanisms circulating amongst missionary organisations also functioned at the level of the active body. An invidious control was exercised, in often minor and individualised ways, over the gestures, movement, language, and appearance of the women and children under the custodianship of the missionaries. New patients in the zenana hospitals and women coming for industrial work were persuaded to abandon their own clothes in favour of the garments and sheets approved by the missionaries. The industrial workers also came under censure for their language, 'compared to which, Mr Goudie says, English swearing is clean. Now any of the newcomers using it are instantly checked by those who have been longer in the class' (*A Visit to Ikkadu* c.1910: 3). Miss Emilie Posnett, when attending a child with typhoid at her home in a nearby village was deeply disturbed to find her, 'not robed in the little white nightdress of an English child, but in gay satin garments of the most brilliant hue' (Posnett c.1910: 3).

In the Madras Presidency, missionary insistence that lower caste women wear blouses led to the 'upper cloth riots', which continued sporadically for over 30 years in the mid-nineteenth century (Warner-Ellis 1883: 29). Such attempts to impose control served to produce a docile body, useful, manipulable, and more intelligible to the missionaries, bringing Indian women, lower as well as high caste, closer to the universalised ideal, whilst simultaneously emphasising their Otherness, or as Homi Bhabha puts it, 'Anglicised but emphatically not English' (Bhabha 1984: 128).

The industrial work, established to provide a livelihood for women, many of whom had been forced to leave their homes when they had converted to Christianity, was largely lace work and fine embroidery. Again, the point of such work seems to lie not so much in its direct utility, but more in that it rescued the women from potential 'associations so depraving' (*Our Missionary Year 1899–1900* 1900: 153) and provided disciplined activity and a controlled environment to protect them from moral harm and such dangers as possession by Satan (Evans 1906: 12).[7]

> It is difficult to estimate what must be the value not only to the women themselves, but through them to their children, of so many hours spent every day in an atmosphere that is clean, physically and morally. (*A Visit to Ikkadu* c.1910: 3)

And further, Miss Spencer of the Zenana Missionary Society Industrial Home wrote to Miss Smith,

> In the home I'm trying to find husbands for the women! One lady calls my circular an 'SOS for husbands' which it is; for three have taken matters into their own hands and again entangled their lives, and my SOS are the only hope of keeping eleven others straight. (Spencer 1929)

Through these and other techniques, the movement, bodies and language of such women were brought within a disciplinary power/knowledge regime. This terrain, in situating women not always as victims in need of rescue but increasingly as weak, at times morally licentious and blameworthy individuals, provided new and fertile ground for furthering the project of mapping and remapping the sexual body.

As part of this counter discourse which became ever more insistent, the issue of motherhood was a central concern. The maternal role of Indian women was seen as a source of moral degeneracy, responsible for the ills of the Indian family and for 'that mighty force of *maternal influence* which perpetuates all the foolish and base institutions and degrading tenets of Hinduism or Islam' (Matthews 1880: 11).

In popular missionary writings and pamphlets of the time, the concerns about 'our eastern sisters' were matched by numerous stories detailing not simply well-rehearsed issues such as child marriage and widowhood but the failure of Indian women to care properly for their children and to hold families together. The journal *India's Women* published by CMS in the 1880s is a major source of such views. Miss Rose laments, 'I heard always and read so much of the "strong love" of the "gentle Hindu mother" but my experience had been that they are devoid of true love and are not "gentle"' (Rose 1885: 315). In an earlier edition, the story of Maya, a dying Indian child, is recounted. As she lay, 'tossing in pain and restlessness', the author asks of her mother, 'Did no lingering touch of motherhood linger in her breast? Alas! heathen mothers are taught thus to forsake their own offspring by their cruel religion.' ('The Girls Union' 1883: 281–2). The amoral influence exercised by mothers (specifically on sons), the growing number of orphans being cared for on mission stations, parents who were prepared to sell or give away their daughters to the missionaries, husbands who deserted their wives, wives who ran away from their husbands, all served to add weight to the argument that the Indian family was in a state of terminal decay. And responsibility for this was shifting from the discursively constructed incompetent masculinity of Indian men to the newly perceived inadequacy of the Indian mother.[8]

In the colonial view the closed world of purdah was a breeding ground of ill-health, both in the immediate sense and in its supposed effect on the physical and moral health of future generations. The missionaries, 'who will see in the Zenanas of India a moral Crimea; women who see in their Indian sisters worthy objects of loving care' (Hasell 1862), espoused the aim of securing their release both from purdah and from the clutches of heathenism. Their methods most often focused on the offer of the material benefits associated with education or medical aid. And at the point of intersection of these two vectors, motherhood was emerging as the centre of the discursive field.

Where once the construction of Indian women as indissolubly tied to their sexuated bodies had originally focused on their identities as virgins, as wives, and as widows, the dominant representation of the twentieth century was to be as mothers. The period of British power did undoubtedly see some individual state interventions which worked to the advantage of limited numbers of women, but the primary motivation throughout was to extend and consolidate colonial power. In rereading the authorised British maps, rather than the contours of resistance, we have shown that the potentially divergent discourses in play functioned to legitimise a series of regulatory practices directly grounded on the bodies of women, and indirectly privileging the political interests of what Irigaray would call the hom(m)osexual economy (Irigaray 1985). In its colonial form, the mapping of the body was an essential and essentialist move in the power/knowledge configurations of the state.

NOTES

1. Whilst most of the material we will present addresses women, see also Janaki Nair (1990) and Mrinalini Sinha (1987) who have identified a repeated move within colonial discourse to gender Indian men as female.
2. See, for example, Uma Chakravarti, 'Whatever Happened to the Vedic *Dasi*? Orientalism, Nationalism and a Script for the Past' and Lata Mani, 'Contentious Traditions: The Debate on *Sati* in Colonial India', both in Sangari and Vaid (1989).
3. For stories of women and children who are *rescued* by the missionaries from illness and idolatry, see *Homes of the East*; *India's Women*; *Picture Papers for Children*, all from CMS/CEZMS and *The Harvest Field* and other Women's Auxiliary Publications from the Methodist Missionary Society.
4. By 1910, the Protestant Missions in south India had 349 unmarried foreign women missionaries working in the field, and 431 wives of male missionaries. (SIMA 1910).
5. Antoinette Burton has detailed the ways in which British feminists of the late 19th century represented themselves as agents of salvation for their Indian sisters and, through a feminised imperial rule, contingent upon British women getting the vote, as the 'saviours of the very empire itself' (Burton 1991: 48).
6. In her article, 'Contentious traditions: the debate in colonial India' Lata Mani has drawn a similar conclusion, '... women are neither the subjects nor objects, but rather the ground of the discourse on sati. ... women themselves are marginal to the debate' (1989: 117). Also, though it is not part of our focus, it should be marked that the competing and resistant discourses of higher status Indian males, which attempted to justify and strengthen the heroic Aryan tradition, were similarly organised around women's bodies.
7. The CMS Archives and the Methodist Archives hold numerous pamphlets which contain details of industrial work for women. See Spencer (1929); Evans (1906); *The History & Aims of CEZMS* (c.1950); *Our Missionary Year 1899–1900* (1900).
8. Again, the popular papers of the missionary societies are a prime source of these stories.

REFERENCES

Archive Sources:

KP: SJLSC Knowsley Pamphlets, Sydney Jones Library Special Collection, University of Liverpool.

CMS Church Missionary Society and Church of England Zenana Missionary Society Archives, Special Collections, Main Library, University of Birmingham.

MA: JRA Methodist Archives, John Rylands Archives, John Rylands Library, University of Manchester.

Bhaba, H. (1984) 'Of Mimicry and Man: The Ambivalence of Colonial Discourse' *October* 28 (Spring): 125–33.

Burton, A. (1990) 'The White Woman's Burden. British Feminists and the Indian Woman, 1865–1915' *Women's Studies International Forum* 13, (4): 295–308.

Burton, A. (1991) 'British Imperial Suffragism and "Global Sisterhood" 1900–1915' *Journal of Women's History* 3, (2): 46–81.

Chakravarti, U. (1989) 'Whatever Happened to the Vedic *Dasi*? Orientalism, Nationalism and a Script for the Past' in *Recasting Women in India: Essays in Colonial History*, ed. Kumkum Sangari and Sudesh Vaid, New Delhi: Kali for Women.

Clarke, H. (1857) 'Prospectus of Simla Railway Company' cited in *Colonization, Defence & Railways in Our Indian Empire*, London: John Wede. (*KP: SJLSC*)

Drummond, A. L. and J. Bulloch (1975) *The Church in Victorian Scotland*, Edinburgh: St Andrews Press.

Evans, J. A. (1906) *Hitherto ... Helped. A Record of God's Dealings with His Children at Baranagore*, Woking: The Gresham Press. (*CMS*)

Foucault, M. (1975) *Discipline and Punish*, London: Allen Lane.

Foucault, M. (1990) *The History of Sexuality, vol 1*, London: Penguin Books.

'The Girls' Union' (1883) *India's Women* 3.17 (Sep/Oct): 281–2. (*CMS*)

The Harvest Field, A Missionary Magazine, (1895–1910) Wesleyan Mission Press. (*MA: JRA*)

Hasell, Rev. S. (1862) cited in 'Sowing and Reaping' (1880) *India's Women* Preparatory Number: 20. (*CMS*)

History and Aims of CEZMS, (c.1950) Collected Papers, Unpublished. (*CMS*)

Homes of the East, (1910–24) CMS. (*CMS*)

Irigaray, L. (1985) *This Sex Which Is Not One*, Ithaca: Cornell UP.

Jewitt-Robinson, Rev E. (1860) *The Daughters of India: Their Social Condition, Religion, Literature, Obligations and Prospects*, Edinburgh: Thomas Murray & Son. (*MA: JRA*)

Mani, L. (1989) 'Contentious traditions: the debate on *Sati* in colonial India' in Sangari and Vaid.

Matthews, J. E. (1880) 'Editorial: our object and our work' *India's Women* Preparatory Number: 11–13. (*CMS*)

Nair, J. (1990) 'Uncovering the Zenana: visions of Indian womanhood in English-women's writings, 1813–1940' *Journal of Women's History* 2 (1): 8–34.

Our Missionary Year 1899–1900. Being a Popular Report on the Work of the Wesleyan Missionary Society, (1900) London: Wesleyan Missionary Society. (*MA: JRA*)

'Our suffering sisters in India and China' (1911) *Homes of the East* 28 (1 Jan). (*CMS*)

Peggs, Rev. J. (1843) *A Cry from the Ganges. The Present State of Exposure of the Sick*, London: Ward & Co. (*MA: JRA*)

Peggs, Rev. J. (1844) *The Infanticide's Cry to Britain. The Present State of Infanticide in India*, London: Ward & Co. (*MA: JRA*).

Picture Papers for Children (c1910) CEZMS. (*CMS*)

Posnett, E. (c.1910) *Won by Medical Work*, Women's Auxiliary of the Wesleyan Methodist Missionary Society. (*MA: JRA*)

Rose, A. (1885) 'Sowing and reaping' *India's Women* 5 (November): 315. (*CMS*)

Said, E. (1978) *Orientalism: Western Conceptions of the Orient*, London: Penguin.

Sangari, K. and S. Vaid, (eds) (1989) *Recasting Women in India: Essays in Colonial History*, New Delhi: Kali for Women.

SIMA (1910) 'Table of statistics of protestant missions, South India' *The Harvest Field* 30.

Sinha, M. (1987) 'Gender and imperialism: colonial policy and the ideology of moral imperialism in late nineteenth-century Bengal' in *Changing Men*, ed. M. Kimmel, London: Sage Publications.

Spencer, M. (1929) Personal Letter from CEZM Industrial Home, Bangalore to Miss Smith, 10 July. (*CMS*)

A Visit to Ikkadu, or With the Doctor on Her Rounds (c.1910) Women's Auxiliary of the Wesleyan Methodist Missionary Society. (*MA: JRA*)

Warner-Ellis, H. (1883) *Our Eastern Sisters & Their Missionary Helpers*, London: Religious Tract Society. (*MA: JRA*)

Weitbrecht, Mrs (c.1878) *The Women of India and Zenana and Educational Work among them*, London: Indian Female Normal School and Instruction.

Weitbrecht, Mrs (1880) 'To our friends' *India's Women*, Preparatory Number, CMS.

'The women of Bengal' (1881) *India's Women* 1 (3)(May/June): 114. (*CMS*)

Women's Work on the Mission Field – Jubilee Number 1858–1908 (1908) Women's Auxiliary of the Wesleyan Methodist Missionary Society. (*MA: JRA*)

6.6

WOMAN, NATION AND NARRATION IN *MIDNIGHT'S CHILDREN*

Nalini Natarajan

In Salman Rushdie's *Midnight's Children*, the midnight of Indian independence is represented through refraction of the colors of the Indian flag onto national celebrations (extravagant 'saffron rockets' and 'green sparkling rain') and the bodies of women giving birth: 'green-skinned', 'whites of eyes ... shot with saffron' (*MC*, 132).[1] We may note significant juxtapositions and identities: woman's pain with communal joy, human with national birth, woman's body *as* the national tricolor flag.

Gender and Nation

The scene illustrates the centrality of gender in the space of the social imaginary that constitutes 'nation' while indicating the dissimilar elements that comprise the collectivity of nationalism. The two women whose ordeal in labor is represented in national colors are from the more marginalized sections (by the dominant middle-class Hindu ethos) of Indian society: a Muslim woman and a humble street singer. The text provides an occasion for introducing my concerns in this essay: the spectacle or visual effect of woman as it shapes the national imaginary, the way woman functions as sign in the imagining of community, the relation of these aspects of woman as sign and spectacle (as figurations of the beloved, mother, and daughter) to the failure of the secularization project that 'Indian' culture generally, and the Bombay cinema industry in particular, envisaged for itself in the early decades of Indian independence.

From: I. Grewal and C. Kaplan (eds), *Scattered Hegemonies: Postmodernity and Transnational Feminist Practices*, Minneapolis: University of Minnesota Press, 1994.

[. . .]

WOMAN AS SIGNIFIER FOR NATION

Woman functioned as a signifier in many ways in the contrary dialectic of stasis and change in the imagining of India. It was required not only to imagine one out of many, an operation requiring a relinquishing of the caste-based hierarchies to a pan-Indian modernity, but also to render this one ontologically stable, an effort that inevitably privileged the dominant cultural group – broadly speaking, the Hindu middle classes.[2] In reading narrative against this paradoxical cultural effort, I look at woman in three moments in nationalism: (1) the movement from regional to national in the 'modernizing' process; (2) the threat of communal or civil rupture within the body politic; and (3) the rise of fundamentalism. Woman's body is a site for testing out modernity, in the first moment; in the second, as 'Bharat Mata' or 'Mother India', a site for mythic unity in the face of fragmentation; and in the third, as 'daughter of the nation', a site for countering the challenge posed by 'Westernization', popularly read as 'women's liberation'.

WOMAN AND NATION IN *MIDNIGHT'S CHILDREN*: THE SCOPIC AND THE CIVIC

> A nation which had never previously existed was about to win its freedom, catapulting us into a world which ... was ... quite imaginary; into a mythical land, a country which would never exist except by the efforts of a phenomenal collective will – except in a dream we all agreed to dream – it was a mass fantasy. (*MC*, 129–30)

The text I am reading traces the fortunes of a Muslim family in complex allegorical relation to the fate of the nation. Woman occupies a minor role in the narrative, but my argument foregrounds her marginality as a strategy of reading. I have already indicated the symbiotic connection between *Midnight's Children* and Bombay cinema. The representation of women is a startling instance of the connection, and it is possible to read the text as parodying Bombay cinema's use of women. At the same time, the text's own use of woman's body as signifier for nation implicates it within a critique of male-dominated culture.

My first argument links the scopic (that which is seen) with the civic. Synecdoche, the imagination of a whole from its parts, essential to nation construction, also becomes the way woman is perceived in *Midnight's Children*. The first national subject textualized in Rushdie is the German-educated doctor Aadam Aziz. He returns home with a void in his head – European scepticism has destroyed his faith in 'Islam' and in 'India'. The void becomes the space of desire. His reintegration occurs over the body of a woman patient, who later becomes his wife. Because she is in purdah (veiled) she is shown to him through holes in a sheet. As he treats her in parts he begins to imagine her as whole. This coincides with his imagining a 'whole' Indian identity for himself, instead of his regional Kashmiri one.

The synecdochic process of discovery or construction of Naseem Aziz is a camera technique familiar in Bombay cinema. The camera focuses on the heroine's body part by part. On one level, this defers to the censors; on another, it leaves the job of construction to the male hero and the audience. The popular film *Mere Mehboob* (1963), for instance, depicts the hero's discovery (which is also imaginative construction) of the woman he loves.[3] He, like Aadam Aziz, has only seen her in parts – walking fully veiled on a university campus. The woman's covering is a sign of orthodox values (here the national culture exploits the titillations of Islamic restrictions on women as presented by many producers who work in Bombay cinema) while the university campus is a signifier for modernity. In another film, *Pakeezah* (1971), a viewing of the heroine's feet takes place on a train.[4] The university and the train are both symbols and spaces of modernization and integration. The audience participates at once in the potential for female viewing offered by a modernizing India, as well as the retreat into traditional taboos that monitor the revealing of the female body. Such representations demarcate the space of desire as male. This imaginary uncovering/covering of woman becomes the site for national self-definition, a site where the contradictory facets of the national ideology are played out. Woman should fulfil the individual male psychic need for scopic/sexual gratification and yet be the figurehead for national culture, guarded by the censors. The contradictions in woman's position as spectacle may be seen in the social dramas of contemporary India – bride viewings by eligible males, the spectacles of lavish marriages financed by the fathers of brides, bride burnings caused by 'inadequate' dowry, the resurgent spectacle of widow-sacrifice, or sati. In each case, the body of woman becomes a focus for the symbolisms of cultural and religious reaction.

While male discursive dismembering (such as Saleem's mutilated finger or bruised scalp) symbolizes national rupture, the representation of woman as parts in both fictional and filmic discourse provides an occasion for imagining wholeness. Although she is a symbol for wholeness, her own integrity remains secondary. The text announces within brackets '(he has told her to come out of purdah)', the punctuation indicating that woman's freedom is an aside in the narrative of nationalism. When the couple move to Amritsar, Naseem Aziz finds herself cruelly exposed to the multiplicity of the Subcontinent as Aziz sets fire to her purdah veils:

> Buckets are brought; the fire goes out; and Naseem cowers on the bed as about thirty-five Sikhs, Hindus and untouchables throng in the smoke-filled room. (*MC*, 33)

Imaginative construction of woman's body is a metaphor for constructing national identity from regional, and the exposure of woman's body is a signal for the melting pot of secular modernity. The text represents this modernity as a sexual threat for women – note the connotations of 'bed', 'throng'. For in the political context of decolonization, modernity is required of Indian women. An

Oedipal trace could be observed here – we recall that Aadam Aziz's mother also came out of purdah in order to finance her son's education. Aadam Aziz demands of his wife after the purdah-burning: 'Forget about being a good Kashmiri girl. Start thinking about being a modern Indian woman' (*MC*, 33). Women are required to shed their traditional inhibitions; their reluctance to do this could indicate disjunctive articulations in the discourse of nationalism, which claims to construct one out of many. 'You, or what?' says Naseem at Aziz's request that she come out of purdah. 'You want me to walk naked in front of strange men' (*MC*, 33).

Incidentally, we may note the difference to woman in Western representations. In *Alice Doesn't*, de Lauretis quotes Mulvey's account of the paradigmatic film narrative where woman is first object of the collective male gaze and then reserved for the hero's eyes alone.[5] The movement of woman as scopic object between public and private spheres is mediated by the wider sociohistorical processes that affect gendering and by the specific anxieties of nationalism. Rushdie's text reveals how women are tied into the process of middle-class homogenization as India modernizes. As part and parcel of the new 'nationhood' and its economic, social, and cultural coordinates, woman becomes an index of the erosion of discrete regional and caste cultures in the movement from regionalism to modernity. The class anxieties that imposed a 'new kind of segregation' on women in the nineteenth century are modified so that women may emerge in public.[6] Thus in Rushdie's text, woman moves from man's individual gaze to the collective gaze of many. But this emergence into the public gaze is as problematic as women's seclusion.

Frantz Fanon has discussed the politics behind the veil in the colonizer's attempt to decimate the colonized culture. The battle to end purdah in the colonial context is inflected by the colonizer's wish to 'rescue' the colonized woman from the 'backward' colonized male.[7] Here in decolonization, the newly independent male demands what he resisted during colonialism, or conceded grudgingly in response to British accusations of 'backwardness'. In both cases, the uncovering of women's bodies is related more to the politics of men's power relations than any interest in female subjectivity.

Rushdie's text ironizes the formation of the national bourgeois imaginary through the relations of Aadam Aziz and Naseem. Although Bombay cinema retains for the male the scopic advantage, and consistently portrays woman as an entity to be discovered and protected in the formation of a new India of patriarchally monitored 'progress' for woman, Rushdie's text unseats these confidences in the spectator/subject. For Aadam Aziz's attempt at mental construction fails – he misapprehends Naseem Aziz. Synecdoche allows a space for the imagined object to assert its autonomy. When the whole is assembled it turns out to be very different from the sum of its parts. Naseem Aziz emerges as the stronger partner in the relationship, defying her husband's desires by becoming fat and refusing to do his sexual bidding. Communication between the couple is forever curtailed through Naseem's silence; her body promises not cognitive wholeness, but rupture.

'MOTHER INDIA'

Naseem Aziz's daughter Amina, though conventionally unattractive, is sexually precocious: she steals her older sister's financé. For the early chapters of her appearance, she lives underground with her fugitive lover – a textual absence. Like Naseem, unmarried women exist only partially in discourse. When she surfaces again, as wife and mother-to-be, national imagining goes side by side with Partition riots. In the Delhi sections of *Midnight's Children* the imagined India reappears in the visual space of the bioscope – the Dilli Dekho machine that shows children the collage of a unified India:

> Inside the peepshow of Lifafa Dass were pictures of the Taj Mahal, and Meenakshi temple, and the holy Ganges ... untouchables being touched; educated persons sleeping in large numbers on railway lines. (*MC*, 84)

We may note here the voyeuristic terminology: peepshow, the sight of untouchables being touched, of the educated homeless ... the exclusions that help the bourgeois Indian child still dream of Indian unity. Meanwhile, in the geographical space of Delhi, attempts at unity are shown to be futile. A Hindu revivalist group (the Ravana gang) terrorizes Muslims, and the Muslim crowd turns on the lone Hindu Lifafa Dass. Once again, the spectacle of woman's body, Amina Sinai provides a national image and averts a riot:

> 'Listen', my mother shouted, 'Listen well, I am with child. I am a mother who will have a child and I am giving this man my shelter. Come on now, if you want to kill, kill a mother also and show the world what men you are!' (*MC*, 86)

Woman as spectacle of motherhood once again evokes dreams of unity and wholeness. Woman here is the dream of unified India, and her unborn child its hypothetical citizen. In pregnancy she is the symbol of the wholeness of the men themselves. Motherhood, which could be a privileged site for women and also a potential challenge to patriarchal systems through its admitting of, in Kristeva's terms, an 'otherness within the self', is appropriated for nationalistic purposes. As Klaus Theweleit has said, 'woman is an infinite untrodden territory of desire which at every stage of historical deterritorialization, men in search of material for utopias have inundated with their desires.' He further adds that it is the lure of a freer existence that marks this territory of desire and is most often indulged in by men in search of power rather than those already dominant.[8]

How does the figure of Mother cement nation? She suggests common mythic origins. Like the land (which gives shelter and 'bears'), she is eternal, patient, essential. National claims have always been buttressed by claims to the soil. The linking of 'Mother' with land gains strength from Sita, who was the daughter of Mother Earth. During moments of 'national' resurgence, the land is figured as a woman and a mother. In the era of militant Hindu resurgence in the late nineteenth century, Bankim Chandra Chatterjee's *Anand Math* captured the figuration through its famous slogan 'Vande Mataram,' Victory to the Mother. The film version of this work expressed the role of woman in the euphoria of a newly independent India. Thus, 'Mother India' is an enormously powerful cultural signifier, gaining strength not only from atavistic memories from the Hindu epics, Sita, Sati Savitri, Draupadi, but also its use in moments of national (typically conflated with Hindu) cultural resurgence.

Figuring woman/mother as nation also suggests another of the sustaining analogies of the myth of nation. In an analysis of the foundational fictions of Latin America, Doris Sommer speaks of how the analogy of family helped to represent marriage between the different racial groups that comprised Latin American nations.[9] In India, however, exogamous marriage (across regions, religions, castes, and subcastes) was not a historical reality. Hence the analogy of nation as family could only lead to the appropriation and invisibility of minority groups in the hegemonic Hindu national narrative.

Let us now see how this appropriation of the maternal body as the 'imaginary site where meaning (or life) is generated,'[10] which excludes women from being meaning makers in their own right, is cemented by film. We have said that Hindi film took upon itself the task of covering the fissures in Indian society through an 'India' it represented for its viewers. This 'India' was best captured by its 'values' figured in woman. Mother as presiding over the link between nation and land/family found its classic expression in a film of the late fifties. I refer to *Mother India*, a film released in 1957, still said to be screened in India every day of the year. Here the nation gains its strength and validity from its metonymic identification of woman with land and family. In *Mother India*, the mother Radha works the land as a serf and is exploited by the forces of capital in Sukhilala the moneylender. She works the land and provides for her family after the death of her husband. And she stands for woman celebrated in the mythic Hindu narratives – of Sita in the *Ramayana* and Draupadi in the *Mahabbarata*, who encounter privations (and in Sita's case rejection and expulsion) in the service of their hero-husbands. Her younger son Birju joins the dacoits in order to avenge his family's ruin. This has been read as an allegory of radical action.[11] But the mother kills her own son, using her moral authority within the family and the nation to uphold the law, making the figure a force for conservatism.

In conflating the maternal with the national, the film extinguishes the heterogeneity of Indian women in favor of the Hindu model. The potential of different cultural formations to interrupt one another and reduce the tendency to privilege man's version of woman over historical women is thereby lost. This stereotyping of Hindu Woman as Mother India gives great impetus to the Hindu fundamentalist project, and makes woman's body the very site of fundamentalism.

Feminist criticism has stressed this need to distinguish between woman as sign and women as historical subjects in their own right.[12] Here feminist perception intersects that of critics of Hollywood cinema who argue that by relying on the primacy of the visual, cinema manages to efface real women in its representation of the image of woman. The image privileges what is present over that which is absent. As Christian Metz has said, the mastery of technique in cinema 'underlines and denounces the lack on which the whole arrangement is based (the absence of the object, replaced by its reflection), an exploit which consists at the same time of making this absence forgotten.'[13]

This disguised lack in cinema has a communal as well as sexual dimension in the case of the film *Mother India*. In this archetypal film of nationalism, the Muslim identity of the actress who played the recognizably Hindu character symbolizing the nation is at once appropriated and emptied of significance. The main actress who played Mother India was the Muslim actress Nargis, and she has always been associated in the minds of the public with Mother India. Her marriage to the Hindu Sunil Dutt, who played her son in the film, cemented her image as Mother India. The cultural message of the film has always been seen as Hindu, with its echoes of Radha, Parvati, Sita, with all of the traditional self-sacrificing virtues ascribed to these women.[14] We have, then, a nationalist articulation of Hindu religion and culture focusing on the figure of a Muslim actress.

Woman, symbol of Hindu nationalism, covers real *women* in India, heterogeneous, various, of many castes, religions, and geographical regions. As spectacle on screen, the regional identity of the actress is usually subsumed under the hegemonizing cultural sway of the Hindi heartland, the Indo-Gangetic plain. The metonymic representation of all women – whatever their cultural identity – as Hindu women is a recurring feature in Hindi film. Indeed, a large majority of popular film actresses have in fact come from the 'minority' sections, but rarely are these sections the subject of the national film industry. Because in Indian popular culture, attention is focused on the person of the actor or actress (gossip magazines about these characters constitute a major popular discourse), such metonymic representations obscure, at the level of individual actor, the fissures in Indian society, and may work to appropriate minority groups.

Read against the cultural politics of the Bombay film industry, the spectacle of Amina as Mother India in *Midnight's Children* yields interesting ironies. The conflations of mother with origin, land, family, and Rule of Law, upheld in the Bombay cinema, are exposed in Rushdie's text. Amina is mother but not to her own son, sharing the raising of Vanitha, the street singer's son, with Mary Pereira. The text is much more interested in maternal betrayal. When Saleem's telepathic gifts give him an inside view of women, he uncovers maternal adultery. But this adultery can be read in terms of national anxiety, suggesting the text's complicity with the imaginary of Bombay cinema. His mother, auntie Pia, and Leela Sabarmati become the collective scapegoat for the emergence of militancy and national heroism, especially significant in the context of the Indo-

Pakistani war of 1965. Once again it is the expulsion of Sita enacted. The graphic scene linking women's purity with national events is conveyed textually through the letter Salim sends to a Commander Sabarmati. He cuts out words from national newspapers relating current events to phrase a letter of warning. Once again, over the body and morality of woman, national events take shape. There is a public unanimity in the reaction to Commander Sabarmati's murder of this wife ('We knew a Navy man wouldn't stand for it' [*MC*, 314]). The ironies are obvious – woman's shame, dispensable in the urge to modernize, becomes a mystified area once the crisis has passed, and India, from being victim, is now the aggressor and victor in subcontinental politics (the two wars with Pakistan, the second over the creation of Bangladesh, established this position for India).

WOMAN IN POSTINDEPENDENCE ANXIETIES

In *Midnight's Children*, the dream image of woman as embodying the desire for nation becomes subjected to greater ironies even as male desire (represented first in Saleem's erotic attraction to his Auntie Pia and then to his sister, Jamila Singer) continues to provide the narrative's impetus. Each time this desire is deflated. Women recur as different kinds of bodies: the body in adultery, the body aging.

But the myth of nation is in fast decay, especially in the context of partition and war, and with it the dream image of woman. Male dismembering – as Saleem loses first hearing, then hair, then finger – is symptomatic of deep national anxiety. But this, too, happens in relation to the anxiety aroused by women. The description of the other children of midnight allows the narrator to imagine many versions of women who are no longer idealized or a dream, but victims of the brutal realities of poverty. There is Sundari, the beggar, whose face is slashed because her beauty blinds people, but now 'was earning a healthy living' and because of her story 'received more alms than any other member of her family' (*MC*, 236), and Parvati the witch, who 'stood mildly amid gasping crowds while her father drove spikes through her neck' (*MC*, 239). In contemporary India, poverty drives women to present a different kind of spectacle. Wee Willie Winkie's wife Vanitha, the street singer, is a common sight in India's big cities, her rags barely covering the body that in other circumstances is so mystified a site. In poverty, woman's shame is dispensable.

And woman's shame is the cornerstone of Islamic fundamentalism. The status of woman as a sign constantly subordinated to male-dictated contexts is demonstrated in the transformation of Saleem's once uninhibited sister the Brass Monkey into Jamila Singer when the family moves to Pakistan. The Brass Monkey, like the Monkey God Hanuman in the *Ramayana*, has entered the world of corruption (read as Western influence on women). Thus she is friendly with the 'hefty' Europeans of Walsingham School and plots with them to discredit the young boys. Later she (temporarily) embraces Christianity. But she can hold her own against these girls – she defeats Evie Burns in a street fight. Saleem's narrative adopts the male-oriented rhetoric of the nation – women are the electorate to be wooed by those in power (*MC*, 221).

(Woman becomes the site of the East–West cultural battle so often depicted in Bombay cinema. The classic example in this genre is the film *Purab aur Paschim* (East–West) (1970), where Indian values for women are reiterated over European. The 'Westernized' heroine (played by the actress Saira Bano) smokes, wears miniskirts, and is reformed into Indian womanhood by her love for the hero (played by Manoj Kumar, a recognized 'nationalist' filmmaker),[15] *Purab aur Paschim* uses woman to vent cultural anxiety in the wake of war and migration. The film targets Indian immigrants in Britain (metonymically represented by women who wear short dresses, smoke, and drink) and the nationalist message of the film is the containment of the threat to national culture (once again represented by Hindu ideals for women) from diasporic Indian populations. In other films of the seventies heroines are similarly reformed or punished for daring convention (*Thodisi Bewafai* and *Do Anjane*, for example).

So too, in *Midnight's Children*, the taming of Jamila Singer, which now involves not the exhibition but the extinguishing of woman as spectacle. As a child, the Brass Monkey was at the center of spectacle – she set fire to shoes. In the cosmopolitan world of Bombay her exuberance was irrepressible, but now, in fundamentalist Pakistan, she is captive to the Pakistani nationalist rhetoric and its view of women. She becomes martyr to the idea of nation. The wheel has come full circle when Jamila's voice, dream/imaginary of the Pakistani nation, is dissociated from her body. Heard by all on the Voice of Pakistan, she is placed during public appearances behind a perforated sheet; 'this was how the history of our family once again became the fate of a nation . . . being the new daughter-of-the-nation, her character began to owe more to the most strident aspects of the national persona than to the child-world of her monkey years' (*MC*, 375). As inspirer of men's souls, she must hide her body: 'Jamila, daughter, your voice will be a sword for purity; it will be the weapon with which we shall cleanse men's souls' (*MC*, 376). Because 'no city which locks its women away is ever short of whores', Saleem Sinai acts on his country's dual view of woman, as saint and whore. While the sister he desires sings of holiness and hides her body, Saleem's lusts drive him to 'women of the street' – latrine cleaners, Tai Bibi, whore of strange odors, and eventually Padma, the muscular pickle maker, whose hairy and strong forearms fascinate him.

Midnight's Children represents and ironizes not only the dream image of woman really servicing the psychic needs of the male subject constructed at the time of decolonization, but also her flip side, fat, gross, dirty but strong, as, from the male narrator's point of view, the dream of nation turns to nightmare in the wake of Indo-Pakistani postcolonial history. The text thus demonstrates woman's body's continued exploitation as a sign (albeit not a fixed one – the role of shame, and of spectacle, for instance, keeps changing), and the shifting space it occupies in the tentative process of decolonization and nation forming imaginatively represented in *Midnight's Children*. Reading Rushdie against Bombay cinema reveals gender as a trope in the narrative imagining of nation. This analysis hopes to reveal how all narratives imagining nation – sophisticated postmodern as well as mass cultural – collude in the engendering of nation as male through their representation of the female body. Thus, though Rushdie's representation parodies this engendering, moments in the text are complicit with Bombay cinema's signifying practices on women.

NOTES

1. Quotations from *Midnight's Children* are cited in the text with page numbers in parentheses using the abbreviation *MC*: Salman Rushdie, *Midnight's Children* (New York: Avon, 1980).
2. Throughout this essay I refer to 'Hindu middle classes' not as an essentialized religious or cultural group, but as a construct of a cultural production relying on recognizable Hindu symbolisms. For the recent political use of Hinduism, see Tapan Basu, Pradip Datta, Sumit Sarkar, Tanika Sarkar, and Sambuddha Sen, *Khaki Shorts, Saffron Flags* (Delhi: Orient Longman, 1993). By 'modernization' I mean the postindependence changes caused by increased intranational mobility of the professional and clerical sectors, and the legal changes in women's status with the widespread education and visibility of women. These are distinct from the role of women in modernization during reform and nationalism. See Partha Chatterjee, 'The Nation and Its Women', in *The Nation and Its Fragments: Colonial and Postcolonial Histories* (Princeton: Princeton University Press, 1993), 116–58. See Immanuel Wallerstein's discussion of 'modernity' and Westernization, 'Culture as the ideological Battleground of the Modern World System', *Global Culture*, ed. Mike Featherstone (London, Newbury Park, and New Delhi: Sage, 1990), 45.
3. Made by H. S. Rawail, this film was very popular in the sixties, chiefly because of its music. The sixties was a decade of euphoric nationalism, centering in the early half around the figure of Nehru and fuelled by the war with China and the two wars with Pakistan.
4. Although released in 1971, this is essentially a film of the late fifties as it took twenty years to complete. In mood and representation of women, it echoes the earlier period of filmmaking. The fifties and sixties were notable for the preponderance of Muslim themes. See Hameeduddin Mahmood, *The Kaleidoscope of Indian Cinema* (New Delhi: East West Press, 1974), 84.
5. Teresa de Lauretis, *Alice Doesn't: Feminism, Semiotics, Cinema* (Bloomington: Indiana University Press, 1984), 139.
6. See Kumkum Sangari and Sudesh Vaid, 'Recasting Women: An Introduction', in *Recasting Women: Essays in Colonial History*, ed. Sangari and Vaid (New Brunswick, NJ: Rutgers University Press, 1990), 10–11.
7. Frantz Fanon, *Studies in a Dying Colonialism* (New York: Monthly Review Press, 1959), 35–67.
8. Klaus Theweleit, *Male Fantasies* (Minneapolis: University of Minnesota Press, 1987), 294.
9. Doris Sommer, 'The Foundational Fictions of Latin America', *Nation and Narration*, ed. Homi Bhabha (New York: Routledge, 1990), 71–98.
10. Mary Jacobus, *Body/Politics: Women and the Discourses of Science*, ed. Mary Jacobus, Evelyn Fox Keller, and Sally Shuttleworth (New York and London: Routledge, 1990), 7.
11. Vijay Mishra, 'The Texts of "Mother India"', *Kunapipi* 11, 1 (1989): 119–37, 134.
12. Mary Poovey, 'Speaking of the Body: Mid-Victorian Constructions of Female Desire', in *Body/Politics*, 29.

13. Mary Ann Doane, 'Technology, Representation, and the Feminine,' in *Body/Politics*, 170–76.
14. Mishra, 'Texts', 125–26.
15. The director is noted for his interest in nationalist themes. The film in question deals with the protection of Indian values for women in an era of change. To assess the nationalist mood at the time, it is useful to note that this era was framed by the two wars with Pakistan – 1965 and 1971 (Sarcar, *Indian Cinema*, 147).

SECTION 7
PERFORMING THE BODY

Surveillance society
look up theorists.

INTRODUCTION

The name of this section could be taken to herald a move into feminist engagements with the arts, but what we have in mind is rather more complex. As feminist theory has moved away from the idea of a fixed and given body, there is increasing interest not just in corporeal construction from the outside as it were, but also in how we are constrained by and/or choose to perform our own bodies. The section looks at the construction of embodied identity in both a discursive and material sense, and addresses a series of broadly related questions: how are bodies and identities mutually constructed through performance? what counts as performing the body? how far is it possible to manipulate and play with bodily performance? can we perform ourselves as other to 'normal' assignations of sex, race, physical ability to good political effect? and implicit to all of these, though not always theorised, who is watching? The trajectory of surveillance/self-surveillance is never far away. For feminist theory, the art of performance is most often read as inextricably intertwined with performativity, while performativity itself has come to occupy the space of becoming that is common to us all.

The relationship between the notion of performance and the concept of performativity is, then, a difficult one to untangle. In an everyday sense it is easiest to think of performance as some kind of deliberative act that intends at very least to mimic what one is not. There is then a clear indication of a foundational self that puts on an appearance, which, however convincing it may be, remains on the surface, external to an unaltered person underneath. If we see an accomplished conference speaker at work, for example, it may make good to sense to ask what she's like 'as a person'. At the same time, and insofar as performance suggests the possibility of deception, it is as a staging that may confound some prior truth, that may leave us uncertain as to what is 'real' and what is not. It is precisely these two implications – of a foundational self and a foundational truth – that the term performativity contests. It implies first that there is no pre-existing self who performs – as Judith Butler puts it, something can be called performative 'in the sense that it constitutes as an effect the very subject that it appears to express' (1991: 24). And second, it implies that any difficulty in separating performance from reality is not a passing confusion but a reflection of the fundamental lack of any grounds of truth. For Butler, in particular, performing the body *fabricates* identity in all sorts of ways, of which gender is one important component.

Not all the selections understand performing the body in the same way as Butler, but all are to some extent concerned with how far we are compelled, as opposed to free, to take on the characteristics that we do. In *Gender Trouble*, from which we offer an extract here, Butler seems to imply that the subversion of embodied norms through the agency of performing differently, deliberately trangressing expectations – as for example in drag, or Straub's account of theatrical cross-dressing – is an intentional and realisable possibility. Her subsequent book, *Bodies that Matter* (1993), was at pains to stress, however, that our choices are highly constrained. Even when an identity is non-normative, it institutes its own internal codes of practice, as Shildrick and Price show in relation to the disabled body, whose performativity is set in the context of multiple disciplinary practices. Moreover, the cycle of power and resistance, familiar from Foucauldian theory, is highly relevant here, such that the refusal of quasi-compulsory practices can be temporarily transgressive but in turn sets up new normativities. The pieces by Fen Coles on women's body-building, and Kathy Davis on cosmetic surgery, for example, may exemplify just such a process, regardless of the participants' own intentions and desires. In a very different way to the other extracts, Heather Findlay's psychoanalytic piece playfully theorises the lesbian dildo debates, in which performativity is mediated by phantasy and fetishism. The relationship between meaning, doing and being is especially acute in her account which nicely captures the controversy that would-be transgressive performance generates.

Though it is always possible to give an account of individual agency, that agency is neither transparent, and nor is it established once and for all. Performativity is the process of becoming – 'kinda subversive and kinda hegemonic' in Eve Sedgwick's words – that is the very condition of embodiment.

REFERENCES AND FURTHER READING

Acker, Kathy (1993) 'Against Ordinary Language. The Language of the Body' in A. and M. Kroker (eds), *The Last Sex: Feminism and Outlaw Bodies*, Basingstoke: Macmillan Press.

Butler, Judith (1991) 'Imitation and Gender Insubordination' in Diana Fuss (ed.), *Inside/Out: Lesbian Theories, Gay Theories*, London: Routledge.

Fuss, Diana (1992) 'Fashion and the Homospectatorial Look', *Critical Inquiry* 18: 712–37.

Grosz, Elizabeth (1994) 'The Body as Inscriptive Surface', Chapter 6 in *Volatile Bodies: Towards a Corporeal Feminism*, Bloomington: Indiana University Press.

Hart, Lynda (1995) 'Blood, Piss and Tears: the Queer Real', *Textual Practice*, 9 (1): 55–66.

Ian, Marcia (1995) 'How Do You Wear Your Body?' in Monica Dorenkamp and Richard Henke (eds), *Negotiating Lesbian and Gay Subjects*, London: Routledge.

Marcus, Susan (1992) 'Fighting Bodies, Fighting Words: A Theory and Politics of Rape Prevention' in Judith Butler and Joan Scott (eds), *Feminist Theorize the Political*, London: Routledge.

Phelan, Peggy (1997) *Mourning Sex: Performing Public Memories*, London: Routledge.

Quinby, Lee (1994) 'Eu(jean)ics: The Fashion in Power' in *Anti-apocalypse: Exercises in Genealogical Criticism*, Minneapolis: Minnesota University Press.

St Martin, Leena and Garvey, Nicola (1996) 'Women's Bodybuilding: Feminist Resistance or Femininity's Recuperation?' in *Body & Society* 2 (4): 45–57.

Urla, Jacqueline and Swedlund, Allan (1995) 'The Anthropometry of Barbie' in Jennifer Terry and Jacqueline Urla (eds), *Deviant Bodies*, Bloomington: Indiana University Press.

Wilson, Elizabeth (1992) 'Fashion and the Postmodern Body' in Leslie Ash and Elizabeth Wilson (eds), *Chic Thrills*, London: Pandora.

7.1

BODILY INSCRIPTIONS, PERFORMATIVE SUBVERSIONS

Judith Butler

[. . .]

The redescription of intrapsychic processes in terms of the surface politics of the body implies a corollary redescription of gender as the disciplinary production of the figures of fantasy through the play of presence and absence on the body's surface, the construction of the gendered body through a series of exclusions and denials, signifying absences. But what determines the manifest and latent text of the body politic? What is the prohibitive law that generates the corporeal stylization of gender, the fantasied and fantastic figuration of the body? We have already considered the incest taboo and the prior taboo against homosexuality as the generative moments of gender identity, the prohibitions that produce identity along the culturally intelligible grids of an idealized and compulsory heterosexuality [see Chapter 2, *Gender Trouble*]. That disciplinary production of gender effects a false stabilization of gender in the interests of the heterosexual construction and regulation of sexuality within the reproductive domain. The construction of coherence conceals the gender discontinuities that run rampant within heterosexual, bisexual, and gay and lesbian contexts in which gender does not necessarily follow from sex, and desire, or sexuality generally, does not seem to follow from gender – indeed, where none of these dimensions of significant corporeality express or reflect one another. When the disorganization and disaggregation of the field of bodies disrupt the regulatory fiction of heterosexual coherence, it seems that the expressive model loses its descriptive force. That

From: J. Butler, *Gender Trouble*, New York: Routledge, 1990.

regulatory ideal is then exposed as a norm and a fiction that disguises itself as a developmental law regulating the sexual field that it purports to describe.

According to the understanding of identification as an enacted fantasy or incorporation, however, it is clear that coherence is desired, wished for, idealized, and that this idealization is an effect of a corporeal signification. In other words, acts, gestures, and desire produce the effect of an internal core or substance, but produce this *on the surface* of the body, through the play of signifying absences that suggest, but never reveal, the organizing principle of identity as a cause. Such acts, gestures, enactments, generally construed, are *performative* in the sense that the essence or identity that they otherwise purport to express are *fabrications* manufactured and sustained through corporeal signs and other discursive means. That the gendered body is performative suggests that it has no ontological status apart from the various acts which constitute its reality. This also suggests that if that reality is fabricated as an interior essence, that very interiority is an effect and function of a decidedly public and social discourse, the public regulation of fantasy through the surface politics of the body, the gender border control that differentiates inner from outer, and so institutes the 'integrity' of the subject. In other words, acts and gestures, articulated and enacted desires create the illusion of an interior and organizing gender core, an illusion discursively maintained for the purposes of the regulation of sexuality within the obligatory frame of reproductive heterosexuality. If the 'cause' of desire, gesture, and act can be localized within the 'self' of the actor, then the political regulations and disciplinary practices which produce that ostensibly coherent gender are effectively displaced from view. The displacement of a political and discursive origin of gender identity onto a psychological 'core' precludes an analysis of the political constitution of the gendered subject and its fabricated notions about the ineffable interiority of its sex or of its true identity.

If the inner truth of gender is a fabrication and if a true gender is a fantasy instituted and inscribed on the surface of bodies, then it seems that genders can be neither true nor false, but are only produced as the truth effects of a discourse of primary and stable identity. In *Mother Camp: Female Impersonators in America*, anthropologist Esther Newton suggests that the structure of impersonation reveals one of the key fabricating mechanisms through which the social construction of gender takes place.[1] I would suggest as well that drag fully subverts the distinction between inner and outer psychic space and effectively mocks both the expressive model of gender and the notion of a true gender identity. Newton writes:

> At its most complex, [drag] is a double inversion that says, 'appearance is an illusion'. Drag says [Newton's curious personification] 'my "outside" appearance is feminine, but my essence "inside" [the body] is masculine.' At the same time it symbolizes the opposite inversion; 'my appearance "outside" [my body, my gender] is masculine but my essence "inside" [myself] is feminine.'[2]

Both claims to truth contradict one another and so displace the entire enactment of gender significations from the discourse of truth and falsity.

The notion of an original or primary gender identity is often parodied within the cultural practices of drag, cross-dressing, and the sexual stylization of butch/femme identities. Within feminist theory, such parodic identities have been understood to be either degrading to women, in the case of drag and cross-dressing, or an uncritical appropriation of sex-role stereotyping from within the practice of heterosexuality, especially in the case of butch/femme lesbian identities. But the relation between the 'imitation' and the 'original' is, I think, more complicated than that critique generally allows. Moreover, it gives us a clue to the way in which the relationship between primary identification – that is, the original meanings accorded to gender – and subsequent gender experience might be reframed. The performance of drag plays upon the distinction between the anatomy of the performer and the gender that is being performed. But we are actually in the presence of three contingent dimensions of significant corporeality: anatomical sex, gender identity, and gender performance. If the anatomy of the performer is already distinct from the gender of the performer, and both of those are distinct from the gender of the performance, then the performance suggests a dissonance not only between sex and performance, but sex and gender, and gender and performance. As much as drag creates a unified picture of 'woman' (what its critics often oppose), it also reveals the distinctness of those aspects of gendered experience which are falsely naturalized as a unity through the regulatory fiction of heterosexual coherence. *In imitating gender, drag implicitly reveals the imitative structure of gender itself – as well as its contingency*. Indeed, part of the pleasure, the giddiness of the performance is in the recognition of a radical contingency in the relation between sex and gender in the face of cultural configurations of causal unities that are regularly assumed to be natural and necessary. In the place of the law of heterosexual coherence, we see sex and gender denaturalized by means of a performance which avows their distinctness and dramatizes the cultural mechanism of their fabricated unity.

The notion of gender parody defended here does not assume that there is an original which such parodic identities imitate. Indeed, the parody is *of* the very notion of an original; just as the psychoanalytic notion of gender identification is constituted by a fantasy of a fantasy, the transfiguration of an Other who is always already a 'figure' in that double sense, so gender parody reveals that the original identity after which gender fashions itself is an imitation without an origin. To be more precise, it is a production which, in effect – that is, in its effect – postures as an imitation. This perpetual displacement constitutes a fluidity of identities that suggests an openness to resignification and recontextualization; parodic proliferation deprives hegemonic culture and its critics of the claim to naturalized or essentialist gender identities. Although the gender meanings taken up in these parodic styles are clearly part of hegemonic, misogynist culture, they are nevertheless denaturalized and mobilized through their parodic recontextualization. As imitations which effectively displace the meaning of the original, they imitate the myth of originality itself. In the place of an original identification which serves as a determining cause, gender identity might be reconceived as a personal/cultural history of received meanings subject to a set of imitative practices which refer laterally to other imitations and which, jointly, construct the illusion of a primary and interior gendered self or parody the mechanism of that construction.

According to Fredric Jameson's 'Postmodernism and Consumer Society', the imitation that mocks the notion of an original is characteristic of pastiche rather than parody:

> Pastiche is, like parody, the imitation of a peculiar or unique style, the wearing of a stylistic mask, speech in a dead language: but it is a neutral practice of mimicry, without parody's ulterior motive, without the satirical impulse, without laughter, without that still latent feeling that there exists something *normal* compared to which what is being imitated is rather comic. Pastiche is blank parody, parody that has lost it humor.[3]

The loss of the sense of 'the normal', however, can be its own occasion for laughter, especially when 'the normal', 'the original' is revealed to be a copy, and an inevitably failed one, an ideal that no one *can* embody. In this sense, laughter emerges in the realization that all along the original was derived.

Parody by itself is not subversive, and there must be a way to understand what makes certain kinds of parodic repetitions effectively disruptive, truly troubling, and which repetitions become domesticated and recirculated as instruments of cultural hegemony. A typology of actions would clearly not suffice, for parodic displacement, indeed, parodic laughter, depends on a context and reception in which subversive confusions can be fostered. What performance where will invert the inner/outer distinction and compel a radical rethinking of the psychological presuppositions of gender identity and sexuality? What performance where will compel a reconsideration of the *place* and stability of the masculine and the feminine? And what kind of gender performance will enact and reveal the performativity of gender itself in a way that destabilizes the naturalized categories of identity and desire.

If the body is not a 'being', but a variable boundary, a surface whose permeability is politically regulated, a signifying practice within a cultural field of gender hierarchy and compulsory heterosexuality, then what language is left for understanding this corporeal enactment, gender, that constitutes its 'interior' signification on its surface? Sartre would perhaps have called this act 'a style of being', Foucault, 'a stylistics of existence'. And in my earlier reading of Beauvoir, I suggest that gendered bodies are so many 'styles of the flesh'. These styles all never fully self-styled, for styles have a history, and those histories condition and limit the possibilities. Consider gender, for instance, as *a corporeal style*, an 'act', as it were, which is both intentional and performative, where '*performative*' suggests a dramatic and contingent construction of meaning.

Wittig understands gender as the workings of 'sex', where 'sex' is an obligatory injunction for the body to become a cultural sign, to materialize itself in obedience to a historically delimited possibility, and to do this, not once or twice, but as a sustained and repeated corporeal project. The notion of a 'project', however, suggests the originating force of a radical will, and because gender is a project which has cultural survival as its end, the term *strategy* better suggests the situation of duress under which gender performance always and variously occurs. Hence, as a strategy of survival within compulsory systems, gender is a performance with clearly punitive consequences. Discrete genders are part of what 'humanizes' individuals within contemporary culture; indeed, we regularly punish those who fail to do their gender right. Because there is neither an 'essence' that gender expresses or externalizes nor an objective ideal to which gender aspires, and because gender is not a fact, the various acts of gender create the idea of gender, and without those acts, there would be no gender at all. Gender is, thus, a construction that regularly conceals its genesis; the tacit collective agreement to perform, produce, and sustain discrete and polar genders as cultural fictions is obscured by the credibility of those productions – and the punishments that attend not agreeing to believe in them; the construction 'compels' our belief in its necessity and naturalness. The historical possibilities materialized through various corporeal styles are nothing other than those punitively regulated cultural fictions alternately embodied and deflected under duress.

Consider that a sedimentation of gender norms produces the peculiar phenomenon of a 'natural sex' or a 'real woman' or any number of prevalent and compelling social fictions, and that this is a sedimentation that over time has produced a set of corporeal styles which, in reified form, appear as the natural configuration of bodies into sexes existing in a binary relation to one another. If these styles are enacted, and if they produce the coherent gendered subjects who pose as their originators, what kind of performance might reveal this ostensible 'cause' to be an 'effect'?

In what senses, then, is gender an act? As in other ritual social dramas, the action of gender requires a performance that is *repeated*. This repetition is at once a reenactment and reexperiencing of a set of meanings already socially established; and it is the mundane and ritualized form of their legitimation.[4] Although there are individual bodies that enact these significations by becoming stylised into gendered modes, this 'action' is a public action. There are temporal and collective dimensions to these actions, and their public character is not inconsequential; indeed, the performance is effected with the strategic aim of maintaining gender within its binary frame – an aim that cannot be attributed to a subject, but, rather, must be understood to found and consolidate the subject.

Gender ought not to be construed as a stable identity or locus of agency from which various acts follow; rather, gender is an identity tenuously constituted in time, instituted in an exterior space through a *stylized repetition of acts.* The effect of gender is produced through the stylization of the body and, hence, must be understood as the mundane way in which bodily gestures, movements, and styles of various kinds constitute the illusion of an abiding gendered self. This formulation moves the conception of gender off the ground of a substantial model of identity to one that requires a conception of gender as a constituted *social temporality.* Significantly, if gender is instituted through acts which are internally discontinuous, then the *appearance of substance* is precisely that, a constructed identity, a performative accomplishment which the mundane social audience, including the actors themselves, come to believe and to perform in the mode of belief. Gender is also a norm that can never be fully internalized; 'the internal' is a surface signification, and gender norms are finally phantasmatic, impossible to embody. If the ground of gender identity is the stylized repetition of acts through time and not a seemingly seamless identity, then the spatial metaphor of a 'ground' will be displaced and revealed as a stylized configuration, indeed, a gendered corporealization of time. The abiding gendered self will then be shown to be structured by repeated acts that seek to approximate the ideal of a substantial ground of identity, but which, in their occasional *dis*continuity, reveal the temporal and contingent groundlessness of this 'ground'. The possibilities of gender transformation are to be found precisely in the arbitrary relation between such acts, in the possibility of a failure to repeat, a de-formity, or a parodic repetition that exposes the phantasmatic effect of abiding identity as a politically tenuous construction.

If gender attributes, however, are not expressive but performative, then these attributes effectively constitute the identity they are said to express or reveal. The distinction between expression and performativeness is crucial. If gender attributes and acts, the various ways in which a body shows or produces its cultural signification, are performative, then there is no preexisting identity by which an act or attribute might be measured; there would be no true or false, real or distorted acts of gender, and the postulation of a true gender identity would be revealed as a regulatory fiction. That gender reality is created through sustained social performances means that the very notions of an essential sex and a true or abiding masculinity or femininity are also constituted as part of the strategy that conceals gender's performative character and the performative possibilities for proliferating gender configurations outside the restricting frames of masculinist domination and compulsory heterosexuality.

Genders can be neither true nor false, neither real nor apparent, neither original nor derived. As credible bearers of those attributes, however, genders can also be rendered thoroughly and radically *incredible.*

NOTES

1. See the chapter 'Role Models' in Esther Newton, *Mother Camp: Female Impersonators in America* (Chicago: University of Chicago Press, 1972).

2. Ibid., p. 103.
3. Fredric Jameson, 'Postmodernism and Consumer Society', in *The Anti-Aesthetic: Essays on Postmodern Culture*, ed. Hal Foster (Port Townsend, WA: Bay Press, 1983), p. 114.
4. See Victor Turner, *Dramas, Fields and Metaphors* (Ithaca: Cornell University Press, 1974). See also Clifford Geertz, 'Blurred Genres: The Refiguration of Thought', in *Local Knowledge, Further Essays in Interpretive Anthropology* (New York: Basic Books, 1983).

7.2

THE GUILTY PLEASURES OF FEMALE THEATRICAL CROSS-DRESSING

Kristina Straub

A curious shift in theatrical cross-dressing took place in late seventeenth-century England. For a variety of complex reasons still being explored by some of our most interesting critics of sexuality and gender in the theater[1] the tradition of boys playing women's parts on the stage became at best an outmoded fashion and at worst an unacceptable breach of gender boundaries for contemporary audiences. At the same time, of course, women entered the acting profession to play women's characters. A growing capacity to perceive gender ambiguity and to find it troubling did not, however, result in the prohibiting of female theatrical cross-dressing, as it did in the limitation of the masculine version to travesty. The cross-dressed actress came into a fashion that lasted, not without changes, throughout the century. Whereas obvious travesty was crucial to the acceptance of male cross-dressing on the early eighteenth-century stage (the actor must be seen as a bad parody of femininity), it seems to have become so for female cross-dressers only in the second half of the century. At mid-century, commentary on female theatrical cross-dressers suggests that the ambiguity which was then intolerable when associated with a male was in fact part of the fun of seeing women in breeches. By the end of the century, discourse about the cross-dressed actress is both more condemnatory of the practice (which was still, however, tolerated) and more insistent that female cross-dressing, like the male, was mere travesty, an obvious parody which left gender boundaries unquestioned.

From: J. Epstein and K. Straub (eds), *Body Guards. The Cultural Politics of Gender Ambiguity*, London: Routledge, 1991.

The cross-dressed actress of the early to mid-eighteenth-century seems to constitute an historical possibility for pleasure in sexual and gender ambiguities. This possibility calls into question the naturalness of an economy of spectatorial pleasure that works on the premise of rigid boundaries between categories of gender and sexuality – male/female, hetero/homosexual. It asks us to unpack dominant constructions of the commodified feminine spectacle as unambiguously oppositional and other to a spectatorial, consuming male gaze.[2] Pat Rogers argues that the eighteenth-century female theatrical cross-dresser was part of a specular economy that fits our present assumptions of an objectified femininity in binary opposition to a masculine gaze. While Rogers is correct up to a point, I will argue here that the commodification of the cross-dressed actress was in fact a good deal more complicated in its audience appeal and that, hence, the cross-dressed actress is less a confirmation than a challenge to modern assumptions about the gendering of spectacle.

Popular discourse about the cross-dressed actress suggests that in mid-century England the female theatrical cross-dresser did not fit the constructions of gender and sexuality that would seem to render natural two key concepts in the modern sex/gender system: (1) the subjugation of a feminine spectacle to the dominance of the male gaze, and (2) the exclusive definition of feminine sexual desire in terms of its relation to masculine heterosexual desire. The autobiographical *Narrative of the Life of Mrs Charlotte Charke* (Charke 1745), a notorious account of an actress who carried her masquerades in male clothes into her life off the boards, is a particularly powerful example of how the cross-dressed actress might function in a way discursively at odds with these two concepts.

The obvious reason for dressing actresses in men's clothes at the end of the seventeenth century is virtually the same as one of the reasons for putting women on the stage at all: conventionally attractive female bodies sell tickets. Judith Milhous documents the popularity of female cross-dressing on the stage during the Restoration and links it to economic competition between the licensed London theaters.[3] Sometimes competing managers did not stop with casting single roles, but rather, as James Wright's history documents, presented whole plays 'all by women, as formerly all by men' (cliii). The same incentive seems to have motivated John Mossop to cast the actress Catley as Macheath during the keen competition between theaters in Dublin at mid-century (Hitchcock 1788 II: 135), and Tate Wilkinson, as a manager, clearly had receipts in mind when, in the 1795 *Wandering Patentee*, he refers to actresses who look well in the 'small clothes'. One might say truthfully that the cross-dressing actress was, in the last instance, economically motivated, an example of the commodification of women in emergent capitalist society. But this commodification itself cannot be reduced to the specularization of women within the structuring principle of the masculine observer and the feminine spectacle, an epistemological and psychological principle which may have been only emergent at this moment in history. The commodification of the cross-dressed actress embodied ideological contradiction about the nature of feminine identity and sexuality; contemporary discourse about this commodification yields a complex picture of what, exactly, is being marketed.

First, as long as the cross-dressed actress was 'packaged' as a commodity for the pleasure of her audience, responses to her suggest that her marketability had as much to do with a playfully ambiguous sexual appeal as with the heterosexually defined attractions of her specularized feminine body. Duality is part of the sexual appeal of the cross-dressed Margaret Woffington, one of the eighteenth century's most written-about cross-dressers. Woffington's 1760 memoir records the reaction of both sexes to her first appearance as Harry Wildair in *The Constant Couple*; the men, it is said, were charmed, and the 'Females were equally well pleased with her acting as the Men were, but could not persuade themselves, that it was a Woman that acted the Character' (*Memoirs* 22). Woffington's ambiguity is presented as a commodity in which the audience takes great pleasure:

> When first in Petticoats you trod the Stage,
> Our Sex with Love you fir'd, your own with Rage!
> In Breeches next, so well you play'd the Cheat,
> The pretty Fellow, and the Rake compleat—
> Each sex, were then, with different Passions mov'd,
> The Men grew envious, and the Women loved.
>
> (Victor 1761 III: 4–5)

Similarly, a 1766 memoir of James Quinn declares that 'it was a most nice point to decide between the gentlemen and the ladies, whether she [Woffington] was the finest woman, or the prettiest fellow' (67–8).

Some of the pleasure afforded by the spectacle of the cross-dressed actress arose, then, out of the doubling of sexual attraction. The actress in men's clothes appeals to both men and women – at least as a specular commodity. This double appeal, however, depends on its containment as a theatrical commodity; off-stage in the 'real' world, the cross-dressed actress was usually represented in the tradition of women who dressed as men for the arousal and/or gratification of heterosexual male desire (Dekker and van de Pol 1989: 54). Popular stories about the adventures of actresses who cross-dressed out of role nearly always bracket their ambiguous sexual appeal in a narrative that privileges both an exclusively heterosexual desire and the actress' function therein. *The Comforts of Matrimony* reports that Susannah Maria Cibber cross-dressed off the stage in order to facilitate her amour with William Sloper (24). Susanna Centlivre's quasi-mythical affair with Anthony Hammond was allegedly carried out by her posing as his younger 'Cousin Jack' and living with Hammond while he was a student at Oxford (Mottley 1747: 185–6; Dibden 1747–1800 *Complete History* II: 313). Catherine Galendo draws on this tradition when she accuses Sarah Siddons of learning the role of Hamlet in order to have the excuse to take fencing lessons from, and consequently to seduce, her husband Mr Galendo (16).

In another story, Woffington is said to have caused her female rival for a male lover's affections to fall in love with her, but the narrative sets this 'confusion' of gender roles and sexual object choice in the context of the actress' efforts to foil her lover's infidelities. These narratives effectively define the 'safe' limits of the cross-dressed actress' ambiguous appeal; as a specular commodity, the gender and sexual confusion associated with the actress could be a source of pleasure as long as it did not contaminate or compromise dominant narratives of heterosexual desire.

Eighteenth-century discourse about the cross-dressed actress evinces an unqualified pleasure in sexual confusion only when that confusion is 'for sale' as a theatrical commodity, an obviously artificial construction. A slightly guilty pleasure in the sexual and gender confusion embodied in the cross-dressed actress is balanced against recuperations of her gender-bending within dominant sexual ideologies. This recuperative discourse often reveals, however, the very anxieties that it seeks to allay. By the mid-eighteenth century, the cross-dressed actress is the focus of rules and strictures that seek to confine her pleasurable ambiguities of sex and gender to a narrowly defined phenomenon, an illusion to be bought and kept within the marketplace of the theater. This discourse of containment voices a 'monstrosity' with which the sexually ambiguous actress flirts and reveals an incipient threat to heterosexual male dominance implicit within the pleasurable ambiguity of the cross-dressed actress.

Despite the pervasiveness and popularity of the custom, commentators on cross-dressing actresses express an uneasiness about this phenomenon that can be read in their desire, by mid-century, to contain it within decorous rules, to defuse the threat of this otherwise entertaining ambiguity by referring it to a 'natural' standard or social 'law'. This discourse of containment also speaks, however, the fears that it seeks to allay. Besides framing the pleasure afforded by the cross-dressed actress within proscriptive rules, it articulates, I would argue, the sources of anxiety about that pleasure. First, it gives voice to fears about the stability and certainty of an emergent dominant definition of masculine sexuality as it is reflected (and refracted) in the cross-dressed female's image. Second, it speaks fears that femininity may itself exceed the limits of privacy and domesticity that would seem, in dominant ideologies, to define it. In short, the cross-dressed actress is seen as threatening to the construction of a stable oppositional relationship between male and female gender and sexuality.

Biographical discourse about cross-dressing actresses becomes increasingly condemnatory by the early nineteenth century. James Boaden refers unequivocally to 'vile and beastly transformations' in 1825 (*Kemble* II: 334); his repulsion at female theatrical transvestism seems to reflect the emergent consciousness of a category for 'deviate' female sexuality. But even as early as 1761 one can read an incipient uneasiness with the ideological effects of female theatrical transvestism. Benjamin Victor, a London theater prompter and manager of one of the Dublin theater during the age of Garrick, reveals even more clearly what is veiled in Boaden's Victorian prose on the subject: fears about the integrity and 'naturalness' of feminine gender identity. His discussion of Woffington's famous role of Sir Harry Wildair candidly states the economic motive – 'she always conferred a Favour upon the Managers whenever she changed her Sex, and filled their Houses' – the cross-dressed actress as pleasurable and lucrative commodity. But he sets limits on this form of commodification: 'And now, ye fair ones of the Stage, it will not be foreign to the Subject, to consider whether it is proper for you . . . to perform the Characters of Men. I will venture in the Name of all sober, discreet, sensible, Spectators . . . to answer *No*! there is something required *so much beyond the Delicacy of your Sex*, to arrive at the Point of Perfection, that, if you hit it, you may be condemned as a Woman, and if you do not, you are injured as an Actress' (my emphasis, II: 4; 5–6). To masquerade as a man, and to do it too well, is to enter a no-woman's-land 'beyond' femininity, to exceed the limits of 'delicacy'.

Furthermore, this trespassing beyond the limits of femininity may lead to ambiguities in sexual object choice that Victor plainly finds disturbing as he continues to address his hypothetical 'fair ones of the Stage': 'supposing you are formed in Mind, and Body . . . like the Actress in Question – for she had Beauty, Shape, Wit, and Vivacity, equal to any theatrical Female in any Time, and capable of any Undertaking in the Province of Comedy, nay of deceiving, and warming into Passion, any of her own Sex, if she had been unknown, and introduced as a young Baronet just returned from his Travels – but still, I say, admirable and admired as she was in this Part, I would not have any other Female of the Stage attempt the Character after her' (III: 6). Victor's nervousness focuses on the sexual as well as gender ambiguity of Woffington in male dress; she is clearly an object of desire, but the question of who might desire her is carefully hedged in the hypothetical – '*if* she had been unknown'. Finally, however, he does not, like Boaden, condemn actresses in male dress, but rather retreats from the hypothetical threat of Woffington as ambiguous sexual object to the relative safety of theatrical tradition and rule: 'the wearing of Breeches merely to pass for a Man, as is the Case in many Comedies, is as far as the Metamorphosis ought to go, and indeed, more than some formal Critics will allow of; but that custom is established into a Law, and as there is great Latitude in it, it should not be in the least extended – when it is, you *o'erstep the Modesty of Nature*, and when that is done, whatever may be the Appearance within Doors, you will be injured by Remarks and Criticisms without' (III; 7). Victor seems to have in mind characters, such as Rosalind, whose masquerades as men are clearly represented *as* masquerades to the audience. By designating female transvestism as a specific form of theatrical illusion, Victor limits the threat of Woffington's ambiguous attractions; but he goes beyond this designation to demand that the illusion be presented as an illusion – further distancing the threat – and containing the ambiguity of the cross-dressed actress within a decorum of theatrical representation.

Later writers on the theater apparently want to exclude ambiguity altogether, not being content with containing it within rules. When ambiguity resists erasure, they often simply insist, all logic and evidence to the contrary, that it confirms the very gender definitions it would seem to problematize. William Cooke writes in 1804 that female cross-dressing should not leave room for any ambiguity. 'Where a woman . . . personates a man *pro tempore* . . . the closer the imitation is made, the more we applaud the performer, but always in the knowledge that the object before us is *a woman assuming the character of a man*' (*Macklin* 12.6). Unlike Victor, however, who allows for ambiguous sexual attraction in Woffington's impersonation, Cook dismisses even the possibility of the successful illusion of a woman assuming a masculine position in the sexual economy: 'when this same woman totally usurps the male character, and we are left to try her merits merely as a man, without making the least allowance for the imbecilities of the other sex, we may safely pronounce, there is no woman, nor ever was a woman, who can fully supply this character. There is such a *reverse* in all the habits and modes of the two sexes, acquired from the very cradle upwards, that it is next to an impossibility for the one to resemble the other so as totally to escape detection' (*Macklin* 126). By 1800, theater historians and biographers tend to exclude the ambiguity that Victor barely admits within the 'rules' of cross-dressing. Boaden writes of Siddons' Hamlet (which he never saw), that 'the unconstrained motion would be wanting for the most part; modesty would be sometimes rather untractable in the male habit, and the conclusion at last might be, "were she *but* man, she would exceed all that man has ever achieved in Hamlet"' (*Memoirs of Mrs Siddons* I: 283). Of Dorothy Jordon playing the part of William in Brooke's *Rosina*, Boaden confidently asserts that 'Did the lady really look like a man, the coarse *androgynus* would be hooted from the stage' (*Life of Mrs Jordan* I: 46), and reads Woffington's famous Wildair through his observations of Jordan:

> When Woffington took it up, she did what she was not aware of, namely, that the audience permitted the actress to *purify* the character, and enjoyed the language from a woman, which might have disgusted from a man speaking before women . . . I am convinced that no creature there supposed it [Woffington a man] for a moment: it was the *travesty*, seen throughout, that really constituted the charm of your performance, and rendered it not only gay, but innocent. And thus it was with Mrs Jordan, who, however beautiful in her figure, stood confessed a perfect and decided woman; and courted, intrigued, and quarrelled, and cudgelled, in whimsical imitation of the ruder man. (*Life of Mrs Jordan* I: 127)

Rather incredibly, Boaden writes of Woffington's trespasses into the discourse of a rake as if they were attempts to clean up scenes in which the male character talks of sex to a woman. The cross-dressed actress is made to serve the dominant construction of separate spheres for men and women even as she would seem to trespass against that separation.

This erasure of sexual ambiguity is more characteristic of biographical texts late in the century. At mid-century, attempts to contain female cross-dressing within rules of tradition or 'nature' tend more often to name what they seek to exclude. *The Actor* (1750) explicitly targets the sexual content of Woffington's impersonation as objectionable: 'We see women sometimes act the parts of men, and in all but love we approve them. Mrs Woffington pleases in Sir Harry Wildair in every part, except where she makes love; but there no one of the audience ever saw her without disgust; it is insipid, or it is worse; it conveys no ideas at all, or very hateful ones; and either the insensibility, or the disgust we conceive, quite break in upon the delusion' (202). The choice between 'no ideas at all, or very hateful ones' sums up the two-pronged threat implied in female cross-dressing, a threat that is both expressed and contained by the biographical discourse we have been examining. First, the spectacle of women representing masculine sexuality summons the threat of a nondominant, nonauthoritative – even impotent – masculinity. Women assuming masculine sexual perogative are 'insipid', and their love-making 'conveys no ideas at all'; they are, in short, failed men, and, as such, would seem to offer no threat whatsoever to masculine sexuality. But, as I will argue, the 'castrated' figure of the cross-dressed actress is also capable of holding a mirror up to masculinity that reflects back an image of castration that cannot be entirely controlled by the mechanisms of projection. When the actress puts on masculine sexuality, even as she functions as its object of desire, she opens possibilities for challenging the stability and authority of that sexuality.

Second, the actress in male dress summons up, in the very act of specularizing the feminine object of desire, the 'hateful' idea of a feminine sexual desire that exceeds the limits of 'normal' heterosexual romantic love. The 'Tommy', as Trumbach says, is a category for 'deviate' feminine sexuality that seems to emerge late in the eighteenth century. Homophobia can certainly be seen, in chillingly recognizable forms, in Cook's denials of sexual or gender ambiguities in cross-dressed performances or in Boaden's disgust at 'vile and beastly transformations'. But even as early as mid-century, discourse about female theatrical cross-dressing evinces an awareness that feminine sexuality and gender identity could stray beyond the boundaries that were coming to define the sphere of feminine feeling and behavior as man's commensurate and oppositional other. The cross-dressed actress points to a feminine desire in excess of this role, and in this case that desire is explicitly sexual. The cross-dressed actress threatened the apparent naturalness and stability of what was becoming dominant gender ideology by suggesting a feminine sexuality that exceeded the heterosexual role of women.

[. . .]

NOTES

1. Much of this work goes on in Shakespeare studies. See, for instance, Greenblatt's 'Fiction and Friction' in *Shakespearean Negotiations*, 66–93. Lisa Jardine's *Still Harping on Daughters* also takes up questions about the reception of boy actors in Renaissance England (9–36). Stephen Orgel's 'Nobody's Perfect' explores reasons for the endurance of the boy actors well into the seventeenth century.
2. The main body of work in which the influential concept of the female spectacle and

the male gaze is developed was done by a group of feminist film theorists associated with *Screen*. See, for instance, Laura Mulvey's 'Visual Pleasure and Narrative Cinema' and Mary Anne Doane's 'Film and the Masquerade' for classic formulations of this concept.

3. Milhous sees female cross-dressing as a competitive strategy against the more expensive-to-stage audience draw of elaborate settings and spectacles (93).

REFERENCES

Boaden, James (1831) *The Life of Mrs Jordan*, 2 vols, London: Edward Bull.

Boaden, James (1827) *Memoirs of Mrs Siddons*, 2 vols, London: Henry Colburn.

Boaden, James (1825) *Memoirs of the Life of John Philip Kemble, Esq*, 2 vols, London: Longman, Hurst, Rees, Orme, Brown, and Green.

Charke, Charlotte (1756) *The History of Henry Dumont, Esq: and Miss Charlotte Evelyn*, London: H. Slater, Jun., and S. Whyte.

Charke, Charlotte (1755) *A Narrative of the Life of Mrs Charlotte Charke*, Ed. Leonard R. N. Ashley, Gainesville, FL: Scholars' Facsimiles and Reprints, 1969.

Cibber, Colley (1968) *An Apology for the Life of Mr Colley Cibber*, Ed. B. R. S. Fone, Ann Arbor: University of Michigan Press.

Comforts of Matrimony; Exemplified in the Memorable Case and Trial, Lately heard Upon an Action brought by Theo——s C——r against – S——, Esq; for Criminal Conversation with the Plaintiff's Wife. London: Sam Baker, 1739.

Cook, William (1804) *Memoirs of Charles Macklin, Comedian*, London: James Asperne.

Dekker, Rudolf and Lotte C. van de Pol (1989) *The Tradition of Female Transvestism in Early Modern Europe*, New York: St Martin's Press.

Dibden, Charles, (1797–1800) *A Complete History of the English Stage*, 5 vols, London.

Doane, Mary Ann (1982) 'Film and the Masquerade: Theorizing the Female Spectator' *Screen* 23: 74–88.

Galendo, Catherine (1809) *Mrs Galendo's Letter to Mrs Siddons: Being a Circumstantial Detail of Mrs Siddon's Life for the last seven years*, London.

Greenblatt, Stephen (1988) *Shakespearian Negotiations: The Circulation of Social Energy in Renaissance England*, Berkeley: University of California Press.

Hitchcock, Robert (1788) *An Historical View of the Irish Stage*, 2 vols, Dublin: R. Marchbank.

Jardine, Lisa (1983) *Still Harping on Daughters: Women and Drama in the Age of Shakespeare*, Sussex: Harvester Press.

Life of Mr James Quinn, Comedian, The. London: S. Bladon, 1766.

Memoirs of the Celebrated Mrs Woffington, Interpersed with Several Theatrical Anecdotes; The Amours of Many Persons of the First Rank; and some Interesting Characters Drawn from Real Life, London, 1760.

Milhous, Judith (1979) *Thomas Betterton and the Management of Lincoln's Inn Fields 1695–1707*, Carbondale: Southern Illinois University Press.

Mottley, John (1747) *A Complete List of all the English Dramatic Poets*, London: W. Reeve.

Mulvey, Laura (1975) 'Visual Pleasure and Narrative Cinema' *Screen* 16: 6–18.

Orgel, Stephen (1989) '"Nobody's Perfect": Or Why Did the English Stage Take Boys for Women?' *South Atlantic Quarterly* 88.1: 7–30.

Rogers, Pat (1982) 'The Britches Part' *Sexuality in Eighteenth-Century Britain*, Ed. Paul-Gabriel Bouce, Manchester: Manchester University Press, 244–58.

Smith, Sidonie (1987) *The Poetics of Women's Autobiography: Marginality and the Fictions of Self-Representation*, Bloomington: Indiana University Press.

Victor, Benjamin (1761) *The History of the Theatres of London and Dublin From the Year 1730 to the Present Time*, 2 vols, London: T. Davies, R. Griffits, T. Becket, and P. A. de Hond, G. Woodfall, J. Coote, G. Kearsley.

Wilkinson, Tate (1790) *The Wandering Patentee: or a History of the Yorkshire Theatres, from 1770 to the present time*, 4 vols, York: Wilson, Spence, and Mawman.

Wright, James (1699) *Historia Histrionica: An Historical Account of the English Stage*, London: G. Groon, for William Hawes.

7.3

BREAKING THE BOUNDARIES OF THE BROKEN BODY

Margrit Shildrick and Janet Price

Theories of postmodernism and the tool of deconstruction are not often associated with the kinds of substantive issues with which an everyday living feminism concerns itself. They may be fine – though never less than controversial – for laying bare the construction of knowledge, or for posing new theorisations of the subject, but it is less obviously clear how they might contribute to an issue-based feminist politics, such as might surround women's health. What we intend to talk about in this article is, nonetheless, just one such area. Our topic is disability; and we want simultaneously to hold in mind the experience of disability as an experience of a supposedly 'broken' body, and disability as precisely one of those transgressive categories that demands that we rethink not simply the boundaries of the body, but equally those between sameness and difference, and indeed self and other. We want to bring together some hard practical concerns with the (con)textual interplay of postmodernism; to show, as Eve Kosofsky Sedgwick says of her own experience of cancer, how 'deconstruction can offer crucial resources of thought for survival under duress . . .' (1994: 12). [. . .] And where postmodernism can come to the aid of a feminism attempting to end its own past indifference is in two ways: first in deconstructing all and every identity, and second in laying bare the ways in which the body itself is constructed and maintained as disabled.

It is not that we think there is no distinction to be made between those women who experience themselves as disabled, and others who are able-bodied, but that, from a postmodern perspective, those are always provisional

From: *Body & Society* (1996) 2 (4): 93–113.

and insecure categories which can never be entirely separate. What we are contesting are the fixed dichotomies – of health/illness, able-bodied/disabled, whole/broken, them and us, and so on – that constitute the very ground of our embodied selves. Indeed, it is through appeal to those hierarchical and apparently stable binaries that we are able to maintain a sense of definition, of the boundaries between sameness and difference, and thus of safety, bodily integrity and (self)identity. Tom Shakespeare identifies the problem as one of the privileged term of the binary: '… it is not us it is non-disabled people's embodiment which is the issue: disabled people remind non-disabled people of their own vulnerability' (Shakespeare 1994: 297). Nevertheless, it is our contention, that although the exclusionary/othering process is usually seen as the prerogative of the dominant, the 'able-bodied', the same move is made in reverse in radical disability politics. As we shall show, the boundaries of what constitutes disabled identity is policed every bit as strongly against its others.

[…]

DISCIPLINARY PRACTICES[1]

Following Foucault's problematisation of the human body and of the epistemology surrounding it, we may see the body not as the point of departure for a bio-psycho-social science of health and illness, but as the very locus of knowledge production. Foucault's concern is with medicine as a disciplinary regime through which the embodied subject is inscribed and brought into being, and with the circulation of power/knowledge as the indivisible condition of discourse. In other words, notions of health, of physical ability, are not absolutes, nor pre-given qualities of the human body, but function both as norms and as practices of regulation and control that produce the bodies they govern.

[…]

While we would agree that Foucault's analysis is flawed by his gender omissions, his deconstructive approach to the episteme of the body and to power is a stepping stone of great significance to a specifically feminist contestation of the politics of disability. It should not be forgotten, of course, that while the body is always marked by gender, it is crossed too and mediated by a variety of other categories such as class, ethnicity, sexual preference and indeed (dis)ability which may both bind and separate women and men. Above all power circulates in the procedures of normalisation by which on the one hand the body is inscribed with meaning (the intelligible body) and on the other rendered manageable (the useful/manipulable body) (Foucault 1977: 136). Together these two modes constitute the docile body which 'may be subjected, used, transformed, and improved' (Foucault 1977: 136). Insertion of bodies into systems of utility – be they at the service of capitalism or patriarchy – devolves on forms of power that are localised over the singular body, and that rely not on brute force but on quasi-voluntary acquiescence. The disciplinary and regulatory techniques practised on the body exemplify the productive nature of power in that they not only set up systems of control, but call forth new desires and institute new normativities. In this, medical science is exemplary in that it constitutes the individual in terms of a series of norms, while at the same time inviting the subject to produce truths about herself. It is not simply that 'the female body became a medical object *par excellence*' (Foucault 1988: 115), but that the external gaze is complemented by a complex mesh of techniques of self-surveillance and confession.

Though the clinical encounter is a paradigmatic site for the technologies of the body which both shape and control, in the modern welfare state the effects of healthcare as a disciplinary regime can extend into other most private and personal aspects of life. The demand to know intimate details about the individual is a common feature of state bureaucracy, but is nowhere more apparent than in the transaction between the welfare claimant and the multifarious over-seeing benefit agencies. In recent years, the trend in Great Britain has been towards various forms of self-certification to replace in-person interview and examination; but, far from liberating the claimant from an authoritarian and intrusive situation, the locus of power/knowledge has merely shifted to equally or additionally onerous forms of surveillance. The gaze now cast over the subject body is that of the subject herself. What is demanded of her is that she should police her own body, and report in intricate detail its failure to meet standards of normalcy; that she should render herself in effect transparent. At the same time, the capillary processes of power reach ever deeper into the body, multiplying the norms of function/dysfunction. As with confession, everything must be told, not by coercive extraction, but 'freely' offered up to scrutiny. The subject is made responsible, and thus all the more cautious and manageable, for her own success in obtaining state benefit. And, should benefit be withheld, then it may be attributed to a failure of reportage as much as to a denying external agency.

These particular modes of disciplinary practice are exemplified with great clarity in the procedures surrounding the benefit currently known as Disability Living Allowance (hereafter DLA).[2] As a benefit directed towards a state of being that affects both men and women of all ages, classes, sexualities, ethnic groups and so on, DLA might seem to illustrate the general operation of power/knowledge in and over the body without specific relevance to gender. We have chosen it as an example, however, precisely because of the way in which disability imbricates conceptually with the wider issue of the existential disablement of the female body in western society. Our concern is not simply that female bodies are the privileged target of disciplinary practices, but also that state-defined disability mirrors the phenomenological experience of women generally. Given that all women are positioned in relation to and measured against an inaccessible body ideal, in part determined by a universalised male body, the experience of female disablement as such may be seen as the further marginalisation of the already marginal. Where all women's experience of their corporeal integrity is generally under threat or inadequately addressed (Young 1990), then those who are additionally defined as disabled may find their bodily experience even more likely to be invalidated (Wendell 1992). In relation then to the 'whole' body of phenomenology, women with disabilities may be seen as doubly dis-abled.

None of this is intended to imply any pre-existing strategy to position 'empirical' disability as a peculiarly feminine condition, and, indeed, male claimants of DLA are subjected for the most part to the same extraordinary procedures as female ones. Nonetheless, insofar as the category of disability is constructed through such practices, it is – and this is a point we would want to make about broken bodies in general – as a condition that is en-gendered as feminine in terms of its implied dependency and passivity. Bearing in mind that the docile bodies produced by disciplinary techniques are an effect in every instance of power/knowledge, what is additionally striking about the shifting and heterogeneous set of conditions named as disability is that in its construction the disciplinary process is laid bare. And, moreover, that heterogeneity, paradigmatically exemplified by the multiple states that fall into the diagnosis ME, is itself masked in the production of a regulatory category that operates as a homogeneous entity – disability – within the social body. Despite the emphasis given to what appear to be very singular determinations of a state of disability, it is in the very gestures of differentiation and individuation – as exemplified by the innumerable subdivisions of the questions posed on the DLA form – that the claimant is inserted into patterns of normalisation which grossly restrict individuality. Ultimately, what the technologies of the body effect, while appearing to incite the singular, is a set of co-ordinated and managed differences.

For the specific benefit of DLA, intended for those who need help with 'personal care' or 'getting around', self-assessment plays a particularly large part in the claims procedure. Nonetheless, the limits of reliability of non-authorative discourse are marked in that the subject's own report must be supplemented by statements from two other people who will be most usually health care professionals. In other words, the gaze is multi-perspectival. What is remarkable about the claims pack (Benefits Agency, 1993) sent to potential claimants is its sheer volume, in which four pages of initial notes are followed by 28 pages of report, the vast majority of which consists of a detailed self-analysis of personal behaviour. The introductory instructions are quite clear about what is expected from claimant self-surveillance: '(t)he more you can tell us, the easier it is for us to get a clear picture of the type of help you need' (Section 2:1); and they suggest: 'keep a record for a day or two of how your illness or disability affects you' (Section 2, 1).

In focusing on singular behaviour, the state sponsored model of disability promotes individual failing above any attention to environmental and social factors. The DLA pack rigidly constructs and controls the definitional parameters of what constitutes disability in such a way that those who need to place themselves within that definition are obliged to take personal responsibility in turning a critical gaze upon their own bodies. The claimant is constrained to answer questions not just on her general capacity to successfully negotiate the everyday processes of washing, dressing, cooking and so on, but on the minutiae of functional capacity at every differential stage, and moreover at differential frequencies. The implication of such a demand is that disability is a fixed and unchanging state which pre-exists its observation. In contradistinction, our point is that not only is disability a fluid set of conditions but that the body itself is always in process. Yet again, ME specifically contests the possibility of predictable performance. A single page (see Figure 1) illustrates the extraordinary complexity and detail in which the claimant is expected to freely confess to her own bodily inadequacy. The questions for each discrete function follow a similar format and many, like those on toilet needs, are duplicated to establish night-time behaviour as well. What this amounts to is an astounding display of power/knowledge and of the capacity to proliferate discourse in accordance with Foucault's dictum: 'the exercise of power creates and causes to emerge new objects of knowledge and accumulates new bodies of information' (1980: 51). No area of bodily functioning escapes the requirement of total visibility, and further, the ever more detailed subdivision of bodily behaviour into a set of discontinuous functions speaks to a fetishistic fragmentation of the embodied person.

In the absence of any sufficient justification that could arise from the declared intentions of the welfare process itself – to provide financial help with personal care or getting around – one must assume that the extent of the benefit agency's 'need' to know is indeed an expression of the power/knowledge complex that underwrites the modern social body. In the section on cooking a main meal, for example, the claimant is asked to distinguish between the inability to use a cooker and the inability to cope with hot pans (*DLA 580*: 16); while in the toilet needs section, women are subjected to a supplementary gaze that requires them to report on their difficulties around menstruation: 'Tell us as much as you can ...' (*DLA 580*: 12). The welfare claimant is controlled not by a display of external coercion but by continuous surveillance and by the insistent demand for a personal accounting that fits the rigidly constructed parameters of disability. The subject herself effects a normalising judgment on her own modes of being by submitting to what Foucault calls a power that 'produces domains of objects and rituals of truth' (1977: 194). Moreover, she acts not as a pre-existent bounded being, but constructs her very selfhood in the process of normalisation. In terms of the *DLA* claim form, she produces herself as a disabled subject. What this display of the productivity of power signals is how control of the social body is effected through disciplinary technologies targeted on the individual body.

Where Foucault was concerned primarily to deconstruct the power relations between the singular, but universalised, body and a series of institutional forms – the prison, the clinic, the school – and to expose the symbiotic links between the individual disciplinary practices and the manipulation of population, feminists have been constrained to emphasise that disciplinary economies are gendered. The interplay of power and knowledge produces difference in just such a way that the bodies of women are the ground on which male hegemony and, at least in part, the power of the state in the service of capitalism are elaborated.[3] For all of us, the polymorphous forms of domination to which we are subjected are frequently masked so as to appear freely chosen, that is,

About help with personal care

■ Part 2 Help you need - **during the day** - continued

▪ **Coping with your toilet needs**

Roughly how many days a week do you need help	No help needed	1 to 3 days	4 to 5 days	6 to 7 days
getting to the toilet ?	☐	☐	☐	☐
using the toilet ?	☐	☐	☐	☐
using something like a commode, bedpan or bottle instead of the toilet ?	☐	☐	☐	☐
coping with incontinence of the bladder ?	☐	☐	☐	☐
coping with incontinence of the bowel ?	☐	☐	☐	☐
using a colostomy bag ?	☐	☐	☐	☐
using incontinence aids, pads or nappies ?	☐	☐	☐	☐

How many times a day do you need help coping with your toilet needs ?

Roughly how many minutes do you need help for each time ?

Please tell us about any equipment that you use to help you with your toilet needs.
This could be rails by the toilet, a special toilet seat, or something like this.

Tell us as much as you can about the help you need coping with your toilet needs.
The more you can tell us, the easier it is for us to get a clear picture of the type of help you need.
For example, if you are a woman you may need help coping with your periods.

12

Figure 1
Source: Benefits Agency, *DLA* section 2: 12.

expressive of personal desire or consented to as necessary for individual or social good. What is not always apparent is that those goods and desires circulate within a system of normativities which, although never inevitable, imposes nonetheless a powerful urge to behave in certain ways, to mark out the boundaries of the proper. Indeed, the efficacy of disciplinary practices may be greatest when they appear not as external demands on the individual but as self-generated and self-policed behaviours. These internalised procedures constitute what Foucault calls the technologies of the self. In other words, the objectifying gaze of the human sciences, which fragments and divides the body against itself, has its counterpart in a personal in-sight, which equally finds the body untrustworthy and in need of governance. Moreover, each form of surveillance incites the other, and renders its subjects wholly transparent.

While it is clear that diverse groups of women, including those classed officially as disabled, are marginalised by many operations, like slimming and keep fit, which are directed at 'whole' bodies, the point remains that all women are positioned *vis-à-vis* an inaccessible body ideal. The reiteration of the technologies of power speak to a body that remains always in a state of pre-resolution, whose boundaries are never secured. Indeed, repetition indicates its own necessary failure to establish any stable body, let alone an ideal one. In the phallocentric order, the female body, whether disabled or not, can never finally answer to the discursive requirements of femininity but remains caught in an endless cycle of bodily fetishisation. In other words, it is a body that always exceeds control.

What we would suggest is that it is precisely that which escapes femininity, the embodied but gender resistant female subject, which provides the moment of contestation. The claim is not that the bodies of women are ever outside the relations of power/knowledge, but that there is potential slippage between what is possible for them and what is required of them by even the most adaptive patriarchal state. Just as disciplinary power incites certain practices in which external expectations are internalised in forms of self-surveillance, so too, those same practices may ground resistance. One example might be the way in which wheelchair athletes subvert expectations of weakness by consistently outperforming non-disabled runners in marathon races.

Given, then, that the construction and quasi-stabilisation of the disabled body is achieved through the continual procedures of both internal and external disciplinary power, might that indicate how a resistant feminism could respond? Foucault's insistence on the absolute interconnectedness of power and resistance – (t)here is no power without potential refusal or revolt' (1988: 84) – indicates the disabled body is never simply passive. But is the struggle ever a successful one, or rather what would constitute success amid the relentless relations of power? As an individual with ME, I do not understand our analysis of *DLA* as a 'personally liberatory experience' but as resistance which 'continually seeks to uncover the constitute mechanisms of truth and knowledge as they construct and position the individual within social and scientific fields' (Shildrick 1997). Foucault himself speaks always of local and discontinuous points of resistance, and recommends the recovery of 'disqualified, illegitimate knowledges' (1980: 83). Perhaps, for feminists and for those others concerned with disability politics, that might point to the obscured histories of bodies. If we can demonstrate that what has been naturalised as the truth of the body is merely the discontinuous outcome of a complex series of normalisations, in which health care has been pre-eminently implicated, then it becomes possible to dissolve devalued identities and theorise new constructions of embodiment. In contesting the universal signification of the living body our aim should be to acknowledge the plurality of possible constructions and the multiple differences which exceed imposed normativities.

For all that, however, final self-identity eludes the embodied self, for the boundaries which organise us into definable categories are in any case discursively unstable, such that constant reiteration is needed to secure them. Just as we perform our sexed and gendered identities, and must constantly police the boundaries between sameness and difference, so too the 'purity' of the 'healthy' body must be actively maintained and protected against its contaminated others – disease, disability, lack of control, material and ontological breakdown. As Diana Fuss puts it:

> Deconstruction dislocates the understanding of identity as self-presence and offers, instead, a view of identity as difference. To the extent that identity always contains the spector of non-identity within it, the subject [and we would stress here the embodied subject] is always divided and identity is always purchased at the price of exclusion of the Other, the repression or repudiation of non-identity. (1989: 102–3)

In a reworking of the separation of self and other, there can be no understanding of, for example, able-bodied, unless there is already an implicit distinction being made that to be able-bodied is not to be disabled. Yet because able-bodied carries within it the trace of the other – a trace which must be continually suppressed if able-bodied is to carry a delimited meaning – such closure is not possible. To deconstruct binary difference, then, to point up all those oppositional categories which begin to undo themselves at the very moment of defining identity through exclusion, disrupts both ontological and corporeal security. In other words, the spectre of the other always already haunts the selfsame: it is the empty wheelchair that generates dis-ease in the fully mobile.

Though the disability movement has both challenged and reconceived the relationship between able-bodied and disabled, its flirtation with identity politics precludes any understanding of how those categories are complicit with one another, and of how each might be radically destabilised. Interestingly, however, one strategy recently advocated in disablement politics is to push the 'healthy' majority to a recognition that they are merely temporarily able bodies (TABs). Although that is intended to mark no more than the material precariousness of health, the notion can be extended to provide just that thoroughgoing critique of health/ill-health, able-bodied/disabled that a poststructuralist approach would demand. The regulatory and disciplinary regimes which impose and maintain normative standards of bodily and mental well-being are necessary precisely because of the inherent leakage and instability of those categories, because the spectre of the other lurks within the selfsame.

A radical politics of disability, then, might disrupt the compulsory character of the norms of abled and disabled, not by pluralising the conditions of disability, as the notion of TABs intends, but rather by exposing the failure of those norms to ever fully contain or express their ideal standards (Butler 1993: 237). And it should be stressed that while disabled and ill people – those whose bodies are deemed as broken – are labelled as other, they do not escape the regulatory apparatus of norms, but are forced to negotiate a set complementary to those of able-bodiedness. In illness and disability, what can be called performative acts – that is the corporeal signs, gestures, claims and desires elicited in embodied subjects – serve no less to produce effects of identity, coherence, control and normativity.

But the very need for the *reiteration* of the regulatory process, through which the materialisation of bodies is compelled, simultaneously destabilises the body, revealing that which exceeds the norm. The discontinuities continually break through, opening up a gap between bodily form, appearance, function and ability: the deaf person who can hear you perfectly, till you turn your back on them; the woman who uses a wheelchair and has just qualified as an aerobics instructor; the visually impaired woman who greets you in the evening on the street but cannot see you in the light of day. These disruptions speak not to the apparent limits of an impaired body, but rather of a break with the normative identities of those who are blind, deaf, disabled, and so on. For just such reasons, performativity may evade normalisation and move instead into transgressive resistance. Speaking of her experience of breast cancer, Eve Sedgwick writes of the 'performativity of a life threatened ... by illness' and of herself 'hurling her major energies outward to inhabit the very farthest of the loose ends where representation, identity, gender, sexuality and the body can't be made to line up neatly together' (1994: 12).

ME AND PERFORMATIVITY

We want now to return to look more specifically at the ways in which the woman with ME, the 'ME sufferer', has been produced through the reiteration of regulatory norms which have materialised the ill and/or disabled body, and in particular at the discursive practices of the medical system and self-help groups which have produced differing reinscriptions of the bodies of those with ME.

[...]

What is at stake for coherent diagnostic closure is that the body should be constructed either as stable or as predictably changeable. Professional theorisation relies, therefore, upon that body being open to investigation – to the invasive waves of ultrasound and the electrical impulses used in muscle tests, to the needle drawing blood, and to the structured psychological questionnaire – and to revealing 'truths' in standardised ways. The patterns of ritualised medical exchange that form the clinical examination permeate the boundaries of the body in fixed and certain ways, to reveal information that might explain and reinscribe a stabilised, though ill, body. The medical gaze seeks to establish an empirical, material 'broken' body but, as with the DLA assessment, the individual is incited at the same time to become an instrument of that ever-extending gaze. She turns it on herself, carrying out her own self-regulation, thus becoming complicit in the process of constituting herself as embodied subject.

A questionnaire produced by the Department of Medicine of the University of Liverpool – for completion by the 'patient' – is an example of how such self-surveillance operates. Through the reiteration of a series of implicit criteria contained in the questions, norms for ME are established. Twenty-six pages of questions cover everything from fatigue and pain, to sex life, sleep patterns, emotional state and social contacts. We can look at one example. A series of questions on pain, which attempt to categorise it and relate it to functional ability – levels of walking, sleeping, having sex, socialising – is followed by a silhouette of the human figure which the patient is asked to complete by precisely mapping where and how she experiences pain (University of Liverpool 1993). The act of marking it down, making it visible, serves to bring a sense of coherence. A pattern emerges through the questions. In answering the questionnaire, those with ME become a part of the process of the inscription of the body, establishing norms without which organisation of the condition has to start afresh each time.

[. . .]

The performative acts of those with ME have also played a major role in the reinscription of the body and the reformation of the identity of ME 'sufferers'. Many have joined self-help groups which play an increasingly important role in the lives of many disabled people and those with illnesses, and which, in working from a range of different political positions, tend to offer, as a minimum, support and information. For some with ME they have provided a vital source of support in the face of dismissal by the medical profession and denial by family or friends of a condition that has caused major changes in individual lives. However, groups such as the ME Association, in order to continue having any coherent role as a self-help organisation, are constrained to express a stable or core identity which materialises the normative subject of the 'ME sufferer'. Although self-help groups see themselves as breaking out of the conformity of the medical model, they do not necessarily offer a radical critique.

In the case of a typical self-help group for lesbians with ME, one aspect of joining involved the completion of a form about my experience of illness: how long, for example, had I been ill, what were my symptoms, what was my background and my interests. The newsletter published summaries each month from new members, clearly providing support and a sense of shared experience to other members. Through the reiteration, repetition and categorisation of symptoms in this and other self-help forums, a normalising process occurs through which some symptoms are accepted as indicative of illness, others not. The process of monitoring remissions, relapses and other changes has become a necessary act for the person with ME, such that the very instability of the condition has become incorporated as a stable feature of a new norm. Members are constrained to 'perform' their illness or disability in ways that fit in with the norms adopted by the self-help group. But the point here is not so much that the performativity of disability is something a subject may freely choose, but that disability itself 'is performative in the sense that it constitutes as an effect the very subject that it appears to express' (Butler 1991: 24). At an uncontested individual level, the deployment of norms offers a fantasy of control, a way of pinning down, categorising and assuming the ability to manage a condition which constantly escapes attempts at closure, which continually produces new symptoms, or which returns to previous symptoms after a respite of weeks or months. They offer a way of adapting to the functioning of a body which suddenly runs out of energy, leaving you stranded in the middle of a shopping expedition, or half-way up a flight of stairs, or in the middle of making love. They create an illusion of mastery, never completely absent but never totally achieved.

In campaigning ME groups, the normative identity claimed relies on the recognition of a physically induced impairment, and there is strong resistance to any suggestion of non-somatic causation. The ME Association in particular perceives itself as having made a major breakthrough in achieving recognition for ME in the International Classification of Diseases, *the* definitive list of recognised conditions. The belief in an underlying viral illness is an article of faith in most ME self-help groups, and there is no place for the woman who has similar symptoms, but relates them all to stress. Part of the operative norm is that of the tragic victim, suffering from a physical illness and yet dismissed by the medical profession and by society. The 1993 Annual Report of the ME Association entitled *The Burden of Proof* reiterates this norm both visually – through the cover image of a drooping, wheelchair-bound figure – and in its contents, which focus on research undertaken into the viral origins and pathological markers of ME. This process of reinscription of the body of illness, through knowledge circulated and legitimised by self-help groups, acts as a strong cohesive link in the maintenance and success of such groups.

The ME Association's imagery – and this is true of some other groups – mirrors conventional media representations in which 'disabled people are portrayed as helpless, needy victims of illness and the plucky, courageous and brave cripple fighting adversity' (Ralph 1993: 9). In contrast, other disability rights groups have strongly critiqued such negative accounts (Corbett 1994; Shakespeare 1994). Nonetheless, however much self-help and campaigning groups have challenged the medical and charity models, familiar in cultural representations, their intention and achievement has been only to replace one set of functional norms with another. Either way, what matters is a stable identity and the discovery of the 'true' self.

> The denial of our reality ... is a suffocation of what makes us exist as unique individuals. It disempowers and weakens us. To gain a proud label, we need to fight this denial and use the language of our 'actual' identity. (Corbett 1994: 347)

What a deconstructionist approach suggests, on the contrary, is that all identities are constantly shifting and developing, both through resistance to existing norms, and through the incitement of new norms. But the process is never complete, and nor is there any final truth of the body at which to aim.

The task of the postmodernist feminist is to lay bare and contest the discursive construction of all seemingly stable categories. It is – to brazenly paraphrase Judith Butler writing on heterosexuality and homosexuality – as though disability 'secures its self-identity and shores up its ontological boundaries by protecting itself from what it sees as the continual predatory encroachments of its contaminated other', ability (1991: 2). Now that, of course, is a reversal of the usual relationship between ability and disability, but our point is that *both* categories are concerned to police their boundaries. [...] Indeed, the failure of feminism in general to respond adequately to issues around disability must surely reflect the difficulty of thinking beyond the binary of sameness and difference. By deconstructing both the regulatory processes of normalisation which map out the divisions between bodies, and by contesting the stability of the able-bodied/disabled subject herself, we hope to break down the boundaries of the broken body. There are neither homogenous categories, nor fully self-present individuals. In rethinking difference in terms of irreducible and multiple differences, we advocate not liberal tolerance but a radical openness to the disruptive otherness *within*.

NOTES

1. Parts of this section are taken from Shildrick (1997) *Leaky Bodies and Boundaries: Feminism, Postmodernism and (Bio)ethics*.
2. The analysis we make of DLA is equally true of the new Incapacity Benefit, introduced by the Department of Social Security in 1995 to replace Invalidity Benefit.
3. For a detailed analysis of how such a move operates with regard to the colonial state, see Price and Shildrick (1995).

REFERENCES

Baudrillard, Jean (1988) 'Consumer Society', in Mark Poster (ed.), *Jean Baudrillard: Selected Writings*, Stanford, CT: Stanford University Press.

Begum, Nasa (1992) 'Disabled Women and the Feminist Agenda' *Feminist Review* 40: 70–84.

Benefits Agency (1993) *DLA 580 Claims Pack*, London: Dept. of Social Security.

Butler, Judith (1990) *Gender Trouble: Feminism and the Subversion of Identity*, London: Routledge.

Butler, Judith (1991) 'Imitation and Gender Insubordination' in Diana Fuss (ed.), *Inside/Out: Lesbian Theories, Gay Theories*, London: Routledge.
Butler, Judith (1993) *Bodies That Matter: On the Discursive Limits of 'Sex'*, London: Routledge.
Corbett, J. (1994) 'A Proud Label: Exploring the Relationship between Disability Politics and Gay Pride' *Disability and Society* 9 (3): 343–57.
Foucault, M. (1977) *Discipline and Punish: The Birth of the Prison*, trans. A. Sheridan, London: Allen Lane.
Foucault, M. (1979) *History Of Sexuality, Vol. 1*, trans. R. Hurley, London: Allen Lane.
Foucault, M. (1980) *Power/Knowledge: Selected Interviews and Other Writings (1977–1984)*, ed. Colin Gordon, Brighton: Harvester Press.
Foucault, M. (1988) 'Power and Sex' in *Politics Philosophy Culture: Interviews and Other Writings (1977–1984)*, ed. Lawrence D. Kritzman, London: Routledge.
Fuss, D. (1989) *Essentially Speaking: Feminism, Nature and Difference*, London: Routledge.
ME Association (1993) *The Burden of Proof, Myalgic Encephalomyelitis Association Annual Review (1992–1993)*, Stanford le Hope: ME Association.
Price, Janet and Margrit Shildrick (1995) 'Mapping the Colonial Body: Sexual Economies and the State in Colonial India', in T. Foley, L. Pilkington, S. Ryder and E. Tilley (eds), *Gender and Colonialism*, Galway: Galway University Press.
Ralph, S. (1993) 'Charity Advertising on British TV' unpublished paper presented to the Group on the Status of Persons with Disability of the Association for Education in Journalism & Mass Communication.
Sedgwick, Eve Kosofsky (1994) *Tendencies*, London: Routledge.
Shakespeare, Tom (1994) 'Culture Representation of Disabled People: Dustbins for Disavowal?' *Disability and Society* 9 (3): 283–99.
Shildrick, Margrit (1997) *Leaky Bodies and Boundaries: Feminism, Postmodernism and (Bio)ethics*, London: Routledge.
University of Liverpool (1993) *Investigations into Chronic Fatigue Questionnaire*, Liverpool: Department of Medicine, University of Liverpool.
Wendell, Susan (1992) 'Toward a Feminist Theory of Disability' in Helen Bequaert Holmes and Laura M. Purdy (eds), *Feminist Perspectives in Medical Ethics*, Bloomington: Indiana University Press.
Young, Iris Marion (1990) *Throwing Like a Girl and Other Essays in Feminist Philosophy and Social Theory*, Bloomington: Indiana University Press.

7.4

FEMININE CHARMS AND OUTRAGEOUS ARMS

Fen Coles

What might muscles on women have to do with strength? Fen Coles shows the lengths to which the bodybuilding industry is prepared to go to hide, excuse and sexualise these threatening muscular developments and to keep female bodybuilders feminine.

This article developed out of a course which looked at representations of the body, particularly the female body, within patriarchal culture. I decided to look closely at the female bodybuilder, the way the bodybuilding industry and patriarchal ideology in general tamper with her, and her final contest appearance which is an odd mixture of muscle and make-up.

The female bodybuilder raises some important questions for feminists because of the ways in which she challenges traditional ideas about femininity. Patriarchal ideology depends on and enforces the idea that sex, gender and sexuality come together 'naturally' as a package, i.e. you are born female, therefore you must naturally be feminine and heterosexual. The female bodybuilder, however, particularly during her performance in competitions, challenges all these ideas together, demonstrating that femininity, heterosexuality and even the female body are constructs.

TAMING THE BEAST

In a bid to apologise for and to soften female muscle, repeated strategies are employed to ensex, engender and heterosexualise the female bodybuilder's disturbing physique. This muscle is repeatedly adorned, restricted and confined.

From: *Trouble and Strife* (1994/95) Winter, 29/30: 67–72.

This very compulsion seems, however, to fail from the outset. Highly visible on the contest stage, pumped, flexed, and increasingly taking up space with each competition, the female bodybuilder is a threatening sight. Ironically, patriarchy's attempt to feminise her muscle makes her appearance all the more unusual and unsettling.

Laurie Schulze describes the domestification of the female bodybuilder as a repeated reining in of her, 'emphasising certain features, suppressing others, and papering over contradictions'. Danae Clark describes this 'taming' process again in a fascinating essay which compares lesbian style and female bodybuilders, demonstrating the ways in which patriarchal ideology strives to homogenise and collapse both into a non-threatening sameness fit for heterosexual society. This is carried out on the female bodybuilder's body by dressing her up in feminine markers and by dressing down her muscle.

DRESSING DOWN/UP MUSCLE

If she complies with such dress codes, the female bodybuilder may be allowed to pass in heterosexual culture. Such 'passing strategies' include covering her muscles with long sleeves and declining to flex them in public. The undisplayed muscle is further promoted by women's magazines. Within these pages, the very women who appear to have surmounted the 'natural' body, reassure the female consumer that her biology, typically her lack of testosterone, will actually prevent her from developing an 'overly' muscular physique. Even were she to possess the 'unique super genes' (or, we might add, the steroids) required for competitive bodybuilding, she may be comforted in the knowledge that, unpumped and unflexed, she will appear as 'normal' as the girl next door. As Laurie Schulze says, the message is that 'these muscles are a difference that won't make a difference'.

Bodybuilding magazines, intent on a respectable public image, could not make this point clearer. On the cover of each magazine, the conflation of sex, gender and sexuality is confirmed and guaranteed. In almost every instance, a male bodybuilder grins at the consumer, flexed, erect, hard, oiled and pumped to full performance level. Dwarfed by this muscled monstrosity, often enfolded in his arms, stands what we are led to believe is a female bodybuilder. The woman featured is in fact usually a model. Always in profile, in a classic 'tits n' arse' shot, curving into the male's peak body, this woman performs her difference, setting off his supreme masculinity/maleness and stamping their heterosexuality.

When the overt display of muscle is called for at competition time and the near naked body staged, feminine props/apologies are called in to reinscribe and 'renaturalise' the female bodybuilder. Instructed to 'get feminine or get out of competitive bodybuilding' (a competition judge's own words), Bev Francis soon received plenty of beauty tips and, went about her transformation (which included plastic surgery).

Feminising muscle places the greatest emphasis on that single part of the body which seems unmarked by muscle, the face. The female bodybuilder's makeover must compensate as much for the effects of pre-contest diuretics (the face appears pinched and overdrawn) and steroids (increased facial hair) as for her bulked-up body.

On such a body, a subtle hint of femininity would achieve nothing. Instead it must be high glamour, overdone and overplayed. Bright, heavy make up and abundant (preferably bouffant) hair all play their part in this. Although short hair highlights the shoulder width required to compete, competitive female bodybuilders almost always have long hair (tied back at competition time) with their power suits, echoing the eighties message that women in business shouldn't combine short hair with shoulder pads. The hair is also almost invariably bleached blonde. (Out of the nine top winners at the 1988 Ms Olympia contest, seven were blonde.) Manuals for competitors are full of suggestions for bleaching, perming, tinting and complete makeovers. The overall display of the head combines whiteness and youthfulness, the supreme femme.

Fitting her into the feminine mould, the female bodybuilder is further fitted out at competition time in tiny, brightly coloured string bikinis. Materials include lycra and leather, adding to the body a superimposed fetishistic quirk. Photo shoots and posing routines are further camped up with outlandish props and costumes. Examples include Kimberly-Anne Jones, famous for her bondage poses (dressed up in leather and carrying a whip) and Andrulla Blanchette, repeatedly photographed in her ripped fishnet tights (and always with her boyfriend).

Lastly, feminine apologies are further extended when female bodybuilders refer in interviews to their softer selves. A 1994 *Options* article reassures us that, despite external appearances, Kimberly-Anne enjoys her Garfield pyjamas and Beverly Hahn weeps at an animal in pain.

The female bodybuilder's proclamation of sexual difference and feminine identity is further problematised by the likely disappearance of her breasts. Help is at hand, however. Although working out body fat cancels out 'feminine curves', the breasts can at least be rescued in the form of breast implants. While calf implants are banned in bodybuilding, generous allowances are made for the return of the female breast. Ironically (although conveniently for her domesticators), the female bodybuilder's busting up compromises and couches some of the muscle groups (serratus/intercostal/abdominal) required for definition at a competitive level. By tacking breasts onto female muscle, the female bodybuilder might be seen to make a particularly forceful move towards reinscribing herself within the site of the male gaze and male fetishism.

SEXUALISING MUSCLE

Perhaps more than any other sport, women's bodybuilding has been subjected to media 'dyke-baiting'. As Laurie Schulze (1990) comments, 'patriarchy and homophobia combine in complex ways to link female bodybuilders with lesbianism'. Indeed the ways in which the muscular woman and the lesbian distress the heterosexual system are closely linked, and they therefore receive similar responses from the dominant culture. Both the muscular woman and the butch lesbian are scorned for wanting to look like men or wanting to be men. Both are therefore often perceived to distort traditional gender appearances as well as gendered behaviour (they are seen as 'inappropriately' aggressive, for example). Accused of wanting to look like a man, and therefore necessarily of lesbianism, the female bodybuilder's sexuality is marked as excessive, disruptive and, worse, indifferent to men.

Such charges have resulted in a sustained move to sexualise and wedge female muscle within a heterosexual frame. Her muscle is sold as sexy accoutrement and her heterosexuality confirmed to the point of exhaustion. This obsession is apparent in bodybuilding magazines in which women and men are repeatedly shown training in couples and marketing products with their arms entwined. Again, magazine covers typically feature couples enjoying each other's well oiled bodies.

As if to finally guarantee the female bodybuilder's successful performance within heterosexuality, adverts for the 'Kegelcisor' or vaginal barbell crop up regularly in these magazines, promising to 'intensify sensation during intercourse' and to 'restore vaginal tightness'. This is itself a rather extreme example of bodybuilding's general promotion which links improved sexual stamina to weight training. It is, however, largely male performance which is highlighted. Any increased sexual endurance in the female bodybuilder is clearly tailored, 'tightened', 'restored' for his sexual enjoyment.

Making the female bodybuilder sexy, translates her into 'molten beauty', 'rosy-cheeked cutie' and 'blond bombshell'. Her workouts sell as '10 steps to a sexy waist' and tips on 'sexy shoulders' (all quotes from *Muscle and Fitness* and *BodyPower*). Above all, her body has a more traditional function within the merging discourses of sexuality, femininity and consumerism (although it should be emphasised that like any other woman, she is commodified both within and outside of the market). Her entry into the sport industry over the last decade has prompted her use as spectacle, her body capitalised on and commodified to sell a range of products. Within this advertising, her body is tightly monitored in order that the sex which sells is unmistakedly female. Consequently, although they may appear under the caption of female bodybuilder, the women who demonstrate gym equipment by mounting or straddling it in a white lace bodysuit, are usually models. Likewise, it is this vital-statistics muscleless figure which is turned out in the astounding array of pornographic videos and posters on offer in bodybuilding magazines. Again, posing as female bodybuilders, these curvaceous women perform *Nude Boxing*, *Cindy V Chris*, *Mistress of Muscle*, *Muscle Melons* and *Steel Innocence* (examples from *MuscleMag*).

STRAINING THE HOUR-GLASS

Struggling with all these attempts to recast female muscle within patriarchal regulations, the industry's governing bodies and judges continue to be locked in confusing contradictions. In a bid for mass appeal and in a panic that female mass might swamp vital sex/gender signs, female bodybuilding competitions have often been judged along lines of conventional attractiveness rather than of size. The sport is singular in that the achievement of maximum strength may be rewarded with minimum gain in terms of ranking.

Uncertainties and anxieties abound at competition time: whether to opt for the glamour, grace and showbiz style of the beauty contest (in which many see female bodybuilding's origins), or the 'pure' bodybuilding contest; whether to rank maximum beefiness highest or to praise a basic hour-glass shape joined with shapely 'feminine' muscles, thereby maintaining a 'feminine mystique' (competitor's words).

Female bodybuilders appear no more decided. Thus while one woman is typically heard explaining 'why my muscles had to go' in order to feel 'normal' (quoted in *MuscleMag*), there are a growing number of female competitors who have not only no interest in pandering to conventions (in terms of muscle size), but who actively enjoy the confusion that their bulk provokes.

What is clear from this constant squeezing of muscle into a feminine package is that the limits of this package are constantly being strained. As Robert Duff and Lawrence Hong (1984) comment, 'the concept of muscularity is relative and is rapidly changing as the sport progresses'. Moreover, there is a definite rebellious whisper in the ranks which continues to demand attention: 'We are saying, who are you to tell us what we should look like? We're saying it with our bodies' (Kimberly-Anne Jones).

IM/PROPER MUSCLE

The increasing number of women entering the bodybuilding industry and their expanding body mass suggest that the female bodybuilder is not a passing phenomenon despite the confusion she engenders. For in spite of (and, indeed, because of) every endeavour to adorn and contort her, her sabotage of gender norms has simply meant that she 'clamours harder to be looked at, to be evaluated and to be discussed', (Drorbaugh 1993).

That her domestification continues to fail is evidenced by the perpetual labelling of her by the public. She has been called grotesque, perverse, obscene; disparaged for not being a 'real' woman, looking like a man/dyke. More commonly, she has been branded freak, hermaphrodite, transvestite, gender impersonator (butch in a frock?). Ironically it is the very means by which her muscles are dressed down/up which guarantees her disturbing potential as cross-gendered: 'within these passing strategies are the very seeds of resistance' (Clark). In other words, as we shall see, attempts to feminise the female bodybuilder – the strategies which allow her to 'pass' in patriarchal culture – become instead the means by which she resists any traditional reading of her as 'feminine'. 'The average person cannot understand why any woman would want to look like a man' (*Options* 1994).

What is it about muscle which insists on its bearer being irrevocably male?

Taken into the cultural arena, muscle is highly gendered, the embodiment of a discourse which states that 'to be an adult male is distinctly to occupy space, to have a physical presence in the world' (Morgan 1993). This space is secured by men viewing muscle as legitimately theirs. Muscle is therefore 'naturally' taken on/given to the male. As Richard Dyer (1982) says, muscles have been traditionally understood as symbolic of male power and just as patriarchy forces us to see power as something which belongs 'naturally' to men, so muscle is seen to be a biological given in men. Natural and real only to him, the representation of muscle is reducible only to him, ratifying the conflation of sex and gender.

The discussions around steroid use amongst bodybuilders is similarly gendered. Within the bodybuilding industry, the potentially fatal damage caused by steroid use is downplayed. Instead, the view of such chemicals as 'true gender benders' (*Muscle & Fitness*) is promoted. Overwhelmingly, however, it is women users who are scolded for 'gender bending' (even though steroid use by men can result in testicular atrophy and the development of breasts, these side effects are almost invariably hushed up). A discourse of appropriateness/inappropriateness abounds again. Testosterone is popularly and incorrectly understood to be the male sex hormone and is therefore (in both its natural and synthetic form) deemed proper only to the male. He is seen to add on/take on further that which was always conferred on him, both testosterone and its effects, bulk muscle. The female bodybuilder's use, by contrast, is deemed improper and monstrous.

Within this discourse of naturalised sex and gender, not only muscles and 'femininity' but muscles and woman are exclusive categories. The female bodybuilder would seem then to be an impossible term. Nearly naked on the contest stage, she offers up her natural body dressed up in someone else's sex signs, not taken on, not added on, but put on. Ultimately, the female bodybuilder's muscles constitute a kind of drag.

FEMALE MALE IMPERSONATORS/MALE FEMALE IMPERSONATORS

It is precisely through the ideas of drag and cross-dressing that the female bodybuilder's subversiveness can be measured. Many of the ideas we receive about the categories of male/female, masculine/feminine, set up the male/masculine as the real in our culture: 'Men are real. Women are "made up"' (MacCannell and MacCannell 1987). If men and masculinity are seen as natural and authentic categories, then the idea of 'performing' masculinity would seem to be a contradiction in terms.

When we see men impersonating women in drag acts, the audience interprets this act as one which exposes women/femininity as artifice or construction. We are less used to seeing this act in reverse. What happens when we see a woman impersonating men/masculinity, categories which patriarchy has taught us are real? We must surely, as Elizabeth Drorbaugh has argued, interpret men/masculinity similarly as constructions. The effect of this is that 'faith in the real may begin to break down' (Drorbaugh 1993).

The female male impersonator shows us then that the 'real' is in fact a lie. This demonstration is undoubtedly subversive for, as Alisa Solomon says, 'what confers male privilege if not some intangible aura of masculinity – and how potent, how sure is that quality when women can put it on as easily as hats and tails?' This act of impersonation is even more subversive when it is performed by the female bodybuilder.

Certainly when the female bodybuilder appears at competition time, we do indeed see something resembling an actual stage performance. Glammed up, oiled and engaged in a posing routine, her display takes on an air of high theatricality. Flexing, she camps up and puts on male muscle. Crucially, however, her cross-dress does not disappear once she is off stage. It may be covered or played down, but it isn't an instantly removable power suit. For this reason, the female bodybuilder performs the transgressive potential of cross-dressing in a particularly radical way – her challenge to traditional ideas of sex and gender is not a costume (like hats and tails) which she can take off after the show; this challenge appears on her body.

The female bodybuilder's challenge does not rest here, however, for, were she simply to perform masculinity, she could once again be returned to the familiar, returned to 'wanting to look like a man'. Indeed it is the very strategies which seek to make her more comfortable, less 'like a man' which further constitute her disturbance. As Yvonne Tasker observes, 'it is precisely the femininity of the female bodybuilder that destabilises her relationship to the supposedly secure categories of sex, sexuality and gender.'

Again, as the other, the non-man, the non-real, women are 'made up', lending itself to the suggestion that 'femininity is always drag, no matter who paints on the nail varnish and mascara' (Solomon 1993). Female bodybuilders in particular are made to 'dramatise their sexuality' (MacCannell and MacCannell 1987), and adopt extreme trappings of the artificial sex: 'The female muscled body is so dangerous that the proclamation of gender must be made very loudly indeed' (MacGinn and Mansfield 1993). Lavish and theatrical make up, blondeness, fetishistic garb and the occasionally tacked on breasts are all part of this noisy announcement: 'the artificial nature of femininity ... most ... grotesquely demonstrated' (Solomon 1993).

The final result resembles in many ways a feminine caricature or, more precisely, a male female impersonator. Accordingly, as both the *Options* article and Laurie Schulze point out, the female bodybuilder has frequently been read by her spectator as a male transvestite.

SEEING IS DISBELIEVING

Both as male female impersonator and female male impersonator, the female bodybuilder disrupts clear gender norms. According to Judith Butler, dressing up in gender trappings can expose that whatever has been deemed appropriate/proper to one sex, has only been 'improperly installed as the effect of a compulsory system'.

In other words, while certain things have traditionally been seen to belong 'naturally' to one sex or the other, drag shows us that the allocation of these things has less to do with what is 'naturally appropriate' and more to do with the effect of a system, in this case patriarchal ideology, which has ruled from the outset what is 'properly' male/female, masculine/feminine. To take this further, Judith Butler (1991), suggests that drag reveals that there is nothing inherently natural about gender. Instead, gender is a kind of impersonation/ imitation. If we see gender as an imitation, then drag cannot be seen to imitate any original or 'authentic' gender. For example, we might understand a drag act as one which copies masculinity or femininity. But we know that masculinity and femininity are themselves artificial constructions. We can therefore take our understanding further and see drag as an imitation of an imitation.

Drag can therefore radically undo traditional ideas of a natural gender/ natural sex. Through drag, the supposedly fixed binaries of masculinity/ femininity, male/female overlap and fall apart. All categories are exposed as false, artificial.

It is precisely this disruption which is undertaken by the female bodybuilder. Spectacular on her stage, seeing her is disbelieving, for finally she cannot be accommodated on either side of any binary. Enacting a double impersonation, her 'female' body fills out a masculine body drag, laced with super-feminine embellishments. The spectator cannot resolve what she 'ought' to be – a woman – and what she appears to be: the impossible juxtaposition of feminine/masculine, female/male, femme/butch, female impersonator/male impersonator.

REFERENCES

Butler, Judith (1991) 'Imitation and Gender Insubordination' in Diana Fuss, ed., *Inside/ Out: Lesbian Theories, Gay Theories*, Routledge.

Drorbaugh, Elizabeth (1993) 'Sliding Scales: Notes on Storme DeLarverie and the Jewel Box Revue, the cross-dressed woman on the contemporary stage, and the invert' in Lesley Ferris, ed., *Crossing The Stage: Controversies on Cross-Dressing*, Routledge.

Duff, Robert and Hong, Lawrence (1984) 'Self-Images of Women Bodybuilders', in *Sociology of Sport Journal* 14.

Dyer, Richard (1982) 'Don't Look Now', in *Screen* 23:3/4.

Ferris, Lesley (ed.) (1993) *Crossing The Stage: Controversies on Cross-Dressing*, Routledge.

MacCannell, Dean and MacCannell, Juliet Flower (1987) 'The Beauty System' in Nancy Armstrong and Leonard Tennenhouse, eds, *The Ideology of Conduct*, Methuen.

MacGinn, Barbara and Mansfield, Alan (1993) 'Pumping Irony: The Muscular and the Feminine' in Sue Scott and David Morgan, eds, *Body Matters: Essays on the Sociology of the Body*, Falmer.

Morgan, David (1993) 'You Too Can Have A Body Like Mine: Reflections on the Male Body and Masculinities' in Sue Scott and David Morgan, eds, *Body Matters: Essays on the Sociology of the Body*, Falmer.

Schulze, Laurie (1990) 'On The Muscle', in Jane Gaines and Charlotte Herzog, eds, *Fabrication*, Routledge.

Scott, Sue and Morgan, David (eds) (1993) *Body Matters: Essays on the Sociology of the Body*, Falmer.

Solomon, Alisa (1993) 'It's Never Too Late To Switch: Crossing Towards Power' in Lesley Ferris, ed., *Crossing The Stage: Controversies on Cross-Dressing*, Routledge.
Tasker, Yvonne (1993) *Spectacular Bodies: Gender, Genre and the Action Cinema*, Routledge.

7.5

'MY BODY IS MY ART': COSMETIC SURGERY AS FEMINIST UTOPIA?

Kathy Davis

In August 1995, the French performance artist Orlan was invited to give a lecture at a multimedia festival in Amsterdam.[1] Orlan has caused considerable furore in the international art world in recent years for her radical body art in which she has her face surgically refashioned before the camera. On this particular occasion, the artist read a statement about her art while images of one of her operations flashed on the screen behind her. The audience watched as the surgeon inserted needles into her face, sliced open her lips, and, most gruesomely of all, severed her ear from the rest of her face with his scalpel. While Orlan appeared to be unmoved by these images, the audience was clearly shocked. Agitated whispers could be heard and several people left the room. Obviously irritated, Orlan interrupted her lecture and asked whether it was 'absolutely necessary to talk about the pictures *now*' or whether she could proceed with her talk. Finally one young woman stood up and exclaimed: 'You act as though it were not *you*, up there on the screen'.[2]

This may seem like a somewhat naive reaction. Good art is, after all, about shifting our perceptions and opening up new vistas. That this causes the audience some unease goes without saying. Moreover, the young woman's reaction is not directed at Orlan the artist who is explaining her art, but rather at Orlan the woman who has had painful surgery. Here is a woman whose face has been mutilated and yet discusses it intellectually and dispassionately. The audience is squirming and Orlan is acting as though she were not directly involved.

From: K. Davis (ed.), *Embodied Practices: Feminist Perspectives on the Body*, London: Sage, 1997.

I became interested in Orlan (and the reactions she evokes) as a result of my own research on women's involvement in cosmetic surgery (Davis 1995). Like many feminists, I was deeply troubled by the fact that so many women willingly and enthusiastically have their bodies altered surgically despite considerable hardship and risk to themselves. While I shared the commonly held feminist view that cosmetic surgery represented one of the more pernicious horrors inflicted by the medical system upon women's bodies, I disliked the concomitant tendency among feminists to treat the recipients as nothing more than misguided or deluded victims. In an attempt to provide a critical analysis of cosmetic surgery which did not undermine the women who saw it as their best option under the circumstances, I conducted in-depth interviews with women who had had or were planning to have some form of cosmetic surgery. They had undergone everything from a relatively simple ear correction or a breast augmentation to – in the most extreme case – having the entire face reconstructed. Since the research was conducted in the Netherlands where cosmetic surgery was included in the national health care package, my informants came from diverse socioeconomic backgrounds. Some were professional women or academics, others were cashiers or home-helps and some were full-time housewives and mothers. Some were married, some single, some heterosexual, some lesbian. They ranged in age from a 17-year-old school girl whose mother took her in for a breast augmentation, to a successful, middle-aged business woman seeking a face lift in order to 'fit into the corporate culture'.

These women told me about their history of suffering because of their appearance, how they decided to have their bodies altered surgically, their experiences with the operation itself and their assessments of the outcome of the surgery. While their stories involved highly varied experiences of embodiment as well as different routes towards deciding to have their bodies altered surgically, they invariably made cosmetic surgery viewable as an understandable and even unavoidable course of action in light of their particular biographical circumstances. I learned of their despair, not because their bodies were not beautiful, but because they were not ordinary – 'just like everyone else'. I listened to their accounts of how they struggled with the decision to have cosmetic surgery, weighing their anxieties about risks against the anticipated benefits of the surgery. I discovered that they were often highly ambivalent about cosmetic surgery and wrestled with the same dilemmas which have made cosmetic surgery problematic for many feminists. My research gave a central role to women's agency, underlining their active and lived relationship with their bodies and showing how they could knowledgeably choose to have cosmetic surgery. While I remained critical of the practice of cosmetic surgery and the discourse of feminine inferiority which it sustains, I did not reject it as an absolute evil, to be avoided at any cost. Instead I argued for viewing cosmetic surgery as a complex dilemma: problem and solution, symptom of oppression and act of empowerment, all in one.

Given my research on cosmetic surgery, I was obviously intrigued by Orlan's surgical experiments. While I was fascinated by her willingness to put her body under the knife, however, I did not immediately see what her project had to offer for understanding why 'ordinary' women have cosmetic surgery. On the contrary, I placed Orlan alongside other contemporary women artists who use their bodies to make radical statements about a male-dominated social world: Cindy Sherman's inflatable porno dolls with their gaping orifices, Bettina Rheim's naked women in their exaggerated sexual posings, or Matuschka's self-portraits of her body after her breast has been amputated. It came as a surprise, therefore, when my research was continually being linked to Orlan's project. Friends and colleagues sent me clippings about Orlan. At lectures about my work, I was invariably asked what I thought about Orlan. Journalists juxtaposed interviews with me and Orlan for their radio programmes or discussed us in the same breath in their newspaper pieces. Our projects were cited as similar in their celebration of women's agency and our insistence that cosmetic surgery was about more than beauty.[3] We were both described as feminists who had gone against the feminist mainstream and dared to be politically incorrect. By exploring the empowering possibilities of cosmetic surgery, we were viewed as representatives of a more nuanced and – some would say – refreshing perspective on cosmetic surgery.

These reactions have increasingly led me to reconsider my initial belief that Orlan's surgical experiments have nothing to do with the experiences of women who have cosmetic surgery. In particular, two questions have begun to occupy my attention. The first is to what extent Orlan's aims coincide with my own; that is, to provide a feminist critique of the technologies and practices of the feminine beauty system while taking women who have cosmetic surgery seriously. The second is whether Orlan's project can provide insight into the motives of the run-of-the-mill cosmetic surgery recipient.

In this article, I am going to begin with this second question. After looking at Orlan's performances as well as how she justifies them, I consider the possible similarities between her surgical experiences and the surgical experiences of the women I spoke with. I then return to the first question and consider the status of Orlan's art as feminist critique of cosmetic surgery – that is, as a utopian revisioning of a future where women reappropriate cosmetic surgery for their own ends. In conclusion, I argue that – when all is said and done – surgical utopias may be better left to art than to feminist critique.

ORLAN'S BODY ART

Orlan came of age in the 1960s – the era of the student uprisings in Paris, the 'sexual revolution' and the emergence of populist street theatre. As visual artist, she has always used her own body in unconventional ways to challenge gender stereotypes, defy religion and, more generally, to shock her audience (Lovelace 1995). For example, in the 1960s, she displayed the sheets of her bridal trousseau stained with semen to document her various sexual encounters, thereby poking fun at the demands for virgin brides in France. In the

1970s, she went to the Louvre with a small audience and pasted a triangle of her own pubic hair to the voluptuously reclining nude depicted in the *Rape of Antiope* – a hairless body devoid of subjecthood, a mere object for consumption. In the 1980s, Orlan shocked Parisian audiences by displaying her magnified genitals, held open by means of pincers, with the pubic hair painted yellow, blue and red (the red was menstrual blood). A video camera was installed to record the faces of her viewers who were then given a text by Freud on castration anxiety.

Her present project in which she uses surgery as a performance is, by far, her most radical and outrageous. She devised a computer-synthesized ideal self-portrait based on features taken from women in famous works of art: the forehead of Da Vinci's *Mona Lisa*, the chin of Botticelli's *Venus*, the nose of Fountainebleau's *Diana*, the eyes of Gérard's *Psyche* and the mouth of Boucher's *Europa*. She did not choose her models for their beauty, but rather for the stories which are associated with them. Mona Lisa represents transsexuality for beneath the woman is – as we now know – the hidden self-portrait of the artist Leonardo Da Vinci; Diana is the aggressive adventuress; Europa gazes with anticipation at an uncertain future on another continent; Psyche incorporates love and spiritual hunger; and Venus represents fertility and creativity.

Orlan's 'self-portraits' are not created at the easel, but on the operating table. The first took place on 30 May 1987, the artist's 40th birthday and eight more have taken place since then. Each operation is a 'happening'. The operating theatre is decorated with colourful props and larger-than-life representations of the artist and her muses. Male striptease dancers perform to music. The surgeons and nurses wear costumes by top designers and Orlan herself appears in net stockings and party hat with one breast exposed. She kisses the surgeon ostentatiously on the mouth before lying down on the operating table. Each performance has a theme (like 'Carnal Art', 'This is My Body, This is My Software', 'I Have Given My Body to Art', 'Identity Alterity'). Orlan reads philosophical, literary or psychoanalytic texts while being operated on under local anaesthesia. Her mood is playful and she talks animatedly even while her face is being jabbed with needles or cut ('producing', as she puts it, 'the image of a cadaver under autopsy which just keeps speaking').[4]

All of the operations have been filmed. The seventh operation-performance in 1993 was transmitted live by satellite to galleries around the world (the theme was omnipresence) where specialists were able to watch the operation and ask questions which Orlan then answered 'live' during the performance. In between operations, Orlan speaks about her work at conferences and festivals throughout the world where she also shows photographs and video clips of her operations. Under the motto 'my body is my art', she has collected souvenirs from her operations and stored them in circular, plexi-glass receptacles which are on display in her studio in Ivry, France. These 'reliquaries' include pieces of her flesh preserved in liquid, sections of her scalp with hair still attached, fat cells which have been suctioned out of her face, or crumpled bits of surgical gauze drenched in her blood. She sells them for as much as 10,000 francs, intending to continue until she has 'no more flesh to sell'.

Orlan's performances require a strong stomach and her audiences have been known to walk out midway through the video. The confrontation of watching the artist direct the cutting up of her own body is just too much for many people to bear. Reactions range from irritation to – in Vienna – a viewer fainting.[5] While Orlan begins her performances by apologizing to her audience for causing them pain, this is precisely her intention. As she puts it, art has to be transgressive, disruptive and unpleasant in order to have a social function. ('Art is not for decorating apartments, for we already have plenty of that with aquariums, plants, carpets, curtains, furniture ...').[6] Both artist and audience need to feel uncomfortable so that 'we will be forced to ask questions'.

For Orlan, the most important question concerns 'the status of the body in our society and its future ... in terms of the new technologies'.[7] The body has traditionally been associated with the innate, the immutable, the god given or the fated-ness of human life. Within modernist science, the body has been treated as the biological bedrock of theories on self and society – the 'only constant in a rapidly changing world' (Frank 1990: 133). In recent years, this view has become increasingly untenable. The body – as well as our beliefs about it – is subject to enormous variation, both within and between cultures. Postmodern thinkers have rejected the notion of a biological body in favour of viewing bodies as social constructions. Orlan's project takes the postmodern deconstruction of the material body a step further. In her view, modern technologies have made any notion of a natural body obsolete. Test-tube babies, genetic manipulation and cosmetic surgery enable us to intervene in nature and develop our capacities in accordance with our needs and desires. In the future, bodies will become increasingly insignificant – nothing more than a 'costume', a 'vehicle', something to be changed in our search 'to become who we are'.[8]

The body of which Orlan speaks is a female body. Whereas her earlier work explored gender stereotypes in historical representations of the female body, her present project examines the social pressures which are exercised upon women through their bodies – in particular, the cultural beauty norms. At first glance, this may seem contradictory, since the goal of her art is to achieve an 'ideal' face. Although she draws upon mythical beauties for inspiration, she does not want to resemble them. Nor is she particularly concerned with being beautiful. Her operations have left her considerably less beautiful than she was before. For example, in operation seven she had silicone implants inserted in her temples (the forehead of Mona Lisa), giving her a slightly extraterrestrial appearance. For her next and last operation, she has planned 'the biggest nose physically possible' – a nose which will begin midway up her forehead. Thus, while Orlan's face is an ideal one, it deviates radically from the masculinist ideal of feminine perfection. Her ideal is radically non-conformist. It does not make us aware of what we lack. When we look at Orlan, we are reminded that we can use our imagination to become the persons we want to be.

Good

Orlan's project explores the problem of identity. Who she is, is in constant flux or, as she puts it, 'by wanting to become another, I become myself'. 'I am a bulldozer: dominant and aggressive ... but if that becomes fixed it is a handicap ... I, therefore, renew myself by becoming timid and tender ...'.[9] Her identity project is radical precisely because she is willing to alter her body surgically in order to experiment with different identities. What happens to the notion of 'race', she wonders, if I shed my white skin for a black one?[10] Similarly, she rejects gender as a fixed category when she claims: 'I am a woman-to-woman transsexual act'. However, Orlan's surgical transformations – unlike a sex-change operation – are far from permanent. In this sense, Orlan's art can be viewed as a contribution to postmodern feminist theory on identity.[11] Her face resembles Haraway's (1991) cyborg – half-human, half-machine – which implodes the notion of the natural body. Her project represents the postmodern celebration of identity as fragmented, multiple and – above all – fluctuating and her performances resonate with the radical social constructionism of Butler (1990, 1993) and her celebration of the transgressive potential of such performativity.

For Orlan, plastic surgery is a path towards self-determination – a way for women to regain control over their bodies. Plastic surgery is one of the primary arenas where 'man's power can be most powerfully asserted on women's bodies', 'where the dictates of the dominant ideology ... become ... more deeply embedded in female ... flesh'.[12] Instead of having her body rejuvenated or beautified, she turns the tables and uses surgery as a medium for a different project. For example, when Orlan's male plastic surgeons balked at having to make her too ugly ('they wanted to keep me cute'), she turned to a female feminist plastic surgeon who was prepared to carry out her wishes. The surgical performances themselves are set up to dispel the notion of a sick body, 'just an inert piece of meat, lying on the table'.[13] Orlan designs her body, orchestrates the operations and makes the final decision about when to stop and when to go on. Throughout the surgery, she talks, gesticulates and laughs. This is her party and the only constraint is that she remain in charge. Thus, while bone breaking might be desirable (she originally wanted to have longer legs), it had to be rejected because it would have required full anaesthesia and, therefore, have defeated the whole purpose of the project. Orlan has to be the creator, not just the creation; the one who decides and not the passive object of another's decisions.

ART AND LIFE

I now want to return to the issue which I raised at the outset of this article: namely, the puzzling fact that my research is continually being associated with Orlan's art. As one journalist noted after reading my book: the only difference between Orlan and the majority of women who have cosmetic surgery is one of degree. Orlan is just an extreme example of what is basically the same phenomenon: women who have cosmetic surgery want to be 'their own Pygmalions'.[14]

At first glance, there are, indeed, similarities between Orlan's statements about her art and how the women I interviewed described their reasons for having cosmetic surgery. For example, both Orlan and these women insisted that they did not have cosmetic surgery to become more beautiful. They had cosmetic surgery because they did not feel at home in their bodies; their bodies did not fit their sense of who they were. Cosmetic surgery was an intervention in identity. It enabled them to reduce the distance between the internal and external so that others could see them as they saw themselves.[15] Another similarity is that both Orlan and the women I spoke with viewed themselves as agents who, by remaking their bodies, remade their lives as well. They all rejected the notion that by having cosmetic surgery, they had allowed themselves to be coerced, normalized or ideologically manipulated. On the contrary, cosmetic surgery was a way for them to take control over circumstances over which they previously had had no control. Like Orlan, these women even regarded their decision to have cosmetic surgery as an oppositional act: something they did for themselves, often at great risk and in the face of considerable resistance from others.

However, this is where the similarities end. Orlan's project is not about a real-life problem; it is about art. She does not use cosmetic surgery to alleviate suffering with her body, but rather to make a public and highly abstract statement about beauty, identity and agency. Her body is little more than a vehicle for her art and her personal feelings are entirely irrelevant. When asked about the pain she must be experiencing, she merely shrugs and says: 'Art is a dirty job, but someone has to do it.'[16] Orlan is a woman with a mission: she wants to shock, disrupt convention and provoke people into discussing taboo issues. 'Art can and must change the world, for that is its only justification.'[17]

This is very different from the reasons the women I spoke with gave for having cosmetic surgery. Their project is a very private and personal one. They want to eliminate suffering which has gone beyond what they feel they should have to endure. They are anxious about the pain of surgery and worried about the outcome. They prefer secrecy to publicity and have no desire to confront others with their decisions. While their explanations touch on issues like beauty, identity and agency (although not necessarily using those words), they are always linked to their experiences and their particular life histories. Their justification for having cosmetic surgery is necessity. It is the lesser of two evils, their only option under the circumstances. They do not care at all about changing the world; they simply want to change themselves.

Thus, cosmetic surgery as art and cosmetic surgery in life appear to be very different phenomena. I, therefore, might conclude that there is little resemblance between Orlan's surgical experiences and those of most women who have cosmetic surgery, after all. Orlan's celebration of surgical technologies seems to have little in common with a project like my own, which aims to provide a feminist critique of cosmetic surgery. Consequently, comparisons between my research and Orlan's project can only be regarded as superficial or premature.

But perhaps this conclusion is overhasty. After all, it was never Orlan's intention to understand the surgical experiences of 'ordinary' women. Nor is it her intention to provide a feminist polemic against the unimaginable lengths to which women will go to achieve an ideal of beauty as defined by men. Hers is not a sociological analysis which explicitly attacks the evils of cosmetic surgery and its pernicious effects on women (Lovelace 1995). Nevertheless, her project is an implicit critique of the dominant norms of beauty and the way cosmetic surgery is practised today. It belongs to the tradition of feminist critique which imaginatively explores the possibilities of modern technology for the empowerment of women. As such, Orlan's project might be viewed as an example of a feminist utopia.

COSMETIC SURGERY AS FEMINIST UTOPIA

Feminists have often envisioned a future where technology has been seized by women for their own ends. Take, for example, Shulamith Firestone's *Dialectic of Sex* (1970) in which she fantasizes a world in which reproductive technology frees women from the chores and constraints of biological motherhood. In a similar vein, the novelist Marge Piercy depicts a feminist utopia in *Woman on the Edge of Time* (1976) where genetic engineering has erased sexual and 'racial' differences, thereby abolishing sexism and racism.[18]

More recently, the feminist philosopher Kathryn Morgan (1991) applies the notion of utopia to cosmetic surgery. She claims that refusal may not be the only feminist response to the troubling problem of women's determination to put themselves under the knife for the sake of beauty. There may, in fact, be a more radical way for feminists to tackle the 'technological beauty imperative'.

She puts forth what she calls 'a utopian response to cosmetic surgery': that is, an imaginary model which represents a desirable ideal that because of its radicality is unlikely to occur on a wide scale (Morgan 1991: 47). Drawing upon feminist street theatre, on the one hand, and postmodern feminist theory – most notably Judith Butler's (1990) notion of gender as performance – on the other, Morgan provides some imaginative, if somewhat ghoulish, examples of cosmetic surgery as feminist utopia.

For example, she envisions alternative 'Miss . . .' pageants where the contestants compete for the title 'Ms Ugly'. They bleach their hair white, apply wrinkle-*inducing* creams or have wrinkles *carved into* their faces, have their breasts pulled *down* and *darken* their skin. (Morgan 1991: 46). Or, she imagines 'beautiful body boutiques' where 'freeze-dried fat cells', 'skin velcro', magnetically attachable breasts complete with nipple pumps, and do-it-yourself sewing kits with pain-killers and needles are sold to interested customers.

These 'performances' can be characterized as a feminist critique of cosmetic surgery for several reasons.

First, they unmask both 'beauty' and 'ugliness' as cultural artefacts rather than natural properties of the female body. They valorize what is normally perceived as ugly, thereby upsetting the cultural constraints upon women to comply with the norms of beauty. By actually undergoing mutations of the flesh, the entire notion of a natural body – that linchpin of gender ideology – is destabilized.

Second, these surgical performances constitute women as subjects who use their feminine body as a site for action and protest rather than as an object of discipline and normalization. These parodies mock or mimic what is ordinarily a source of shame, guilt or alienation for women. Unlike the 'typical' feminine disorders (anorexia, agoraphobia or hysteria) which are forms of protest where women are victims, Morgan's actions require '*healthy*' (*sic*) women who already 'have a feminist understanding of cosmetic surgery' (Morgan 1991: 45).

Third, by providing a travesty of surgical technologies and procedures, these performances magnify the role that technology plays in constructing femininity through women's bodies. At the same time, they usurp men's control over these technologies and undermine the power dynamic which makes women dependent on male expertise (Morgan 1991: 47). Performances show how technology might be reappropriated for feminist ends.

Morgan acknowledges that her surgical utopias may make her readers a bit queasy or even cause offence. However, this is as it should be. It only shows that we are still in the thrall of the cultural dictates of beauty and cannot bear to imagine women's bodies as ugly. Anyone who feels that such visions go 'too far' must remind herself that she has merely become anaesthetized to the mutilations which are routinely performed on women by surgeons every day (Morgan 1991: 46–7). Where the 'surgical fix' is concerned, 'shock therapy' is the only solution.

DOES COSMETIC SURGERY CALL FOR A UTOPIAN RESPONSE

The attractions of a utopian approach to cosmetic surgery are considerable. It enables feminists to take a stand against the cultural constraints upon women to be beautiful and dramatically exposes the excesses of the technological fix. It destabilizes many of our preconceived notions about beauty, identity and the female body and it provides a glimpse of how women might engage with their bodies in empowering ways. However, most important of all – and I believe this is why such approaches appeal to the feminist imagination – it promises the best of both worlds: a chance to be critical of the victimization of women without having to be victims ourselves.

While I am entertained and intrigued by the visions put forth by Morgan and enacted by Orlan, I must admit that they also make me feel profoundly uneasy. This unease has everything to do with my own research on cosmetic surgery. On the basis of what women have told me, I would argue that a utopian response to cosmetic surgery does not just open up radical avenues for feminist critique; it also limits and may even prevent this same critique. It is my contention that there are, at least, four drawbacks.

First, a utopian response discounts the suffering which accompanies any cosmetic surgery operation. One of the most shocking aspects of Orlan's performances is that she undergoes surgery which is clearly painful and yet shrugs off the pain ('Of course, there are several injections and several grimaces ... but I just take painkillers like everyone else')[19] or explains that the audience feels more pain looking at the surgery than she does in undergoing it. ('Sorry to have made you suffer, but know that I do not suffer, unlike you ...'.)[20] This nonchalance is belied by the postoperative faces of the artist – proceeding from swollen and discoloured to, several months later, pale and scarred. Whether a woman has her wrinkles smoothed out surgically or carved in has little effect on the pain she feels during the surgery. Such models, therefore, presuppose a non-sentient female body – a body which feels no pain.[21]

Second, a utopian response discounts the risks of cosmetic surgery. Technologies are presented as neutral instruments which can be deployed to feminist ends. Both Orlan and Morgan describe surgery as conceived, controlled and orchestrated by the autonomous feminine subject. She has the reins in her hand. However, even Orlan has had a 'failed' operation: one of her silicone implants wandered and had to be reinserted – this time not in front of the video camera. Such models overstate the possibilities of modern technology and diminish its limitations.

Third, a utopian response ignores women's suffering with their appearance. The visions presented by both Orlan and Morgan involve women who are clearly unaffected by the crippling constraints of femininity. They are not dissatisfied with their appearance as most women are; nor, indeed, do they seem to care what happens to their bodies at all. For women who have spent years hating their excess flesh or disciplining their bodies with drastic diets, killing fitness programmes or cosmetic surgery, the image of 'injecting fat cells' or having the breasts 'pulled down' is insulting. The choice of 'darkened skin' for a feminist spectacle which aims to 'valorize the ugly' is unlikely to go down well with women of colour. At best, such models negate their pain. At worst, they treat women who care about their appearance as the unenlightened prisoners of the beauty system who are more 'culturally scripted' than their artistic sisters.

Fourth, a utopian response discounts the everyday acts of compliance and resistance which are part of ordinary women's involvement in cosmetic surgery. The surgical experiments put forth by Orlan and Morgan have the pretension of being revolutionary. In engaging in acts which are extraordinary and shocking, they not only entertain and disturb, but also distance us from the more mundane forms of protest.[22] It is difficult to imagine that cosmetic surgery might entail *both* compliance *and* resistance. The act of having cosmetic surgery involves going along with the dictates of the beauty system, but also refusal – refusal to suffer beyond a certain point. Utopian models privilege the flamboyant, public spectacle as feminist intervention and deprivilege the interventions which are part of living in a gendered social order.

In conclusion, I would like to return to the young woman I mentioned at the beginning of this chapter. At first glance, her reaction might be attributed to her failure to appreciate the radicality of Orlan's project. She is apparently unable to go beyond her initial, 'gut level' response of horror at the pictures and consider what Orlan's performances have to say in general about the status of the female body in a technological age. She is just not sophisticated enough to benefit from this particular form of feminist 'shock therapy'.

However, having explored the 'ins' and 'outs' of surgical utopias, I am not convinced that this is how we should interpret her reaction. Her refusal to take up Orlan's invitation may also be attributed to concern. She may feel concern for the pale woman before her whose face still bears the painful marks of her previous operations. Or she may be concerned that anyone can talk so abstractly and without emotion about something which is so visibly personal and painful. Or she may simply be concerned that in order to appreciate art, she is being required to dismiss her own feelings.

Her concern reminds us of what Orlan and, indeed, any utopian approach to cosmetic surgery leaves out: the sentient and embodied female subject, the one who feels concern about herself and about others. As feminists in search of a radical response to women's involvement in cosmetic surgery, we would do well to be concerned about this omission as well.

NOTES

1. This festival was organized by Triple X which puts on an annual exhibition including theatre, performance, music, dance and visual art. I would like to thank Peter van der Hoop for supplying me with information about Orlan. I am indebted to Willem de Haan, Suzanne Phibbs and the participants of the postgraduate seminar 'Gender, Body, Love', held at the Centre for Women's Research in Oslo, Norway in May 1996 for their constructive and insightful comments.
2. *De Groene Amsterdammer* (23 August 1995).
3. See, for example, a recent article by Xandra Schutte in *De Groene Amsterdammer* (13 December 1995) or 'Passages and Passanten' (VPRO Radio 5, 17 November 1995).
4. Quoted in Reitmaier (1995: 8).
5. *Falter* (1995, No. 49: 28).
6. Quoted in Reitmaier (1995: 7).
7. See Reitmaier (1995: 8).
8. Quoted in Tilroe (1996: 17).
9. *Actuel* (January 1991: 78).
10. Obviously, Orlan has not read John Howard Griffin's (1961) *Black Like Me* in which a white man chronicles his experiences of darkening his skin in order to gain access to African-American life in the mid-1950s. For him, becoming the racial Other was a way to understand the material and bodily effects of racism – an experiment which was anything but playful and ultimately resulted in the author's untimely death from skin cancer. See Awkward (1995) for an excellent discussion of such experiments from a postmodern ethnographic perspective.
11. While Orlan has been cited as a model for postmodern feminist critiques of identity, her project is, in some ways, antithetical to this critique. She celebrates a notion of the sovereign, autonomous subject in search of self which is much more in line with Sartre's existentialism than poststructuralist theory *à la* Butler. See, for example, the debate between Butler and others in Benhabib et al. (1995).
12. Quoted in Reitmaier (1995: 9).
13. *De Volkskrant* (5 June 1993).
14. *De Groene Amsterdammer* (13 December 1995: 29).

15. Quoted in Reitmaier (1995: 8).
16. Quoted in Reitmaier (1995: 10).
17. Quoted in Reitmaier (1995: 7).
18. See José van Dyck (1995) for an excellent analysis of feminist utopias (and dystopias) in debates on the new reproductive technologies.
19. Quoted in Reitmaier (1995: 10).
20. Statement given at performance in Amsterdam.
21. This harks back to the notion that women – particularly working-class women and women of colour – do not experience pain to the same degree that affluent, white women and men do. This notion justified considerable surgical experimentation on women in the last century. See, for example, Dally (1991).
22. It could be argued that in the context of the art business where success depends upon being extraordinary, Orlan is simply complying with convention. This would make her no more, but also no less, revolutionary than any other woman who embarks upon cosmetic surgery.

REFERENCES

Awkward, Michael (1995) *Negotiating Difference. Race, Gender, and the Politics of Positionality*, Chicago and London: The University of Chicago Press.

Benhabib, Seyla, Judith Butler, Drucilla Cornell and Nancy Fraser (1995) *Feminist Contentions. A Philosophical Exchange*, New York and London: Routledge.

Butler, Judith (1990) *Gender Trouble: Feminism and the Subversion of Identity*, New York: Routledge.

Butler, Judith (1993) *Bodies that Matter. On the Discursive Limits of 'Sex'*, New York: Routledge.

Dally, Ann (1991) *Women Under the Knife*, London: Hutchinson Radius.

Davis, Kathy (1995) *Reshaping the Female Body. The Dilemma of Cosmetic Surgery*, New York: Routledge.

Firestone, Shulamith (1970) *The Dialectic of Sex. The Case for Feminist Revolution*, New York: Bantam.

Frank, Arthur (1990) 'Bringing Bodies Back In: A Decade Review'. *Theory, Culture and Society* 7: 131–62.

Griffin, John Howard (1961) *Black Like Me*, New York: Signet.

Haraway, Donna J. (1991) *Simians, Cyborgs, and Women. The Reinvention of Nature*, London: Free Association Books.

Lovelace, Cary (1995) 'Orlan: Offensive Acts', *Performing Arts Journal* 49: 13–25.

Morgan, Kathryn Pauly (1991) 'Women and the Knife: Cosmetic Surgery and the Colonization of Women's Bodies' *Hypatia* 6(3): 25–53.

Piercy, Marge (1976) *Woman on the Edge of Time*, New York: Fawcett Crest.

Reitmaier, Heidi (1995) '"I Do Not Want to Look Like … " Orlan on becoming-Orlan', *Women's Art* 5 June: 5–10.

Tilroe, Anna (1996) *De huid van de kameleon. Over hedendaagse beeldende kunst*, Amsterdam: Querido.

Van Dyck, José (1995) *Manufacturing Babies and Public Consent*, London: Macmillan.

7.6

'FREUD'S FETISHISM' AND THE LESBIAN DILDO DEBATES

Heather Findlay

From the pages of lesbian porn magazines to the meetings of the Modern Language Association, a highly organized discourse has developed around a rather unlikely object: the dildo. No other sex toy has generated the quantity or quality of discussion among mostly urban, middle-class, white lesbians than the dildo.[1] What interests me about this discourse is that a number of subcultural products (advertisements, erotic fiction, the sex toys themselves) have consistently drawn from a set of familiar conventions, thus constituting a kind of shared fantasy about lesbian dildo use. Like all fantasy, this one no doubt occludes more than it reveals about the reality of lesbian desire, whatever that may be. My focus, however, is on what the French might call the *mise-en-scène* of a particular dildo fantasy and its relation to the issues raised in Freud's 1927 essay on fetishism.[2] By analysing the dildo in conjunction with Freud's text, I hope to shed light not only on the dildo debates and the feminist 'sex wars' of which they are a part but also on the (perhaps paradoxical) relevance to lesbians of the psychoanalytic theory of fetishism, a 'perversion' which Freud – as a number of his feminist readers have discussed in some detail[3] – claims to be exclusively male. My aim is not simply to apply psychoanalysis to lesbian sexuality but also to do what is in a sense more difficult: to reread Freud from the perspective of lesbian theory and practice and to unravel those points at which his text may be as 'symptomatic' as the behavior it attempts to describe.

From: *Feminist Studies* (1992) Fall, 18 (3).

THE DILDO WARS

To date, discourse among lesbians over the dildo has been marked by a debate divided roughly into two camps. On the one hand, some lesbians have debunked the dildo and its notorious cousin the strap-on, calling them 'male-identified'. The most colorful, if noncanonical, spokeswomen for this position have been published in the 'Letters' column of the lesbian pornographic magazine *On Our Backs*. These letters are written in protest of the fact that, as Colleen Lamos puts it, 'The dildo is clearly a matter of intense interest and a focus of fetishistic desire for the readers' and publishers of the magazine.[4] For example, Daralee and Nancy, two self-described 'outrageous and oversexed S/M dykes' from Birmingham, Alabama, confess that they are nonetheless puzzled about the current lesbian romance with the dildo.

> What's the deal with women 'portraying' themselves as equipped with penises? We can't figure out how this could be erotic to a woman-identified-woman. If they want a dick, why are they with a woman wearing a dildo and not a man? Don't misunderstand us; we're heavily into penetration ... but this whole life-like dildo market is baffling to us.[5]

The authors' distaste for dildos, especially 'lifelike' ones, is based on the conviction that a dildo represents a penis and is therefore incompatible with 'woman-identified' sexuality. In fact, the letter's reference to women-identified-women situates it within a larger, radical feminist critique of sex,[6] including (as an editor at *off our backs* put it) 'games that rely on paraphernalia [and] roles'.[7] Ironically enough for our kinky Birmingham correspondents, the critique of the dildo has developed in tandem with radical feminist attacks on butch-femme and sadomasochism such as Sheila Jeffreys's, which hold that both practices reproduce a 'heteropatriarchy' based on masculine and feminine sex roles.[8]

On the other hand, some lesbians have argued that dildos do not represent penises; rather, they are sex toys that have an authentic place in the history of lesbian subculture. This argument can be detected in Joan Nestle's writings on butch-femme sexuality[9] or, more explicitly, in the columns of self-made lesbian sexologist Susie Bright, a.k.a. Susie Sexpert. 'The facts about dildos', she assures us, 'aren't as controversial as their famous resemblance to the infamous "penis" and all that *it* represents.' In an attempt to downplay the 'political, social, and emotional connotations of dildos', Bright writes that 'a dildo can be a succulent squash, or a tender mold of silicon. Technically, it is any device you use for the pleasure of vaginal or anal penetration. ... Penises can only be compared to dildos in the sense that they take up space'.[10] As a mere 'device', a sex toy that can be vegetable or mineral, a dildo has only a remote, sheerly ontological relationship to the penis: like it, the dildo exists, it 'takes up space' in the vagina or anus. In sum, as part of their challenge to what they see as the repressive sexual politics of radical feminism, Bright and other 'prosex' feminists have defended the dildo by downplaying its referentiality, by denying that the dildo represents a penis.

Among other things, we might accuse the antidildo camp of having an unsophisticated understanding of representation. Daralee and Nancy's letter, for example, assumes that a dildo not only represents but is also the same thing as 'a dick'. This assumption announces itself in the quotation marks Daralee and Nancy put around 'portraying'; by doubting the difference a representation can make, the authors affirm that wearing a dildo is, quite literally, equipping oneself with a penis.[11] To Bright's credit, her comments shift the burden of suspicion from the representation to the thing represented: if the lesbians from Birmingham put 'portraying' in quotation marks, Bright quotes and italicizes 'the infamous "penis" and all *it* represents', thereby referring to the organ itself as if it were somebody else's peculiar idea – as if, in other words, it were already a signifier. Bright may, however, fall prey to her own brand of literalism: is it possible for a dildo to stand, as it were, only for itself? In the face of 'this whole life-like dildo market' so baffling to the Birmingham lesbians – or, for that matter, the popularity of huge Black dildos so troubling to *Black Lace* editor Alycee J. Lane[12] – is it possible to insist that dildos are strictly nonrepresentational, that they are not (to quote Lane) the 'location' of both sexual and racial 'terror and desire'?

ENTER FREUD WITH FETISH

Putting these questions aside for the moment, I would like to point out that, regardless of the speakers' feelings about psychoanalysis, the dildo debate may be said to revolve around the question of whether the dildo is a fetish in Freud's definition of the term. At the beginning of 'Fetishism', Freud tells a parable about how the fetish reveals itself to be a penis replacement. According to the story, fetishism is the little boy's response to the castration anxiety he experiences upon first seeing his mother's genitals.

> What happened, therefore, was that the boy refused to take cognizance of the fact of his having perceived that a woman does not possess a penis. No, that could not be true: for if a woman had been castrated, then his own possession of a penis was in danger; and against that there rose in rebellion the portion of his narcissism which Nature has, as a precaution, attached to that particular organ.

Freud gives his story a happy ending, at least for the boy: 'To put it more plainly: the fetish is a substitute for the woman's [the mother's] penis that the little boy once believed in and – for reasons familiar to us – does not want to give up.'[13] Because the fetish represents a penis, Freud argues, it allows the subject to maintain, despite evidence to the contrary, that castration is not a danger. In fact, it allows him to maintain that castration has not happened at all. I will go on to say more about this passage, but for the moment we should note that the question of whether the dildo is a penis replacement lies at the very heart of the debate I summarized above. Critics of the dildo claim that it is a penis replacement, and its proponents claim that it is not. The debate over dildo use, in other words, is a debate about the politics of fetishism.

This may seem surprising, considering that Freud's theory of fetishism seems inhospitable to lesbian experience, so exclusive is its phallo- and hetero-centrism. The most glaring example of this is Freud's generic fetishist himself, who is a straight man.[14] Moreover, dildos may not be fetish objects, technically speaking. In the fetishism essay, Freud explains that fetishes, due to unconscious censorship, most often take their form from whatever the little boy sees *just before* he witnesses the spectacle of his mother's castration: fur and velvet symbolize the mother's pubic hair, undergarments 'crystallize the moment of undressing', and so on. Fetishistic desire exemplifies how (if I may refer to the oft-quoted formula) the unconscious is like a language: the fetishist finds himself representing his love object by means of contiguous associations, that is, by metonymy. Thus, to the extent that fetishes most often refer indirectly to the penis (that is, they tend not to be obvious phallic symbols) dildos may not be 'true' fetishes. Freud's essay, in fact, does not discuss them at all.

[...]

More often than not, dildo ads will deny that dildos are penis substitutes. One ad from Eve's Garden, a feminist sex boutique in New York, ends with this terse message to anyone who might be wondering: 'Finally, we should add that we don't think of dildos as imitation penises'; they are 'sexual accessories, not substitutes'. Upon reading this kind of disclaimer, we may wonder if perhaps the lady doth protest too much. In the end, lesbians with uncooperative sexual tastes still market and purchase politically incorrect dildos which are shaped like penises and named after mythological patriarchs, as in the 'Adam I' or the 'Jupiter II'. Even the conventions of English usage undermine the ad's nonsensical explanation, 'We've designed [our dildos] to be factually and aesthetically pleasing.' If dildos do not owe their allure to the fact that they represent penises, then how are they 'factually pleasing'? For that matter, who ever says that anything is 'factually pleasing'? In sum, dildo ads may cause us to conclude with Elizabeth A. Grosz that, in the case of lesbianism, 'it may be possible to suggest some connection' between women and fetishism after all[15] and that the maleness of fetishism is determined more by one's subject position than by biological gender.

Yet we might want to pay close attention to this recurrent anxiety – peculiar to lesbian dildo fetishism – over conflating a representation (a dildo) with reality (a penis). For one thing, this anxiety is totally absent in gay male fetishism, where dildos (and their purveyors) strive overtly to replicate penises, even particular penises, like porn star Jeff Stryker's. The makers of the Jeff Stryker model guarantee that their dildos are cast in a mould taken directly from Stryker's erect member. Exact replication, even a kind of indexical representation, is celebrated among gay men – perhaps all too uncritically.

Lesbians, on the other hand, have marketed a series of dildos which, in an obvious attempt to break the association between a piece of silicon and a penis, are shaped like dolphins, ears of corn, and even the Goddess. This urge to steer away from realism stems from the fact that these feminist dildo suppliers and their customers are suspicious of conflating a representation with reality, especially in the case of a phallus. One of the most important and repeated gestures of feminist critique has been to show how patriarchy asserts itself by making precisely this conflation, by collapsing the difference between a symbol and a real body organ.[16] In this light, being penetrated by the Goddess amounts to a defiant response to the patriarchal law that symbols of pleasure and potency always refer back to, or be somehow proper to, men's bodies. Like good Lacanians, these feminists are busily and happily disarticulating the phallus from the penis.

THE CUNT WE WISH DID NOT EXIST

But the problem of substitution is not the only issue in lesbian dildo use; nor is it the only function of the fetish for Freud. In his essay, Freud specifies further that the fetish is a representation not of just any 'chance penis' but of the *mother's* penis, the penis that the boy child once imagines his mother to have had but then discovers unhappily that she has lost. Faced with the 'fact' of his mother's castration, and thus the possibility of his own, the little boy replaces this image with a fetish which 'remains as a token of triumph over the threat of castration and a protection against it.'[17] At first glance, lesbian dildo fetishism seems to have nothing to do with mothers, castrated or not. But in Freud's account, the fetish does not refer to a real woman; it represents an imaginary, phallic one. In light of this, is not Joan Nestle's bedildoed lover in her short story, 'My Woman Poppa',[18] the fetishist's primal love object, the phallic woman the little boy imagines his mother to be? A more dramatic example may be a sequence from Fatale Video's pornographic collection *Clips*.[19] In this episode, set in a ranch house-style living room complete with a sonorous television, a recliner, and copies of the *Wall Street Journal*, a very femme Fanny kneels in front of her shirtless butch lover Kenny and watches as Kenny unzips her pants to reveal her pinkish-brown, 'realistic' dildo. Is Fanny not taking pleasure in the fetishistic fantasy of a woman with breasts *and* a penis? Because Fanny is in the position of the spectator, is our gaze on the scene fetishistic as well? Indeed, Fanny and the spectator rehearse Freud's formula for fetishism: despite evidence to the contrary, she insists (as we also do?) upon the fantasy that the penis is (still) there.

Fetishism occurs as well on the level of the video's narrative. After Kenny unzips her pants, Fanny takes her to the couch. Although the camera centers on Fanny as she sucks Kenny's dildo, we are also shown that Fanny is stimulating her lover's clitoris by rubbing her crotch with her shoulder. As she brings Kenny to orgasm, even though we know that the pleasure is clitoral, for the sake of the fantasy we believe it is phallic. In terms of recent reformulations of Freud's theory of fetishism, pleasure in this sequence conforms to Slavoj

Žižek's version of fetishistic logic: she knows very well what she is doing, but she is doing it anyway.[20]

As an aside, we might also raise the issue of the dildo harness, which is sometimes the object of as much fetishistic attention as the toy it is designed to hold. Indeed, the harness may approximate the classical Freudian fetish more closely than the dildo, precisely because, as I suggest above, the fetish usually refers metonymically, rather then metaphorically, to the mother's missing penis. For the sake of testing Freud's paradigm we might say that the harness, which few lesbian fetishists ever imagine being made out of anything but the blackest of black leather, gains its allure from the fact that it is – like a woman's crotch – fleshy and dark. Even more significant may be the gaping hole at the front. Inserting and removing a dildo from this hole rehearses, in a literal manner, the traumatic primal experience of Freud's little fetishist: now she has it, now she doesn't.

Yet this function of the fetish, its use as a protective device against the mother's castration, is its most troubling one. How much do lesbians have in common with Freud's little fetishist, who believes in the 'fact' that his mother – and thus all women – are lacking what he possesses, if only precariously? This question is a pressing one because, as a consequence of his belief in women's castration, the fetishist is deeply misogynist. He understands sexual difference simply in terms of women's deficiency. More concretely, Freud insists that after the fetishist represses the terrifying scenario of his mother's lack, 'an aversion, which is never lacking in any fetishist, to the real female genitals remains a *stigma indelebile* of the repression which has taken place.'[21] In the case of lesbian dildo fetishism, if lesbians take up the position of Freud's straight, male fetishist, how much is their pleasure accompanied by this aversion – which is supposedly never absent in any fetishist – to women and their bodies?

As light as the kind of butch-femme play in 'My Woman Poppa' or *Clips* may seem, more than one lesbian has suffered from the darker undercurrents of fetishistic desire. In an article entitled 'Sex, Lies, and Penetration: A Butch Finally 'Fesses Up', Jan Brown details what for her is the painful by-product of butch-femme roleplaying and of the dildo that constitutes its fetishistic center. 'We butches', she writes,

> have a horror of the pity fuck. We cannot face the charity of the mercy orgasm or the thought of contempt in our partner's eyes when we have allowed them to convince us that they really do want to touch us, take us, that they really do want to reach behind our dick and into the cunt we both wish did not exist.[22]

Brown's account reverses the usual subject/object relation and suggests that a subtle objectification is at work not of the feminine but of the masculine partner in butch-femme play. What Brown refers to sneeringly as 'the sacred myth of the stone butch', a myth held in place by her holy fetish object, turns out to be a defensive construct that both partners use to veil a lack, 'the cunt we both wish did not exist.'

If, for Brown, the dildo allows lesbians to circumvent the question of the cunt, some feminists of color are concerned that it serves a similar function *vis-à-vis* racial difference. Alycee J. Lane, for example, writes that a friend once accused her of having 'race issues' because she owns a dildo which, as Lane explains, is 'six inches, rubbery, cheap and *mauve*.' Lane defends her anatomically incorrect dildo in the tradition of 'prosex' feminism: 'a dildo is a dildo, not a dick ... The only thing that mattered, really, was the way my g-spot got worked.' Yet Lane remains unsatisfied with this explanation, especially after a trip to her local sex shop reveals that 'flesh-colored' dildos are actually cream-colored 'What does it *mean*', she ponders, 'when white hegemony extends to the production of dildos?' and that brown dildos are hard to get because, as the cashier explains, 'They sell rather quickly, you know.' She is directed to another bin. 'I turned and looked. They were not dildos; they were *monstrosities*. Twenty-four inches and thick as my arm. "Big Black Dick" said the wrapper. ... I looked around for some "Big White Dick" or even "Big Flesh Colored Dick." No luck. And I seriously doubt that *they* were in high demand.' Lane's experience changes her mind about the referentiality of the dildo. 'Race,' she concludes, 'permeates American culture. A sex toy easily becomes the location for racial terror and desire, because sex itself is that location.'[23]

In precisely what way, however, does a monstrous black dildo function as a signifier of racial terror and desire? When Freud claims that, in the heterosexual encounter, the fetishist's toy allows him to circumvent castration (the fetish 'endows women with the attribute which makes them acceptable as sexual objects'),[24] he suggests that the fetish alleviates the fact that (to cite Jacques Lacan's dictum) there is no relation between the sexes. Similarly, it seems that the black dildo fetish can 'make acceptable' a specifically racial lack – the lack, that is, under white hegemony of a relation between the races. As in the case of Freud's fetishist and his 'woman', the big black dildo allows whites to carry on a relation with Blacks which is, in reality, no relation at all. If sex *itself* is the location of racial terror and desire, we might say that the more general (and apparently lucrative) sexual fantasy of Black superpotency – a fantasy that Jackie Goldsby accuses lesbians of sharing[25] – is another, powerful cultural fetish that allows us to circumvent the Real of racial disintegration. Behind the big black dick, in other words, is the race we wish did not exist.

THE CHANCE PENIS

I am not suggesting that all lesbian fetishism ends in this kind of aversion. Even if we go back to Freud and reread, we see that the fetishistic perspective may not be monolithic. Freud's fetishist, in fact, is guilty of some rather reckless logic, and his triumph over women's castration is not as quick and easy as it may appear. Here, in full, is the passage I have already quoted:

> When now I announce that the fetish is a substitute for the penis, I shall certainly create disappointment; so I hasten to add that it is not a substi-

> tute for any chance penis, but for a particular and quite special penis that had been extremely important in early childhood but had later been lost. ... What happened, therefore, was that the boy refused to take cognizance of the fact of his having perceived that a woman does not possess a penis. No, that could not be true: for if a woman had been castrated, then his own possession of a penis was in danger. ...

First off, I am struck by the sense of urgency in Freud's rhetoric, produced perhaps by his identification with the frightened little boy. This identification is rendered most obvious when Freud speaks for the little boy and ventriloquizes his response to seeing his mother's genitals: 'No, that could not be true: for if a woman had been castrated ...' and so on. Indeed, Freud 'hastens' to erect a completed theoretical framework in the place of a lack, the 'disappointment' he fears he will 'certainly create' in his reader, in the same way that the little boy erects a fetish in the place of his mother's supposed castration. Emphasizing the uncanny way in which his theory doubles that of the little boy's, Freud hastens to complete it by adding, as we know, the notion of the mother's lack.

Because both Freud and his exemplary boy are in such a rush to posit the mother's castration, we might wonder if, contrary to Freud's insistence, the penis in question *may be* a 'chance penis'. In other words, perhaps it is not necessarily the mother's and perhaps somebody or something else might trigger the first blow to the child's narcissism. Actually, Freud describes rather circuitously the consequences of the little fetishist's sight of his mother's genitals: 'what happened, therefore, was that the boy refused to take cognizance of the fact of his having perceived that a woman does not possess a penis.' At this crucial moment, Freud does not write that the fetishist represses the fact of women's castration but, rather, 'the fact of his having perceived' that women are castrated. The idea of women's lack, in other words, is *already* part of the boy's defensive reaction.

If everyone were not hastening to erect defenses around whatever it is we fear to lose, perhaps we might pause to consider that, in a different context, such a loss might be perceived or represented as something other than women's 'fault'. For example, in the context of a developed lesbian subculture, might the butch figure represent a different sort of lack, such as the absence of men like fathers and male lovers? And is it possible to imagine this lack as something other than deficiency, as something other than what my mother had in mind when she pleaded with me to 'try more men'? Ideally, we might reenvision this lack as cultural difference in all its complexity. In Freud's account, this is exactly what the classic fetishistic perspective fails to do when, in its simplicity, it reduces difference to an economy of the same (to cite Jacques Derrida's formula). Faced with the alterity of the mother's body, the little boy interprets it in terms of his own body: like him, she either has or does not have a penis.

I would like to suggest a third position in the debate on lesbian dildos. If, in answer to the question of whether dildos represent penises, one camp says 'yes' and the other says 'no', perhaps a third position might be 'yes, but . . .' This kind of affirmation/negation is, as Sarah Kofman reminds us, the logic of fetishism itself as an 'undecidable compromise'.[26] In her reading of the fetishism essay, Kofman points out that Freud's subjects choose fetishes which, in the final analysis, both deny *and* affirm women's castration. For example, in the case of his patient who enjoys cutting women's hair, Freud discovers that the fetish 'contains within itself the two mutually incompatible assertions: "the woman has still got a penis" and "my father has castrated the woman".'[27] This undecidable compromise is also, I believe, what Fatale's video *Clips* performs, and the way that it does it is through parody: the goofy background music, the stereotypical middle-class living room, Kenny's beer drinking, even Fanny's virginal white lingerie. Through the duplicitous strategy of parody, the video audaciously inserts lesbianism into a context which still refers lovingly to the standard components of bourgeois heterosexuality – this is, after all, a video produced and consumed primarily by lesbians from more or less nuclear, middle-class families. The representation of Kenny's dildo is also inflected through parody; the video pokes fun at the association between dildo and penis, and at the same time it acknowledges and exploits its erotic power. If, as Judith Butler has proposed, such 'parodic redeployment[s] of power' may have a specifically political value for lesbians,[28] we might add that parody is also a fundamentally fetishistic strategy. The makers of *Clips*, for example, know very well that what they are doing is phallocentric but, with a subversive laugh, they are doing it anyway.

[. . .]

NOTES

1. See Colleen Lamos's excellent essay on pornographic representations of the dildo, '*On Our Backs:* Taking on the Phallus' (unpublished manuscript of a talk delivered at the Annual Convention of the Modern Language Association, San Francisco, 28 Dec. 1991), forthcoming in *The Lesbian Postmodern* (Columbia University Press). In general, a number of papers at the 1991 Annual Lesbian and Gay Studies Conference, including my essay and the panel entitled 'Flaunting the Phallus' in which it appeared, attest to the curious increase of scholarly interest in lesbianism and the phallus, epitomized perhaps by Judith Butler's 'The Lesbian Phallus and the Morphological Imaginary' (*differences* 4 [Spring 1992]: 133–71). Recent interventions by women of color in the dildo debate include Jackie Goldsby's 'What It Means to Be Colored Me', *Outlook* 9 (Summer 1990): 15; and Alycee J. Jane's editorial on dildos, 'What's *Race* Got to Do with It?' in the Black lesbian pornographic magazine *Black Lace* (Summer 1991): 21.
2. Sigmund Freud, 'Fetishism', in *The Standard Edition of the Complete Psychological Works of Sigmund Freud*, ed. James Strachey, 24 vols. (London: Hogarth, 1964), 21: 153–61. All further references to Freud are from this volume. For an excellent summary of French psychoanalysis's emphasis on the 'syntax' of fantasy, see Elizabeth Cowie, 'Fantasia', in *The Woman in Question*, ed. Parveen Adams and Elizabeth Cowie (Cambridge: Massachusetts Institute of Technology Press, 1990), 149–96.
3. Several of his feminist readers have discussed Freud's exclusion of women from fetishism. See Naomi Schor, 'Female Fetishism: The Case of George Sand', in *The*

Female Body in Western Culture: Contemporary Perspectives, ed. Susan Suleiman (Cambridge: Harvard University Press, 1985), 363–72; and Elizabeth A. Grosz's 'Lesbian Fetishism?' *differences*. 3 (Summer 1991): 39–54.

4. Lamos, 3.
5. Daralee and Nancy, 'Letters' column *On Our Backs* (Winter 1989): 5.
6. See Anne Koedt et al., eds, *Radical Feminism* (New York: Quadrangle, 1973). Daralee and Nancy's 'women-identified-women' is a catchphrase, of course, originating in part from the Radicalesbian manifesto, 'The Woman-Identified Woman', reprinted in *Radical Feminism*, 240–5. For a recent discussion of the radical feminist critique of sexuality, see Alice Echols's *Daring to Be Bad: Radical Feminism in America, 1967–1975* (Minneapolis: University of Minnesota Press, 1989).
7. See Fran Moira's review of a panel at the 1982 Barnard sexuality conference, 'lesbian sex mafia (1 s/m) speak out', *off our backs* 12 (June 1982): 24.
8. Sheila Jeffreys, 'Butch and Femme: Now and Then', in *Not a Passing Phase: Reclaiming Lesbians in History, 1840–1985*, ed. Lesbian History Group (London: Women's Press, 1989), 178.
9 See, for example, 'The Fem Question' in *Pleasure and Danger: Exploring Female Sexuality*, ed. Carole S. Vance (London: Pandora, 1989), 233; and Joan Nestle's 'My Woman Poppa' in *The Persistent Desire: A Femme-Butch Reader*, ed. Joan Nestle (Boston: Alyson, 1992), 348–51.
10. Susie Bright, *Susie Sexpert's Lesbian Sex World* (Pittsburgh: Cleis Press, 1990), 19.
11. Similarly, radical feminist attacks on butch-femme can be accused of failing to distinguish between gender stereotyping and 'gender performance' in Judith Butler's sense of the term (see Judith Butler, *Gender Trouble: Feminism and the Subversion of Identity* [New York: Routledge, 1989], 137–9); that is, between butch-femme as a duplication of masculine/feminine sex roles, and butch-femme as a performative, 'erotic statement' (Joan Nestle, *A Restricted Country* [Ithaca: Firebrand Books, 1986], 100).
12. Lane, 21.
13. Freud, 153, 152–3.
14. He is, at least, not a homosexual. Fetishism, according to Freud, 'saves the fetishist from being a homosexual by endowing women with the attribute which makes them acceptable as sexual objects' (154). In other words, the fetish object displaces the sight of the horrifying female sex organ and thus 'saves' the subject from having to seek sex with other men. Interestingly, Freud pauses at this moment to admit he remains puzzled as to why a 'great majority' of men are heterosexual, considering the fact that 'probably no male human being is spared the fright of castration at the female genital.' See also 155.
15. Grosz, 51.
16. See, for example, Jane Gallop's writings on the penis/phallus distinction, in particular her chapter, 'Beyond the Phallus', in *Thinking through the Body* (New York: Columbia, 1988), 119–33.
17. Freud, 154.
18. Nestle, 'My Woman Poppa', 348–50.
19. 'When Fanny Liquidates Kenny's Stocks', in *Clips*, dir. Nan Kinney and Debi Sundahl, Fatale Video [1988].
20. See Slavoj Žižek, *The Sublime Object of Ideology* [London: Verso, 1989), 28–33.
21. Freud, 154.
22. Jan Brown, 'Sex, Lies, and Penetration: A Butch Finally "Fesses Up", *Outlook 7* (Winter 1990): 34; reprinted in Nestle, *A Persistent Desire*, 410–15.
23. Lane, 21.
24. Freud, 154.
25. See Jackie Goldsby's critique of Susie Bright's video presentation, 'All Girl Action: A History of Lesbian Erotica' (1990). Responding in part to Bright's defense of Russ

Meyer's *Vixen*, which portrays a white woman getting 'ram[med]' by a Black woman with a 'larger-than-life white dildo', Goldsby concludes that 'as Bright's lecture presented it, lesbian eroticism – its icons, its narratives, its ideologies – is white' (15).
26. Sarah Kofman, *The Enigma of Woman: Woman in Freud's Writings*, trans. Catherine Porter (Ithaca: Cornell University Press, 1985), 88.
27. Freud, 157.
28. Butler, 124.

COPYRIGHT ACKNOWLEDGEMENTS

Grateful acknowledgement is made to the following sources for permission to reproduce material in this book that has previously been published elsewhere. Every effort has been made to trace copyright holders but if any have inadvertently been overlooked, the editors and publishers will be pleased to make the necessary arrangements at the first opportunity.

WOMAN AS BODY?

1.1 Londa Schiebinger, 'Theories of Gender and Race' from *Nature's Body: Gender in the Making of Modern Science*, 1993, Beacon Press, Boston, Massachusetts, USA, with permission from the author. Copyright © 1993 Londa Schiebinger.
1.2 Elizabeth Spelman, 'Woman as Body: Ancient and Contemporary Views'. This article is reprinted from *Feminist Studies*, 8 (1): 109–31, Spring 1982, by permission of the publisher, Feminist Studies, Inc., c/o Department of Women's Studies, University of Maryland, College Park, MD 20742.
1.3 Lynda Birke, 'Biological Sciences' from *Companion to Feminist Philosophy*, eds Alison Jagger and Iris Young, 1998. Reprinted with permission from Blackwell Publishers, Oxford, UK and the author.
1.4 Felly Nkweto Simmonds, 'My Body, Myself. How Does a Black Woman Do Sociology?' from *Black British Feminism: A Reader*, ed. Heidi Safia Mirza, 1997. Reprinted with permission from Routledge, London, UK and the author.
1.5 Helen Marshall, 'Our Bodies Ourselves. Why We Should Add Old Fashioned Empirical Phenomenology to the New Theories of the Body'. Reprinted from *Women's Studies International Forum*, 19 (3): 253–65, 1996, with permission from Elsevier Science.

SEXY BODIES

2.1 Luce Irigaray, 'When Our Lips Speak Together' from *Signs: Journal of Women in Culture and Society*, 6, (1): 69–79, 1980. Translated by Carolyn Burke. Reprinted with permission from University of Chicago Press and the author. Copyright © 1980 by the University of Chicago.

2.2 Jeanette Winterson, 'The Nose' and 'Taste' from *Written on the Body*, 1992, Jonathan Cape, London, UK and A. Knopf, New York.

2.3 Evelynn M. Hammonds, 'Toward a Genealogy of Black Female Sexuality: The Problematic of Silence'. Copyright © 1997. From *Feminist Genealogies, Colonial Legacies, Democratic Futures*, eds Jacqui Alexander and Chandra Talpade Mohanty. Reproduced by permission of Routledge Inc.

2.4 Lynne Segal, 'Body Matters: Cultural Inscriptions' from *Straight Sex: The Politics of Pleasure*, 1994, Virago Press, London, UK.

2.5 Barbara Creed, 'Lesbian Bodies: Tribades, Tomboys and Tarts' from *Sexy Bodies: The Strange Carnalities of Feminism*, eds. Elizabeth Grosz and Elspeth Probyn, 1995. Repinted by permission of Routledge, London, UK and the author.

2.6 Judith Halberstam, 'F2M: The Making of Female Masculinity' from *The Lesbian Postmodern*, ed. Laura Doan. Copyright © 1994. Columbia University Press, New York. Reprinted with permission of the publisher.

2.7 Juniper Wiley, 'NO BODY is Doing It: Cybersexuality' from *Body & Society*, 1 (1): 151–60, 1995. Reprinted by permission of Sage Publications, London, UK and the author. Copyright © Sage Publications Ltd 1995.

2.8 Sue-Ellen Case, 'The Hot Rod Bodies of Cybersex' from *The Domain-Matrix*, 1996, Indiana University Press, Bloomington, Indiana, USA.

BODIES IN SCIENCE AND BIOMEDICINE

3.1 Audre Lorde, 'A Burst of Light: Living with Cancer' from *A Burst of Light. Essays by Audre Lorde*, 1988. Reproduced by permission of Firebrand Books, Ithaca, New York, USA. Copyright © Audre Lord.

3.2 Eve Kosofsky Sedgwick, *Tendencies*, pp. 12–15. Copyright © 1994, Duke University Press. Reprinted with permission.

3.3 Ludmilla J. Jordanova, 'Natural facts: A Historical Perspective on Science and Sexuality' from *Nature, Culture and Gender*, eds. Carol MacCormack and Marilyn Strathern, 1980. Reprinted by permission of Cambridge University Press, Cambridge, UK and the author.

3.4 Anne Fausto-Sterling, 'Hormonal Hurricanes: Menstruation, Menopause, and Female Behaviour'. Excerpts, as submitted, from Chapter 4 from *Myths of Gender: Biological Theories about Women and Men* by Anne Fausto-Sterling. Copyright © 1986 by Basic Books, Inc. Reprinted by permission of Basic Books, a subsidiary of Perseus Books Group, PLC.

3.5 Emily Martin, 'The Egg and the Sperm: How Science has Constructed a Romance Based on Stereotypical Male–Female Roles' from *Signs: Journal of Women in Culture and Society*, 16, (3): 485–501, 1991. Reprinted by permission of University of Chicago Press and the author. Copyright © 1991 by the University of Chicago.

3.6 Jana Sawicki, 'Disciplining Mothers: Feminism and the New Reproductive Technologies'. Copyright © 1991. From *Disciplining Foucault: Feminism, Power and the Body* by Jana Sawicki. Reproduced by permission of Routledge Inc.

3.7 Donna Haraway, 'The Biopolitics of Postmodern Bodies: Determinations of Self in Immune System Discourse' from *differences: A Journal of Feminist Cultural Studies*, 1 (1): 3–43, 1989. Indiana University Press, Indiana, USA.

AFTER THE BINARY

4.1 Denise Riley, 'Bodies, Identities, Feminisms', from *'Am I That Name?' Feminism and the Category of 'Women' In History*, 1988. Reprinted with the permission of Macmillan Press Ltd, Basingstoke, UK and University of Minnesota Press, Minneapolis, USA. Copyright © Denise Riley 1988.

4.2 Moira Gatens, 'Power, Bodies and Difference' from *Destabilising Theory*,

1992, Polity Press, Cambridge, UK. Reprinted with permission of Moira Gatens.
4.3 Judith Butler, Excerpt from 'Introduction', *Bodies that Matter*. Copyright © 1993. Reproduced by permission of Routledge, Inc.
4.4 Susan Bordo, 'Feminism, Foucault and the Politics of the Body' from *Up Against Foucault*, ed. Carol Ramazanoglu, 1993. Reprinted by permission of Routledge, London, UK and the author.
4.5 Trinh T. Minh-ha, 'Write Your Body' and 'The Body in Theory' from *Woman, Native, Other. Writing Postcoloniality and Feminism*, 1989. Indiana University Press, Bloomington, Indiana, USA.
4.6 Elizabeth Grosz, 'The Body' from *Feminism and Psychoanalysis: A Critical Dictionary*, ed. Elizabeth Wright, 1992, Blackwell Publishers, Oxford, UK. Copyright © Basil Blackwell Ltd.

ALTER/ED BODIES

5.1 Anne Balsamo, 'Forms of Technological Embodiment: Reading the Body in Contemporary Culture' from *Cyberspace/Cyberbodies/Cyberpunk*, eds M. Featherstone and R. Burrows, 1996, Reprinted by permission of Sage Publications, London, UK and the author. Copyright Sage Publications Ltd, 1996.
5.2 Rosi Braidotti, 'Signs of Wonder and Traces of Doubt: On Teratology and Embodied Differences' from *Between Monsters, Goddesses and Cyborgs*, eds Nina Lykke and Rosi Braidotti, 1996, Zed Press, London, UK.
5.3 Pratibha Parmar and Alice Walker, 'Interview' from *Warrior Marks: Female Genital Mutilation and the Sexual Blinding of Women*, 1993, Jonathan Cape, London, UK and A. Knopf, New York.
5.4 Awa Thiam, *Speak Out Black Sisters*, Pluto Press, 1986. Trans by Dorothy S. Blair. London, UK.
5.5 Bibi Bakare-Yusuf, 'The Economy of Violence: Black Bodies and the Unspeakable Terror' from *Gender and Catastrophe*, ed. Ronit Lentin, 1997. Reprinted with permission of Zed Press, London, UK and the author.
5.6 Susan Wendell, 'Feminism, Disability and Transcendence of the Body'. Copyright © 1996. From *The Rejected Body* by Susan Wendell. Reproduced by permission of Routledge, Inc.

BODYSPACEMATTER

6.1 Christine Battersby, 'Her Body/Her Boundaries' from *The Phenomenal Woman*, 1998. Reproduced by permission of Polity Press, Cambridge, UK; Routledge, Inc., New York, USA, and the author. Copyright © 1998 Christine Battersby.
6.2 Gillian Rose, 'Women and Everyday Spaces' from *Feminism and Geography: The Limits of Geographical Knowledge*, 1993, Polity Press, Cambridge, UK, and University of Minnesota Press, Minneapolis, USA. Copyright © 1993 Gillian Rose.
6.3 Andrea Benton Rushing, 'Surviving Rape. A Morning/Mourning Ritual' from *Theorizing Black Feminisms: The Visionary Pragmatism of Black Women*, eds Stanlie M. James and Abena Busia, 1993. Reproduced by permission of Routledge, London, UK and the author.
6.4 Elizabeth Grosz, 'Bodies–Cities'. Copyright © 1995. From *Space, Time and Perversion. Essays on the Politics of Bodies* by Elizabeth Grosz. Reproduced by permission of Routledge, Inc. and Elizabeth Grosz.
6.5 Janet Price and Margrit Shildrick, 'Mapping the Colonial Body: Sexual Economies and the State in Colonial India' from *Gender and Colonialism*, eds T. Foley, L. Pilkington, S. Ryder and E. Tilley, 1995, Galway University Press, Galway, Ireland.
6.6 Nalini Natarajan, 'Woman, Nation and Narration in *Midnight's Children*' from *Scattered Hegemonies: Postmodernity and Transnational Feminist Practices*, eds Inderpal Grewal and Caren Kaplan, 1994, University of Minnesota Press, Minneapolis, USA. Copyright © 1994 by the Regents of the University of Minnesota.

PERFORMING THE BODY

7.1 Judith Butler, 'Bodily Inscriptions; Performative Subversions'. Copyright © 1990. From *Gender Trouble* by Judith Butler. Reproduced by permission of Routledge, Inc.

7.2 Kristina Straub, 'The Guilty Pleasures of Female Theatrical Cross-Dressing and the autobiography of Charlotte Clarke'. Copyright © 1991. From *Body Guards. The Cultural Politics of Gender Ambiguity*, eds Julia Epstein and Kristina Straub. Reproduced by permission of Routledge, Inc.

7.3 Margrit Shildrick and Janet Price, 'Breaking the Boundaries of the Broken Body' from *Body & Society*, 2 (4): 93–113, 1996. Reprinted by permission of Sage Publications, London, UK. Copyright © Sage Publications Ltd, 1996.

7.4 Fen Coles, 'Feminine Charms and Outrageous Arms' from *Trouble and Strife*, 29/30: 67–72, 1994/95.

7.5 Kathy Davis, '"My Body is my Art": Cosmetic Surgery as Feminist Utopia?' from *Embodied Practices. Feminist Perspectives on the Body*, ed. Kathy Davis, 1997. Reprinted by permission of Sage Publications, London, UK. Copyright © Sage Publications Ltd, 1997.

7.6 Heather Findlay, '"Freud's Fetishism" and the Lesbian Dildo Debates'. This article is reprinted from *Feminist Studies*, 18(3): 563–75, Fall 1992, by permission of the publisher, Feminist Studies, Inc., c/o Department of Women's Studies, University of Maryland, College Park, MD 20742.

SUBJECT INDEX

NAME INDEX